CAMBRIDGE LIBRARY COLLECTION

Books of enduring scholarly value

Physical Sciences

From ancient times, humans have tried to understand the workings of the world around them. The roots of modern physical science go back to the very earliest mechanical devices such as levers and rollers, the mixing of paints and dyes, and the importance of the heavenly bodies in early religious observance and navigation. The physical sciences as we know them today began to emerge as independent academic subjects during the early modern period, in the work of Newton and other 'natural philosophers', and numerous sub-disciplines developed during the centuries that followed. This part of the Cambridge Library Collection is devoted to landmark publications in this area which will be of interest to historians of science concerned with individual scientists, particular discoveries, and advances in scientific method, or with the establishment and development of scientific institutions around the world.

Mathematical and Physical Papers

William Thomson, first Baron Kelvin (1824–1907), is best known for devising the Kelvin scale of absolute temperature and for his work on the first and second laws of thermodynamics, though throughout his 53-year career as a mathematical physicist and engineer at the University of Glasgow he investigated a wide range of scientific questions in areas ranging from geology to transatlantic telegraph cables. The extent of his work is revealed in the six volumes of his *Mathematical and Physical Papers*, published from 1882 until 1911, consisting of articles that appeared in scientific periodicals from 1841 onwards. Volume 4, published in 1910, includes articles from the period 1867–1906. Themes covered in this book examine issues relating to water, such as hydrodynamics, tidal theory and deep sea ship waves.

Cambridge University Press has long been a pioneer in the reissuing of out-of-print titles from its own backlist, producing digital reprints of books that are still sought after by scholars and students but could not be reprinted economically using traditional technology. The Cambridge Library Collection extends this activity to a wider range of books which are still of importance to researchers and professionals, either for the source material they contain, or as landmarks in the history of their academic discipline.

Drawing from the world-renowned collections in the Cambridge University Library, and guided by the advice of experts in each subject area, Cambridge University Press is using state-of-the-art scanning machines in its own Printing House to capture the content of each book selected for inclusion. The files are processed to give a consistently clear, crisp image, and the books finished to the high quality standard for which the Press is recognised around the world. The latest print-on-demand technology ensures that the books will remain available indefinitely, and that orders for single or multiple copies can quickly be supplied.

The Cambridge Library Collection will bring back to life books of enduring scholarly value (including out-of-copyright works originally issued by other publishers) across a wide range of disciplines in the humanities and social sciences and in science and technology.

Mathematical and Physical Papers

VOLUME 4

LORD KELVIN
EDITED BY JOSEPH LARMOR

CAMBRIDGE UNIVERSITY PRESS

Cambridge, New York, Melbourne, Madrid, Cape Town,
Singapore, São Paolo, Delhi, Tokyo, Mexico City

Published in the United States of America by Cambridge University Press, New York

www.cambridge.org
Information on this title: www.cambridge.org/9781108029018

This edition first published 1910
This digitally printed version 2011

ISBN 978-1-108-02901-8 Paperback

MATHEMATICAL

AND

PHYSICAL PAPERS

CAMBRIDGE UNIVERSITY PRESS
London: FETTER LANE, E.C.
C. F. CLAY, Manager

Edinburgh: 100, PRINCES STREET
Berlin: A. ASHER AND CO.
Leipzig: F. A. BROCKHAUS
New York: G. P. PUTNAM'S SONS
Bombay and Calcutta: MACMILLAN AND CO., Ltd.

MATHEMATICAL

AND

PHYSICAL PAPERS

VOLUME IV

HYDRODYNAMICS AND GENERAL DYNAMICS

BY THE RIGHT HONOURABLE

SIR WILLIAM THOMSON, BARON KELVIN

O.M., P.C., G.C.V.O., LL.D., D.C.L., SC.D., M.D., ...

PAST PRES. R.S., FOR. ASSOC. INSTITUTE OF FRANCE,

GRAND OFFICER OF THE LEGION OF HONOUR, KT PRUSSIAN ORDER *POUR LE MÉRITE*,

CHANCELLOR OF THE UNIVERSITY OF GLASGOW

FELLOW OF ST PETER'S COLLEGE, CAMBRIDGE

ARRANGED AND REVISED WITH BRIEF ANNOTATIONS BY

SIR JOSEPH LARMOR, D.Sc., LL.D., SEC. R.S.

LUCASIAN PROFESSOR OF MATHEMATICS IN THE UNIVERSITY OF CAMBRIDGE

AND FELLOW OF ST JOHN'S COLLEGE

CAMBRIDGE:

AT THE UNIVERSITY PRESS

1910

Cambridge:
PRINTED BY JOHN CLAY, M.A.
AT THE UNIVERSITY PRESS

PREFACE.

IN the first three volumes of this reprint the papers were numbered consecutively from I to CIV, those which had already been collected in the volume of "Papers on Electrostatics and Magnetism" (Macmillans, 1872; reprinted 1884), being inserted only in title with a reference to their place in that volume. Lord Kelvin himself began the preparation of a Volume IV., and material which had been standing in type to form the first sheets was ultimately printed off as an Appendix, pp. 569—593, to the volume of "Baltimore Lectures" published in 1904; but the numbering of these papers is in error on account of additions afterwards made to Volume III., as mentioned below. In the volume of "Baltimore Lectures" he also reprinted, either in the text or as Appendices, a considerable number of later papers connected with the Dynamical Theory of Light. Moreover at the end of Volume III. of the "Mathematical and Physical Papers" (1890) he inserted a collection of papers then only just written, on the relations of the ether to electrodynamic propagation, stimulated thereto by the phenomena of alternating currents in cables, and of Hertzian electric waves in space, both then undergoing exploration. Also a considerable number of the less abstract papers have been selected and reprinted without regard to date in the three volumes of "Popular Lectures and Addresses" (Macmillans: Vol. I., Constitution of Matter, 1889; Vol. II., Geology and General Physics, 1894; Vol. III., Navigational Affairs, 1891).

It appeared to the Editor, when he was requested by Lady Kelvin to take charge of the completion of the collected edition of Lord Kelvin's work, that in consequence of the variety of procedure in these partial reprints, any attempt at continuing the previous numbering of the papers must be abandoned. It seemed, moreover, that a clearer view of the sequence of Lord Kelvin's scientific activity could be obtained by classifying the subject-matter under a number of broad headings, collecting together under each head in chronological order the material that belonged to it, and at the same time making the record practically complete by including titles of other papers with references to the places where they had already been republished. This procedure has been carried back in time far enough to bring it into connexion with Lord Kelvin's own chronological scheme of his earlier work, as reprinted in the previous volumes of this collection. In order to secure the convenience of fuller continuity under various headings, where it could readily be obtained, it has been thought advisable to reprint some pages already included in the "Baltimore Lectures" and elsewhere.

The difficulties involved in this rearrangement of the material were lightened by the use of two important bibliographies of Lord Kelvin's work. For the papers up to 1884, the published volumes of the Royal Society's Catalogue of Scientific Papers were available; and for the remaining period access was obtained, through the courtesy of Prof. McLeod, to the titles now prepared for the continuation of that catalogue. For the latter half of the present volume and all the next, the very complete bibliography of 661 titles appended to Prof. Silvanus Thompson's *Life of Lord Kelvin* has kindly been made available for the same purpose. In that list the cross-references to reprints or abstracts of the various papers are remarkably full; yet such is the complexity of the material that considerable research has been required to establish the relations between the various entries. A considerable proportion of the list consists of titles of verbal communications, often historically interesting, made to learned Societies, where nothing, or at most only an accidental press abstract, has been published; and in that respect it is of course more complete than the list of substantial publications which alone is given here and is summarised in the table of Contents.

The greater part of the present volume is taken up by papers on the subject of Hydrodynamics, the theory of the motions of fluids. The pages 1—230 form a connected reprint of the papers on Vortex Motion, such as has for a long time been a *desideratum*. Few subjects in modern physical mathematics can vie with this one in originality and elegance, and we may add in difficulty of development; and the great names of Helmholtz and Kelvin are inseparably connected in regard to it. The path-breaking memoir of the former takes rank among the classical examples of elegant and final mathematical formulation of the abstract essentials of a physical phenomenon, carrying on the reader in its sweep and compelling his ready assent; while the keen geometrical intuition and the objective presentation of Lord Kelvin's more fragmentary efforts towards the fuller development of the subject, made under the stimulus of his romantic conception of an atomic theory based on vortical motion in a perfect fluid, admits the student in a manner behind the scenes, where new knowledge is being forged and prepared by a master for incorporation in the formal scheme of science. Hardly any discipline is more informing for the training of a physical mathematician than an intermission of the engrossing logical details of algebraic analysis, in favour of such direct and tentative intuitional siege of the suggestions of order presented by natural phenomena, as is recorded in these papers.

The next section, pp. 231—269, contains Lord Kelvin's contributions to the Dynamical Theory of the Tides, in which his original aim was to restore the impugned authority of Laplace's analysis; but at the same time, by the exhibition of many new features and new points of view, he initiated the extensive modern development and improvement of this beautiful theory. These dynamical papers thus form a fit complement to their author's work in initiating, and guiding, the complete practical achievement of tidal prediction, through the formal harmonic analysis of the actual periodic motions which constitute the tides of the irregular terrestrial oceans as we know them. In regard to this section the advice of Sir George Darwin has been available; and he has kindly contributed the historical note on p. 269.

Then follows a section, pp. 270—456, on Waves on Water. This has been, on its physical side, a subject pre-eminently British ever since its mathematical machinery was formulated by Cauchy and Poisson,—perhaps mainly owing to the requirements of the scientific engineers who developed, first the navigation of canals, and afterwards the designing of ships, with a view to diminishing the drain on the propelling energy which arises from the production of waves. Lord Kelvin's knowledge of the sea, acquired as a yachtsman, led him directly to investigations of the effect of wind and current, in which he has been followed by Lord Rayleigh, and, mainly on the meteorological side, by Helmholtz; while the search for the *rationale* of the frictional resistance to ships, and the experiments of Osborne Reynolds on the demarcation between smooth and turbulent flow, prompted difficult investigations in viscous flow which still remain incomplete. The mode in which the motion at the front and rear of a limited regular train of waves spreads itself out, has features which are important also for the understanding of the advance of a beam of radiation into a dispersive medium; while the regular trains of standing undulations established in a current, by flow over a submerged obstacle, elucidate a mode of genesis of wavy motion which may also find application in meteorological atmospheric phenomena. A main feature in this section is the graphical representations of results, and thanks are due to the Royal Society of Edinburgh for providing *clichés* of the numerous diagrams.

Then follows a section on General Dynamics, pp. 457—531, in which are collected various fragmentary papers, beginning with the thorny question of the partition of thermal molecular energy, and ramifying into applications of the Principle of Action and other dynamical methods to the subject of periodic orbits, now fundamental in dynamical astronomy, and to the graphical solution of dynamical problems. As following naturally on this subject, the volume is completed by a brief section on Elastic Propagation, pp. 532—560, which is little more than a chronological list of titles of papers, mainly of optical and electrical interest, on propagation and reflexion in ordinary elastic solid media and in electric cables, which have been republished already in other volumes. The special subject of propagation through the ether, considered in its more modern electric connexions, has been reserved for the next volume.

The final volume (V.) of the Mathematical and Physical Papers will contain papers, now arranged ready for press, on Thermodynamics, Cosmical and Geological Physics, Electrodynamics and Electrolysis, Molecular and Crystalline Theory, Radioactivity and Electrionic Theory, with perhaps some addresses and other miscellaneous scientific matter.

The Editor desires to acknowledge much expert assistance which has greatly lightened his task. Many of the proof sheets have been looked through by Mr W. J. HARRISON, Fellow of Clare College, Cambridge. In the correction of the latter half of the volume, the vigilance of Mr GEORGE GREEN, who was Lord Kelvin's scientific secretary for the later years of his life, and who thus brought special knowledge to bear, has ensured the correction of many small oversights, in addition to more important points which are explicitly mentioned in footnotes. While Prof. W. M^cF. ORR, F.R.S., whose assistance was specially invoked in connexion with the difficult topics treated on p. 330, has kindly examined in proof all the subsequent sheets of the volume, and has also supplied most of the list of *errata* belonging to the earlier part. For general advice relating to various matters, the Editor is under obligation to Lord RAYLEIGH and Prof. HORACE LAMB, and to Dr J. T. BOTTOMLEY. In the reprint, obvious minor errors and misprints have been corrected without mention; but all important changes have been referred to in footnotes inserted by the Editor, which are enclosed in square brackets.

Thanks are due, as always, to the officials of the Cambridge University Press for the excellence of their work, and their unfailing courtesy.

J. L.

ST JOHN'S COLLEGE, CAMBRIDGE.
March, 1910.

CONTENTS.

HYDRODYNAMICS.

THEORY OF THE TIDES.

WAVES ON WATER.

GENERAL DYNAMICS.

ELASTIC PROPAGATION.

CORRIGENDA

Page 10, line 15 from foot, *for* moment *read* momentum

,, 73, Cf. Lamb's *Hydrodynamics*, §§ 129, 130

,, 102, first eq., *for* τ *read* T

,, 136, line 8 from foot, *for* diminishing *read* increasing

,, 136, ,, 2 ,, *for* $-$ *read* $+$

,, 141, ,, 3, *for* pp. 97—109 *read* pp. 109—116

,, 143, eq. (8′), *for* $\frac{1}{r^2}\frac{d^2h}{rd\theta^2}$ *read* $\frac{1}{r^2}\frac{d^2h}{d\theta^2}$

,, 143, line 10 from foot, *omit* in *after* Using

,, 144, ,, 5, *for* θ *read* a

,, 156, ,, 8, *for* $i=1$ *read* $i=0$

,, 160, ,, 12, *for* 4 *read* 2

,, 165, footnote, *read* [*supra*, p. 1]

,, 175, line 7, *omit* every. Cf. Lamb's *Hydrodynamics*, § 164

,, 186, see footnotes, p. 334

,, 255, eq. (3) and (4), *delete* r

,, 255, ,, (6), *for* r *read* r^2

,, 256, ,, (8), (9), (11), (12), *for* r *read* r^2

,, 305, ,, (14), *for* $2^{\frac{3}{2}}$ in last expression *read* 2

,, 317 seq. Reference may be made to W. M. Hicks, *Brit. Assoc. Report*, 1885, Address to Section A, p. 517, also p. 930 : also to same author, *Phil. Trans.* Vol. 192 (1898), p. 33, on "Spiral or Gyrostatic Vortex Aggregates."

HYDRODYNAMICS

1. On Vortex Atoms.

[*Proceedings of the Royal Society of Edinburgh*, Vol. VI, pp. 94—105; reprinted in *Phil. Mag.* Vol. XXXIV, 1867, pp. 15—24.]

After noticing Helmholtz's admirable discovery of the law of vortex motion in a perfect liquid—that is, in a fluid perfectly destitute of viscosity (or fluid friction)—the author said that this discovery inevitably suggests the idea that Helmholtz's rings are the only true atoms. For the only pretext seeming to justify the monstrous assumption of infinitely strong and infinitely rigid pieces of matter, the existence of which is asserted as a probable hypothesis by some of the greatest modern chemists in their rashly-worded introductory statements, is that urged by Lucretius and adopted by Newton—that it seems necessary to account for the unalterable distinguishing qualities of different kinds of matter. But Helmholtz has proved an absolutely unalterable quality in the motion of any portion of a perfect liquid in which the peculiar motion which he calls "Wirbelbewegung" has been once created. Thus any portion of a perfect liquid which has "Wirbelbewegung" has one recommendation of Lucretius's atoms —infinitely perennial specific quality. To generate or to destroy "Wirbelbewegung" in a perfect fluid can only be an act of creative power. Lucretius's atom does not explain any of the properties of matter without attributing them to the atom itself. Thus the "clash of atoms," as it has been well called, has been invoked by his modern followers to account for the elasticity of gases. Every other property of matter has similarly required an assumption of specific forces pertaining to the atom. It is as easy (and as improbable—not more so) to assume whatever specific forces may be required in any portion of matter which possesses the "Wirbelbewegung," as in a solid indivisible piece of matter; and hence the Lucretius atom has no *prima facie* advantage over the Helmholtz atom. A magnificent display of smoke-rings, which he recently had the pleasure of witnessing in Professor Tait's lecture-room, diminished by one the number of assumptions required to explain the properties of matter on the hypothesis that all bodies are composed of vortex atoms in a perfect homogeneous liquid. Two smoke-rings were frequently seen to bound obliquely from one another, shaking violently from the effects of the shock. The result was very similar to that observable in two large india-rubber rings striking one another in the air. The elasticity of each smoke-ring seemed no further from perfection than might be expected in a solid india-rubber ring of the same shape, from what we know of the viscosity of india-rubber. Of course this kinetic elasticity of form is perfect elasticity for vortex rings in a perfect liquid. It is at least as good a beginning as the "clash of atoms" to account for the elasticity of gases. Probably the beautiful investigations of D. Bernoulli, Herapath, Joule, Krönig, Clausius, and Maxwell, on the various thermodynamic properties of gases, may have all the positive assumptions they have been obliged to make, as to mutual forces between two atoms and kinetic energy acquired by individual atoms or molecules, satisfied by vortex rings, without requiring any other property in the matter whose motion composes them than inertia and incompressible occupation of space. A full mathematical investigation of the mutual action between two vortex rings of any given magnitudes and velocities passing one another in any two lines, so directed that they never come nearer one another than a large multiple of the diameter of either, is a perfectly solvable mathematical problem; and the novelty of the circumstances contemplated presents difficulties of an exciting character. Its solution will become the foundation of the proposed new kinetic theory of gases. The possibility of founding a theory of elastic solids and liquids on the dynamics of more closely-packed vortex atoms may be reasonably anticipated. It may be remarked in connexion with this anticipation, that the mere title of Rankine's paper on "Molecular Vortices," communicated to the Royal Society of Edinburgh in 1849 and 1850, was a most suggestive step in physical theory.

Diagrams and wire models were shown to the Society to illustrate knotted or knitted vortex atoms, the endless variety of which is infinitely more than sufficient to explain the varieties and allotropies of known simple bodies and their mutual affinities. It is to be remarked that two ring atoms linked together or one knotted in any manner with its ends meeting, constitute a system which, however it may be altered in shape, can never deviate from its own peculiarity of multiple continuity, it being impossible for the matter in any line of vortex motion to go through the line of any other matter in such motion or any other part of its own line. In fact, a closed line of vortex core is literally indivisible by any action resulting from vortex motion.

The author called attention to a very important property of the vortex atom, with reference to the now celebrated spectrum-analysis practically established by the discoveries and labours of Kirchhoff and Bunsen. The dynamical theory of this subject, which Professor Stokes had taught to the author of the present paper before September 1852, and which he has taught in his lectures in the University of Glasgow from that time forward, required that the ultimate constitution of simple bodies should have one or more fundamental periods of vibration, as has a stringed instrument of one or more strings, or an elastic solid consisting of one or more tuning-forks rigidly connected. To assume such a property in the Lucretius atom, is at once to give it that very flexibility and elasticity for the explanation of which, as exhibited in aggregate bodies, the atomic constitution was originally assumed. If, then, the hypothesis of atoms and vacuum imagined by Lucretius and his followers to be necessary to account for the flexibility and compressibility of tangible solids and fluids were really necessary, it would be necessary that the molecule of sodium, for instance, should be not an atom, but a group of atoms with void space between them. Such a molecule could not be strong and durable, and thus it loses the one recommendation which has given it the degree of acceptance it has had among philosophers; but, as the experiments shown to the Society illustrate, the vortex atom has perfectly definite fundamental modes of vibration, depending solely on that motion the existence of which constitutes it. The discovery of these fundamental modes forms an intensely interesting problem of pure mathematics. Even for a simple Helmholtz ring, the analytical difficulties which it presents are of a very formidable character, but certainly far from insuperable in the present state of mathematical science. The author of the present communication had not attempted, hitherto, to work it out except for an infinitely long, straight, cylindrical vortex. For this case he was working out solutions corresponding to every possible description of infinitesimal vibration, and intended to include them in a mathematical paper which he hoped soon to be able to communicate to the Royal Society. One very simple result which he could now state is the following. Let such a vortex be given with its section differing from exact circular figure by an infinitesimal harmonic deviation of order i. This *form* will travel as waves round the axis of the cylinder in the same direction as the vortex rotation, with an angular velocity equal to $(i-1)/i$ of the angular velocity of this rotation. Hence, as the number of crests in a whole circumference is equal to i, for an harmonic deviation of order i there are $i-1$ periods of vibration in the period of revolution of the vortex. For the case $i=1$ there is no vibration, and the solution expresses merely an infinitesimally displaced vortex with its circular form unchanged. The case $i=2$ corresponds to elliptic deformation of the circular section; and for it the period of vibration is, therefore, simply the period of revolution. These results are, of course, applicable to the Helmholtz ring when the diameter of the approximately circular section is small in comparison with the diameter of the ring, as it is in the smoke-rings exhibited to the Society. The lowest fundamental modes of the two kinds of transverse vibrations of a ring, such as the vibrations that were seen in the experiments, must be much graver than the elliptic vibration of section. It is probable that the vibrations which constitute the incandescence of sodium-vapour are analogous to those which the smoke-rings had exhibited; and it is therefore probable that the period of each vortex rotation of the atoms of sodium-vapour is much less than $\frac{1}{525}$ of the millionth of the millionth of a second, this being approximately the period of vibration of the yellow sodium light. Further, inasmuch as this light consists of two sets of vibrations coexistent in slightly different periods, equal approximately to the time just stated, and of as nearly as can be perceived equal intensities, the sodium atom must have two fundamental modes of vibration, having those for their respective periods, and being about equally excitable by such forces as the atom experiences in the incandescent vapour. This last condition renders it probable that the two fundamental modes concerned are approximately similar (and not merely different orders of different series chancing to concur very nearly in their periods of vibration). In an approximately circular and uniform disk of elastic solid the fundamental modes of transverse vibration, with nodal division into quadrants, fulfil both the conditions. In an approximately circular and uniform ring of elastic solid these conditions are fulfilled for the flexural vibrations in its plane, and also in its transverse vibrations perpendicular to its own plane. But the circular vortex ring, if created with one part somewhat thicker than another, would not remain so, but would experience longitudinal vibrations round its own circumference, and could not possibly have two fundamental modes of vibration similar in character and approximately equal in period. The same assertion may, it is probable*, be practically extended to any atom consisting of a single vortex ring, however involved, as illustrated by those of the models shown to the Society which consisted of only a single wire knotted in various ways. It seems, therefore, probable that the sodium atom may not consist of a single vortex line; but it may very probably consist of two approximately equal vortex rings passing through one another like two links of a chain. It is, however, quite certain that a vapour consisting of such atoms, with proper volumes and angular velocities in the two rings of each atom, would act precisely as incandescent sodium-vapour acts—that is to say, would fulfil the "spectrum test" for sodium.

The possible effect of change of temperature on the fundamental modes cannot be pronounced upon without mathematical investigation not hitherto executed; and therefore we cannot say

* *Note*, April 26, 1867.—The author has seen reason for believing that the sodium characteristic might be realized by a certain configuration of a single line of vortex core, to be described in the mathematical paper which he intends to communicate to the Society.

that the dynamical explanation now suggested is mathematically demonstrated so far as to include the very approximate identity of the periods of the vibrating particles of the incandescent vapour with those of their corresponding fundamental modes at the lower temperature at which the vapour exhibits its remarkable absorbing-power for the sodium light.

A very remarkable discovery made by Helmholtz regarding the simple vortex ring is that it always moves, relatively to the distant parts of the fluid, in a direction perpendicular to its plane, towards the side towards which the rotatory motion carries the inner parts of the ring. The determination of the velocity of this motion, even approximately, for rings of which the sectional radius

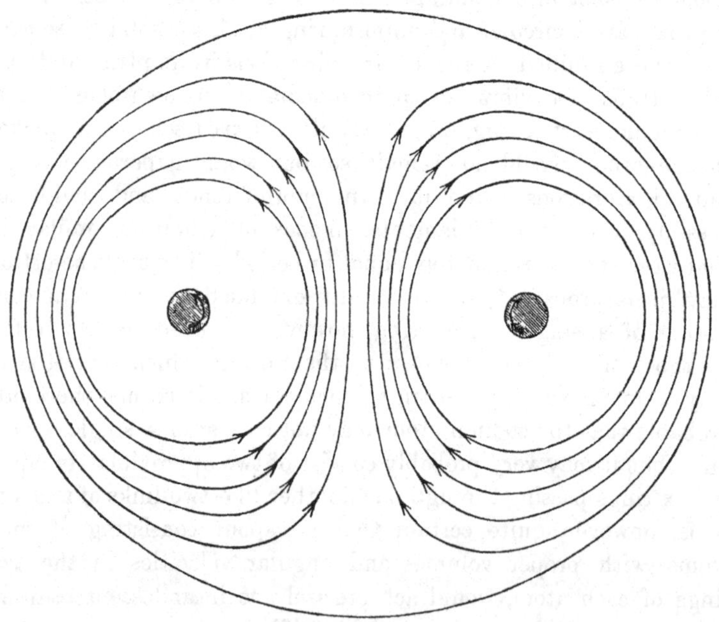

is small in comparison with the radius of the circular axis, has presented mathematical difficulties which have not yet been overcome*. In the smoke-rings which have been actually observed, it seems to be always something smaller than the velocity of the

* See, however, note added to Professor Tait's translation of Helmholtz's paper (*Phil. Mag.* 1867, vol. xxxiii. Suppl.), where the result [see *infra*, p. 67] of a mathematical investigation which the author of the present communication has recently succeeded in executing is given.

fluid along the straight axis through the centre of the ring; for the observer standing beside the line of motion of the ring sees, as its plane passes through the position of his eye, a convex* outline of an atmosphere of smoke in front of the ring. This convex outline indicates the bounding surface between the quantity of smoke which is carried forward with the ring in its motion and the surrounding air which yields to let it pass. It is not so easy to distinguish the corresponding convex outline behind the ring, because a confused trail of smoke is generally left in the rear. In a perfect fluid the bounding surface of the portion carried forward would necessarily be quite symmetrical on the anterior and posterior sides of the middle plane of the ring. The motion of the surrounding fluid must be precisely the same as it would be if the space within this surface were occupied by a smooth solid; but in reality the air within it is in a state of rapid motion, circulating round the circular axis of the ring with increasing velocity on the circuits nearer and nearer to the ring itself. The circumstances of the actual motion may be imagined thus:—Let a solid column of india-rubber, of circular section, with a diameter small in proportion to its length, be bent into a circle, and its two ends properly spliced together so that it may keep the circular shape when left to itself; let the aperture of the ring be closed by an infinitely thin film; let an impulsive pressure be applied all over this film, of intensity so distributed as to produce the definite motion of the fluid, specified as follows, and instantly thereafter let the film be all liquified. This motion is, in accordance with one of Helmholtz's laws, to be along those curves which would be the lines of force, if, in place of the india-rubber circle, were substituted a ring electromagnet†; and the velocities at different points

* The diagram represents precisely the convex outline referred to, and the lines of motion of the interior fluid carried along by the vortex, for the case of a double vortex consisting of two infinitely long, parallel, straight vortices of equal rotations in opposite directions. The curves have been drawn by Mr D. M'Farlane, from calculations which he has performed by means of the equation of the system of curves, which is

$$\frac{y^2}{a}=\frac{2x}{a}\cdot\frac{N+1}{N-1}-\left(1+\frac{x^2}{a^2}\right), \text{ where } \log_e N=\frac{x+b}{a}.$$

The proof will be given in the mathematical paper which the author intends to communicate in a short time to the Royal Society of Edinburgh.

† That is to say, a circular conductor with a current of electricity maintained circulating through it.

are to be in proportion to the intensities of the magnetic forces in the corresponding points of the magnetic field. The motion, as has long been known, will fulfil this definition, and will continue fulfilling it, if the initiating velocities at every point of the film perpendicular to its own plane be in proportion to the intensities of the magnetic force in the corresponding points of the magnetic field. Let now the ring be moved perpendicular to its own plane in the direction *with* the motion of the fluid through the middle of the ring, with a velocity very small in comparison with that of the fluid at the centre of the ring. A large approximately globular portion of the fluid will be carried forward with the ring. Let the velocity of the ring be increased; the volume of fluid carried forward will be diminished in every diameter, but most in the axial or fore-and-aft diameter, and its shape will thus become sensibly oblate. By increasing the velocity of the ring forward more and more, this oblateness will increase, until, instead of being wholly convex, it will be concave before and behind, round the two ends of the axis. If the forward velocity of the ring be increased until it is just equal to the velocity of the fluid through the centre of the ring, the axial section of the outline of the portion of fluid carried forward will become a lemniscate. If the ring be carried still faster forward, the portion of it carried with the india-rubber ring will be itself annular; and, relatively to the ring, the motion of the fluid will be backwards through the centre. In all cases the figure of the portion of fluid carried forward and the lines of motion will be symmetrical, both relatively to the axis and relatively to the two sides of the equatorial plane. Any one of the states of motion thus described might of course be produced either in the order described, or by first giving a velocity to the ring and then setting the fluid in motion by aid of an instantaneous film, or by applying the two initiative actions simultaneously. The whole amount of the impulse required, or, as we may call it, the effective momentum of the motion, or simply the momentum of the motion, is the sum of the integral values of the impulses on the ring and on the film required to produce one or other of the two components of the whole motion. Now it is obvious that as the diameter of the ring is very small in comparison with the diameter of the circular axis, the impulse on the ring must be very small in comparison with the impulse on the film, unless the velocity given to the ring is much greater than that given to the central parts of the film. Hence, unless the velocity given to the ring is so very great as to reduce the volume of the fluid carried forward with it to something not incomparably greater than the volume of the solid ring itself, the momenta of the several configurations of motions we have been considering will exceed by but insensible quantities the momentum when the ring is fixed. The value of this momentum is easily found by a proper application of Green's formulæ. Thus the actual momentum of the portion of fluid carried forward (being the same as that of a solid of the same density moving with the same velocity), together with an equivalent for the inertia of the fluid yielding to let it pass, is approximately the same in all these cases, and is equal to a Green's integral expressing the whole initial impulse on the film. The equality of the effective momentum for different velocities of the ring is easily verified without analysis for velocities not so great as to cause sensible deviations from spherical figure in the portion of fluid carried forward. Thus in every case the length of the axis of the portion of the fluid carried forward is determined by finding the point in the axis of the ring at which the velocity is equal to the velocity of the ring. At great distances from the plane of the ring that velocity varies, as does the magnetic force of an infinitesimal magnet on a point in its axis, inversely as the cube of the distance from the centre. Hence the cube of the radius of the approximately globular portion carried forward is in simple inverse proportion to the velocity of the ring, and therefore its momentum is constant for different velocities of the ring. To this must be added, as was proved by Poisson, a quantity equal to half its own amount, as an equivalent for the inertia of the external fluid; and the sum is the whole effective momentum of the motion. Hence we see not only that the whole effective momentum is independent of the velocity of the ring, but that its amount is the same as the magnetic moment in the corresponding ring electromagnet. The same result is of course obtained by the Green's integral referred to above.

The synthetical method just explained is not confined to the case of a single circular ring specially referred to, but is equally applicable to a number of rings of any form, detached from one another, or linked through one another in any way, or to a single line knotted to any degree and quality of "multiple continuity," and joined continuously so as to have no end. In every possible such case the motion of the fluid at every point, whether of the vortex core or of the fluid filling all space round it, is perfectly determined by Helmholtz's formulæ when the shape of the core is given. And the synthetic investigation now explained proves that the effective momentum of the whole fluid motion agrees in magnitude and direction with the magnetic moment of the corresponding electromagnet. Hence, still considering for simplicity only an infinitely thin line of core, let this line be projected on each of three planes at right angles to one another. The areas of the plane circuit thus obtained (to be reckoned according to De Morgan's rule when autotomic, as they will generally be) are the components of momentum perpendicular to these three planes. The verification of this result will be a good exercise on "multiple continuity." The author is not yet sufficiently acquainted with Riemann's remarkable researches on this branch of analytical geometry to know whether or not all the kinds of "multiple continuity" now suggested are included in his classification and nomenclature.

That part of the synthetical investigation in which a thin solid wire ring is supposed to be moving in any direction through a fluid with the free vortex motion previously excited in it, requires the diameter of the wire at every point to be infinitely small in comparison with the radius of curvature of its axis and with the distance of the nearest of any other part of the circuit from that point of the wire. But when the effective moment of the whole fluid motion has been found for a vortex with infinitely thin core, we may suppose any number of such vortices, however near one another, to be excited simultaneously; and the whole effective momentum in magnitude and direction will be the resultant of the momenta of the different component vortices each estimated separately. Hence we have the remarkable proposition that the effective momentum of any possible motion in an infinite incompressible fluid agrees in direction and magnitude with the magnetic moment of the corresponding electromagnet in Helmholtz's theory. The author hopes to give the mathematical formulæ expressing and proving this statement in the more detailed paper, which he expects soon to be able to lay before the Royal Society.

The question early occurs to any one either observing the phenomena of smoke-rings or investigating the theory,—What conditions determine the size of the ring in any case? Helmholtz's investigation proves that the angular vortex velocity of the core varies directly as its length, or inversely as its sectional area. Hence the strength of the electric current in the electromagnet, corresponding to an infinitely thin vortex core, remains constant, however much its length may be altered in the course of the transformations which it experiences by the motion of the fluid. Hence it is obvious that the larger the diameter of the ring for the same volume and strength of vortex motions in an ordinary Helmholtz ring, the greater is the whole kinetic energy of the fluid, and the greater is the momentum; and we therefore see that the dimensions of a Helmholtz ring are determinate when the volume and strength of the vortex motion are given, and, besides, either the kinetic energy or the momentum of the whole fluid motion due to it. Hence if, after any number of collisions or influences, a Helmholtz ring escapes to a great distance from others and is then free, or nearly free, from vibrations, its diameter will have been increased or diminished according as it has taken energy from, or given energy to, the others. A full theory of the swelling of vortex atoms by elevation of temperature is to be worked out from this principle.

Professor Tait's plan of exhibiting smoke-rings is as follows:—A large rectangular box, open at one side, has a circular hole of 6 or 8 inches diameter cut in the opposite side. A common rough packing-box of 2 feet cube, or thereabout, will answer the purpose very well. The open side of the box is closed by a stout towel or piece of cloth, or by a sheet of india-rubber stretched across it. A blow on this flexible side causes a circular vortex ring to shoot out from the hole on the other side. The vortex rings thus generated are visible if the box is filled with smoke. One of the most convenient ways of doing this is to use two retorts with their necks thrust into holes made for the purpose in one of the sides of the box. A small quantity of muriatic acid is put into one of these retorts, and of strong liquid ammonia into the other. By a spirit-lamp applied from time to time to one or other of these retorts, a thick cloud of sal-ammoniac is readily maintained in the inside of the box. A curious and interesting experiment may be made with two boxes thus arranged, and placed either side by side close to one another or facing one another so as to project smoke-rings meeting from opposite directions—or in various relative positions, so as to give smoke-rings proceeding in paths inclined to one another. at any angle, and passing one another at various distances. An interesting variation of the experiment may be made by using clear air without smoke in one of the boxes. The invisible vortex rings projected from it render their existence startlingly sensible when they come near any of the smoke-rings proceeding from the other box.

2. On Vortex Motion.

[*Transactions of the Royal Society of Edinburgh*, Vol. xxv. 1869, pp. 217—260. Read 29th April, 1867.]

(§§ 1—59 recast and augmented 28th August to 12th November, 1868.)

1. The mathematical work of the present paper has been performed to illustrate the hypothesis, that space is continuously occupied by an incompressible frictionless liquid acted on by no force, and that material phenomena of every kind depend solely on motions created in this liquid. But I take, in the first place, as subject of investigation, a finite mass of incompressible frictionless* fluid completely enclosed in a rigid fixed boundary.

2. The containing vessel may be either *simply* or *multiply continuous*†. And I shall frequently consider solids surrounded by the liquid, which also may be either simply or multiply continuous. It will not be necessary to exclude the supposition that any such solid may touch the outer boundary over some finite area, in which case it is *not* surrounded by the liquid; but each such solid, whether surrounded by the liquid or not, and whether moveable or fixed, must be considered as a part of the whole boundary of the liquid.

3. Let the whole fluid be given at rest, and let no force, except pressure from the containing vessel, or from the surfaces of solids immersed in it, ever act on any part of it. Let there be any number of solids, perfectly incompressible, and of the same density as the fluid; but either perfectly rigid, or more or less

* A frictionless fluid is defined as a mass continuously occupying space, whose contiguous portions press on one another everywhere exactly in the direction perpendicular to the surface separating them.

† Helmholtz—*Ueber Integrale der hydrodynamischen Gleichungen, welche den Wirbelbewegungen entsprechen*, Crelle (1858); translated by Tait in *Phil. Mag.* 1867, I. Riemann—*Lehrsätze aus der Analysis situs, &c.*, Crelle (1857). See also § 58, below.

flexible, with perfect or imperfect elasticity. Some of these may at times be supposed to lose rigidity, and become perfectly liquid; and portions of the liquid may be supposed to acquire rigidity, and thus to constitute solids. Let the solids act on one another with any forces, pressures, frictions, or mutual distant actions, subject only to the law of "action and reaction." Let motions originate among them, and in the liquid, either by the natural mutual actions of the solids or by the arbitrary application of forces to them during some limited time. It is of no consequence to us whether these forces have reactions on matter outside the containing vessel, so that they might be called "natural forces" in the present state of science (which admits action and reaction at a distance); or are applied arbitrarily by supernatural action without reaction. To avoid circumlocution, and, at the same time, to conform to a common usage, we shall call them *impressed forces.*

4. From the homogeneousness as to density of the contents of the fixed bounding vessel, it follows that the centre of inertia of the whole system of liquid and solids immersed in it remains at rest; in other words, the integral momentum of the motion is zero. Hence (Thomson and Tait's *Natural Philosophy*, § 297) the time integral of the sum of the components of *pressure on the containing vessel,* parallel to any fixed line, is equal to the time-integral of the sum of the components of *impressed forces* parallel to the same line. This equality exists, of course, at each instant during the action of the impressed forces, and continues to exist for the constant values of their time integrals, after they have ceased. Thus, in the subsequent motion of the solids, and of the fluids compelled to yield to them, whatever pressure may come to act on the containing vessel, whether from the fluid or from some of the solids coming in contact with it, the components of this pressure, parallel to any fixed line, summed for every element of the inner surface of the vessel, must vanish for every interval of time during which no impressed forces act. If, for example, one of the solids strikes the containing vessel, there will be an impulsive pressure of the fluid over all the rest of the fixed containing surface, having the sum of its components parallel to any line, equal and contrary* to the corresponding component of the

* I shall use the word *contrary* to designate merely directional opposition; and reserve the unqualified word *opposite,* to signify *contrary and in one line.*

impulsive pressure of the solid on the part of this surface which it strikes [see § 8, and consider oblique impulse of an inner moving solid, on the fixed solid spherical boundary]. *But, after the impressed forces cease to act, and as long as the containing vessel is not touched by any of the solids, the integral amount of the component of fluid pressure on it, parallel to any line, vanishes.*

5. If now forces be applied to stop the whole motion of fluid and solids [as (§ 62) is done, if the solids are brought to rest by forces applied to themselves only], the time integrals of the sums of the components of these forces, parallel to any stated lines, *may or may not in general be equal and contrary* to the time integrals of the corresponding sums of components of the initiating impressed forces (§ 3). But we shall see (§§ 19, 21) that *if the containing vessel be infinitely large, and all of the moving solids be infinitely distant from it during the whole motion,* there must be not merely the equality in question between the time integrals of the components in contrary directions of the initiating and stopping impressed forces, but there must be (§ 21) *completely equilibrating opposition between the two systems.*

6. To avoid circumlocution, henceforth I shall use the unqualified term *impulse* to signify a system of impulsive forces, to be dealt with as if acting on a rigid body. Thus the most general impulse may be reduced to an impulsive force, and couple in plane perpendicular to it, according to Poinsot; or to two impulsive forces in lines not meeting, according to his predecessors. Further, I shall designate by *the impulse of the motion at any instant,* in our present subject, the system of impulsive forces on the moveable solids which would generate it from rest; or any other system which would be equivalent to that one if the solids were all rigid and rigidly connected with one another, as, for instance, the Poinsot resultant impulsive force and minimum couple. The line of this resultant impulsive force will be called the *resultant axis of the motion,* and the moment of the minimum couple (whose plane is perpendicular to this line) will be called the *rotational moment* of the motion.

7. But, having thus defined the terms I intend to use, I must, to warn against errors that might be fallen into, remark that the momentum of the whole motions of solids and liquid is *not* equal to what I have defined as *the impulse,* but (§ 4) is equal

to zero; being the force-resultant of "the impulse" and the impulsive pressure exerted on the liquid by the containing vessel during the generation of the motion: and that the moment of momentum of the whole motion round the centre of inertia of the contents of the vessel is *not* equal to the *rotational moment*, as I have defined it, but is equal to the moment of the couple constituted by "the impulse" and the impulsive pressure of the containing vessel on the liquid. It must be borne in mind that however large, and however distant all round from the moveable solids, the containing vessel may be, it exercises a finite influence on the momentum and moment of momentum of the whole motion within it. But if it is infinitely large, and infinitely distant all round from the solids, it does so by infinitely slow motion through an infinitely large mass of fluid, and exercises no finite influence on the finite motion of the solids or of the neighbouring fluid. This will be readily understood, if for an instant we suppose the rigid containing vessel to be not fixed, but quite free to move as a rigid body without mass. The momentum of the whole motion will then be not zero, but exactly equal to the force-resultant of the impulse on the solids; and the moment of momentum of the whole motion round the centre of inertia will be precisely equal to the resultant impulsive couple found by transposing the constituent impulsive forces to this point after the manner of Poinsot. But the finite motion of the immersed solids, and of the fluid in their neighbourhood which we shall call the *field of motion*, will not be altered by any finite difference, whether the containing vessel be held fixed or left free, provided it be infinitely distant from them all round. It is, therefore, essentially indifferent whether we keep it fixed or let it be free. The former supposition is more convenient in some respects, the latter in others; but it would be inconvenient to leave any ambiguity, and I shall adhere (§ 1) to the former in all that follows.

8. To further illustrate the impulse of the motion, and its resultant impulsive force and couple, according to the previous definitions, as distinguished from the momentum, and the moment of momentum, of the whole contents of the vessel, let the vessel be spherical. Its impulsive pressure on the liquid will always be reducible to a single resultant in a line through its centre, which (§ 4) will be equal and contrary to the force-resultant of "the

impulse"; and, therefore, with it will constitute in general a couple. The resultant, of this couple and the couple-resultant of the impulse, will be equal to the moment of momentum of the whole motion round the centre of the sphere (which is the centre of inertia). But if the vessel be infinitely large, and infinitely distant all round from the moveable solids, the moment of momentum of the whole motion is irrelevant; and what is essentially important, is the impulse and its force and couple-resultants, as defined above.

9. The following way of stating (§§ 10, 12), and proving (§§ 11—15), a fundamental proposition in fluid motion will be useful to us for the theory of the impulse, whether of the moveable solids we have hitherto considered or of vortices.

10. The moment of momentum of every spherical portion of a liquid mass in motion, relatively to the centre of the sphere, is always zero, if it is so at any one instant for every spherical portion of the same mass.

11. To prove this, it is first to be remarked, that the moment of momentum of that part of the liquid which at any instant occupies a certain fixed spherical space can experience no change, at that instant (or its *rate* of change vanishes at that instant), because the fluid pressure on it (§ 1), being perpendicular to its surface, is everywhere precisely towards its centre. Hence, if the moment of momentum of the matter in the fixed spherical space varies, it must be by the moment of momentum of the matter which enters it not balancing exactly that of the matter which leaves it. We shall see later (§§ 20, 17, 18) that this balancing is vitiated by the entry of either a moving solid, or of some of the liquid, if any there is, of which spherical portions possess moment of momentum, into the fixed spherical space; but it is perfect under the condition of § 10, as will be proved in § 15.

12. First, I shall prove the following purely mathematical lemmas; using the ordinary notation u, v, w for the components of fluid velocity at any point (x, y, z).

Lemma (1). The condition (last clause) of § 10 requires that $udx + vdy + wdz$ be a complete differential*, at whatever instant and through whatever part of the fluid the condition holds.

* This proposition was, I believe, first proved by Stokes in his paper "On the Friction of Fluids in Motion, and the Equilibrium and Motion of Elastic Solids," *Cambridge Philosophical Transactions*, 14th April, 1845.

Lemma (2). If $u\,dx + v\,dy + w\,dz$ be a complete differential of a single valued function of x, y, z, through any finite space of the fluid, at any instant, the condition of § 10 holds through that space at that instant.

13. The following is Stokes' proof of Lemma (1):—First, for any motion whatever, whether subject to the condition of § 10 or not, let L be the component moment of momentum round OX of an infinitesimal sphere with its centre at O. Denoting by $\iiint$ integration through this space we have

$$L = \iiint (wy - vz)\,dx\,dy\,dz \quad \text{.................(1).}$$

Now let $(dw/dx)_0$, $(dw/dy)_0$, &c. denote the values at O of the differential coefficients. We have, by Maclaurin's theorem,

$$w = x\left(\frac{dw}{dx}\right)_0 + y\left(\frac{dw}{dy}\right)_0 + z\left(\frac{dw}{dz}\right)_0,$$

and so for v. Hence, remembering that $(dw/dx)_0$, &c. are constants for the space through which the integration is performed, we have

$$\iiint dx\,dy\,dz\,wy$$

$$= \left(\frac{dw}{dx}\right)_0 \iiint xy\,dx\,dy\,dz + \left(\frac{dw}{dy}\right)_0 \iiint y^2 dx\,dy\,dz + \left(\frac{dw}{dz}\right)_0 \iiint zy\,dx\,dy\,dz.$$

The first and third of the triple integrals vanish, because every diameter of a homogeneous sphere is a principal axis; and if A denote moment of momentum of the spherical volume round its centre, we have for the second

$$\iiint y^2\,dx\,dy\,dz = \tfrac{1}{2}A.$$

Dealing similarly with vz in the expression for L, we find

$$L = \tfrac{1}{2}A\left[\left(\frac{dw}{dy}\right)_0 - \left(\frac{dv}{dz}\right)_0\right] \quad \text{.................(2).}$$

But L must be zero according to the condition of § 10; and, therefore, as the centre of the infinitesimal sphere now considered may be taken at any point of space through which this condition holds at any instant, we must have, throughout that space,

$$\left.\begin{aligned} \frac{dw}{dy} - \frac{dv}{dz} &= 0 \\ \text{and similarly} \qquad \frac{du}{dz} - \frac{dw}{dx} &= 0 \\ \frac{dv}{dx} - \frac{du}{dy} &= 0 \end{aligned}\right\} \quad \text{.......................(3);}$$

which proves Lemma (1).

14. To prove Lemma (2), let

$$u=\frac{d\phi}{dx},\quad v=\frac{d\phi}{dy},\quad w=\frac{d\phi}{dz}\quad \text{..............(4)};$$

and let L denote the component moment of momentum round OX, through any spherical space with O in centre. We have [(1) of § 13],

$$L=\iiint dxdydz\,(wy-vz)\quad \text{................(5)},$$

$\iiint$ denoting integration through this space (not now infinitesimal). But by (4)

$$yw-vz=\left(y\frac{d}{dz}-z\frac{d}{dy}\right)\phi=\frac{d\phi}{d\psi}\text{..............(6)};$$

if $d/d\psi$ denote differentiation with reference to ψ, in the system of co-ordinates x, ρ, ψ, such that

$$y=\rho\cos\psi,\quad z=\rho\sin\psi\quad \text{................(7)}.$$

Hence, transforming (5) to this system of co-ordinates, we have

$$L=\iiint dxd\rho\,\rho d\psi\frac{d\phi}{d\psi}\quad \text{....................(8)}.$$

Now, as the whole space is spherical, with the origin of co-ordinates in its centre, we may divide it into infinitesimal circular rings with OX for axis, having each for normal section an infinitesimal rectangle with dx and $d\rho$ for sides. Integrating first through one of these rings, we have

$$dxd\rho\,\rho\int_0^{2\pi}\frac{d\phi}{d\psi}d\psi,$$

which vanishes, because ϕ is a single-valued function of the co-ordinates. Hence $L=0$, which proves Lemma (2).

15. Returning now to the dynamical proposition, stated at the conclusion of § 11; for the promised proof, let R denote the radial component velocity of the fluid across any element, $d\sigma$, of the spherical surface, situated at (x, y, z); and let u, v, w be the three components of the resultant velocity at this point; so that

$$R=u\frac{x}{r}+v\frac{y}{r}+w\frac{z}{r}\quad \text{....................(9)}.$$

The volume of fluid leaving the hollow spherical space across $d\sigma$ in an infinitesimal time dt is $Rd\sigma\,.\,dt$, and the moment of momentum of this moving mass round the centre has, for component round OX,

$$(wy - vz)\, Rd\sigma\, dt.$$

Hence, if L denote the component of the moment of momentum of the whole mass within the spherical surface at any instant, t, we have (§ 11),

$$\frac{dL}{dt} = \iint (wy - vz)\, R\, d\sigma \text{ ....(10).}$$

Now, using Lemma (1) of § 12, and the notation of § 14, we have

$$wy - vz = \frac{d\phi}{d\psi},$$

and, by (9),

$$R = \frac{d\phi}{dr},$$

where d/dr denotes rate of variation per unit length perpendicular to the spherical surface, that is differentiation with reference to r, the other two co-ordinates being directional relatively to the centre. Hence, using ordinary polar co-ordinates, r, θ, ψ, we have

$$\frac{dL}{dt} = r^2 \iint \frac{d\phi}{dr} \frac{d\phi}{d\psi} \sin\theta\, d\theta\, d\psi \text{ ....(11).}$$

But the "equation of continuity" for an incompressible liquid (being

$$\frac{du}{dx} + \frac{dv}{dy} + \frac{dw}{dz} = 0),$$

gives* $\nabla^2 \phi = 0$, for every point within the spherical space; and therefore [Thomson and Tait, App. B]

$$\phi = S_0 + S_1 r + S_2 r^2 + \&\text{c.} \text{ ....(12),}$$

a converging series, where S_0 denotes a constant, and S_1, S_2, &c., surface harmonics of the orders indicated.

Hence $$R = \frac{d\phi}{dr} = S_1 + 2r S_2 + 3r^2 S_3 + \&\text{c.} \text{ ....(13).}$$

And it is clear from the synthesis of the most general surface harmonic, by zonal, sectorial, and tesseral harmonics [Thomson and Tait, § 781], that $dS_i/d\psi$ is a surface harmonic of the same

* By ∇^2 we shall always understand $d^2/dx^2 + d^2/dy^2 + d^2/dz^2$.

order as S_i*: from which [Thomson and Tait, App. B (16)], it follows that,

$$\iint S_i \frac{dS_{i'}}{d\psi} \sin\theta\, d\theta\, d\psi = 0,$$

except when $i' = i$. But this is true also when $i' = i$ because

$$S_i \frac{dS_i}{d\psi} = \tfrac{1}{2} \frac{d\,(S_i^2)}{d\psi},$$

and therefore, as in § 14, the integration for ψ, from $\psi = 0$ to $\psi = 2\pi$ gives zero. Hence (11) gives

$$\frac{dL}{dt} = 0.$$

This and § 11 establish § 10.

16. Lemma (1) of § 11, and § 10 now proved, show that in any motion whatever of an incompressible liquid, whether with solids immersed in it or not, $udx + vdy + wdz$ is always a complete differential through any portion of the fluid, for which it is a complete differential at any instant, to whatever shape and position of space this portion may be brought in the course of the motion. This is the ordinary statement of the fundamental proposition of fluid motion referred to in § 9, which was first discovered by Lagrange. (For another proof see § 60.) I have given the preceding demonstration, not so much because it is useful to look at mathematical structures from many different points of view, but (§ 19) because the dynamical considerations and the formulæ I have used are immediately available for establishing the theory of the impulse (§§ 3—8), of which a

* This follows, of course, from the known analytical theorem that the operations ∇^2 and $(y\, d/dz - z\, d/dy)$ are commutative, which is proved thus :—

By differentiation we have

$$\frac{d^2\left(y\dfrac{d\phi}{dz}\right)}{dy^2} = y\frac{d^2}{dy^2}\frac{d\phi}{dz} + 2\frac{d}{dy}\frac{d\phi}{dz};$$

and therefore, since $d/dy\; d\phi/dz = d/dz\; d\phi/dy$,

$$\nabla^2\left(y\frac{d\phi}{dz} - z\frac{d\phi}{dy}\right) = y\nabla^2\left(\frac{d\phi}{dz}\right) - z\nabla^2\left(\frac{d\phi}{dy}\right) = \left(y\frac{d}{dz} - z\frac{d}{dy}\right)\nabla^2\phi,$$

or

$$\nabla^2\left(y\frac{d}{dz} - z\frac{d}{dy}\right)\phi = \left(y\frac{d}{dz} - z\frac{d}{dy}\right)\nabla^2\phi,$$

ϕ being any function whatever. Hence, if $\nabla^2\phi = 0$, we have

$$\nabla^2\left(y\frac{d\phi}{dz} - z\frac{d\phi}{dy}\right) = 0.$$

fundamental proposition was stated above (§ 5). To prove this proposition (in § 19) I now proceed.

17. Imagine any spherical surfaces to be described round a moveable solid or solids immersed in a liquid. The surrounding fluid can only press (§ 1) perpendicularly; and therefore when any motion is (§ 3) generated by impulsive forces applied to the solids, the moment round any diameter of the momentum of the matter within the spherical surface at the first instant, must be exactly equal to the moment of those impulsive forces round this line. And the moment round this line, of the momentum of the matter in the space between any two concentric spherical surfaces is zero, provided neither cuts any solid, and provided that, if there are any solids in this space, no impulse acts on them.

18. Hence, considering what we have defined as "the impulse of the motion," (§ 6), we see that its moment round any line is equal to the moment of momentum round the same line, of all the motion within any spherical surface having its centre in this line, and enclosing all the matter to which any constituent of the impulse is applied. This will still hold, though there are other solids not in the neighbourhood, and impulses are applied to them: provided the moments of momentum of those only which are within S are taken into account, and provided none of them is cut by S.

19. The statements of § 11, regarding fluid occupying at any instant a fixed spherical surface, are applicable without change to the fluids and solids occupying the space bounded by S, because of our present condition, that no solid is cut by S. Hence every statement and formula of § 15, as far as equation (11), may be now applied to the matter within S; but instead of (12) we now have [Thomson and Tait, § 736], if we denote by T_1, T_2, &c., another set of surface spherical harmonics,

$$\phi = \left.\begin{array}{l} S_0 + S_1 r + S_2 r^2 + \&c. \\ \qquad + T_1 r^{-2} + T_2 r^{-3} + \&c. \end{array}\right\} \quad \text{............(14)*},$$

for all space between the greatest and smallest spherical surface concentric with S, and having no solids in it, because through all

* There is no term T_0/r, because this would give, in the integral of flow across the whole spherical surface, a finite amount of flow out of or into the space within, implying a generation or destruction of matter.

this space, § 16, and the equation of continuity prove that $\nabla^2\phi = 0$. Hence, instead of (13), we now have

$$\left.\begin{aligned} R = \frac{d\phi}{dr} &= S_1 + 2rS_2 + 3r^2S_3, \text{ \&c.} \\ &-\frac{2}{r^3}T_1 - \frac{3}{r^4}T_2 - \frac{4}{r^5}T_3 + \text{\&c.} \end{aligned}\right\} \quad \text{.........(15).}$$

Hence finally

$$\frac{dL}{dt} = \sum_{i=0}^{i=\infty} \iint \left[iS_i \frac{dT_i}{d\psi} - (i+1)\, T_i \frac{dS_i}{d\psi} \right] \sin\theta \, d\theta \, \psi \quad \text{...(16).}$$

Now if, as assumed in § 5, neither any moveable solids, nor any part of the boundary exist within any finite distance of S all round; S_1, S_2, &c., must each be infinitely small: and therefore (16) gives $dL/dt = 0$. This proves the proposition asserted in § 5: because a system of forces cannot have zero moment round every line drawn through any finite portion of space, without having force-resultant and couple-resultant each equal to zero.

20. As the rigidity of the solids has not been taken into account, all or any of them may be liquefied (§ 3) without violating the demonstration of § 19. To save circumlocutions, I now define a *vortex* as a portion of fluid having any motion that it could not acquire by fluid pressure transmitted through itself from its boundary. Often, merely for brevity, I shall use the expression *a body* to denote either a solid or a vortex, or a group of solids or vortices.

21. The proposition thus proved may be now stated in terms of the definitions of § 6, which were not used in § 5, and so becomes simply this:—*The impulse of the motion of a solid or group of solids or vortices and the surrounding liquid remains constant as long as no disturbance is suffered from the influence of other solids or vortices, or of the containing vessel.*

This implies, of course (§ 6), that the magnitudes of the force-resultant and the rotational moment of the impulse remain constant, and the position of its axis invariable.

22. In Poinsot's system of the statics of a rigid body we may pass from the resultant force and couple along and round the central axis to an equal resultant force along the parallel line through any point, and a greater couple the resultant of the former (or minimum) couple, and a couple in the plane of the two parallels, having its moment equal to the product of their distance

into the resultant force. So we may pass from the force-resultant and rotational moment of the impulse along and round its axis, to an equal force-resultant and greater moment of impulse, by transferring the former to any point, Q, not in the axis (§ 6) of the motion. This greater moment is (§ 18) equal to the moment of momentum round the point Q, of the motion within any spherical surface described from Q as centre, which encloses all the vortices or moving solids.

23. Hence a group of solids or vortices which always keep within a spherical surface of finite radius, or a single body, moving in an infinite liquid, can have no permanent average motion of translation in any direction oblique to the direction of the force-resultant of the impulse, if there is a finite force-resultant. For the matter within a finite spherical surface enclosing the moving bodies or body, cannot have moment of momentum round the centre increasing to infinity.

24. But there may be motion of translation when the force-resultant of the impulse vanishes; and there will be, for example, in the case of a solid, shaped like the screw-propeller of a steamer, immersed in an infinite homogeneous liquid, and set in motion by a couple in a plane perpendicular to the axis of the screw.

25. And when the force-resultant of the impulse does *not* vanish, there may be no motion of translation, or there may even be translation in the direction opposite to it. Thus, for example, a rigid ring, with cyclic motion, established (§ 63) through it, will, if left at rest, remain at rest. And if at any time urged by an impulse in either direction in the line of the force-resultant of the impulse of the cyclic motion, it will commence and continue moving with an average motion of translation in that direction; a motion which will be uniform, and the same as if there were no cyclic motion, when the ring is symmetrical. If the translatory impulse is contrary to the cyclic impulse, but less in magnitude, the translation will be contrary to the whole force-resultant impulse.

If the translatory impulse is equal and opposite to the cyclic impulse, there will be translation with zero force-resultant impulse—another example of what is asserted in § 24. In this case, if the ring is plane and symmetrical, or of any other shape such that the cyclic motion (which, to fix ideas, we have supposed given first, with the ring at rest) must have had only a force-resultant, and no rotational moment, we have a solid moving with a uniform motion of translation through a fluid, and both force and couple resultant of the whole impulse zero.

26. From §§ 21 and 4, we see that, however long the time of application of the impressed forces may be—provided only that, during the whole of it, the solid or group of solids has been at an infinite distance from all other solids and from the containing vessel—the time integrals of the impressed forces parallel to three fixed axes, and of their moments round these lines, are equal to the six corresponding components of "the impulse" (§ 6).

27. If two groups, at first so far asunder as to exercise no sensible influence on one another, come together, the "impulse" of the whole system remains unchanged by any disturbance each may experience from the other, whether by impacts of the solids, or through motion and pressure of the surrounding fluid; and (§ 6) it is always reducible to the force-resultant along the central axis, and the minimum couple-resultant, of the two impulses reckoned as if applied to one rigid body. The same holds, of course, if one group separates into two so distant as to no longer exert any sensible influence on one another.

28. Hence whatever is lost of impulse perpendicular to a fixed plane, or of component rotational movement round a fixed line, by one group through collision with another, is gained by the other.

29. Two of the moveable solids, or two groups, will be said to be *in collision* when, having been so far asunder as not to disturb one another's motions sensibly, they are so near as to do so. This disturbance will generally be supposed to be through fluid pressure only, but impacts of solids on solids may take place during a collision.

30. We are now prepared to investigate (§§ 30, 31, 32) the influence of a fixed solid on the impulse of a moveable solid, or of a vortex, or of a group of solids or vortices, passing near it, thus—If during such collisions or separations as are considered in §§ 27, 28, forces are impressed on any one or more of the solids, their alteration of the whole impulse is (§ 26) to be reckoned by adding to each of its rectangular components the time integral of the

corresponding component of these impressed forces. Now, let us suppose such forces to be impressed on any one of the moveable solids as shall keep it at rest. These forces are zero as long as no moving solid is within a finite distance. But if a moving solid or vortex, or group of solids or vortices, passes near the fixed solid, the change of pressure due to the motion of the fluid will tend to move it, and the impression of force on it becomes necessary to keep it fixed. Let $d\sigma$ be an element of its surface; (x, y, z) the co-ordinates of the centre of this element; α, β, γ the inclinations of the normal at (x, y, z) to the three rectangular axes; and p the fluid pressure at time t, and point (x, y, z). The six components of force and couple required to hold the body fixed at time t, are

$$\left.\begin{array}{l}\iint d\sigma.\cos\alpha.p,\ \iint d\sigma.\cos\beta.p,\ \iint d\sigma.\cos\gamma.p;\\ \iint d\sigma(y\cos\gamma - z\cos\beta)p,\ \iint d\sigma(z\cos\alpha - x\cos\gamma)p,\ \iint d\sigma(x\cos\beta - y\cos x)p\end{array}\right\} \quad \text{.........(1).}$$

If in these expressions we substitute

$$\int pdt \quad \text{..............................(2),}$$

in place of p ($\int dt$ denoting a time integral from any era of reckoning before the disturbance became sensible, up to time t, which may be any instant during the collision, or after it is finished), we have the changes in the corresponding components of the impulse up to time t, provided there has been no impact of moveable solid on the fixed solid.

31. Let now the "velocity potential" (as we shall call it, in conformity with a German usage which has been adopted by Helmholtz), be denoted by ϕ; that is (§ 16), let ϕ be such a function of (x, y, z, t) that

$$u = \frac{d\phi}{dx},\quad v = \frac{d\phi}{dy},\quad w = \frac{d\phi}{dz} \text{.................(3),}$$

and let $\dot{\phi}$ (or $d\phi/dt$) denote its rate of variation per unit of time at any instant t, for the point (x, y, z) regarded as fixed.

Also, let q denote the resultant fluid velocity, so that

$$q^2 = u^2 + v^2 + w^2 = \frac{d\phi^2}{dx^2} + \frac{d\phi^2}{dy^2} + \frac{d\phi^2}{dz^2} \quad \text{..........(4).}$$

The ordinary hydrodynamical formula gives

$$p = \Pi - \dot{\phi} - \tfrac{1}{2}q^2 \text{........................(5);}$$

where Π denotes the constant pressure in all sensibly quiescent parts of the fluid.

32. The constant term Π disappears from p in each of the integrals (1) of § 30, because a solid is equilibrated by equal pressure around. And in the time integral (2), we have

$$\int \dot{\phi}\, dt = \phi \quad \text{.........................(6)};$$

and therefore if (XYZ) (LMN) denote the changes in the force- and couple-components of the impulse produced by the collision up to time t, we have

$$\left.\begin{array}{l} X = -\iint d\sigma \cos\alpha \left(\phi + \tfrac{1}{2}\int q^2 dt\right), \quad Y = \&c., \quad Z = \&c. \\ L = -\iint d\sigma \left(y\cos\gamma - z\cos\beta\right)\left(\phi + \tfrac{1}{2}\int q^2 dt\right), \quad M = \&c., \quad N = \&c. \end{array}\right\} \quad \text{.........(7)}.$$

But because the fluid is quiescent in the neighbourhood of the fixed body when the moving body or group of bodies is infinitely distant from it; it follows that before the commencement and after the end of the collision we have $\phi = 0$ at every point of the surface of the fixed body. Hence, for every value of t representing a time after the completion of the collision, the preceding expressions become

$$\left.\begin{array}{l} X = -\tfrac{1}{2}\iint d\sigma \cos\alpha \int q^2 dt, \quad Y = \&c., \quad Z = \&c. \\ L = -\tfrac{1}{2}\iint d\sigma \left(y\cos\gamma - z\cos\beta\right)\int q^2 dt, \quad M = \&c., \quad N = \&c. \end{array}\right\} \text{...(8)};$$

which express that *the integral change of impulse experienced by a body or group of bodies, in passing beside a fixed body without striking it, may be regarded as a system of impulsive attractions towards the latter, everywhere in the direction of the normal, and amounting to* $\tfrac{1}{2}\int q^2 dt$ *per unit of area.* But it must not be forgotten that the term ϕ in the expression [§ 31 (5)] for p produces, as shown in § 30 (1), an influence *during the collision*, the integral effect of which only disappears from the expression [§ 32 (7)] for the impulse *after the collision is completed;* that is (§ 29) after the moving system has passed away so far as to leave no sensible fluid motion in the neighbourhood of the fixed body.

33. Hence, and from § 23, we see that when there is no impact of moving solid against the fixed body, and when the moving solid or group of solids passes altogether on one side of the fixed body, the direction of the translation will be deflected, as if there were, on the whole, an *attraction towards* the fixed body,

or a *repulsion from it*, according as (§ 25) the translation is in the direction of the impulse or opposite to it. For, in each case, the impulse is altered by the introduction of an impulse *towards* the fixed body upon the moving body or bodies as they pass it; and (§ 23) the translation before and after the collision is always along the line of the impulse, and is altered in direction accordingly. This will be easily understood from the diagrams, where in each case B represents the fixed body, the dotted line $ITT'I'$ and arrow-heads II', the directions of the force-resultant of the impulse at successive times, and the full arrow-heads TT', the directions of the translation.

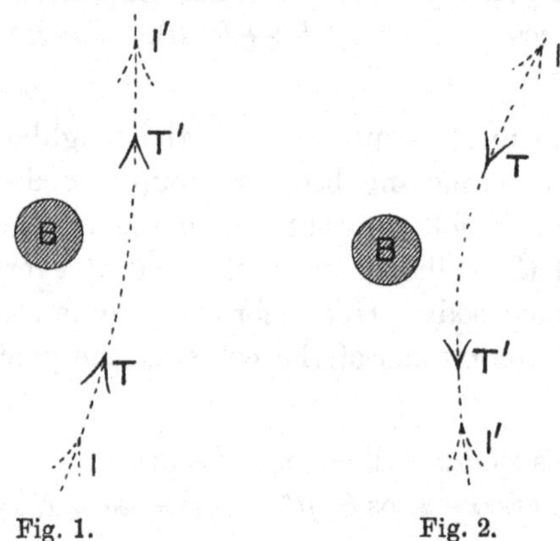

Fig. 1. Fig. 2.

All ordinary cases belong to the class illustrated by fig. 1. The case of a rigid ring, with cyclic motion (§ 25) established round it as core, belongs to the class illustrated by fig. 2, if the ring be projected through the fluid in the direction perpendicular to its own plane, and contrary to the cyclic motion through its centre.

34. When (§ 66) we substitute vortices for the moving solids, we shall see (§ 67) that the translation is probably always in the direction *with* the impulse. Hence, as illustrated by fig. 1, there is always the deflection, as if by *attraction*, when a group of vortices pass all on one side of a fixed body. This is easily observed, for a simple Helmholtz ring, by sending smoke-rings on a large scale, according to Professor Tait's plan, in such directions as to pass very near a convex fixed surface. An ordinary 12-inch globe, taken off its bearings and hung by a thin cord, answers very well for the fixed body.

35. The investigation of §§ 30, 31, 32 is clearly applicable to a vortex or a moving body, or to a group of vortices or moving bodies, which keep always near one another (§ 23), passing near a projecting part of the fixed boundary, and being, before and after this collision (§ 29), at a very great distance from every part of the fixed boundary. Thus a Helmholtz ring, projected so as to pass near a projecting angle of two walls, shows a deflection of its course, as if caused by attraction towards the corner.

36. In every case the force-resultant of the impulse is, as we shall presently see (§ 37), determinate when the flow of the liquid across every element of any surface completely enclosing the solids or vortices is given; but not so, from such data, either the axis (§ 6) or the rotational moment, as we see at once by considering the case of a solid sphere (which may afterwards be supposed liquefied) set in motion by a force in any line not through the centre, and a couple in a plane perpendicular to it. For this line will be the "axis," and the impulsive couple will be the rotational moment of the whole motion of the solid and liquid. But the liquid, on all sides, will move exactly as it would if the impulse were merely an impulsive force of equal amount in a parallel line through the centre of the sphere, with therefore this second line for "axis" and zero for rotational moment. For illustration of rotational moment remaining latent in a liquid (with or without solids) until made manifest by actions, tending to alter its axis, or showing effects of centrifugal force due to it, see § 66 and others later.

37. The component impulse in any direction is equal to the corresponding component momentum of the mass enclosed within the surface S, containing all the places of application of the impulse, together with that of the impulsive pressure outwards on this surface. But as the matter enclosed by S (whether all liquid or partly liquid and partly solid) is of uniform density, its momentum will be equal to its mass multiplied into the velocity of the centre of gravity of the space within the surface S supposed to vary so as to enclose always the same matter, and will therefore depend solely on the normal motion of S; that is to say, on the component of the fluid velocity in the direction of the normal at every point of S. And the impulsive fluid pressure, corresponding to the generation of the actual motion from rest, being the time

integral of the pressure during the instantaneous generation of the motion, is (§§ 31, 32) equal to $-\phi$, the velocity potential; which (§ 61) is determinate for every point of S, and of the exterior space when the normal component of the fluid motion is given for every point of S. Hence the proposition asserted in § 36. Denoting by $d\sigma$ any element of S; N the normal component of the fluid velocity; α the inclination to OX, of the normal drawn *outwards* through $d\sigma$; and X the x-component of the impulse; we have for the two parts of this quantity considered above, and its whole value, the following expressions; of which the first is taken in anticipation from § 42—

$$\left.\begin{array}{ll} x\text{-momentum of matter, within } S, & =\iint Nx\,d\sigma \quad \text{(8) of § 42} \\ x\text{-component of impulsive pressure} & \\ \quad \text{on } S\text{, outwards,} & =-\iint\phi\cos\alpha\,d\sigma \end{array}\right\}\ldots(1),$$

$$X=\iint(Nx-\phi\cos\alpha)\,d\sigma \quad \ldots\ldots(2).$$

It is worthy of remark that this expression holds for the impulse of all the solids or vortices within S, even if there be others in the immediate neighbourhood outside: and that therefore its value must be zero if there be no solids or vortices within S, and N and ϕ are due solely to those outside.

38. If ϕ be the potential of a magnet or group of magnets, some within S and others outside it, and N the normal component magnetic force, at any point of S, the preceding expression (2) is equal to the x-component of the magnetic moment of all the magnets within S, multiplied by 4π. For let ρ be the density of any continuous distribution of positive and negative matter, having for potential, and normal component force, ϕ and N respectively, at every point of S. We have [Thomson and Tait, § 491 (c)] $\rho=-1/4\pi\,\nabla^2\phi$, and therefore

$$\iiint\rho x\,dx\,dy\,dz=-\frac{1}{4\pi}\iiint x\left(\frac{d^2\phi}{dx^2}+\frac{d^2\phi}{dy^2}+\frac{d^2\phi}{dz^2}\right)dx\,dy\,dz\ldots(3).$$

Now, integrating by parts*, as usual with such expressions, we have

* The process here described leads merely to the equation obtained by taking the last two equal members of App. A (1) (Thomson and Tait) for the case $a=1$, $U=\phi$, $U'=x$.

$$\iiint x \frac{d^2\phi}{dx^2} dx dy dz = \iint x \frac{d\phi}{dx} dy dz - \iiint \frac{d\phi}{dx} dx dy dz$$
$$= \iint \left(x \frac{d\phi}{dx} - \phi\right) dy dz.$$

Hence integrating each of the other two terms of (3) once simply, and reducing as usual [Thomson and Tait, App. A (a)] to a surface integral we have

$$\iiint \rho x dx dy dz = -\frac{1}{4\pi} \iint (Nx - \phi \cos \alpha) d\sigma \ldots\ldots\ldots (4);$$

which proves the proposition, and also, of course, that if there be no matter within S, the value of the second member is zero.

39. Hence, considering the magnetic and hydrokinetic analogous systems with the sole condition that at every point of some particular closed surface the magnetic potential is equal to the velocity potential, we conclude that 4π times the magnetic moment of all the magnetism within any surface, in the magnetic system, is equal to the force-resultant of the impulse of the solids or vortices within the corresponding surface in the hydrokinetic system; and that the directions of the magnetic axis and of the force-resultant of the impulse are the same. For the theory of magnetism, it is interesting to remark that indeterminate distributions of magnetism within the solids, or portions of fluid to which initiating forces (§ 3) were applied, or determinate distributions in infinitely thin layers at their surfaces, may be found, which through all the space external to them shall produce the same potential as the velocity-potential, and therefore the same distribution of force as the distribution of velocity through the whole fluid. But inasmuch as when the magnetic force in the interior of a magnet is defined in the manner explained in § 48 (2) of my *Mathematical Theory of Magnetism* *, it is expressible through all space by the differential coefficients of a potential; and, on the contrary, for the kinetic system $udx + vdy + wdz$ is not a complete differential generally through the spaces occupied by the solids, the agreement between resultant force and resultant flow holds only through the space exterior to the magnets and solids in the magnetic and kinetic systems respectively. But if the other definition of resultant force within a magnet [*Math. Theory of*

* *Trans. R. S. Lond.* 1851; or *Thomson's Electrical Papers*, Macmillan, 1869.

Magnetism, § 77, foot-note, and § 78], published in preparation for a sixth chapter "On Electro-magnets" (still in my hands in manuscript, not quite completed), and which alone can be adopted for spaces occupied by non-magnetic matter traversed by electric currents, the magnetic force has not a potential within such spaces; and we shall see (§ 68) that determinate distributions of closed electric currents through spaces corresponding to the solids of the hydrokinetic system can be found which shall give for every point of space, whether traversed by electric currents or not, a resultant magnetic force, agreeing in magnitude and direction with the velocity, whether of solid or fluid, at the corresponding point of the hydrokinetic system. This thorough agreement for all space renders the electro-magnetic analogue preferable to the magnetic; and, having begun with the magnetic analogous system only because of its convenience for the demonstration of § 38, we shall henceforth chiefly use the purely eloctro-magnetic analogue.

40. To prove the formula used in anticipation, in § 37 (1) we must now (§§ 41, 42, 43) find the momentum of the whole matter —fluid, fluid and solid, or even solid alone—at any instant within a closed surface S, in terms of the normal component velocity of the matter at any point of this surface, or, which is the same, the normal velocity of this surface itself, if we suppose it to vary so as to enclose always the same matter.

41. Let V be the volume of the space bounded by any varying closed surface S. As yet we need not suppose V constant. Let $\bar{x}$, $\bar{y}$, $\bar{z}$ be the co-ordinates of the centre of gravity. We have

$$V\bar{x} = \tfrac{1}{2} \iint [x^2 dy\,dz] \quad\ldots\ldots\ldots\ldots\ldots\ldots\ldots(5),$$

where [ ] indicates that the expression within it is to be taken between proper limits for S. Now as S varies with the time, the area through which $\iint dy\,dz$ is taken will in general vary; but the increments or decrements which it experiences at different parts of the boundary of this area, in the infinitely small time dt, contribute no increments or decrements to $\iint [x^2 dy\,dz]$, as we see most easily by first supposing S to be a surface everywhere convex outwards. Hence

$$\frac{d}{dt}\iint [x^2 dy\,dz] = \iint \left[\frac{d\,(x^2)}{dt}\,dy\,dz\right] = 2\iint \left[x\,\frac{dx}{dt}\,dy\,dz\right] \ldots\ldots(6).$$

But if N denote the velocity with which the surface moves in the direction of its outward normal at (x, y, z), we have, in the preceding expression

$$\frac{dx}{dt} = N \sec \alpha$$

if α be the inclination of the outward normal to OX. Hence

$$\frac{d(V\bar{x})}{dt} = \iint [xN \sec \alpha \, dy \, dz].$$

But the conditions as to limits indicated by [ ] are clearly satisfied, if, $d\sigma$ denoting an element of the surface, such that

$$dydz = \cos \alpha \, d\sigma,$$

we simply take $\iint d\sigma$ over the whole surface. Thus we have

$$\frac{d(V\bar{x})}{dt} = \iint xN d\sigma \quad \text{....................(7).}$$

42. In any case in which V is constant, this becomes

$$V\frac{d\bar{x}}{dt} = \iint xN d\sigma \quad \text{....................(8).}$$

If now the varying surface, S, is the boundary of a portion of the matter—fluid or solid—of uniform density unity, with whose motions we are occupied, the x-component momentum of this portion is $V d\bar{x}/dt$; and, therefore, equation (8) is the required (§ 40) expression.

43. The same formulæ (7) and (8) are proved more shortly of course by the regular analytical process given by Poisson* and Green† in dealing with such subjects; thus, in short. Let u, v, w be the components of velocity, of any matter, compressible or incompressible, at any point (x, y, z) within S; and let c denote the value at this point of $du/dx + dv/dy + dw/dz$, so that

$$\frac{du}{dx} = c - \left(\frac{dv}{dy} + \frac{dw}{dz}\right) \quad \text{....................(9).}$$

We have, for the component momentum of the whole matter within S, if of unit density at the instant considered,

$$\iiint u dx dy dz = \iint ux dy dz - \iiint x \frac{du}{dx} dx dy dz \text{......(10).}$$

* *Théorie de la Chaleur*, § 60.

† *Essay on Electricity and Magnetism.*

But by (9)

$$\iiint x\frac{du}{dx}dxdydz = \iiint cx\,dxdydz - \iiint x\left(\frac{dv}{dy}+\frac{dw}{dz}\right)dxdydz,$$

and by simple integrations,

$$\iiint x\left(\frac{dv}{dy}+\frac{dw}{dz}\right)dxdydz = \iint x\,(vdxdz + wdxdy).$$

Using these in (10), and altering the expression to a surface integral, as in Thomson and Tait, App. A (a), we have

$$\begin{aligned}\iiint u\,dxdydz &= \iint x\,(udydz + vdzdx + wdxdy) - \iiint cx\,dxdydz \\ &= \iint xNd\sigma - \iiint cx\,dxdydz \quad \ldots\ldots\ldots\ldots\ldots\ldots\ldots(11),\end{aligned}$$

which clearly agrees with (7).

When this mass is incompressible, we have $c = 0$ by the formula so ill named the equation "of continuity" (Thomson and Tait, § 191), and we fall upon (8).

The proper analytical interpretation of the differential coefficients du/dx, &c., and of the equation of continuity, when, as at the surfaces of separation of fluid and solids, u, v, w are discontinuous functions having abruptly varying values, presents no difficulty.

44. In the theory of the impulse applied to the collision (§ 29) of solids or vortices moving through a liquid, the force-resultant of the impulse corresponds, as we have seen, precisely to the resultant momentum of a solid in the ordinary theory of impact. Some difficulty may be felt in understanding how the zero-momentum (§ 4) of the whole mass is composed; there being clearly positive momentum of solids and fluids in the direction of the impulse in some localities near the place of its application, and negative in others. [Consider, for example, the simple case of a solid of revolution struck by a single impulse in the line of its axis. The fluid moves in the direction of the impulse, before and behind the body, but in the contrary direction in the space round its middle.] Three modes of dividing the whole moving mass present themselves as illustrative of the distribution of momentum through it; and the following propositions (§ 45) with reference to them are readily proved (§§ 46, 47, 48).

45. I. Imagine any cylinder of finite periphery, not necessarily circular, completely surrounding the vortices (or moving solids), and any other surrounding none, and consider the in-

finitely long prisms of variously moving matter at any instant surrounded by these two cylinders. The component momentum parallel to the length of the first is equal to the component of the impulse parallel to the same direction; and that of the second is zero.

II. Imagine any two finite spherical surfaces, one enclosing all the vortices or moving solids, and the other none. The resultant-momentum of the whole matter enclosed by the first is in the direction of the impulse, and is equal to $\frac{2}{3}$ of its value. The resultant-momentum of the whole fluid enclosed by the second is the same as if it all moved with the same velocity, and in the same direction, as at its centre.

III. Imagine any two infinite planes at a finite distance from one another and from the field of motion, but neither cutting any solid or vortex. The component perpendicular to them of the momentum of the matter occupying at any instant the space between them (whether this includes none, some, or all of the vortices or moving solids) is zero.

46. To prove these propositions:—

Consider in either case a finite length of the prism extending to a very great distance in each direction from the field of motion, and terminated by plane or curved ends. Then, the motion being, as we may suppose (§ 61) started from rest by impulsive pressures on the solids [or (§ 66) on the portions of fluid constituting the vortices]; the impulsive fluid pressure on the cylindrical surface can generate no momentum parallel to the length; and to generate momentum in this direction there will be, in case 1, the impressed impulsive forces on the solids, and the impulsive fluid pressures on the ends; but in case 2 there will be only the impulsive fluid pressure on the ends. Now, the impulsive fluid pressures on the ends diminish [§ 50 (15)] according to the inverse square of the distance from the field of motion, when the prism is prolonged in each direction, and are therefore infinitely small when the prisms are infinitely long each way. Whence the proposition I.

47. By using the harmonic expansions § 19 (14), (15), in the several expressions § 37 (1), (2); and the fundamental theorem

$$\iint \mathfrak{S}_i \mathfrak{S}_{i'} d\sigma = 0,$$

of the harmonic analysis [Thomson and Tait, App. B (16)]; and putting $S_i=0$ for one case, and $T_i=0$ for the other; we prove the two parts of Prop. II., § 45, immediately.

48. To prove Prop. III., § 45, the well-known theory of electric images in a plane conductor* may be conveniently referred to. It shows that if N_1 denotes the normal component force at any point of an infinite plane due to any distribution, μ, of matter in the space lying on one side of the plane, a distribution of matter over the plane having $N_1/2\pi$ for surface density at each point exerts the same force as μ through all the space on the other side of the plane, and therefore that the whole quantity of matter in that surface distribution is equal to the whole quantity of matter in μ†. Hence, $\iint d\sigma$ denoting integration over the infinite *plane*,

$$\iint N_1 d\sigma = 0 \quad\text{.........................(12)},$$

if the whole quantity of matter in μ be zero. Hence, if N be the normal force due to matter through space on both sides of the plane, provided the whole quantity of matter on each side separately is zero,

$$\iint N d\sigma = 0 \quad\text{..........................(13)};$$

since N is the sum of two parts, for each of which separately (12) holds. This, translated into hydrokinetics, shows that the whole flow of matter across any infinite plane is zero at every instant when it cuts no solids or vortices. Hence, and from the uniformity of density which (§ 3) we assume, the centre of gravity of the matter between any two infinite fixed parallel planes has no motion in the direction perpendicular to them at any time when no vortex or moving solid is cut by either: which is Prop. III. of § 45 in other words.

* Thomson, *Camb. and Dub. Math. Journal*, 1849; Liouville's *Journal*, 1845 and 1847; or Reprints of *Electrical Papers* (Macmillan, 1869).

† This is verified synthetically with ease, by direct integrations showing (whether by Cartesian or polar plane co-ordinates), that

$$\int_{-\infty}^{\infty}\int_{-\infty}^{\infty}\frac{a\,dy\,dz}{(a^2+y^2+z^2)^{\frac{3}{2}}}=2\pi.$$

And taking d/da of this, we have

$$\int_{-\infty}^{\infty}\int_{-\infty}^{\infty}\frac{(y^2+z^2-2a^2)\,dy\,dz}{(a^2+y^2+z^2)^{\frac{5}{2}}}=0,$$

the synthesis of (12).

49. The integral flow of matter across any surface whatever, imagined to divide the whole volume of the finite fixed containing vessel of § 1 into two parts is necessarily zero, because of the uniformity of density; and therefore the momentum of all the matter bounded by two parallel planes, extending to the inner surface of the containing vessel, and the portion of this surface intercepted between them has always zero for its component perpendicular to these planes, whether or not moving solids or vortices are cut by either or both these planes. But it is remarkable that when any moving solid or vortex is cut by a plane, the integral flow of matter across this plane (if the containing vessel is infinitely distant on all sides from the field of motion), converges to a generally *finite* value, as the plane is extended to very great distances all round from the field of motion, which are still infinitely small in comparison with the distances to the containing vessel; and diminishes from that finite value to zero by another convergence, when the distances to which the plane is extended all round begin to be comparable with, and ultimately become equal to, the distances of the curve in which it cuts the containing vessel. Hence we see how it is that the condition of neither plane cutting any moving solid or vortex is necessary to allow § 45, III. to be stated without reference to the containing vessel, and are reminded that the equality to zero asserted in this proposition is proved in § 48 to be approximated to when the planes are extended to distances all round, which, though infinitely short of the distances to the containing vessel, are very great in comparison with their perpendicular distances from the most distant parts of the field of motion.

50. The convergencies concerned in § 45, I., III., may be analysed thus. Perpendicular to the resultant impulse draw any two planes on the two sides of the field of motion, with all the moving solids and vortices between them, and divide a portion of the space between them into finite prismatic portions by cylindrical (or plane) surfaces perpendicular to them. Suppose now one of these prismatic portions to include all the moving solids and vortices, and without altering the prismatic boundary, let the parallel planes be removed in opposite directions to distances each infinite (or very great) in comparison with the distance of the most distant of the moving solids or vortices. By § 45, I., the momen-

tum of the motion within this prismatic space is (approximately) equal to the force-resultant, I, of the impulse, and that of the motion within any one of the others is (approximately) zero.

But the sum of these (approximately) zero values must, on account of § 45, III., be equal to $-I$, if the portions of the planes containing the ends of the prismatic spaces be extended to distances very great in comparison with the distance between the planes. To understand this, we have only to remark that if ϕ denotes the velocity potential at a point distant D from the middle of the field, and x from a plane through the middle perpendicular to the impulse, we have (§ 53) approximately,

$$\phi = -\frac{Ix}{4\pi D^3}$$

provided D be great in comparison with the radius of the smallest sphere enclosing all the moving solids or vortices. Hence, putting $x = \pm a$ for the two planes under consideration, denoting by A the area of either end of one of the prismatic portions, and calling D *the proper mean distance* for this area, we have (§ 45) for the momentum of the fluid motion within this prismatic space, provided it contains no moving solids or vortices,

$$-2\frac{Ia}{4\pi D^3}$$

This vanishes when A/D^2 is an infinitely small fraction (as a/D is at most unity); but it is finite if A/D^2 is finite, provided a/D be not infinitely small. And its integral value (compare § 48, footnote) converges to $-I$, when the portion of area included in the integration is extended till a/D is infinitely small for all points of its boundary.

51. Both as regards the mathematical theory of the convergence of definite integrals, and as illustrating the distribution of momentum in a fluid, it is interesting to remark that, u denoting component velocity parallel to x, at any point (x, y, z), the integral $\iiint u\,dx\,dy\,dz$, expressing momentum, may, as is readily proved, have any value from $-\infty$ to $+\infty$ according to the portions of space through which it is taken.

52. As a last illustration of the distribution of momentum, let the containing vessel be spherical of finite radius a.

We have, as in § 19,

$$\left.\begin{aligned}\phi = S_0 + S_1 r + S_2 r^2 + \&c. \\ + T_1 r^{-2} + T_2 r^{-3} + \&c.\end{aligned}\right\} \quad \text{.............. (14)},$$

each series converging, provided r is less than a, and greater than the radius of the smallest concentric spherical surface enclosing all the solids or vortices. Now, by the condition that there be no flow across the fixed containing surface, we must have

$$\frac{d\phi}{dr} = 0, \text{ when } r = a \quad \text{.................... (15)},$$

which gives

$$S_i = \frac{i+1}{i}\frac{T_i}{a^{2i+1}} \quad \text{...................... (16)};$$

and (14) becomes

$$\phi = \frac{T_1}{r^2}\left(1 + 2\frac{r^3}{a^3}\right) + \frac{T_2}{r^3}\left(1 + \frac{3}{2}\frac{r^5}{a^5}\right) + \&c. \quad \text{......... (17)}.$$

But [§ 37 (1)] if the whole amount of the x-component of impulsive pressure exerted by the fluid within the spherical surface of radius r, upon the fluid round it be denoted by F, we have

$$F = -\iint \phi \cos\theta \, d\sigma \quad \text{.................... (18)},$$

θ being the inclination to OX of the radius through $d\sigma$. Now $\cos\theta$ is a surface harmonic of the first order, and therefore all the terms of the harmonic expansion, except the first, disappear in the integral, which consequently becomes

$$F = -\left(1 + 2\frac{r^3}{a^3}\right)\iint T_1 \cos\theta \frac{d\sigma}{r^2} \quad \text{........... (19)}.$$

Now let

$$T_1 = -\frac{Ax + By + Cz}{r} \quad \text{................. (20)},$$

this being [Thomson and Tait, App. B, §§ i, j] the most general expression for a surface harmonic of the first order. We have $\cos\theta = x/r$; and therefore (by spherical harmonics, or by the elementary analysis of moments of inertia of a uniform spherical surface),

$$-\iint T_1 \cos\theta \frac{d\sigma}{r^2} = \frac{A}{r^4}\iint x^2 d\sigma = \frac{4\pi A}{3} \quad \text{........ (21)};$$

and (19) becomes

$$F = \left(1 + 2\frac{r^3}{a^3}\right)\frac{4\pi A}{3} \quad \text{................. (22)}.$$

Whence, if X denote the x-momentum of the fluid at any instant in the space between the concentric spherical surfaces of radius r and r',

$$X = -\frac{2}{3}\frac{r^3 - r'^3}{a^3} 4\pi A \quad\text{.....................(23).}$$

If r and r' be each infinitely small in comparison with a, this expression vanishes, as it ought to do, in accordance with § 45, II. But if

$$\left.\begin{array}{ll} & \dfrac{r'}{a} = 0, \text{ and } r = a, \\ \text{it becomes} & X = -\tfrac{2}{3} \,.\, 4\pi A \end{array}\right\} \quad\text{.....................(24),}$$

fulfilling § 4, by showing in the fluid outside the spherical surface of radius r a momentum equal and opposite to that (§ 45, II.) of the whole matter, whether fluid or solid, within that surface.

53. Comparing § 47 and § 52, we see that if X, Y, Z be rectangular components of the force-resultant of the impulse, the term $T_1 r^{-2}$ of the harmonic expansion (14) is as follows:—

$$T_1 r^{-2} = \frac{Xx + Yy + Zz}{4\pi r^3} \quad\text{.................(25),}$$

provided all the solids and vortices taken into account are within a spherical surface whose radius is very small in comparison with the distances of all other vortices or moving solids, and with the shortest distance to the fixed bounding surface.

54. Helmholtz, in his splendid paper on Vortex Motion, has made the very important remark, that a certain fundamental theorem of Green's, which has been used to demonstrate the determinateness of solutions in hydrokinetics, is subject to exception when the functions involved have multiple values. This calls for a serious correction and extension of elementary hydrokinetic theory, to which I now proceed.

55. In the general theorem (1) of Thomson and Tait, App. A, let $\alpha = 1$. It becomes

$$\iiint \left(\frac{d\phi}{dx}\frac{d\phi'}{dx} + \frac{d\phi}{dy}\frac{d\phi'}{dy} + \frac{d\phi}{dz}\frac{d\phi'}{dz}\right) dx\,dy\,dz$$

$$= \iint d\sigma\, \phi \eth \phi' - \iiint dx\,dy\,dz\, \phi \nabla^2 \phi' = \iint d\sigma\, \phi' \eth \phi - \iiint dx\,dy\,dz\, \phi' \nabla^2 \phi$$

$$\text{.........(1),}$$

which is true without exception if ϕ and ϕ' denote any two *single-valued* functions of x, y, z; $\iiint dxdydz$ integration through the space enclosed by any finite closed surface, S; $\iint d\sigma$ integration over the area of this surface; and $\mathfrak{d}$ rate of variation per unit of length in the normal direction at any point of it. This is Green's original theorem, with Helmholtz's limitation added (in italics). The reader may verify it for himself.

56. But if either ϕ or ϕ' is a many-valued function, and the differential coefficients $d\phi/dx$, ..., $d\phi'/dx$, ..., each single-valued, the double equation (1) cannot be generally true. Its first member is essentially unambiguous; but the process of integration by which the second member or the third member is found, would introduce ambiguity if ϕ or ϕ' is many-valued. In one case the first member, though not equal to the ambiguous second, would be equal to the third, provided ϕ' is not also many-valued; and in the other, the first member, though not equal to the third, would be equal to the second, provided ϕ is not many-valued.

For example, let
$$\phi' = \tan^{-1}\frac{y}{x} \qquad\qquad (2),$$

and let S consist of the portions of two planes perpendicular to OZ, intercepted between two circular cylinders having OZ for axis, and the portions of these cylinders intercepted between the two planes. The inner cylindrical boundary excludes from the space bounded by S, the line OZ where ϕ' has an infinite number of values, and $d\phi'/dx$, and $d\phi'/dz$ have infinite values. We have

$$\frac{d\phi'}{dx} = \frac{-y}{x^2+y^2}, \quad \frac{d\phi'}{dy} = \frac{x}{x^2+y^2} \qquad\qquad (3),$$

and at every point of S, $\mathfrak{d}\phi' = 0$. Then, if ϕ be single-valued, there is no failure in the process proving the equality between the first and second members of (1), which becomes

$$\iiint \frac{x\dfrac{d\phi}{dy} - y\dfrac{d\phi}{dx}}{x^2+y^2}\, dxdydz = 0 \qquad\qquad (4).$$

Compare § 14 (6) to end.

The third member of (1) becomes

$$\iint d\sigma \tan^{-1}\frac{y}{x}\mathfrak{d}\phi - \iiint \tan^{-1}\frac{y}{x}\nabla^2\phi\, dxdydz \qquad\qquad (5),$$

which is no result of unambiguous integration of the first member through the space enclosed by S, as we see by examining, in this case, the particular meaning of each step of the ordinary process in rectangular co-ordinates for proving Green's theorem. It is thus seen that we must add to (5) a term

$$2\pi \iint dx\, dz \left(\frac{d\phi}{dy}\right)_{y=0},$$

if in its other terms the value of $\tan^{-1} y/x$ is reckoned continuously round from one side of the plane ZOX to the other: or

$$-2\pi \iint dy\, dz \left(\frac{d\phi}{dx}\right)_{x=0},$$

if the continuity be from one side of ZOY to the other; to render it really equal to the first member of (1). Thus, taking for example, the first form of the added term, we now have for the corrected double equation (1) for the case of $\phi' = \tan^{-1} y/x$, ϕ any single-valued function, and S the surface, composed of the two co-axal cylinders and two parallel planes specified above:

$$\iiint \frac{x\dfrac{d\phi}{dy} - y\dfrac{d\phi}{dx}}{x^2+y^2}\, dx\, dy\, dz = 0 = 2\pi \iint dx\, dz \left(\frac{d\phi}{dy}\right)_{y=0} + \iint d\sigma \tan^{-1}\frac{y}{x}\, \eth\phi$$

$$- \iiint dx\, dy\, dz \tan^{-1}\frac{y}{a} \nabla^2\phi \quad \ldots\ldots\ldots\ldots\ldots\ldots (6).$$

But if we annex to S any barrier stopping circulation round the inner cylindrical core, all ambiguity becomes impossible, and the double equation (1) holds. For instance, if the barrier be the portion of the plane ZOX, intercepted between the co-axal cylinders and parallel planes constituting the S of § 55, so that $\iint d\sigma$ must now include integration over each side of this rectangular area; (6) becomes simply the strict application of (1) to the case in question.

57. The difficulty of the exceptional interpretation of Green's theorem for the class of cases exemplified in §§ 55 and 56, depends on the fact that $\int F ds$ may have different values when reckoned along the lengths of different curves, drawn within the space bounded by S, from a point P to a point Q; ds being an infinitesimal element of the curve, and F the rate of variation of ϕ per unit of length along it. Let PCQ, $PC'Q$ be two curves for which the $\int F ds$ has different values; and let both lie wholly within S.

If we draw any curve from P to Q; make it first coincide with PCQ, and then vary it gradually until it coincides with $PC'Q$; it must in some of its intermediate forms cut the bounding surface S: for we have

$$Fds = \frac{d\phi}{dx}dx + \frac{d\phi}{dy}dy + \frac{d\phi}{dz}dz$$

throughout the space contained within S, and $d\phi/dx, d\phi/dy, d\phi/dz$ are each of them unambiguous by hypothesis; which implies that $\int Fds$ has equal values for all gradual variations of one curve between P and Q, each lying wholly within S. Now, in a simply continuous space, a curve joining the points P and Q may be gradually varied from any curve PCQ to any other $PC'Q$, and therefore if the space contained within S be simply continuous, the difficulty depending on the multiplicity of value of ϕ or ϕ' cannot exist. And however multiply continuous (§ 58) the space may be, the difficulty may be evaded if we annex to S a surface or surfaces stopping every aperture or passage on the openness of which its multiple continuity depends; for these annexed surfaces, as each of them occupies no space, do not disturb the triple integrations (1), and will, therefore, not alter the values of its first member; but by removing the multiplicity of continuity, they free each of the integrations by parts, by which its second or third members are obtained, from all ambiguity. To avoid circumlocution, we shall call β the addition thus made to S; and further, when the space within S is (§ 58) not merely doubly but triply, or quadruply, or more multiply, continuous, we shall designate by β_1, β_2; or $\beta_1, \beta_2, \beta_3$; and so on; the several parts of β required in any case to stop all multiple continuity of the space. These parts of β may be quite detached from one another, as when the multiple continuity is that due to detached rings, or separate single tunnels in a solid. But one part β_1 may cut through part of another, β_2, as when two rings (§ 58, diagram) linked into one another without touching constitute part of the boundary of the space considered. And we shall denote by $\iint d\varsigma$, integration over the surface β, or over any one of its parts, β_1, β_2, &c. Let now P and Q be each infinitely near a point B, of β, but on the two sides of this surface. Let κ denote the value of $\int Fds$ along any curve lying wholly in the space bounded by S, and joining PQ without cutting the barrier; this value being the same for all such curves, and for all positions of B to which it may be brought without leaving β, and without making either P or Q pass through any part of β. That is to say, κ is a single constant when the space is not more than doubly continuous; but it denotes one or other of n constants $\kappa_1, \kappa_2, \ldots \kappa_n$, which may be all different from one another, when the space is n-ply continuous. Lastly, let κ' denote the same element, relatively to ϕ', as κ relatively to ϕ. We find that the first steps of the integrations by parts now introduce, without ambiguity, the additions

$$\Sigma\kappa\iint d\varsigma\mathfrak{d}\phi' \text{ and } \Sigma\kappa'\iint d\varsigma\mathfrak{d}\phi \quad \ldots\ldots\ldots\ldots\ldots(7),$$

to the second and third members of (1): Σ denoting summation of the integrations for the different constituents $\beta_1, \beta_2, \ldots$ of β; but only a single term when the space is (§ 58) not more than doubly continuous. Green's theorem thus corrected becomes

$$\iiint\left(\frac{d\phi}{dx}\frac{d\phi'}{dx}+\frac{d\phi}{dy}\frac{d\phi'}{dy}+\frac{d\phi}{dz}\frac{d\phi'}{dz}\right)dxdydz$$

$$=\iint d\sigma\phi\mathfrak{d}\phi'+\Sigma\kappa\iint d\varsigma\mathfrak{d}\phi'-\iiint\phi\nabla^2\phi'dxdydz$$

$$=\iint d\sigma\phi'\mathfrak{d}\phi+\Sigma\kappa'\iint d\varsigma\mathfrak{d}\phi-\iiint\phi'\nabla^2\phi dxdydz\ldots\ldots(8).$$

58. Adopting the terminology of Riemann, as known to me through Helmholtz, I shall call a finite position of space n-ply continuous when its bounding surface is such that there are n irreconcilable paths between any two points in it. To prevent any misunderstanding, I add (1), that by *a* portion of space I mean such a portion that any point of it may be travelled to from any other point of it, without cutting the bounding surface; (2), that the "paths" spoken of all lie within the portion of space referred to; and (3), that by irreconcilable paths between two points P and Q; I mean paths such, that a line drawn first along one of them cannot be gradually changed till it coincides with the other, being always kept passing through P and Q, and always wholly within the portion of space considered. Thus, when all the paths between any two points are reconcilable, the space is simply continuous. When there are just two sets of paths, so that each of one set is irreconcilable with any one of the other set, the space is doubly continuous; when there are three such sets it is triply continuous, and so on. To avoid circumlocutions, we shall suppose S to be the boundary of a hollow space in the interior of a solid mass, so thick that no operations which we shall consider

shall ever make an opening to the space outside it. A tunnel through this solid opening at each end into the interior space constitutes the whole space doubly continuous; and if more tunnels be made, every new one adds one to the degree of multiple continuity. When one such tunnel has been made, the surface of the tunnel is continuous with the whole bounding surface of the space considered; and in reckoning degrees of continuity, it is of no consequence whether the ends of any fresh tunnel be in one part or another of this whole surface. Thus, if two tunnels be made side by side, a hole anywhere opening from one of them into the other adds one to the degree of multiple continuity. Any solid detached from the outer bounding solid, and left, whether fixed or movable in the interior space, adds to the bounding surface an isolated portion, but does not interfere with the reckoning of multiple continuity. Thus, if we begin with a simply continuous space bounded outside by the inner surface of the supposed external solid, and internally by the boundary of the detached solid in its interior, and if we drill a hole in this solid we produce double continuity. Two holes, or two solids in the interior each with one hole (such as two ordinary solid rings), constitute triple continuity, and so on. A sponge-like solid whose pores communicate with one another, illustrates a high degree of multiple continuity, and it is of no consequence whether it is attached to the external bounding solid or is an isolated solid in the interior. Another type of multiple continuity, that presented by two rings linked in one another, was referred to in § 57.

When many rings are linked into one another in various combinations, there are complicated mutual intersections of the several partial barriers β_1, β_2, ... required to stop all multiple continuity. But without having any portion of the bounding solid detached, as in that case in which one at least of the two rings is loose, we have varieties of multiple continuity curiously different from that illustrated by a single ordinary straight or bent tunnel, illustrated sufficiently by the simplest types, which are obtained by boring a tunnel along a line agreeing in form with the axis of a cord or wire on which a simple knot is tied; and by fixing the two ends of wire with a knot on it to the bounding solid, so that the surface of the wire shall become part of the bounding surface of the space considered, the knot not being pulled tight, and the wire being arranged not to touch itself in any point; or by placing a knotted wire, with its ends united, in the interior of the space. No amount of knotting or knitting, however complex, in the cord whose axis indicates the line of tunnel, complicates in any way the continuity of the space considered, or alters the simplicity of the barrier surface required to stop the circulation. But it is otherwise when a knotted or knitted wire forms part of the bounding solid. A single simple knot, though giving only double continuity, requires a curiously self-cutting surface for stopping barrier: which, in its form of minimum area, is beautifully shown by the liquid film adhering to an endless wire, like the first figure, dipped in a soap solution and removed. But no complication of these types, or of combinations of them with one another, eludes the statements and formulæ of § 57.

Instalment, received Nov.—Dec. 1869 [§ 59—§ 64 (*f*)].

59. I shall now give a dynamical lemma, for the immediate object of preparing to apply Green's corrected theorem (§ 57) to the motion of a liquid through a multiply continuous space. But later we shall be led by it to very simple demonstrations of Helmholtz's fundamental theorems of vortex motion; and shall see that it may be used as a substitute for the common equations of hydrokinetics.

(Lemma.) An endless finite tube* of infinitesimal normal section, being given full of liquid (whether circulating round

* A finite length of tube with its ends done away by uniting them together.

through it, or at rest) is altered in shape, length, and normal section, in any way, and with any speed. The average value of the component velocity of the fluid along the tube, reckoned all round the circuit (irrespectively of the normal section), varies inversely as the length of the circuit.

59 (*a*) To prove this, consider first a single particle of unit mass, acted on by any force, and moving along a smooth guiding curve, which is moved and bent about quite arbitrarily. Let ρ be the radius of curvature, and ξ, η the component velocities of the guiding curve, towards the centre of curvature, and perpendicular to the plane of curvature, at the point P, through which the moving particle is passing at any instant. Let ζ be the component velocity of the particle itself, along the instantaneous direction of the tangent through P. Thus ξ, η, ζ are three rectangular components of the velocity of the particle itself. Let $\mathfrak{Z}$ be the component in the direction of ζ, of the whole force on P. We have, by elementary kinetics,

$$\frac{d\zeta}{dt} = \mathfrak{Z} + \frac{\zeta\xi}{\rho} + \xi\frac{d\xi}{ds} + \eta\frac{d\eta}{ds} \quad \ldots\ldots\ldots\ldots\ldots(1)^*,$$

* This theorem (not hitherto published ?) will be given in the second volume of Thomson and Tait's *Natural Philosophy*. It may be proved analytically from the general equations of the motion of a particle along a varying guide-curve (Walton, *Cambridge Mathematical Journal*, 1842, February); or more synthetically, thus—Let l, m, n be the direction cosines of PT, the tangent to the guide at the point through which the particle is passing at any instant; (x, y, z) the co-ordinates of this point, and $(\dot{x}, \dot{y}, \dot{z})$ its component velocities parallel to fixed rectangular axes. We have

$$\zeta = l\dot{x} + m\dot{y} + n\dot{z}\,; \text{ and } \mathfrak{Z} = l\ddot{x} + m\ddot{y} + n\ddot{z},$$

and from this

$$\frac{d\zeta}{dt} = l\ddot{x} + m\ddot{y} + n\ddot{z} + \dot{l}\dot{x} + \dot{m}\dot{y} + \dot{n}\dot{z} = \mathfrak{Z} + \dot{l}\dot{x} + \dot{m}\dot{y} + \dot{n}\dot{z}.$$

But it is readily proved (Thomson and Tait's *Natural Philosophy*, § 9, to be made more explicit on this point in a second edition) that the angular velocity with which PT changes direction is equal to $\sqrt{(\dot{l}^2 + \dot{m}^2 + \dot{n}^2)}$, and, if this be denoted by ω, that

$$\frac{\dot{l}}{\omega},\ \frac{\dot{m}}{\omega},\ \frac{\dot{n}}{\omega}$$

are the direction cosines of the line PK, perpendicular to PT in the plane in which PT changes direction, and on the side towards which it turns. Hence,

$$\frac{d\zeta}{dt} = \mathfrak{Z} + \kappa\omega$$

if κ denote the component velocity of P along PK. Now, if the curve were fixed we should have $\omega = \zeta/\rho$, by the kinematic definition of curvature (Thomson and Tait, § 5); and the plane in which PT changes direction would be the plane of curvature. But in the case actually supposed, there is also in this plane an additional angular

where ρ denotes the radius of curvature, and $d\xi/ds$, $d\eta/ds$ rates of variation of ξ and η from point to point along the curve at one time.

59 (*b*) Now, instead of a single particle of unit mass, let an infinitesimal portion, μ, of a liquid, filling the supposed endless tube, be considered. Let ϖ be the area of the normal section of the tube in the place where μ is, and δs the length along the tube of the space occupied by it, at any instant; so that (as the density of the fluid is called unity),

$$\mu = \varpi \delta s.$$

Further, let dp/ds denote the rate of variation of the fluid pressure along the tube, so that

$$\mathfrak{Z} = -\varpi \frac{dp}{ds} \delta s.$$

Thus we have, by (1)

$$\frac{d\zeta}{dt} = \frac{\zeta\xi}{\rho} + \xi \frac{d\xi}{ds} + \eta \frac{d\eta}{ds} - \frac{dp}{ds} \quad\text{..................(2).}$$

(*c*) Now, because the two ends of the arc δs move with the fluid, we have, by the kinematics of a varying curve,

$$\frac{d\delta s}{dt} = \frac{d\zeta}{ds} \delta s - \frac{\xi}{\rho} \delta s \quad\text{.....................(3);}$$

and, therefore, $$\frac{d(\zeta\delta s)}{dt} = \frac{d\zeta}{dt} \delta s + \zeta \left(\frac{d\zeta}{ds} \delta s - \frac{\xi}{\rho} \delta s\right) \quad\text{..............(4).}$$

Substituting in this for $d\zeta/dt$ its value by (2), we have

$$\frac{d(\zeta\delta s)}{dt} = \left(\xi \frac{d\xi}{ds} + \eta \frac{d\eta}{ds} - \frac{dp}{ds} + \zeta \frac{d\zeta}{ds}\right) \delta s,$$

or $$\frac{d(\zeta\delta s)}{dt} = \delta\left(\tfrac{1}{2}q^2 - p\right) \quad\text{.....................(5),}$$

velocity equal to $d\xi/ds$, and a component angular velocity in the plane of PT and η, equal to $d\eta/ds$; due to the normal motion of the varying curve. Hence the whole angular velocity ω is the resultant of two components,

$$\frac{\zeta}{\rho} + \frac{d\xi}{ds} \text{ in the plane of } \xi,$$

and $$\frac{d\eta}{ds} \text{ in the plane of } \eta.$$

Hence $$\xi\left(\frac{\zeta}{\rho} + \frac{d\xi}{ds}\right) + \eta \frac{d\eta}{ds} = \kappa\omega,$$

and the formula (1) of the text is proved.

if q denote the resultant fluid velocity; and δ the differences for the two ends of the arc δs. Integrating this through the length of any finite arc P_1P_2 of the fluid, its ends P_1, P_2 moving with the fluid, we have

$$\frac{d\Sigma_1^2(\zeta\delta s)}{dt} = (\tfrac{1}{2}q^2 - p)_2 - (\tfrac{1}{2}q^2 - p)_1 \quad \text{...........(6)},$$

the suffixes denoting the values of the bracketed function, at the points P_2 and P_1, respectively; and Σ_1^2 denoting integration along the arc from P_1 to P_2. Let now P_2 be moved forward, or P_1 backward, till these points coincide, and the arc P_1P_2 becomes the complete circuit; and let Σ denote integration round the whole closed circuit. (6) becomes

$$\frac{d\Sigma(\zeta\delta s)}{dt} = 0 \quad \text{......................(7)};$$

and we conclude that $\Sigma\zeta\delta s$ remains constant, however the tube be varied. This is the proposition to be proved, as the "average velocity" referred to is found by dividing $\Sigma(\zeta\delta s)$ by the length of the tube.

59 (d) The tube, imagined in the preceding, has had no other effect than exerting, by its inner surface, normal pressure on the contained ring of fluid. Hence the proposition* at the beginning

* Equation (6), from which, as we have seen, that proposition follows immediately, may be proved with greater ease, and not merely for an incompressible fluid, but for any fluid in which the density is a function of the pressure, by the method of rectilineal rectangular co-ordinates from the ordinary hydrokinetic equations. These equations are

$$\frac{Du}{Dt} = -\frac{d\varpi}{dx}, \quad \frac{Dv}{Dt} = -\frac{d\varpi}{dy}, \quad \frac{Dw}{Dt} = -\frac{d\varpi}{dz},$$

if D/Dt denote rate of variation per unit of time, of any function depending on a point or points *moving with the fluid;* and $\varpi = \int dp/\rho$, ρ denoting density. In terms of rectilineal co-ordinates we have

$$\zeta\delta s = u\delta x + v\delta y + w\delta z.$$

Hence $$\frac{D(\zeta\delta s)}{Dt} = \frac{Du}{Dt}\delta x + u\frac{D\delta x}{Dt} + \text{\&c.}$$

Now $$\frac{D\delta x}{Dt} = \delta u, \quad \frac{D\delta y}{Dt} = \delta v, \text{ and } \frac{D\delta z}{Dt} = \delta w.$$

These and the kinetic equations reduce the preceding to

$$\frac{D(\zeta\delta s)}{Dt} = u\delta u + v\delta v + w\delta w - \frac{d\varpi}{dx}\delta x - \frac{d\varpi}{dy}dy - \frac{d\varpi}{dz}\delta z = \delta\left[\tfrac{1}{2}(u^2+v^2+w^2) - \varpi\right] \text{...(8)};$$

whence, by Σ integration, equation (6) generalised to apply to compressible fluids.

of § 59 is applicable to any closed ring of fluid forming part of an incompressible fluid mass extending in all directions through any finite or infinite space, and moving in any possible way; and the formulæ (5) and (6) are applicable to any infinitesimal or infinite arc of it with two ends not met. Thus in words—

PROP. (1). *The line-integral of the tangential component velocity round any closed curve of a moving fluid remains constant through all time.*

And, PROP. (2), The rate of augmentation, per unit of time, of the space integral of the velocity along any terminated arc of the fluid is equal to the excess of the value of $\frac{1}{2}q^2-p$, at the end towards which tangential velocity is reckoned as positive, above its value at the other end.

59 (*e*) The condition that $udx+vdy+wdz$ is a complete differential [proved above (§ 13) to be the criterion of irrotational motion] means simply

That the flow [defined § 60 (*a*)] *is the same in all different mutually reconcilable lines from one to another of any two points in the fluid;* or, which is the same thing,

That the circulation [§ 60 (*a*)] *is zero round every closed curve capable of being contracted to a point without passing out of a portion of the fluid through which the criterion holds.*

From Proposition (1), just proved, we see that this condition holds through all time for any portion of a moving fluid for which it holds at any instant; and thus we have another proof of Lagrange's celebrated theorem (§ 16), giving us a new view of its dynamical significance, which [see for example § 60 (*g*)] we shall find of much importance in the theory of vortex motion.

(*f*) But it is only in a closed curve, *capable of being contracted to a point without passing out of space occupied by irrotationally moving fluid,* that the circulation is necessarily *zero,* in irrotational motion. In § 57 we saw that a continuous fluid mass, occupying doubly or multiply continuous space, may move altogether irrotationally, yet so as to have finite circulation in a closed curve $PP'QQ'P$, provided $PP'Q$ and $PQ'Q$ are "irreconcilable paths" between P and Q. *That the circulation must be the same in all mutually reconcilable closed curves* (compare § 57), is an immediate consequence from the now proved [§ 59 (Prop. 2)] equality of the flows [§ 60 (*a*)] in all mutually reconcilable conterminous arcs. For by leaving one part of a closed curve unchanged, and varying the remaining arc continuously, no change is produced in the flow, in this part; and, by repetitions of the process, a closed curve may be changed to any other reconcilable with it.

60. *Definitions and elementary propositions.* (*a*) The line-integral of the tangential component velocity along any finite line, straight or curved, in a moving fluid, is called the flow in that line. If the line is endless (that is, if it forms a closed curve or polygon), the *flow* is called *circulation.* The use of these terms abbreviates the statements of Propositions (2) and (1) of § 59 to the following:—

[§ 59, Prop. (2)]. The rate of augmentation, per unit of time, of the flow in any terminated line which moves with the fluid, is equal to the excess of the value of $\frac{1}{2}q^2 - p$ at the end from which, above its value at the end towards which, positive flow is reckoned.

[§ 59, Prop. (1)]. The circulation in any closed line moving with the fluid, remains constant through all time.

(*b*) If any open finite surface, lying altogether within a fluid, be cut into parts by lines drawn across it, the circulation in the boundary of the whole is equal to the sum of the circulations in the boundaries of the parts. This is obvious, as the latter sum consists of an equal positive and negative flow in each portion of the boundary common to two parts, added to the sum of the flows in all the parts into which the single boundary of the whole is divided.

(*c*) Hence the circulation round the boundaries of infinitesimal areas, infinitely near one another in one plane, are simply proportional to these areas.

(*d*) *Proposition.* Let any part of the fluid rotate as a solid (that is, without changing shape); or consider simply the rotation of a solid. The "circulation" in the boundary of any plane figure moving with it is equal to twice the area enclosed, multiplied by the component angular velocity in that plane (or round an axis perpendicular to that plane). For, taking r, θ to denote polar coordinates of any point in the boundary, A the enclosed area, and

ω the component angular velocity in the plane, and continuing the notation of § 59, we have

$$\zeta = r\omega \frac{rd\theta}{ds},$$

and therefore

$$\Sigma\zeta\delta s = \omega\Sigma r^2 \frac{d\theta}{ds}\delta s = \omega\Sigma r^2\delta\theta = \omega \times 2A.$$

60 (*e*) *Definition.* For a fluid moving in any manner, the circulation round the boundary of an infinitesimal plane area, divided by double the area, is called the *component rotation* in that plane (or round an axis perpendicular to that plane) of the neighbouring fluid.

In this statement, the single word "rotation" is used for *angular velocity of rotation:* and the definition is justified by (*c*) and (*d*); also by § 13 (2) above, applied to (*p*) below. It agrees, in virtue of (*p*), with the definition of rotation in fluid motion given first of all, I believe, by Stokes, and used by Helmholtz in his memorable *Vortex Motion*, also in Thomson and Tait's *Natural Philosophy*, §§ 182 and 190 (*j*).

(*f*) *Proposition.* If ξ, η, ζ be the components of rotation at any point, P, of a fluid, round three axes at right angles to one another, and ω the component round an axis, making with them angles whose cosines are l, m, n,

$$\omega = \xi l + \eta m + \zeta n.$$

To prove this, let a plane perpendicular to the last-mentioned axis cut the other three in A, B, C. The circulation in the periphery of the triangle ABC is, by (*b*), equal to the sum of the circulations in the peripheries PBC, PCA, and PAB. Hence, calling Δ and α, β, γ the areas of these four triangles, we have, by (*e*),

$$\omega\Delta = \xi\alpha + \eta\beta + \zeta\gamma.$$

But α, β, γ are the projections of Δ on the planes of the pairs of the rectangular axes; and so the proposition is proved.

It follows, of course, that the composition of rotations in a fluid fulfils the law of the compositions of angular velocities of a solid, of linear velocities, of forces, &c.

(*g*) Hence, in any infinitesimal part of the fluid, the circulation is zero in the periphery of every plane area passing through a certain line;—the resultant axis of rotation of that part of the fluid. But (*a*) the circulation remains zero in every closed line moving with the fluid, for which it is zero at any time. Hence

60 (*h*) The axial lines [defined (*i*)] move with the fluid.

(*i*) *Definition.* An axial line through a fluid moving rotationally, is a line (straight or curved) whose direction at every point coincides with the resultant axis of rotation through that point.

(*j*) *Proposition.* The resultant rotation of any part of the fluid varies in simple proportion to the length of an infinitesimal arc of the axial line through it, terminated by points moving with the fluid. To prove this, consider any infinitesimal plane area, A, moving with the fluid. Let ω be the resultant rotation, and θ the angle between its axis and the perpendicular to the plane of A. This makes $\omega \cos \theta$ the component rotation in the plane of A; and therefore $A\omega \cos \theta$ remains constant. Now, draw axial lines through all points of the boundary of A, forming a tube whose area of normal section is $A \cos \theta$. The resultant rotation must vary inversely as this area, and therefore (in consequence of the incompressibility of the fluid) directly as the length of an infinitesimal line along the axis.

(*k*) Form a surface by axial lines drawn through all points of any curve in the fluid. The circulation is zero round the boundary of any infinitesimal area of this surface; and therefore (*b*) it is zero round the boundary of any finite area of it.

(*l*) Let the curve of (*k*) be closed, and therefore the surface tubular. On this surface let $ABCA$, $A'B'C'A'$ be any two curves closed round the tube, and ADA' any arc from A to A'. The circulation in the closed path, $ADA'B'C'A'DACBA$, is zero by (*h*). Hence the circulation in $ABCA$ is equal to the circulation in $A'B'C'A'$—that is to say,

The circulations are equal in all circuits of a vortex tube.

(*m*) *Definitions.* An *axial surface* is a surface made up of axial lines. A *vortex tube* is an axial surface through every point of which a finite endless path, cutting every axial line it meets, can be drawn. Any such path, passing just once round, is called *a circuit* or, *the circuit* of the tube. The *rotation of a vortex tube* is the circulation in its circuit. A *vortex sheet* is (a portion as it were of a collapsed vortex tube) a surface on the two sides of which the fluid moves with different tangential component velocities.

60 (n) Draw any surface cutting a vortex tube, and bounded by it. The surface integral of the component rotation round the normal has the same value for all such surfaces; and this common value is what we now call the rotation of the tube.

(o) In an unbounded infinite fluid, an axial tube must be either finite and endless or infinitely long in each direction*. In an infinite fluid with a boundary (for instance, the surface of an enclosed solid), an axial tube may have two ends, each in the boundary surface; or it may have one end in the boundary surface, and no other; or it may be infinitely long in each direction, or it may be finite and endless. In a finite fluid mass, an axial tube may be endless, or may have one end, but, if so, must have another, both in the boundary surface.

(p) *Proposition.* Applying the notation of (f), to axes parallel to those of co-ordinates x, y, z, and denoting, as formerly, by u, v, w, the components of the fluid velocity at (x, y, z), we have

$$\xi=\tfrac{1}{2}\left(\frac{dw}{dy}-\frac{dv}{dz}\right),\quad \eta=\tfrac{1}{2}\left(\frac{du}{dz}-\frac{dw}{dx}\right),\quad \zeta=\tfrac{1}{2}\left(\frac{dv}{dx}-\frac{du}{dy}\right).$$

The proof is obvious, according to the plan of notation, &c., followed in § 13 above.

(q) Hence by (f), (e), and (b)

$$\iint dS\left\{l\left(\frac{dw}{dy}-\frac{dv}{dz}\right)+m\left(\frac{du}{dz}-\frac{dw}{dx}\right)+n\left(\frac{dv}{dx}-\frac{du}{dy}\right)\right\}$$
$$=\int(udx+vdy+wdz),$$

where $\iint dS$ denotes integration over any portion of surface bounded by a closed curve; $\int(udx+\&c.)$ integration round the whole of this curve; and (l, m, n) the direction cosines of any point (x, y, z) in the surface. It is worthy of remark that the equation of continuity for an incompressible fluid does not enter into the

* **Vortex tubes apparently ending in the fluid, for instance, a portion of fluid bounded by a figure of revolution, revolving round its axis as a solid, constitute no exception. Each infinitesimal vortex tube in this case is completed by a strip of vortex sheet and so is endless.**

demonstration of this proposition, and therefore u, v, w may be any functions whatever of x, y, z. In a purely analytical light the result has an important bearing on the theory of the integration of complete or incomplete differentials. It was first given, with the indication of a more analytical proof than the preceding, in Thomson and Tait's *Natural Philosophy*, § 190 (j).

60 (r) Propositions (h) (j) (n) (o) of the present section (§ 60) are due to Helmholtz; and with his integration for associated rotational and cyclic irrotational motion in an unbounded fluid, to be given below, constitute his general theory of vortex motion. (n) and (o) are purely kinematical; (h) and (j) are dynamical.

(s) Henceforth I shall call *a circuit* any closed curve not continuously reducible to a point, in a multiply continuous space. I shall call *different circuits,* any two such closed curves if mutually irreconcilable (§ 58), but different mutually reconcilable closed curves will not be called different circuits.

(t) Thus, $(n+1)$ply continuous space, is a space for which there are n, and only n, different circuits. This is merely the definition of § 58, abbreviated by the definite use of the word circuit, which I now propose. The general terminology regarding simply and multiply continuous spaces is, as I have found since § 58 was written, altogether due to Helmholtz; Riemann's suggestion, to which he refers, having been confined to two-dimensional space. I have deviated somewhat from the form of definition originally given by Helmholtz, involving, as it does, the difficult conception of a stopping barrier*; and substituted for it the definition by reconcilable and irreconcilable paths. It is not easy to conceive the stopping barrier of any one of the first three diagrams of § 58, or to understand its singleness; but it is easy to see that in each of those three cases, any two closed curves drawn round the solid wire represented in the diagrams are reconcilable, according to the definition of this term given in § 58, and

* But without this conception we can make no use of the theory of multiple continuity in hydrokinetics (see §§ 61–63), and Helmholtz's definition is, therefore, perhaps preferable after all to that which I have substituted for it. Mr Clerk Maxwell tells me that J. B. Listing has more recently treated the subject of multiple continuity in a very complete manner in an article entitled "Der Census räumlicher Complexe."—*Königl. Ges. Göttingen*, 1861. See also Prof. Cayley "On the Partition of a Close."—*Phil. Mag.* 1861.

therefore, that the presence of any such solid adds only one to the degree of continuity of the space in which it is placed.

60 (u) If we call a *partition*, a surface which separates a closed space into two parts, and, as hitherto, a *barrier*, any surface edged by the boundary of the space, Helmholtz's definition of multiple continuity may be stated shortly thus:—

A space is $(n+1)$ply continuous if n barriers can be drawn across it, none of which is a partition.

(v) Helmholtz has pointed out the importance in hydrokinetics of many-valued functions, such as $\tan^{-1} y/x$, which have no place in the theories of gravitation, electricity, or magnetism, but are required to express electro-magnetic potentials, and the velocity potentials for the part of the fluid which moves irrotationally in vortex motion. It is, therefore, convenient, before going farther, that we should fix upon a terminology, with reference to functions of that kind, which may save us circumlocutions hereafter.

(w) A function $\phi(x, y, z)$ will be called *cyclic* if it experiences a constant augmentation every time a point P, of which x, y, z are rectangular rectilinear co-ordinates, is carried from any position round a certain circuit to the same position again, without passing through any position for which either $d\phi/dx$, $d\phi/dy$, or $d\phi/dz$ becomes infinite. The value of this augmentation will be called the cyclic constant for that particular circuit. The cyclic constant must clearly have the same value for all circuits mutually reconcilable (§ 58), in space throughout which the three differential coefficients remain all finite.

(x) When the function is cyclic with reference to several different mutually irreconcilable circuits, it is called polycyclic. When it is cyclic for only one set of circuits, it is called monocyclic.

Example.—The apparent area of a circle as seen from a point (x, y, z) anywhere in space, is a monocyclic function of x, y, z, of which the cyclic constant is 4π.

The apparent area of a plane curve of the $(2n)$th degree, consisting of n detached closed (that is finite endless) branches (some of which might be enclosed within others) is an n-cyclic function, of which the n-cyclic constants are essentially equal, being each 4π.

Algebraic equations among three variables (x, y, z) may easily be found to represent tortuous curves, constituting one or more finite, isolated, endless branches (which may be knotted, as shown in the first three diagrams of § 58, or linked into one another, as in the fourth and fifth). The integral expressing what, for brevity, we shall call the *apparent area* of such a curve, is a cyclic function, which, if polycyclic, has essentially equal values for all its cyclic constants. By the *apparent area of a finite endless curve* (tortuous or plane), I mean the *sum of the apparent areas of all barriers edged by it, which we can draw without making a partition.*

It is worthy of notice that every polycyclic function may be reduced to a sum of monocyclic functions.

60 (y) Fluid motion is called *cyclic* unless the circulation is zero in every closed path through the fluid, when it is called *acyclic.* Rotational motion is (e) essentially cyclic.

(z) Irrotational motion may [§ 59 (f)] be either acyclic or cyclic. If cyclic it is *monocyclic* if there is only one distinct circuit, or *polycyclic* if there are several distinct circuits, in which there is circulation. It is *purely cyclic* if the boundary of the space occupied by irrotationally moving fluid is at rest. If the boundary moves and the motion of the fluid is cyclic, it is *acyclic compounded with cyclic.*

61 (a) We are now prepared to investigate the most general possible irrotational motion of a single continuous fluid mass, occupying either simply or multiply continuous space, with, for every point of the boundary, a normal component velocity given arbitrarily, subject only to the condition that the whole volume remains unaltered.

(b) *Genesis of a cyclic motion.* Commencing, as in § 3, with a fluid mass at rest throughout, let all multiplicity of the continuity of the space occupied by it be done away with by temporary barrier surfaces, β_1, β_2... stopping the circuits, as described in § 57. The bounding surface of the fluid, which ordinarily consists of the inner surface of the containing vessel, will thus be temporarily extended to include each side of each of these barriers. Let now, as in § 3, any possible motion be arbitrarily given to the bounding surface. The liquid is consequently set in motion, purely through fluid pressure; and the motion is [§§ 10–15, or 60, 59] throughout irrotational. Hence irrotational motion fulfilling the prescribed surface conditions is possible, and the actual motion is, of course (as the solution of every real problem is), unambiguous. But from this bare physical principle we could not even suspect, what the following simple application of Green's equation proves, that the surface normal velocity at any instant determines the interior motion irrespectively of the previous history of the motion from rest.

61 (c) *Determinacy of irrotational motion in simply continuous space.* In § 57 (1), which is immediately applicable, as the volume is now simply continuous, make $\phi' = \phi$, and put $\nabla^2\phi = 0$, so that ϕ may be the velocity potential of an incompressible fluid. That double equation becomes the following single equation

$$\iiint\left(\frac{d\phi^2}{dx^2} + \frac{d\phi^2}{dy^2} + \frac{d\phi^2}{dz^2}\right) dx\,dy\,dz = \iint d\sigma\phi\mathfrak{d}\phi,$$

where the surface integration $\iint d\sigma$ must now include each side of each of the barrier surfaces $\beta_1, \beta_2 \ldots$. Hence, if $\mathfrak{d}\phi = 0$ for every point of the bounding surface, we must have

$$\iiint\left(\frac{d\phi^2}{dx^2} + \frac{d\phi^2}{dy^2} + \frac{d\phi^2}{dz^2}\right) dx\,dy\,dz = 0,$$

which requires that

$$\frac{d\phi}{dx} = 0, \quad \frac{d\phi}{dy} = 0, \quad \frac{d\phi}{dz} = 0:$$

that is to say, if there is no motion of the boundary surface in the direction of the normal, there can be no motion of the irrotational species in the interior; whence it follows that there cannot be two different internal irrotational motions with the same surface normal component velocities. Thus, as a particular case, beginning with a fluid at rest, let its boundary be set in motion; and brought again to rest at any instant, after having been changed in shape to any extent, through any series of motions. The whole liquid comes to rest at that instant.

A demonstration of this important theorem, which differs essentially from the preceding, and includes what the preceding does not include, a purely analytical proof of the possibility of irrotational motion throughout the fluid, fulfilling the arbitrary surface-condition specified above, as was first published in Thomson and Tait's *Natural Philosophy*, § 317 (3), and is to be given below, with some variation and extension. In the meantime, however, we satisfy ourselves as to the *possibility* of irrotational motions fulfilling the various surface-conditions with which we are concerned, because the surface motions are possible and require the fluid to move, and [§§ 10–15, or § 59] because the fluid cannot acquire rotational motion through fluid pressure from the motion of its boundary; and we go on, by aid of Green's extended formula [§ 57 (7)], to prove the determinateness of the interior motion under conditions now to be specified for multiply continuous space, as we have done by his unaltered formula [§ 57 (1)] for simply continuous space.

62. *Genesis of cyclic irrotational motion.* In the case of motion considered in § 61, the value of the normal component velocity is not independently arbitrary over the whole boundary, but has equal arbitrary values, positive and negative, on the two sides of each of the barriers β_1, β_2, &c. We must now introduce a fresh restriction in order that, when the barriers are liquefied, the motion of the fluid may be irrotational throughout the space thus re-opened into multiple-continuity. For although we have secured that the normal component velocity is equal everywhere on the two sides of each barrier, we have hitherto left the tangential velocity unheeded. If they are not equal on the two sides, and in the same direction, there will be a finite slipping of fluid on fluid across the surface left by the dissolution of the infinitely thin barrier membrane; constituting [§ 60 (m) above], as Helmholtz has shown, a "vortex sheet." The analytical expression of the condition of equality between the tangential velocities is that the variation of the velocity potential in tangential directions shall be equal on the two sides of each barrier. Hence, by integration, we see that the difference between the values of the velocity potential on the two sides must be the same over the whole of each barrier. This condition requires that the initiating pressure be equal over the whole membrane. For, at any time during the instituting of the motion, let p_1, p_2 be the pressures at two points P_1, P_2 of the fluid, and moving with the fluid, infinitely near one another on the two sides of one of the membranes, so that the pressure ϖ, which must be applied to the membrane to produce this difference of fluid pressure on the two sides, is equal to $p_1 - p_2$ in the direction opposed to p_1. And let

ϕ_1, ϕ_2 be the velocity potentials at P_1 and P_2, so that if $\int ds$ denote integration from P_1 to P_2, along any path P_1PP_2 whatever from P_1 to P_2, altogether through the fluid (and therefore cutting none of the membranes), and ζ the component of fluid velocity along the tangent at any point of this curve, we have

$$\int \zeta ds = \phi_2 - \phi_1 \quad \text{.......................(1).}$$

Hence, by (6) of § 59,

$$\frac{d(\phi_2 - \phi_1)}{dt} = \varpi - \tfrac{1}{2}(q_1^2 - q_2^2) \quad \text{..............(2),}$$

where q_1, q_2 denote the resultant fluid velocities at P_1 and P_2. Now, the normal component velocities at P_1 and P_2 are necessarily equal; and therefore, if the components parallel to the tangent plane of the intervening membrane are also equal, we have

$$q_1 = q_2$$

and the preceding becomes

$$\frac{d(\phi_2 - \phi_1)}{dt} = \varpi \quad \text{.........................(3).}$$

But if the tangential component velocities at P_1 and P_2 are not only equal, but in the same direction, $\phi_2 - \phi_1$ must, as we have seen, be constant over the membrane, and therefore ϖ must also be constant.

Suppose now that after pressure has been applied for any time in the manner described, of uniform value all over the membrane at each instant, it is applied no longer, and the membrane (having no longer any influence) is done away with. The fluid mass is left for ever after in a state of motion, which is irrotational throughout, but cyclic. The "circulation" [§ 60 (a)], or the cyclic constant being equal to $\phi_2 - \phi_1$, for every circuit reconcilable with $P_1PP_2P_1$ is given by the equation

$$\phi_2 - \phi_1 = -\int \varpi dt \quad \text{.......................(4),}$$

$\int dt$ denoting a time-integral extended through the whole period during which ϖ had any finite value.

The same kind of operation may be performed, on each of the n barriers temporarily introduced in § 61 to reduce the $(n+1)$fold continuity of the space occupied by the fluid, to simple continuity.

The velocity potential at any point of the fluid will then be a polycyclic function [§ 60 (x)] equal to the sum of the separate values corresponding to the pressure separately applied to the several barriers. Thus we see how a state of irrotational motion, cyclic with reference to every one of the different circuits of a multiply continuous space, and having arbitrary values for the corresponding cyclic constants, or circulations, may be generated. But the proof of the possibility of fluid motion fulfilling such conditions, founded on this planning out of a genesis of it, leaves us to imagine that it might be different according to the infinitely varied choice we may make of surfaces for the initial forms of the barriers, or according to the order and the duration of the applications of pressure to them in virtue of which these figures may be changed more or less, and in various ways, before the initiating pressures all cease; and hitherto we have seen no reason even to suspect the following proposition to the contrary.

63. (PROP.) The motion of a liquid moving irrotationally within an $(n+1)$ply continuous space is determinate when the normal velocity at every point of the boundary, and the values of the circulations in the n circuits, are given.

This is proved by an application of Green's extended formula (7) of § 57, showing, as the simple formula (1) of the same section showed us in § 61 for simply continuous space, that the difference of the velocity potentials of two motions, each fulfilling this condition, is necessarily zero throughout the whole fluid. Let ϕ, ϕ' be the velocity potentials of two motions fulfilling the prescribed conditions, and let

$$\psi = \phi - \phi'.$$

At every point of the boundary (the barriers not included) the prescribed conditions require that $\mathfrak{d}\phi = \mathfrak{d}\phi'$, and therefore $\mathfrak{d}\psi = 0$. Again, the cyclic constants for ϕ' are equal to those for ϕ; those for ψ, being their differences, must therefore vanish. Hence, if the ϕ and ϕ' of § 57 (7) be made equal to one another and to avoid confusion with our present notation we substitute ψ for each, the second members of that double equation vanish, and it becomes simply

$$\iiint \left(\frac{d\psi^2}{dx^2} + \frac{d\psi^2}{dy^2} + \frac{d\psi^2}{dz^2} \right) dx\,dy\,dz = 0\,;$$

which, as before (§ 61), proves that $\psi = 0$, and therefore $\phi' = \phi$; and so establishes our present proposition.

EXAMPLE (1). The solution $\phi = \tan^{-1} y/x$ considered in § 56, fulfils Laplace's equation, $\nabla^2\phi = 0$; and obviously satisfies the surface condition, not merely for the annular space with rectangular meridional section there considered, but for the hollow space bounded by the figure of revolution obtained by carrying a closed curve of any shape round any axis (OZ) not cutting the curve; which, for brevity, we shall in future call a *hollow circular ring*. Hence the irrotational motion possible within a fixed hollow circular ring is such that the velocity potential is proportional to the angle between the meridian plane through any point, and a fixed meridian.

EXAMPLE (2). The solid angle, α, subtended at any point (x, y, z) by an infinitesimal plane area, A, in any fixed position, fulfils Laplace's equation $\nabla^2\alpha = 0$. This well-known proposition may be proved by taking A at the origin, and perpendicular to OX, when we have

$$\alpha = \frac{Ax}{(x^2+y^2+z^2)^{\frac{3}{2}}} = A\frac{d}{dz}\frac{-1}{(x^2+y^2+z^2)^{\frac{1}{2}}} \qquad \text{.........(5)},$$

for which $\nabla^2\alpha = 0$ is verified.

The solid angle subtended at (x, y, z) by any single closed circuit is the sum of those subtended at the same point by all parts into which we may divide any limited surface having this curve for its bounding edge. [Consider particularly curves such as those represented by the first three diagrams of § 58.] Hence if we call ϕ the solid angle subtended at (x, y, z) by this surface, Laplace's equation $\nabla^2\phi$ is fulfilled. Hence ϕ represents the velocity potential of the irrotational motion possible for a liquid contained in an infinite fixed closed vessel, within which is fixed, at an infinite distance from the outer bounding surface, an infinitely thin wire bent into the form of the closed curve in question.

The particular case of this example for which the curve is a circle, presents us with the simplest specimen of cyclic irrotational motion not confined [as that of Example (1) is] to a set of parallel planes. The velocity potential being the apparent area of a circular disc (or the area of a spherical ellipse) is readily found, and shown to be expressible readily in terms of a complete elliptic integral of the third class, and therefore in terms of incomplete elliptic functions of the first and second classes. The equi-potential surfaces are therefore traceable by aid of Legendre's tables. But it is to Helmholtz that we owe the remarkable and useful discovery, that the equations of the *stream lines* (or lines perpendicular to the equi-potential surfaces) are expressible in terms of complete integrals of the first and second classes. They are therefore easily traceable by aid of Legendre's tables. The annexed diagram, of which we shall make much use later, shows these curves as calculated and drawn by Mr Macfarlane from Helmholtz's formula, expressed in terms of rectangular co-ordinates. An improved method of tracing them is described in a note by Mr Clerk Maxwell*, which he has kindly allowed me to append to this paper.

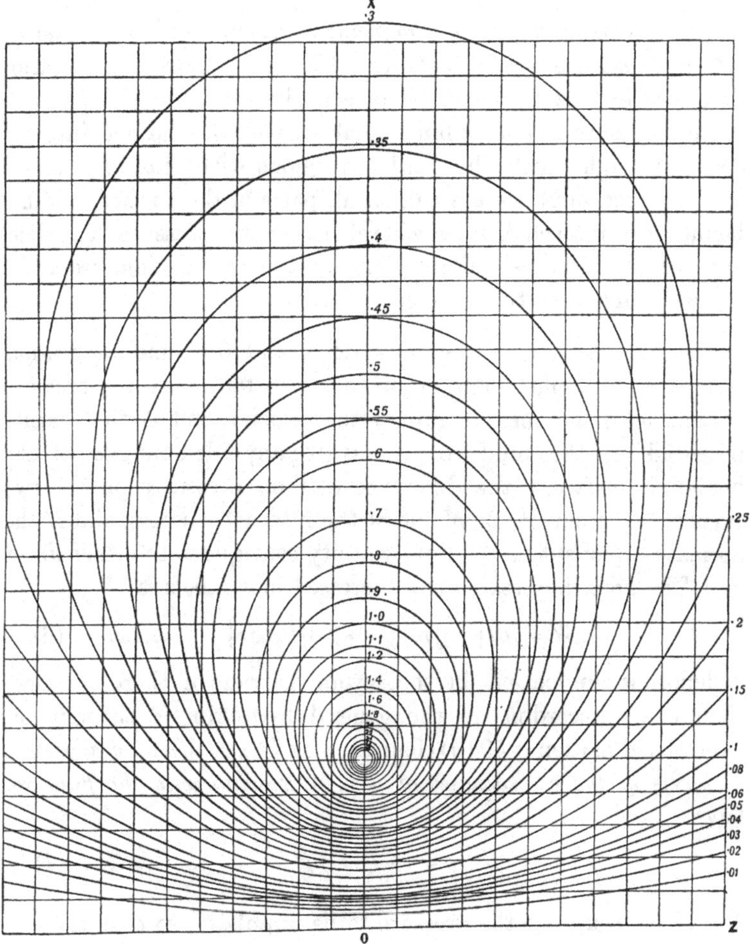

[* Cf. Maxwell's *Electricity and Magnetism*, vol. II.]

EXAMPLE (3). The motion described in Example (2) will remain unchanged outside any solid ring formed by solidifying and reducing to rest a portion of the fluid bounded by stream lines surrounding the infinitely thin wire. Thus we have a solid thick endless wire or bar forming a ring, or an endless knot as illustrated in the first three diagrams of § 59, of peculiar sectional figure depending on the stream lines round the arbitrary curve of Example (2); and the cyclic irrotational motion which, if placed in an infinite liquid it permits, is that whose velocity potential is proportional to the solid angle defined geometrically in the general solution given under Example (2).

64. *Kinetic energy of compounded acyclic and polycyclic irrotational motion—kinetico-statics.* The work done in the operation described in § 62 is calculated directly by summing the products of the pressure into an infinitesimal area of the surface, into the space through which the fluid contiguous with this area moves in the direction of the normal, for all parts of the surface, whether boundary or internal barrier, where the genetic pressure is applied, and for all infinitesimal divisions of the whole time from the commencement of the motion.

(*a*) Let w denote the work done, and $\int dt$ time-integration, from the beginning of motion up to any instant. At any previous instant let p be the pressure, q the velocity, and ϕ the velocity potential, of the fluid contiguous to any element $d\sigma$ of the bounding surface, k the difference of fluid pressures on the two sides of any element, $d\varsigma$, of one of the internal barriers, and N the normal component of the fluid velocity contiguous to either $d\sigma$ or $d\varsigma$. The preceding statement expressed in symbols is

$$W = \int dt\,[-\iint pN d\sigma + \Sigma \iint kN d\varsigma] \quad\text{..................(6)},$$

Σ denoting summation for the several barriers if there are more than one. According to the general hydrokinetic theorem for irrotational motion [§ 59 (6) compare with § 31 (5)], *with ϕ expressed in terms of the co-ordinates of a point moving with the fluid,* we have

$$p = -\frac{d\phi}{dt} + \tfrac{1}{2}q^2 \quad\text{.......................(7)}.$$

Now, let us suppose the pressure to be impulsive, so that there is infinitely little change of shape either of the bounding surface or of the barriers during the time $\int dt$. This will also imply that $d\phi/dt$ is infinitely great in comparison with $\frac{1}{2}q^2$; so that

$$p = -\frac{d\phi}{dt} \quad \text{.........................(8)}.$$

And according to the notation of § 57 we have

$$N = \mathfrak{d}\phi \quad \text{.........................(9)}.$$

Also k is constant over each barrier surface.
Hence (6) becomes

$$W = \int dt \left[\iint \frac{d\phi}{dt} \mathfrak{d}\phi d\sigma + \Sigma k \iint \mathfrak{d}\phi d\varsigma \right] \quad \text{.........(10)}.$$

64 (*b*) The initiating motion of the bounding surface and the pressures on the barriers may be varied quite arbitrarily from the beginning to the end of the impulse; so that the history within that period of the acquisition of the prescribed final velocity may be altogether different, and not even simultaneous, in different parts of the bounding surface. Thus k_1 and k_2 may be quite different functions of t; provided only $\int k_1 dt$ and $\int k_2 dt$ have the prescribed values, which we shall denote by $\mathfrak{k}_1$ and $\mathfrak{k}_2$ respectively.

(*c*) But, for one example, we may suppose ϕ to have at each instant of $\int dt$ everywhere one and the same proportion of its final value; so that if the latter be denoted by Φ, and if we put

$$\frac{\phi}{\Phi} = m \quad \text{.........................(11)},$$

m is independent of co-ordinates of position, but may of course be any arbitrary function of the time. Hence, observing that

$$\int dt\, m \frac{dm}{dt} = \tfrac{1}{2},$$

as the final value of m is 1, (10) becomes

$$W = \tfrac{1}{2} [\iint \Phi \mathfrak{d}\Phi d\sigma + \Sigma \mathfrak{k} \iint \mathfrak{d}\Phi d\varsigma] \quad \text{...........(12)}.$$

(*d*) The second member of this equation doubled agrees with the two equal second members of (7) § 57 with ϕ and ϕ' each made equal to Φ. And the first member of that equation becomes twice the kinetic energy of the whole motion. Hence, when $\phi' = \phi$, and $\nabla^2\phi = 0$, (7) of § 57 expresses the equation of energy for the impulsive generation, of the fluid motion corresponding to velocity potential ϕ, by pressures varying throughout according to the same function of the time; the first member being twice the kinetic energy of the motion generated, and the second twice the work done in the process.

64 (*e*). As another example, let us suppose the initiating pressures to be so applied as first to generate a motion corresponding to velocity potential ϕ, and after that to change the velocity potential from ϕ to $\phi+\phi'$, denoting by ϕ and ϕ' any two functions, such that $\phi+\phi'=\Phi$, and each fulfilling Laplace's equation: and let the augmentation from zero to ϕ, and again from ϕ to $\phi+\phi'$ be uniform through the whole fluid. The work done in the first process, found as above (12),

$$\tfrac{1}{2}\left[\iint\phi\mathfrak{d}\phi\,d\sigma+\Sigma\kappa\iint\mathfrak{d}\phi\,d\varsigma\right]\ \ldots\ldots\ldots\ldots..(13),$$

if κ_1, κ_2, &c., denote the cyclic constants relative to ϕ, as $\mathfrak{k}_1$, $\mathfrak{k}_2$, &c., relatively to Φ, and the additional work done in the second process, similarly found, is

$$\tfrac{1}{2}\left[\iint\phi'(2\mathfrak{d}\phi+\mathfrak{d}\phi')\,d\sigma+\Sigma\kappa'\iint(2\mathfrak{d}\phi+\mathfrak{d}\phi')\,d\varsigma\right]\ldots\ldots(14).$$

(*f*) Now, as we have seen (§ 63) that the actual fluid motion depends at each instant wholly on the normal velocity at each point of the bounding surface and the values of the cyclic constants, it follows that the work done in generating it ought to be independent of the order and law of the acquisition of velocity at the bounding surface, and of the attainment of the values of the several cyclic constants. Hence, the sum of (13) and (14) ought to be equal to (12). But if for Φ in (12) we substitute $\phi+\phi'$, the difference between its value and that of the sum of (13) and (14) is found to be

$$\tfrac{1}{2}\left[\iint(\phi\mathfrak{d}\phi'-\phi'\mathfrak{d}\phi)\,d\sigma+\Sigma\left(\kappa\iint\mathfrak{d}\phi'\,d\varsigma-\kappa'\iint\mathfrak{d}\phi\,d\varsigma\right)\right]\ldots(15);$$

which, being half the difference between the two equal second members of (7) § 57 for the case of

$$\nabla^2\phi=0 \text{ and } \nabla^2\phi'=0,$$

is equal to zero. Hence, the equality of the second members of (7) § 57, constitutes the analytical reconciliation of the equations of energy for different modes of generation of the same fluid motion.

3. THE TRANSLATORY VELOCITY OF A CIRCULAR VORTEX RING.

[*Appended to Prof. Tait's translation of Helmholtz's Memoir on Vortex Motion; Phil. Mag.* XXXIII. 1867, 511—512.]

FOLLOWING as nearly as may be Helmholtz's notation, let g be the radius of the circular axis of a uniform vortex-ring, and a the radius of the section of its core (which will be approximately circular when a is small in comparison with g), the vortex motion being so instituted that there is no molecular rotation in any part of the fluid exterior to this core, and that in the core the angular velocity of the molecular rotation is approximately ω, or rigorously

$$\frac{\omega\chi}{g}$$

for any fluid particle at distance χ from the straight axis.

I find that the velocity of translation is approximately equal to

$$\frac{\omega a^2}{2g}\left(\log\frac{8g}{a}-\frac{1}{4}\right)$$

(quantities of the same order as this multiplied by a/g being neglected).

The velocity of the liquid at the surface of the core is approximately constant and equal to ωa. At the centre of the ring it is $\pi\omega a^2/g$.

If these be denoted by Q and W respectively, and if T be the velocity of translation, we therefore have

$$T=\frac{a}{2g}\left(\log\frac{8g}{a}-\frac{1}{4}\right)Q$$

$$=\frac{\log\frac{8g}{a}-\frac{1}{4}}{2\pi}W.$$

Hence the velocity of translation is very large in comparison with the fluid velocity along the axis through the centre of the ring, when the section is so small that $\log 8g/a$ is large in comparison with 2π. But the velocity of translation is always small in comparison with the velocity of the fluid at the surface of the core, and the more so the smaller is the diameter of the section in comparison with the diameter of the ring.

These results remove completely the difficulty which has hitherto been felt with reference to the translation of infinitely thin vortex-filaments. I have only succeeded in obtaining them since the communication of my mathematical paper (April 29, 1867) to the Royal Society of Edinburgh, but hope to be allowed to add a proof of them to that paper should it be accepted for the Transactions.

4, 5. HYDROKINETIC SOLUTIONS AND OBSERVATIONS*.

[From the *Philosophical Magazine*, Vol. XLII., Nov. 1871, pp. 362—377; reprinted in *Baltimore Lectures*, 1904, pp. 584—601.]

4. ON THE MOTION OF FREE SOLIDS THROUGH A LIQUID.

Part I.—[Acyclic Motion produced by a freely moving Solid.]

THIS paper commences with the following extract from the author's private journal, of date January 6, 1858:—

"Let $\mathfrak{X}$, $\mathfrak{Y}$, $\mathfrak{Z}$, $\mathfrak{L}$, $\mathfrak{M}$, $\mathfrak{N}$ be rectangular components of an impulsive force and an impulsive couple applied to a solid of invariable shape, with or without inertia of its own, in a perfect liquid, and let u, v, w, ϖ, ρ, σ be the components of linear and angular velocity generated. Then, if the *vis viva*† (twice the mechanical value) of the whole motion be, as it cannot but be, given by the expression

$$Q = [u, u]\, u^2 + [v, v]\, v^2 + \ldots + 2\, [v, u]\, vu + 2\, [w, u]\, wu + \ldots \\ + 2\, [\varpi, u]\, \varpi u + \ldots,$$

where $= [u, u]$, $[v, v]$, &c. denote 21 constant coefficients determinable by transcendental analysis from the form of the surface of the solid, probably involving only elliptic transcendentals when the surface is ellipsoidal: involving, of course, the moments of inertia of the solid itself: we must have

$$[u, u]\, u + [v, u]\, v + [w, u]\, w + [\varpi, u]\, \varpi + [\rho, u]\, \rho + [\sigma, u]\, \sigma = \mathfrak{X}, \text{\&c.},$$
$$[u, \varpi]\, u + [v, \varpi]\, v + [w, \varpi]\, w + [\varpi, \varpi]\, \varpi + [\rho, \varpi]\, \rho + [\sigma, \varpi]\, \sigma = \mathfrak{L}, \text{\&c.}$$

* Parts I. and II. from the *Proceedings of the Royal Society of Edinburgh* for 1870–71. Parts III. and IV. from letters to Professor Tait of August 1871. Part V. appended to this communication September 1871.

† Henceforth T, instead of $\frac{1}{2}Q$, is used to denote the "mechanical value," or, as it is now called, the "kinetic energy" of the motion.

If now a continuous force X, Y, Z, and a continuous couple L, M, N, referred to axes fixed in the body, is applied, and if $\mathfrak{X}$, ..., &c. denote the impulsive force and couple capable of generating from rest the motion u, v, w, ϖ, ρ, σ, which exists in reality at any time t; or, merely mathematically, if $\mathfrak{X}$ &c. denote for brevity the preceding linear functions of the components of motion, the equations of motion are as follows:—

$$\left.\begin{aligned} &\frac{d\mathfrak{X}}{dt} - \mathfrak{Y}\sigma + \mathfrak{Z}\rho = X, \quad \frac{d\mathfrak{Y}}{dt} = \text{\&c.} \\ &\frac{d\mathfrak{L}}{dt} - \mathfrak{Y}w + \mathfrak{Z}v - \mathfrak{M}\sigma + \mathfrak{N}\rho = L \\ &\frac{d\mathfrak{M}}{dt} - \mathfrak{Z}u + \mathfrak{X}w - \mathfrak{N}\varpi + \mathfrak{L}\sigma = M \\ &\frac{d\mathfrak{N}}{dt} - \mathfrak{X}v + \mathfrak{Y}u - \mathfrak{L}\rho + \mathfrak{M}\varpi = N \end{aligned}\right\} \quad \text{.........(1).}$$

Three first integrals, when

$$X=0, \quad Y=0, \quad Z=0, \quad L=0, \quad M=0, \quad N=0 \quad \text{...(2),}$$

must of course be, and obviously are,

$$\mathfrak{X}^2 + \mathfrak{Y}^2 + \mathfrak{Z}^2 = \text{const.} \quad \text{.....................(3),}$$

resultant momentum constant;

$$\mathfrak{L}\mathfrak{X} + \mathfrak{M}\mathfrak{Y} + \mathfrak{N}\mathfrak{Z} = \text{const.} \quad \text{..............(4),}$$

resultant of moment of momentum constant; and

$$u\mathfrak{X} + v\mathfrak{Y} + w\mathfrak{Z} + \varpi\mathfrak{L} + \rho\mathfrak{M} + \sigma\mathfrak{N} = Q \quad \text{.........(5)."}$$

These equations were communicated in a letter to Professor Stokes, of date (probably January) 1858, and they were referred to by Professor Rankine, in his first paper on Stream-lines, communicated to the Royal Society of London*, July 1863.

They are now communicated to the Royal Society of Edinburgh, and the following proof is added:—

* These equations will be very conveniently called the Eulerian equations of the motion. They correspond precisely to Euler's equations for the rotation of a rigid body, and include them as a particular case. As Euler seems to have been the first to give equations of motion in terms of co-ordinate components of velocity and force referred to lines fixed relatively to the moving body, it will be not only convenient, but just, to designate as "Eulerian equations" any equations of motion in which the lines of reference, whether for position, or velocity, or moment of momentum, or force, or couple, move with the body or the bodies whose motion is the subject. [The notebook of date 1858, containing this early statement of the theory of the impulse, has been preserved. For developments see Lamb's *Hydrodynamics*.]

Let P be any point fixed relatively to the body; and at time t, let its co-ordinates relatively to axes OX, OY, OZ, fixed in space, be x, y, z. Let PA, PB, PC be three rectangular axes fixed relatively to the body, and (A, X), (A, Y), ... the cosines of the nine inclinations of these axes to the fixed axes OX, OY, OZ.

Let the components of the "impulse"* or generalized momentum parallel to the fixed axes be ξ, η, ζ, and its moments round the same axes λ, μ, ν; so that if X, Y, Z be components of force acting on the solid, in line through P, and L, M, N components of couple, we have

$$\left.\begin{array}{l}\dfrac{d\xi}{dt}=X,\quad \dfrac{d\eta}{dt}=Y,\quad \dfrac{d\zeta}{dt}=Z\\[2ex] \dfrac{d\lambda}{dt}=L+Zy-Yz,\quad \dfrac{d\mu}{dt}=M+Xz-Zx,\quad \dfrac{d\nu}{dt}=N+Yx-Xy\end{array}\right\}\text{(6)}.$$

Let $\mathfrak{X}$, $\mathfrak{Y}$, $\mathfrak{Z}$ and $\mathfrak{L}$, $\mathfrak{M}$, $\mathfrak{N}$ be the components and moments of the impulse relatively to the axes PA, PB, PC moving with the body. We have

$$\left.\begin{array}{l}\xi=\mathfrak{X}(A,X)+\mathfrak{Y}(B,X)+\mathfrak{Z}(C,X)\\ \dots\dots\dots\dots\dots\dots\dots\dots\dots\dots\\ \dots\dots\dots\dots\dots\dots\dots\dots\dots\dots\\ \lambda=\mathfrak{L}(A,X)+\mathfrak{M}(B,X)+\mathfrak{N}(C,X)+\mathfrak{Z}y-\mathfrak{Y}z\\ \dots\dots\dots\dots\dots\dots\dots\dots\dots\dots\\ \dots\dots\dots\dots\dots\dots\dots\dots\dots\dots\end{array}\right\}\dots(7).$$

Now let the fixed axes OX, OY, OZ be chosen coincident with the position at time t of the moving axes PA, PB, PC: we shall consequently have

$$\left.\begin{array}{l}x=0,\quad y=0,\quad z=0\\[1ex] \dfrac{dx}{dt}=u,\quad \dfrac{dy}{dt}=v,\quad \dfrac{dz}{dt}=w\end{array}\right\}\dots\dots\dots\dots\dots\dots(8),$$

$$\left.\begin{array}{l}(A,X)=(B,Y)=(C,Z)=1\\ (A,Y)=(A,Z)=(B,X)=(B,Z)=(C,X)=(C,Y)=0\\[1ex] \dfrac{d(A,Y)}{dt}=\sigma,\quad \dfrac{d(B,X)}{dt}=-\sigma,\quad \dfrac{d(C,X)}{dt}=\rho\\[2ex] \dfrac{d(A,Z)}{dt}=-\rho,\quad \dfrac{d(B,Z)}{dt}=\varpi,\quad \dfrac{d(C,Y)}{dt}=-\varpi\end{array}\right\}\dots(9).$$

Using (7), (8), and (9) in (6), we find (1).

* See "Vortex Motion," § 6, *Trans. Roy. Soc. Edin.* (1868).

One chief object of this investigation was to illustrate dynamical effects of heliçoidal property (that is, right or left-handed asymmetry). The case of complete isotropy, with heliçoidal quality, is that in which the coefficients in the quadratic expression for T fulfil the following conditions.

$$\left.\begin{array}{ll} [u, u]=[v, v]=[w,w] & \text{(let } m \text{ be their common value)} \\ [\varpi,\varpi]=[\rho, \rho]=[\sigma,\sigma] & \quad ,, \quad n \quad ,, \quad ,, \\ [u, \varpi]=[v, \rho]=[w,\sigma] & \quad ,, \quad h \quad ,, \quad ,, \\ \multicolumn{2}{l}{[v, w]=[w,u]=[u, v]=0 ; \ [\rho,\sigma]=[\sigma,\varpi]=[\varpi,\rho]=0} \\ \text{and} & \\ \multicolumn{2}{l}{[u, \rho]=[u, \sigma]=[v, \sigma]=[v,\varpi]=[w,\varpi]=[w,\rho]=0} \end{array}\right\} \ldots(10);$$

so that the formula for T is

$$T=\tfrac{1}{2}\{m(u^2+v^2+w^2)+n(\varpi^2+\rho^2+\sigma^2)+2h(u\varpi+v\rho+w\sigma)\} \quad (11).$$

For this case, therefore, the Eulerian equations (1) become

$$\left.\begin{array}{l} \dfrac{d(mu+h\varpi)}{dt}-m(v\sigma-w\rho)=X, \text{\&c.} \\ \text{and} \quad \dfrac{d(n\varpi+hu)}{dt}=L, \text{\&c.} \end{array}\right\} \ldots(11).$$

[Memorandum:—Lines of reference fixed relatively to the body.]

But inasmuch as (11) remains unchanged when the lines of reference are altered to any other three lines at right angles to one another through P, it is easily shown directly from (6), (7), and (9) that if, altering the notation, we take u, v, w to denote the components of the velocity of P parallel to three fixed rectangular lines, and ϖ, ρ, σ the components of the body's angular velocity round these lines, we have

$$\left.\begin{array}{l} \dfrac{d(mu+h\varpi)}{dt}=X, \text{\&c.} \\ \text{and} \quad \dfrac{d(n\varpi+hu)}{dt}+h(\sigma v-\rho w)=L, \text{\&c.} \end{array}\right\} \ldots(12),$$

[Memorandum:—Lines of reference fixed in space]

which are more convenient than the Eulerian equations.

The integration of these equations, when neither force nor couple acts on the body ($X=0$ &c., $L=0$ &c.), presents no difficulty; but its result is readily seen from § 21 ("Vortex

Motion") to be that, when the impulse is both translatory and rotational, the point P, round which the body is isotropic, moves uniformly in a circle or spiral so as to keep at a constant distance from the "axis of the impulse," and that the components of angular velocity round the three fixed rectangular axes are constant.

An isotropic heliçoid may be made by attaching projecting vanes to the surface of a globe in proper positions; for instance, cutting at 45° each, at the middles of the twelve quadrants of any three great circles dividing the globe into eight quadrantal triangles. By making the globe and the vanes of light paper, a body is obtained rigid enough and light enough to illustrate by its motions through air the motions of an isotropic heliçoid through an incompressible liquid. But curious phenomena, not deducible from the present investigation, will, no doubt, on account of viscosity, be observed.

Part II.—[*Cyclic Motion through a perforated Solid.*]

Still considering only one moveable rigid body, infinitely remote from disturbance of other rigid bodies, fixed or moveable, let there be an aperture or apertures through it, and let there be irrotational circulation or circulations (§ 60, "Vortex Motion") through them. Let ξ, η, ζ be the components of the "impulse" at time t, parallel to three fixed axes, and λ, μ, ν its moments round these axes, as above, with all notation the same, we still have (§ 26, "Vortex Motion")

$$\left.\begin{aligned}\frac{d\xi}{dt}&=X,\ \&c.\\ \frac{d\lambda}{dt}&=L+Zy-Yz,\ \&c.\end{aligned}\right\}\ \ldots\ldots(6)\ \text{(repeated)}.$$

But, instead of for T a quadratic function of the components of velocity as before, we now have

$$T=E+\tfrac{1}{2}\{[u,u]\,u^2+\ldots+2\,[u,v]\,uv+\ldots\}\ \ldots\ldots(13),$$

where E is the kinetic energy of the fluid motion when the solid is at rest, and $\tfrac{1}{2}\{[u,u]\,u^2+\ldots\}$ is the same quadratic as before. The coefficients $[u,\ u]$, $[u,\ v]$, &c. are determinable by a transcendental analysis, of which the character is not at all influenced by the circumstance of there being apertures in the solid. And instead of $\xi = dT/du$, &c., as above, we now have

$$\left.\begin{array}{l} \xi = \dfrac{dT}{du} + Il, \quad \eta = \dfrac{dT}{dv} + Im, \quad \zeta = \dfrac{dT}{dw} + In \\ \lambda = \dfrac{dT}{d\varpi} + I(ny - mz) + Gl, \quad \mu = \&c., \quad \nu = \&c. \end{array}\right\} \ldots\ldots(14),$$

where I denotes the resultant "impulse" of the cyclic motion when the solid is at rest, l, m, n its direction-cosines, G its "rotational moment" (§ 6, "Vortex Motion"), and x, y, z the co-ordinates of any point in its "resultant axis." These, (14) with (13), used in (6) give the equations of the solid's motion referred to fixed rectangular axes. They have the inconvenience of the coefficients $[u, u]$, $[u, v]$, &c. being functions of the angular coordinates of the solid. The Eulerian equations (free from this inconvenience) are readily found on precisely the same plan as that adopted above for the old case of no cyclic motion in the fluid.

The formulæ for the case in which the ring is circular, has no rotation round its axis, and is not acted on by applied forces, though, of course, easily deduced from the general equations (14), (13), (6), are more readily got by direct application of first principles. Let P be such a point in the axis of the ring, and $\mathfrak{C}$, A, B such constants that $\frac{1}{2}(\mathfrak{C}\omega^2 + Au^2 + Bv^2)$ is the kinetic energy due to rotational velocity ω round D, any diameter through P, and translational velocities u along the axis and v perpendicular to it. The impulse of this motion, together with the supposed cyclic motion, is therefore compounded of

$$\text{momentum in lines through } P \left\{\begin{array}{l} Au + I \text{ along the axis,} \\ Bv \text{ perpendicular to axis,} \end{array}\right.$$

and moment of momentum $\mathfrak{C}\omega$ round the diameter D.

Hence if OX be the axis of resultant momentum, (x, y) the co-ordinates of P relatively to fixed axes OX, OY; θ the inclination of the axis of the ring to OX; and ξ the constant value of the resultant momentum, we have

$$\left.\begin{array}{l} \xi\cos\theta = Au + I; \quad -\xi\sin\theta = Bv; \quad \xi y = \mathfrak{C}\omega \\ \text{and} \\ \dot{x} = u\cos\theta - v\sin\theta; \quad \dot{y} = u\sin\theta + v\cos\theta; \quad \dot{\theta} = \omega \end{array}\right\} \ldots(15).$$

Hence for θ we have the differential equation

$$A\mathfrak{C}\frac{d^2\theta}{dt^2}+\xi\left[I\sin\theta+\frac{A-B}{2B}\xi\sin 2\theta\right]=0 \quad\text{......(16)},$$

which shows that the ring oscillates rotationally according to the law of a horizontal magnetic needle carrying a bar of soft iron rigidly attached to it parallel to its magnetic axis.

When θ is and remains infinitely small, $\dot{\theta}$, y, and $\dot{y}$ are each infinitely small, x remains infinitely nearly constant, and the ring experiences an oscillatory motion in period

$$2\pi\sqrt{\frac{B\mathfrak{C}}{[I+(A-B)\dot{x}](I+A\dot{x})}},$$

compounded of translation along OY and rotation round the diameter D. This result is curiously comparable with the well-known gyroscopic vibrations.

5. Influence of Wind and Capillarity on Waves in water supposed frictionless.

Part III.—The Influence of Wind on Waves in water supposed frictionless. (*Letter to Professor Tait, of date August* 16, 1871.)

Taking OX vertically downwards and OY horizontal, let

$$x = h \sin n (y - at) \quad \text{.....................(1)}$$

be the equation of the section of the water by a plane perpendicular to the wave-ridges; and let h (the half wave-height) be infinitely small in comparison with $2\pi/n$ (the wave-length). The x-component of the velocity of the water at the surface is then

$$- nah \cos n (y - at) \quad \text{.....................(2)};$$

and this (because h is infinitesimal) must be the value of $d\phi/dx$ for the point $(0, y)$, if ϕ denote the velocity-potential at any point (x, y) of the water. Now because

$$\frac{d^2\phi}{dx^2} + \frac{d^2\phi}{dy^2} = 0,$$

and ϕ is a periodic function of y, and a function of x which becomes zero when $x = \infty$, it must be of the form

$$P \cos (ny - e) \epsilon^{-nx},$$

where P and e are independent of x and y. Hence, taking $d\phi/dx$, putting $x = 0$ in it, and equating it to (2), we have

$$- Pn \cos (ny - e) = - nah \cos (ny - nat);$$

and therefore $P = ah$, and $e = nat$; so that we have

$$\phi = ah\epsilon^{-nx} \cos n (y - at) \quad \text{.................(3)}.$$

This, it is to be remarked, results simply from the assumptions that the water is frictionless, that it has been at rest, and that its surface is moving in the manner specified by (1).

If the air were a frictionless liquid moving irrotationally, with a constant velocity V at heights above the water (that is to say, values of $-x$) considerably exceeding the wave-length, its velocity potential ψ, found on the same principle, would be

$$(V-\alpha)\, h\epsilon^{nx} \cos n(y-\alpha t) + Vy \text{..............(4).}$$

Let now q denote the resultant velocity at any point (x, y) of the air. Neglecting infinitesimals of the order $(nh)^2$, we have

$$\tfrac{1}{2}q^2 = \tfrac{1}{2}V^2 - V(V-\alpha)\, nh\epsilon^{nx} \sin n(y-\alpha t) \text{ ......(5).}$$

Now, if p denote the pressure at any point (x, y) in the air, and σ the density of the air, we have by the general equation for pressure in an irrotationally moving fluid,

$$C - p = \sigma\left(\frac{d\psi}{dt} + \tfrac{1}{2}q^2 - gx\right) \text{.................(6).}$$

Using (4) and (5) in this and putting $C = \tfrac{1}{2}\sigma V^2$, we find

$$-p = \sigma\{-nh(V-\alpha)^2 \epsilon^{nx} \sin n(y-\alpha t) - gx\} \text{ ......(7).}$$

Similarly if p' denote the pressure at any point (x, y) of the water, since in this case q^2 is infinitesimal, we have

$$-p' = nh\alpha^2\epsilon^{-nx} \sin n(y-\alpha t) - gx \text{ ....(8),}$$

the density of the water being taken as unity.

Now let T be the cohesive tension of the separating surface of air and water. The curvature of the surface at any point (x, y) given by equation (1), being d^2x/dy^2, is equal to

$$-n^2h \sin n(y-\alpha t) \text{ ....(9).}$$

Hence at any point (x, y) fulfilling (1),

$$p - p' = Tn^2h \sin n(y-\alpha t) \text{..............(10);}$$

and by (7) and (8), with for x its value by (1) (which, as h is infinitesimal, only affects their last terms), we have

$$p - p' = h\{n[\sigma(V-\alpha)^2 + \alpha^2] - g(1-\sigma)\} \sin n(y-\alpha t) \text{...(11).}$$

This, compared with (10), gives

$$n[\sigma(V-\alpha)^2 + \alpha^2] - g(1-\sigma) = Tn^2 \text{ ....(12).}$$

Let
$$w = \sqrt{\frac{g(1-\sigma) + Tn^2}{(1+\sigma)n}} \text{ ....(13),}$$

which (being the value of α for $V=0$) is the velocity of propagation of waves with no wind, when the wave-length is $2\pi/n$. Then (12) becomes

$$\frac{\alpha^2 + \sigma(V-\alpha)^2}{1+\sigma} = w^2 \text{.....................(14)},$$

which determines α, the velocity of the same waves when there is wind, of velocity V, in the direction of propagation of the waves. Solving the quadratic, we have

$$\alpha = \frac{\sigma V}{1+\sigma} \pm \left\{w^2 - \frac{\sigma V^2}{(1+\sigma)^2}\right\}^{\frac{1}{2}} \text{..............(15)}.$$

This result leads to the following conclusions:

(1) When $V < w\sqrt{(1+\sigma)}/\sqrt{\sigma}$,

the values of α are positive and negative; that is to say, waves can travel with or against the wind. The positive value is always greater; that is to say, waves travel faster with than against the wind. The velocity of waves travelling *against* the wind is always less than w, the velocity without wind.

(2)* When $V < 2w$, the velocity of waves travelling with the wind is greater than w. When $V = 2w$, the velocity of the waves *with* the wind is undisturbed by the wind; a result obvious without

* The conclusion (2) is incorrect, and is corrected in the reprint in *Baltimore Lectures*, where the results (1), (2), (3), (4) are replaced by the following:

"for given wave-length, $2\pi/n$, the greatest wave-velocity is $w\sqrt{(1+\sigma)}$, which is reached when this is the velocity of the wind. It is interesting to see that with wind of any other speed than that of the waves, and in the direction of the waves, their speed is less. For instance, the wave-speed with no wind, which is w, is less by approximately $\frac{1}{2}\sigma$ of w, (or about 1/1650,) than the speed when the wind is with the waves and of their speed. The explanation clearly is that when the air is motionless relatively to the wave crests and hollows its inertia is not called into play.

"(1) When $\frac{V}{w} = \sqrt{\frac{1+\sigma}{\sigma}} \doteq 28{\cdot}7 \times \left(1 + \frac{1}{1650}\right)$,

one of the values of α is zero, that is to say, static corrugations of wave-length $2\pi/n$, would be equilibrated by wind of velocity

$$w\sqrt{[(1+\sigma)/\sigma]}.$$

But the equilibrium would be unstable.

"(2) When $\frac{V}{w} = \frac{1+\sigma}{\sqrt{\sigma}} \doteq 28{\cdot}7 \times \left(1 + \frac{1}{825}\right)$,

the two values of α are equal.

"(3) When $\frac{V}{w} > \frac{1+\sigma}{\sqrt{\sigma}}$,

both values of α are imaginary, and therefore the wind would blow into spin-drift, waves of length $2\pi/n$ or shorter." Then follows (16) and (16′).

analysis. When $V > 2w$, the velocity of waves travelling with the wind is less than the velocity of the same waves without wind.

(3) When $V > w\sqrt{(1+\sigma)/\sigma}$, waves of such length that w would be their velocity without wind, cannot travel against the wind.

(4) When $V > w(1+\sigma)/\sqrt{\sigma}$, there cannot be waves of so small length as that for which the undisturbed velocity is w, and the equilibrium of the water is essentially unstable. And (13) shows that the minimum value of w is

$$\sqrt{\frac{2\sqrt{gT}(1-\sigma)}{1+\sigma}} \quad\text{.....................(16).}$$

Hence the water with a plane level surface is unstable* if the velocity of the wind exceeds

$$\sqrt{\frac{2\sqrt{gT}(1-\sigma^2)}{\sigma}} \quad\text{.....................(16').}$$

Part IV.—(Letter to Professor Tait, of date August 23, 1871.)

Defining a ripple as any wave on water whose length $< 2\pi\sqrt{(T'/g')}$†, where

$$g' = g\frac{1-\sigma}{1+\sigma}, \qquad T' = \frac{T}{1+\sigma} \quad\text{..............(17),}$$

($\sigma = \cdot 00122$), you always see an exquisite pattern of ripples in front of any solid cutting the surface of water and moving horizontally at any speed, fast or slow. The ripple-length is the smaller root of the equation

$$\frac{2\pi}{\lambda}T' + \frac{\lambda}{2\pi}g' = w^2 \quad\text{.....................(18),}$$

where w is the velocity of the solid. The latter may be a sailing-vessel or a row-boat, a pole held vertically and carried horizontally, an ivory pencil-case, a penknife-blade either edge or flat side foremost, or (best) a fishing-line kept approximately vertical by a lead weight hanging down below water, while carried along at about half a mile per hour by a becalmed vessel. The fishing-line shows both roots admirably; ripples in front, and waves of same velocity (λ the greater root of same equation) in rear. If so fortunate as to be becalmed again, I shall try to get a drawing of the whole pattern, showing the transition at the sides from ripples to waves. When the speed with which

* ["would be unstable even if the air were frictionless." *B.L.* reprint.]

† Which for pure water = 1·7 centim. (see Part V.).

the fishing-line is dragged is diminished towards the critical velocity

$$\sqrt{2\sqrt{g'T'}},$$

which is the minimum velocity of a wave, being [see *Part V.* below] for pure water 23 centims. per second (or 1/2·29 of a nautical mile per hour), the ripples in front elongate and become less curved, and the waves in rear become shorter, till at the critical velocity waves and ripples seem nearly equal, and with ridges nearly in straight lines perpendicular to the line of motion. (This is observation.) It seems that the critical velocity may be determined with some accuracy by experiment thus [see *Part V.* below]:—

Remark that the shorter the ripple-length the greater is the velocity of propagation, and that the moving force of the ripple-motion is partly gravity, but chiefly cohesion; and with very short ripple-length it is almost altogether cohesion, *i.e.* the same force as that which makes a dew drop tremble. The least velocity of frictionless air that can raise a ripple on rigorously quiescent frictionless water is [(16) above]

660 centimetres per second, = 12·8 nautical miles per hour,

(being $\frac{1+\sigma}{\sqrt{\sigma}} \times$ minimum wave-velocity).

Observation shows the sea to be ruffled by wind of a much smaller velocity than this. Such ruffling, therefore, is due to viscosity of the air.

Postscript to Part IV.—(*October* 17, 1871).

The influence of viscosity gives rise to a greater pressure on the anterior than on the posterior side of a solid moving uniformly relatively to a fluid. A symmetrical solid, as for example a globe, moving uniformly through a frictionless fluid, experiences augmentation of pressure in front and behind equally; and diminished pressure over an intervening zone. Observation (as for instance in Mr J. R. Napier's experiments on his "pressure log," for measuring the speed of vessels, and experiments by Joule and myself*, on the pressure at different points of a solid globe exposed to wind) shows that, instead of being increased, the pressure is sometimes actually diminished on the posterior side

* "Thermal Effects of Fluids in Motion," *Royal Society Transactions*, 1860; and *Phil. Mag.* 1860, vol. xx. p. 552.

of a solid moving through a real fluid such as air or water. Wind blowing across ridges and hollows of a fixed solid (such as the furrows of a field) must, because of the viscosity of the air, press with greater force on the slopes facing it than on the sheltered slopes. Hence if a regular series of waves at sea consisted of a solid body moving with the actual velocity of the waves, the wind would do work upon it, or it would do work upon the air, according as the velocity of the wind were greater or less than the velocity of the waves. This case does not afford an exact parallel to the influence of wind on waves, because the surface particles of water do not move forward with the velocity of the waves as those of the furrowed solid do. Still it may be expected that when the velocity of the wind exceeds the velocity of propagation of the waves, there will be a greater pressure on the posterior slopes than on the anterior slopes of the waves; and *vice versâ*, that when the velocity of the waves exceeds the velocity of the wind, or is in the direction opposite to that of the wind, there will be a greater pressure on the anterior than on the posterior slopes of the waves. In the first case the tendency will be to augment the wave, in the second case to diminish it. The question whether a series of waves of a certain height gradually augment with a certain force of wind or gradually subside through the wind not being strong enough to sustain them, cannot be decided offhand. Towards answering it Stokes's investigation of the work against viscosity of water required to maintain a wave*, gives a most important and suggestive instalment. But no theoretical solution, and very little of experimental investigation, can be referred to with respect to the eddyings of the air blowing across the tops of the waves, to which, by its giving rise to greater pressure on the posterior than on the anterior slopes, the influence of the wind in sustaining and maintaining waves is chiefly if not altogether due.

My attention having been called three days ago, by Mr Froude, to Scott Russell's Report on Waves (British Association, York, 1844), I find in it a remarkable illustration or indication of the leading idea of the theory of the influence of wind on waves, that the velocity of the wind must exceed that of the waves, in the following statement:—"Let him [an observer studying the

* *Transactions of the Cambridge Philosophical Society*, 1851 ("Effect of Internal Friction of Fluids on the Motion of Pendulums," Section V.).

surface of a sea or large lake, during the successive stages of an increasing wind, from a calm to a storm] begin his observations in a perfect calm, when the surface of the water is smooth and reflects like a mirror the images of surrounding objects. This appearance will not be affected by even a slight motion of the air, and a velocity of less than half a mile an hour ($8\frac{1}{2}$ in. per sec.) does not sensibly disturb the smoothness of the reflecting surface. A gentle zephyr flitting along the surface from point to point, may be observed to destroy the perfection of the mirror for a moment, and on departing, the surface remains polished as before; if the air have a velocity of about a mile an hour, the surface of the water becomes less capable of distinct reflexion, and on observing it in such a condition, it is to be noticed that the diminution of this reflecting power is owing to the presence of those minute corrugations of the superficial film which form waves of the *third order*. These corrugations produce on the surface of the water an effect very similar to the effect of those panes of glass which we see corrugated for the purpose of destroying their transparency, and these corrugations at once prevent the eye from distinguishing forms at a considerable depth, and diminish the perfection of forms reflected in the water. To fly-fishers this appearance is well known as diminishing the facility with which the fish see their captors. This first stage of disturbance has this distinguishing circumstance, that the phenomena on the surface cease almost simultaneously with the intermission of the disturbing cause, so that a spot which is sheltered from the direct action of the wind remains smooth, the waves of the third order being incapable of travelling spontaneously to any considerable distance, except when under the continued action of the original disturbing force. This condition is the indication of present force, not of that which is past. While it remains it gives that deep blackness to the water which the sailor is accustomed to regard as an index of the presence of wind, and often as the forerunner of more.

"The second condition of wave motion is to be observed when the velocity of the wind acting on the smooth water has increased to two miles an hour. Small waves then begin to rise uniformly over the whole surface of the water; these are waves of the second order, and cover the water with considerable regularity Capillary waves disappear from the ridges of these waves, but are to be found sheltered in the hollows between them, and on the anterior slopes of these waves. The regularity of the distribution of these secondary waves over the surface is remarkable; they begin with about an inch of amplitude, and a couple of inches long; they enlarge as the velocity or duration of the wave increases; by and by conterminal waves unite; the ridges increase, and if the wind increase the waves become cusped, and are regular waves of the *second order*. They continue enlarging their dimensions; and the depth to which they produce the agitation increasing simultaneously with their magnitude, the surface becomes extensively covered with waves of nearly uniform magnitude."

The "Capillary waves" or "waves of the third order" referred to by Russell are what I, in ignorance of his observations on this branch of his subject, had called "ripples." The velocity of 8½ inches (21½ centimetres) per second is precisely the velocity he had chosen (as indicated by his observations) for the velocity of propagation of the straight-ridged waves streaming obliquely from the two sides of the path of a small body moving at speeds of from 12 to 36 inches per second; and it agrees remarkably with my theoretical and experimental determination of the absolute minimum wave-velocity (23 centimetres per second; see *Part V.*). Russell has not explicitly pointed out that his critical velocity of 8½ inches per second was an absolute minimum velocity of propagation. But the idea of a minimum velocity of waves can scarcely have been far from his mind when he fixed upon 8½ inches per second as the minimum of wind that can sustain ripples. In an article to appear in *Nature* on the 26th of this month, I have given extracts from Russell's Report (including part of a quotation which he gives from Poncelet and Lesbros in the memoirs of the French Institute for 1829), showing how far my observations on ripples had been anticipated. I need say no more here than that these anticipations do not include any indication of the dynamical theory which I have given, and that the subject was new to me when *Parts III, IV* and *V* of the present communication were written.

Part V.—Waves under motive power of Gravity and Cohesion jointly, without wind.

Leaving the question of wind, consider (13), and introduce notation of (16), (17) in it. It becomes

$$w^2 = \frac{g'}{n} + T'n \quad \text{.......................(19).}$$

This has a minimum value,

$$\left.\begin{aligned} w^2 &= 2\sqrt{g'T'} \\ \text{when} \qquad n &= \sqrt{\frac{g'}{T'}} \end{aligned}\right\} \quad \text{.......................(20).}$$

In applying these formulæ to the case of air and water, we may neglect the difference between g and g', as the value of σ is about $\frac{1}{820}$; and between T and T', although it is to be remarked that it is T' rather than T that is ordinarily calculated from experiments on capillary attraction. From experiments of Gay-Lussac's it appears that the value of T' is about ·073 of a gramme weight per centimetre; that is to say, in terms of the kinetic unit of force founded on the gramme as unit of mass,

$$T' = g \times {\cdot}073.$$

To make the density of water unity (as that of the lower liquid has been assumed), we must take one centimetre as unit of length. Lastly, with one second as unit of time, we have

$$g = 982;$$

and (18) gives

$$w = \sqrt{982\left(\frac{1}{n} + {\cdot}074 \times n\right)}$$

for the wave-velocity in centimetres per second, corresponding to wave-length $2\pi/n$. When $1/n = \sqrt{{\cdot}073} = {\cdot}27$ (that is, when the wave-length is 1·7 centimetre), the velocity has a minimum value of 23 centimetres per second.

The part of the preceding theory which relates to the effect of cohesion on waves of liquids occurred to me in consequence of having recently observed a set of very short waves advancing steadily, directly in front of a body moving slowly through water, and another set of waves considerably longer following steadily in its wake. The two sets of waves advanced each at the same rate as the moving body; and thus I perceived that there were two different wave-lengths which gave the same velocity of propagation. When the speed of the body's motion through the water was increased, the waves preceding it became shorter, and those in its wake became longer. Close before the cut-water of a vessel moving at a speed of not more than two or three knots* through very smooth water, the surface of the water is marked with an exquisitely fine and regular fringe of ripples, in which several scores of ridges and hollows may be distinguished (and probably counted, with a little practice) in a space extending 20 or 30 centimetres in advance of the solid. Right astern of either a steamer or sailing vessel moving at any speed above four or five knots, waves may generally be seen following the vessel at exactly its own speed, and appearing of such lengths as to verify as nearly as can be judged the ordinary formula

$$l = \frac{2\pi w^2}{g}$$

for the length of waves advancing with velocity w, in deep water. In the well-known theory of such waves, gravity is assumed as the sole origin of the motive forces. When cohesion was thought of at all (as, for instance, by Mr Froude in his important nautical experiments on models towed through water, or set to oscillate to test qualities with respect to the rolling of ships at sea), it was justly judged to be not sensibly influential in waves exceeding 5 or 10 centimetres in length. Now it becomes apparent that for waves of any length less than 5 or 10 centimetres cohesion contributes sensibly to the motive system, and that, when the length is a small fraction of a centimetre, cohesion is much more influential than gravity as "motive" for the vibrations.

The following extract † from part of a letter to Mr Froude, forming part of a communication to *Nature* (to appear on the 26th of this month), describes observations for an experimental determination of the minimum velocity of waves in sea-water:—...

* The speed "one knot" is a velocity of one nautical mile per hour, or 51·5 centimetres per second.

[† Reprinted, p. 88, *infra*.]

6. Ripples and Waves*.

[From *Nature*, Vol. v. 1871, pp. 1—3.]

You have always considered cohesion of water (capillary attraction) as a force which would seriously disturb such experiments as you were making, if on too small a scale. Part of its effect would be to modify the waves generated by towing your models through the water. I have often had in my mind the question of waves as affected by gravity and cohesion jointly, but have only been led to bring it to an issue by a curious phenomenon which we noticed at the surface of the water round a fishing-line one day slipping out of Oban (becalmed) at about half a mile an hour through the water. The speed was so small that the lead kept the line almost vertically downwards; so that the experimental arrangement was merely a thin straight rod held nearly vertical, and moved through smooth water at speeds from about a quarter to three-quarters of a mile per hour. I tried boat-hooks, oars, and other forms of moving solids, but they seemed to give, none of them, so good a result as the fishing-line. The small diameter of the fishing-line seemed to favour the result, and I do not think its roughness interfered much with it. I shall, however, take another opportunity of trying a smooth round rod like a pencil, kept vertical by a lead weight hanging down under water from one end, while it is held up by the other end. The fishing-line, however, without any other appliance proved amply sufficient to give very good results.

What we first noticed was an extremely fine and numerous set of short waves preceding the solid, much longer waves following it right in the rear, and oblique waves streaming off in the usual manner at a definite angle on each side, into which the waves in

* **Extract from a letter to Mr W. Froude, by Sir W. Thomson.**

front and the waves in the rear merged so as to form a beautiful and symmetrical pattern, the tactics of which I have not been able thoroughly to follow hitherto. The diameter of the "solid" (that is to say the fishing-line) being only two or three millimetres, and the longest of the observed waves five or six centimetres, it is clear that the waves at distances in any directions from the solid exceeding fifteen or twenty centimetres, were sensibly unforced (that is to say moving each as if it were part of an endless series of uniform parallel waves undisturbed by any solid). Hence the waves seen right in front and right in rear showed (what became immediately an obvious result of theory) two different wave-lengths with the same velocity of propagation. The speed of the vessel falling off, the waves in rear of the fishing-line became shorter and those in advance longer, showing another obvious result of theory. The speed further diminishing, one set of waves shorten and the other lengthen, until they become, as nearly as I can distinguish, of the same lengths, and the oblique lines of waves in the intervening pattern open out to an obtuse angle of nearly two right angles. For a very short time a set of parallel waves some before and some behind the fishing-line, and all advancing direct with the same velocity, were seen. The speed further diminishing the pattern of waves disappeared altogether. Then slight tremors of the fishing-line (produced for example by striking it above water) caused circular rings of waves to diverge in all directions, those in front advancing at a greater speed relatively to the water than that of the fishing-line. All these phenomena illustrated very remarkably a geometry of ripples communicated a good many years ago to the *Philosophical Magazine* by Hirst, in which, however, so far as I can recollect, the dynamics of the subject were not discussed. The speed of the solid, which gives the uniform system of parallel waves before and behind it, was clearly an absolute minimum wave-velocity, being the limiting velocity to which the common velocity of the larger waves in rear and shorter waves in front was reduced by shortening the former and lengthening the latter to an equality of wave-length.

Taking ·074 of a gramme weight per centimetre of breadth for the cohesive tension of a water surface (calculated from experiments by Gay Lussac, contained in Poisson's theory of capillary attraction, for pure water at a temperature, so far as

I recollect, of about 9° Cent.) and one gramme as the mass of a cubic centimetre, I find, for the minimum velocity of propagation of surface waves, 23 centimetres per second*. The minimum wave velocity for sea-water may be expected to be not very different from this. (It would of course be the same if the cohesive tension of sea water were greater than that of pure water in precisely the same ratio as the density.)

About three weeks later, being becalmed in the Sound of Mull, I had an excellent opportunity, with the assistance of Prof. Helmholtz, and my brother from Belfast, of determining by observation the minimum wave velocity with some approach to accuracy. The fishing-line was hung at a distance of two or three feet from the vessel's side, so as to cut the water at a point not sensibly disturbed by the motion of the vessel. The speed was determined by throwing into the sea pieces of paper previously wetted, and observing their times of transit across parallel planes, at a distance of 912 centimetres asunder, fixed relatively to the vessel by marks on the deck and gunwale. By watching carefully the pattern of ripples and waves, which connected the ripples in front with the waves in rear, I had seen that it included a set of parallel waves slanting off obliquely on each side, and presenting appearances which proved them to be waves of the critical length and corresponding minimum speed of propagation. Hence the component velocity of the fishing-line perpendicular to the fronts of these waves was the true minimum velocity. To measure it, therefore, all that was necessary was to measure the angle between the two sets of parallel lines of ridges and hollows, sloping away on the two sides of the wake, and at the same time to measure the velocity with which the fishing-line was dragged through the water. The angle was measured by holding a jointed two-foot rule, with its two branches, as nearly as could be judged, by the eye, parallel to the sets of lines of wave-ridges. The angle to which the ruler had to be opened in this adjustment was the angle sought. By laying it down on paper, drawing two straight lines by its two edges, and completing a simple geometrical construction with a length properly introduced to represent the measured velocity of the moving solid, the required minimum

* One nautical mile per hour, the only other measurement of velocity, except the French metrical reckoning, which ought to be used in any practical measurement, is 51·6 centimetres per second.

wave-velocity was readily obtained. Six observations of this kind were made, of which two were rejected as not satisfactory. The following are the results of the other four:—

Velocity of Moving Solid.	Deduced Minimum Wave-Velocity.
51 centimetres per second.	23·0 centimetres per second.
38 " "	23·8 " "
26 " "	23·2 " "
24 " "	22·9 " "
Mean	23·22

The extreme closeness of this result to the theoretical estimate (23 centimetres per second) was, of course, merely a coincidence, but it proved that the cohesive force of sea-water at the temperature (not noted) of the observation cannot be very different from that which I had estimated from Gay Lussac's observations for pure water.

I need not trouble you with the theoretical formulæ just now, as they are given in a paper which I have communicated to the Royal Society of Edinburgh, and which will probably appear soon in the *Philosophical Magazine.* If 23 centimetres per second be taken as the minimum speed they give 1·7 centimetres for the corresponding wave-length.

I propose, if you approve, to call ripples, waves of lengths less than this critical value, and generally to restrict the name waves to waves of lengths exceeding it. If this distinction is adopted, ripples will be undulations such that the shorter the length from crest to crest the greater the velocity of propagation; while for waves the greater the length the greater the velocity of propagation. The motive force of ripples is chiefly cohesion; that of waves chiefly gravity. In ripples of lengths less than half a centimetre the influence of gravity is scarcely sensible; cohesion is nearly paramount. Thus the motive of ripples is the same as that of the trembling of a dew drop and of the spherical tendency of a drop of rain or spherule of mist. In all waves of lengths exceeding five or six centimetres, the effect of cohesion is practically insensible, and moving force may be regarded as wholly gravity This seems amply to confirm the choice you have made of dimensions in your models, so far as concerns escaping disturbances due to cohesion.

The introduction of cohesion into the theory of waves explains a difficulty which has often been felt in considering the patterns of standing ripples seen on the surface of water in a finger-glass made to sound by rubbing a moist finger on its lip. If no other levelling force than gravity were concerned, the length from crest to crest corresponding to 256 vibrations per second would be a fortieth of a millimetre. The ripples would be quite undistinguishable without the aid of a microscope, and the disturbance of the surface could only be perceived as a dimming of the reflections seen from it. But taking cohesion into account, I find the length from crest to crest corresponding to the period of $\frac{1}{256}$ of a second to be 1·9 millimetres, a length which quite corresponds to ordinary experience on the subject.

When gravity is neglected the formula for the period (P) in terms of the wave-length (l) the cohesive tension of the surface (T) and the density of the fluid (ρ), is

$$P = \sqrt{\frac{l^3 \rho}{2\pi T}},$$

where T must be measured in kinetic units. For water we have $\rho = 1$, and (according to the estimate I have taken from Poisson and Gay Lussac) $T = 982^* \times \cdot 074 = 73$. Hence for water

$$P = \frac{l^{\frac{3}{2}}}{\sqrt{2\pi \times 73}} = \frac{l^{\frac{3}{2}}}{21 \cdot 4}.$$

When l is anything less than half a centimetre the error from thus neglecting gravity is less than 5 per cent. of P. When l exceeds $5\frac{1}{2}$ centimetres the error from neglecting cohesion is less than 5 per cent. of the period. It is to be remarked that, while for waves of sufficient length to be insensible to cohesion, the period is proportional to the square root of the length, for ripples short enough to be insensible to gravity, the period varies in the sesquiplicate ratio of the length.

WILLIAM THOMSON.

Mr Froude having called my attention to Mr Scott Russell's Report on Waves (British Association, York, 1844) as containing observations on some of the phenomena which formed the subject of the preceding letter to him, I find in it, under the heading

* 982 being the weight of one gramme in kinetic units of force centimetres per second.

"Waves of the Third Order," or "Capillary Waves," a most interesting account of the "ripples" (as I have called them), seen in advance of a body moving uniformly through water; also a passage quoted by Russell from a paper of date, Nov. 16, 1829, by Poncelet and Lesbros*, where it seems this class of waves was first described.

Poncelet and Lesbros, after premising that the phenomenon is seen when the extremity of a fine rod or bar is lightly dipped in a flowing stream, give a description of the curved series of ripples (which first attracted my attention in the manner described in the preceding letter). Russell's quotation concludes with a statement from which I extract the following:—..."on trouve que les rides sont imperceptibles quand la vîtesse est moyennement au dessous de 25 c. per seconde."

Russell gives a diagram to illustrate this law. So far as I can see, the comparatively long waves following in rear of the moving body have not been described either by Poncelet and Lesbros or by Russell, nor are they shown in the plan contained in Russell's diagram. But the curve shown above the plan (obviously intended to represent the section of the water-surface by a vertical plane) gives these waves in the rear as well as the ripples in front, and proves that they had not escaped the attention of that very acute and careful observer. In respect to the curves of the ripple-ridges, Russell describes them as having the appearance of a group of confocal hyperbolas, which seems a more correct description than that of Poncelet and Lesbros, according to which they present the aspect of a series of parabolic curves. It is clear, however, from my dynamical theory that they cannot be accurate hyperbolas; and, as far as I am yet able to judge, Russell's diagram exhibiting them is a very good representation of their forms. Anticipating me in the geometrical determination of a limiting velocity, by observing the angle between the oblique terminal straight ridge-lines streaming out on the two sides, Russell estimates it at 8½ inches (21½ centimetres) per second.

Poncelet and Lesbros's estimate of 25 centimetres per second for the smallest velocity of solid relatively to fluid which gives ripples in front, and Russell's terminal velocity of 21½ centimetres per second, are in remarkable harmony with my theory and

* *Memoirs of the French Institute*, 1829.

observation which give 23 centimetres per second as the minimum velocity of propagation of wave or ripple in water.

Russell calls the ripples in front "forced," and the oblique straight waves streaming off at the sides "free"—appellations which might seem at first sight to be in thorough accordance with the facts of observation, as, for instance, the following very important observation of his own:—

"It is perhaps of importance to state that when, while these forced waves were being generated, I have suddenly withdrawn the disturbing point, the first wave immediately sprang back from the others (showing that it had been in a state of compression), and the ridges became parallel; and, moving on at the rate of $8\frac{1}{2}$ inches per second, disappeared in about 12 seconds."

Nevertheless I maintain that the ripples of the various degrees of fineness seen in the different* parts of the fringe are all properly "free" waves, because it follows from dynamical theory that the motion of every portion of fluid in a wave, and, therefore, of course, the velocity of propagation, is approximately the same as if it were part of an infinite series of straight-ridged parallel waves, provided that in the actual wave the radius of curvature of the ridge is a large multiple of the wave-length, and that there are several approximately equal waves preceding it and following it.

No indication of the dynamical theory contained in my communication to the *Philosophical Magazine*, and described in the preceding letter to Mr Froude, appears either in the quotation from Poncelet and Lesbros, or in any other part of Mr Scott Russell's report; but I find with pleasure my observation of a minimum velocity below which a body moving through water gives no ripples, anticipated and confirmed by Poncelet and Lesbros, and my experimental determination of the velocity of the oblique straight-ridged undulations limiting the series of ripples, anticipated and confirmed by Russell.

* The dynamical theory shows that the length from crest to crest depends on the corresponding component of the solid's velocity. For very fine ripples it is approximately proportional to the reciprocal of the square of this component velocity, and therefore to the square of the secant of the angle between the line of the solid's motion and the horizontal line perpendicular to the ridge of the ripple.

7. On the forces experienced by solids immersed in a moving liquid.

[From the *Proceedings of the Royal Society of Edinburgh for* 1869–70; reprinted with the additions in [ ] in *Papers on Electrostatics and Magnetism*, 1872, pp. 567–571.]

Cyclic irrotational motion* [§ 60 (z)] once established through an aperture or apertures, in a movable solid immersed in a liquid, continues for ever after with circulation or circulations unchanged [§ 60 (a)], however the solid be moved, or bent, and whatever influences experienced from other bodies. The solid, if rigid and left at rest, must clearly continue at rest relatively to the fluid surrounding it to an infinite distance, provided there be no other solid within an infinite distance from it. But if there be any other solid or solids at rest within any finite distance from the first, there will be mutual forces between them, which, if not balanced by proper application of force, will cause them to move. The theory of the equilibrium of rigid bodies in these circumstances might be called Kinetico-statics; but it is in reality a branch of physical statics simply. For we know of no case of true statics in which some if not all of the forces are not due to motion; whether as in the case of the hydrostatics of gases, thanks to Clausius and Maxwell, we perfectly understand the character of the motion, or, as in the statics of liquids and elastic solids, we only know that some kind of molecular motion is essentially concerned. The theorems which I now propose to bring before the Royal Society regarding the forces experienced by bodies mutually influencing one another through the mediation of a moving liquid, though they are but theorems of abstract hydrokinetics, are of some interest in physics as illustrating the great question of the

* The references §§ without farther title are to the author's paper on Vortex Motion, recently published in the *Transactions* (1869), which contains definitions of all the new terms used in the present article. Proofs of such of the propositions now enunciated as require proof are to be found in a continuation of that paper.

18th and 19th centuries:—Is action at a distance a reality, or is gravitation to be explained, as we now believe magnetic and electric forces must be, by action of intervening matter?

I. (Proposition.) Consider first a single fixed body with one or more apertures through it; as a particular example, a piece of straight tube open at each end. Let there be irrotational circulation of the fluid through one or more such apertures. It is readily proved [from § 63 *Exam.* (2)]* that the velocity of the fluid at any point in the neighbourhood agrees in magnitude and direction with the resultant electro-magnetic force, at the corresponding point, in the neighbourhood of an electro-magnet replacing the solid, constructed according to the following specification. The "core," on which the "wire" is wound, is to be of any material having infinite diamagnetic inductive capacity†, and is to be of the same size and shape as the solid immersed in the fluid. The wire is to form an infinitely thin layer or layers, with one circuit going round each aperture. The whole strength of current in each circuit, reckoned in absolute electro-magnetic measure, is to be equal to the circulation of the fluid through that aperture divided by $\sqrt{4\pi}$. The resultant electro-magnetic force at any point will be numerically equal to the resultant fluid velocity at the corresponding point in the hydrokinetic system, multiplied by $\sqrt{4\pi}$.

Thus, considering, for example, the particular case of a straight tube open at each end, let the diameter be infinitely small in comparison with the length. The "circulation" will exceed by but an infinitely small quantity the product of the velocity within the tube into the length. In the neighbourhood of each end, at distances from it great in comparison with the diameter of the tube and short in comparison with the length, the stream lines will be straight lines radiating from the end. The velocity, outwards from one end and inwards towards the other, will therefore be inversely as the square of the distance from the end. Generally at all considerable distances from the ends, the dis-

* Or from Helmholtz's original integration of the hydrokinetic equations.

† Real diamagnetic substances are, according to Faraday's very expressive language, relatively to lines of magnetic force, *worse conductors* than air.

The ideal substance of infinite diamagnetic inductive capacity is a substance which completely *sheds off* lines of magnetic force, or which is perfectly impervious to *magnetic force.*

tribution of fluid velocity will be the same as that of the magnetic force in the neighbourhood of an infinitely thin bar longitudinally magnetised uniformly from end to end.

Merely as regards the comparison between fluid velocity and resultant magnetic forces, Euler's fanciful theory of magnetism is thus curiously illustrated. This comparison, which has been long known as part of the correlation between the mathematical theories of electricity, magnetism, conduction of heat, and hydrokinetics, is merely kinematical, not dynamical. When we pass, as we presently shall, to a strictly dynamical comparison relatively to the mutual force between two hard steel magnets, we shall find the same law of mutual action between two tubes, with liquid flowing through each, but with this remarkable difference, that the forces are opposite in the two cases; unlike poles attracting and like poles repelling in the magnetic system, while in the hydrokinetic there is attraction between like ends and repulsion between unlike ends.

II. (Proposition.) Consider two or more fixed bodies, such as the one described in Prop. I. The mutual actions of two of these bodies are equal, but in opposite directions, to those between the corresponding electro-magnets. The particular instance referred to above shows us the remarkable result, that through fluid pressure we can have a system of mutual action, in which like attracts like with force varying inversely as the square of the distance. Thus, if the exit ends of tubes, open at each end with fluid flowing through them, be placed in the neighbourhood of one another, and the entering ends be at infinite distances, the mutual forces resulting will be simply attractions according to this law. The lengths of the tubes on this supposition are infinitely great, and therefore, as is easily proved from the conservation of energy, the quantities flowing out per unit of time are but infinitesimally affected by the mutual influence. [When any change is allowed in the relative positions of two tubes by which work is done, a *diminution* of kinetic energy of the fluid is produced within the tubes, and at the same time an *augmentation* of its kinetic energy in the external space. The former is equal to double the work done; the latter is equal to the work done; and so the loss of kinetic energy from the whole liquid is simply equal to the work done.]

III. Proposition II. holds, even if one of the bodies considered be merely a solid, with or without apertures; if with apertures, having no circulation through them. In such a case as this the corresponding magnetic system consists of a magnet or electro-magnet, and a merely diamagnetic body, not itself a magnet, but disturbing the distribution of magnetic force around it by its diamagnetic influence. Thus, for example, a spherical solid at rest in the field of motion surrounding a fixed body, through apertures in which there is cyclic irrotational motion, will experience from fluid pressure a resultant force through its centre equal and opposite to that experienced by a sphere of infinite diamagnetic capacity, similarly situated in the neighbourhood of the corresponding electro-magnet. Therefore, according to Faraday's law for the latter, and the comparison asserted in Prop. I., it would experience a force from places of less towards places of greater fluid velocity, irrespectively of the direction of the stream lines in its neighbourhood; a result easily deduced from the elementary formula for fluid pressure in hydrokinetics.

I have long ago shown that an elongated diamagnetic body in a uniform magnetic field tends, as tends an elongated ferromagnetic body, to place its length along the lines of force. Hence a long solid, pivoted on a fixed axis through its middle in a uniform stream of liquid, tends to place its length perpendicularly across the direction of motion; a known result (Thomson and Tait's *Natural Philosophy*, § 335). Again, two globes held in a uniform stream with the lines joining their centres, require force to prevent them from mutually approaching one another. In the magnetic analogue, two spheres of diamagnetic or ferromagnetic inductive capacity repel one another when held in a line at right angles to the lines of force. A hydrokinetic result similar to this for the case of two equal globes, is to be found in Thomson and Tait's *Natural Philosophy*, § 332.

IV. (Proposition.) If the body considered in § III. [be an infinitely small globe*, and] be acted on by force applied so as always to balance the resultant of the fluid pressure, calculated for it according to II. and III. for whatever position it may come to at any time, and if it be influenced besides by any

*[The proposition as originally published, without limitation, is obviously false although that it is so I have only perceived to-day.—Sept., 1872.]

other system of applied forces superimposed on the former, it will move just as it would move under the influence of the latter system of forces alone, were the fluid at rest, except in so far as compelled to move by the body's own motion through it. A particular case of this proposition was first published many years ago, by Professor James Thomson, on account of which he gave the name of "vortex of free mobility" to the cyclic irrotational motion symmetrical round a straight axis. [*Additional, Sept.* 14, 1872.—The same proposition holds for a globe of any dimensions, in a field of fluid motion consisting of circulation or circulations with infinitely fine rigid endless curve or curves for core, and no other rigid body in the liquid. Demonstration to appear in the *Proceedings of the Royal Society of Edinburgh* for 1871–2*.]

* *Proceedings of the Royal Society of Edinburgh*, March 4, 1872 [*infra*, p. 108].

8. On Attractions and Repulsions due to Vibration.

[Extracts from two letters to Prof. F. Guthrie, from the *Philosophical Magazine* for June 1871; reprinted in *Papers on Electrostatics and Magnetism*, 1872, pp. 571—4, §§ 741—3.]

Glasgow, *Nov.* 14*th*, 1870.

I have to-day received the *Proceedings of the Royal Society* containing your paper "On Approach caused by Vibration," which I have read with great interest. The experiments you describe constitute very beautiful illustrations of the known theorem for fluid pressure in abstract hydrokinetics, with which I have been much occupied in mathematical investigations connected with vortex-motion.

741. According to this theorem, the average pressure at any point of an incompressible frictionless fluid originally at rest, but set in motion and kept in motion by solids moving to and fro, or whirling round in any manner, though a finite space of it, is equal to a constant diminished by the product of the density into half the square of the velocity. This immediately explains the attractions demonstrated in your experiments; for in each case the average square of velocity is greater on the side of the card nearest the tuning-fork than on the remote side. Hence obviously the card must be attracted by the fork as you have found it to be; but it is not so easy at first sight to perceive that the square of the average velocity must be greater on the surfaces of the tuning-fork next to the card than on the remote portions of the vibrating surface. Your theoretical observation, however, that the attraction must be mutual, is beyond doubt valid, as we may convince ourselves by imagining the stand which bears the tuning-fork and the card to be perfectly free to move through the fluid. If the card were attracted towards the tuning-fork, and there were

not an equal and opposite force on the remainder of the whole surface of the tuning-fork and support, the whole system would commence moving, and continue moving with an accelerated velocity in the direction of the force acting on the card—an impossible result. It might, indeed, be argued that this result is not impossible, as it might be said that the kinetic energy of the vibrations could gradually transform itself into kinetic energy of the solid mass moving through the fluid, and of the fluid escaping before and closing up behind the solid. But "common sense" almost suffices to put down such an argument, and elementary mathematical theory, especially the theory of momentum in hydrokinetics explained in my article on "Vortex-motion,"* negatives it.

742. The law of the attraction which you observed agrees perfectly with the law of magnetic attraction in a certain ideal case which may be fully specified by the application of a principle explained in a short article [§§ 733—740] communicated to the Royal Society of Edinburgh in February last [1870], as an abstract of an intended continuation of my paper on "Vortex-motion." Thus, if we take as an ideal tuning-fork two globes or disks moving rapidly to-and-fro in the line joining their centres, the corresponding magnet will be a bar with poles of the same name as its two ends and a double opposite pole in its middle. Again, the analogue of your paper disk is an equal and similar diamagnetic of extreme diamagnetic inductive capacity [§ 734]. The mutual force between the magnetic and the diamagnetic will be equal and opposite to the corresponding hydrokinetic force at each instant. To apply the analogy, we must suppose the magnet to gradually vary from maximum magnetization to zero, then through an equal and opposite magnetization back through zero to the primitive magnetization, and so on periodically. The resultant of fluid pressure on the disk is not at each instant equal and opposite to the magnetic force at the corresponding instant, but the average resultant of the fluid pressure is equal to the average resultant of the magnetic force. Inasmuch as the force on the diamagnetic is generally repulsion from the magnet, however the magnet be held, and is unaltered in amount by the reversal of the magnetization, it follows that

* *Transactions of the Royal Society of Edinburgh*, read 29th April, 1867.

the average resultant of the fluid pressure is an attraction on the whole towards the tuning-fork, into whatever position the tuning-fork be turned relatively to it....

Nov. 23, 1870.

743....There are, no doubt, curiously close analogies between some of the circumstances of motion in contiguous fluids of different densities, and the distribution of magnetic force in a field occupied by substances of different inductive capacities. Thus, if in a great space occupied by frictionless incompressible liquid denser in some portions than in others, a solid be suddenly set in motion, the lines of the fluid motion first generated agree perfectly [compare §§ 751—763 below] with the permanent lines of magnetic force in a correspondingly heterogeneous medium under the influence of a bar-magnet, to be substituted for the moveable solid and placed with its magnetic axis in the line of the solid's motion. As to amounts, the fluid velocity multiplied into the density is simply equal to the resultant magnetic force at each point, if the particular definition [the "electromagnetic definition" (§ 517, *Postscript*)] of the resultant magnetic force in a medium of heterogeneous inductive capacity, given in the foot-note to [§ 516 above] § 48 of my paper on the "Mathematical Theory of Magnetism,"* be adopted. But here the analogy ends; the rigidity in virtue of which a solid moveable in a fluid medium differing from it in magnetic inductive capacity keeps its form, does not exist [contrast § 751 below] in the hydrokinetic analogue....

* *Philosophical Transactions*, June 21, 1849. Published in Part I. for 1851.

9. On the Motion of Rigid Solids in a Liquid circulating irrotationally through perforations in them or in a fixed solid.

[From the *Proceedings of the Royal Society of Edinburgh*, reprinted in *Phil. Mag.** Vol. XLV. 1873, pp. 332—345.]

1. Let ψ, ϕ,... be the values, at time t, of generalised co-ordinates fully specifying the positions of any number of solids movable through space occupied by a perfect liquid destitute of rotational motion, and not acted on by any force which could produce it. Some or all of these solids being perforated, let χ, χ', χ'', &c. be the quantities of liquid which from any era of reckoning, up to the time t, have traversed the several apertures. According to an extension of Lagrange's general equations of motion, used in Vol. I. of Thomson and Tait's *Natural Philosophy*, §§ 331—336, proved in §§ 329, 331 of the German translation of that volume, and to be further developed in the second English edition now in the press†, we may use these quantities χ, χ',... as if they were co-ordinates so far as concerns the equations of motion. Thus, although the position of any part of the fluid is not only not explicitly specified, but is actually indeterminate, when ψ, ϕ,... χ, χ',... are all given, we may regard χ, χ'... as specifying all that it is necessary for us to take into account regarding the motion of the liquid, in forming the equations of motion of the solids; so that if ξ, η,..., and Ψ, Φ... denote the generalised components of momentum and of force [Thomson and Tait, § 313 (*a*), (*b*)] relatively to ψ, ϕ,..., and if κ, κ',... K, K'... denote corresponding elements

* The title and first part (§§ 1—13) are new. The remainder (§§ 14, 15) was communicated to the Royal Society at the end of last December.—W. T. September 26, 1872.

† [Under the heading 'ignoration of coordinates.']

relatively to $\chi, \chi' \ldots$, we have (Hamiltonian form of Lagrange's general equations)

$$\left.\begin{array}{ll} \dfrac{d\xi}{dt} + \dfrac{\mathfrak{d}\tau}{d\psi} = \Psi, & \dfrac{d\eta}{dt} + \dfrac{\mathfrak{d}T}{d\phi} = \Phi \ldots \\ \dfrac{d\kappa}{dt} + \dfrac{\mathfrak{d}T}{d\chi} = K, & \dfrac{d\kappa'}{dt} + \dfrac{\mathfrak{d}T}{d\chi'} = K' \ldots \end{array}\right\} \ldots\ldots\ldots\ldots(1),$$

where T denotes the whole kinetic energy of the system, and $\mathfrak{d}$ differentiation on the hypothesis of $\xi, \eta, \ldots \kappa, \kappa' \ldots$ constant.

2. To illustrate the meaning of $\chi, K, \kappa, \chi', \ldots$, let B be one of the perforated solids, to be regarded generally as movable. Draw an immaterial barrier surface Ω across the aperture to which they are related, and consider this barrier as fixed relatively to B. Let N denote the normal component velocity, relatively to B and Ω of the fluid at any point of Ω; and let $\iint d\sigma$ denote integration over the whole area of Ω: then

$$\iint N d\sigma = \dot{\chi} \ldots\ldots\ldots\ldots\ldots\ldots\ldots\ldots(2);$$

and

$$\chi = \int dt \iint N d\sigma \ \ldots\ldots\ldots\ldots\ldots\ldots\ldots(3),$$

which is a symbolical expression of the definition of χ. To the surface of fluid coinciding with Ω at any instant, let pressure be applied of constant value K per unit of area, over the whole area; and at the same time let force (or force and couple) be applied to B equal and opposite to the resultant of this pressure supposed for a moment to act on a rigid material surface Ω rigidly connected with B. The *motive* (that is to say, system of forces) consisting of the pressure K on the fluid surface, and force and couple B as just defined, constitutes the generalised component force corresponding to χ [Thomson and Tait, § 313 (*b*)]; for it does no work upon any motion of B or other bodies of the system if χ is kept constant; and if χ varies work is done at the rate

$$K\dot{\chi} \text{ per unit of time,}$$

whatever other motions or forces there may be in the system. Lastly, calling the density of the fluid unity, let κ denote "circulation"* [V. M. § 60 (*a*)]† of the fluid in any circuit crossing

* Or $\int F ds$ if F denote the tangential component of the absolute velocity of the fluid at any point of the circuit, and $\int ds$ line integration once round the circuit.

† References distinguished by the initials V. M. are to the part already published of the author's paper on Vortex Motion. (*Transactions of the Royal Society of Edinburgh*, 1867–8 and 1868–9.)

β once, and only once: it is this which constitutes the generalised component momentum relatively to χ [Thomson and Tait, § 313 (*e*)]; for (V. M. § 72) we have

$$\kappa = \int_0^t K dt \quad \text{........................(4)},$$

if the system given at rest (or in any state of motion for which $\kappa = 0$) be acted on by the motive K during time t*.

3. The kinetic energy T is, of course, necessarily a quadratic function of the generalised momentum-components, $\xi, \eta, \ldots \kappa, \kappa' \ldots$, with coefficients generally functions of ψ, $\phi, \ldots$, but necessarily independent of $\chi, \chi', \ldots$. In consequence of this peculiarity it is convenient to put

$$T = Q(\xi - \alpha\kappa - \alpha'\kappa' - \&c., \eta - \beta\kappa - \beta'\kappa' - \&c., \ldots) + \mathfrak{Q}(\kappa, \kappa', \ldots) \quad \text{.........(5)},$$

where Q, $\mathfrak{Q}$ denote two quadratic functions. This we may clearly do, because, if i be the number of the variables $\xi, \eta, \ldots$, and j the number of $\kappa, \kappa' \ldots$; the whole number of coefficients in the single quadratic function expressing τ is $\frac{1}{2}(i+j)(i+j+1)$, which is equal to the whole number of the coefficients $\frac{1}{2}i(i+1) + \frac{1}{2}j(j+1)$ of the two quadratic functions, together with the $i\,j$ available quantities $\alpha, \beta, \ldots \alpha', \beta', \ldots, \ldots$

4. The meaning of the quantities $\alpha, \beta, \ldots \alpha', \ldots$ thus introduced is evident when we remember that

$$\frac{dT}{d\xi} = \dot{\psi}, \qquad \frac{dT}{d\eta} = \dot{\phi}, \ldots \qquad \frac{dT}{d\kappa} = \dot{\chi}, \qquad \frac{dT}{d\kappa'} = \dot{\chi}', \ldots \quad \text{...(6)}.$$

For, differentiating (5), and using these, we find

$$\dot{\psi} = \frac{dQ}{d\xi}, \qquad \dot{\phi} = \frac{dQ}{d\eta}, \ldots \quad \text{.................(7)},$$

and using these latter,

$$\dot{\chi} = \frac{d\mathfrak{Q}}{d\kappa} - \alpha\dot{\psi} - \beta\dot{\phi} - \&c., \quad \dot{\chi}' = \frac{d\mathfrak{Q}}{d\kappa'} - \alpha'\dot{\psi} - \beta'\dot{\phi} - \&c., \ldots \quad \text{.........(8)}.$$

Equations (8) show that $-\alpha\dot{\psi}$, $-\beta\dot{\phi}$, $-\alpha'\dot{\psi}$, &c., are the contributions to the flux across Ω, Ω', &c., given by the separate velocity-

* The general limitation, for impulsive action, that the displacements effected during it are infinitely small, is not necessary in this case. Compare § 5 (11), below.

components of the solids. And (7) show that to prevent the solids from being set in motion when impulses $\kappa, \kappa', \ldots$ are applied to the liquid at the barrier surfaces, we must apply to them impulses expressed by the equations

$$\xi = \alpha\kappa + \alpha'\kappa' + \&c., \quad \eta = \beta\kappa + \beta'\kappa' + \&c., \ldots \quad \ldots\ldots(9).$$

5. To form the equations of motion, we have, in the first place,

$$\frac{\mathfrak{d}T}{d\chi} = 0, \quad \frac{\mathfrak{d}T}{d\chi'} = 0, \ldots \quad \ldots\ldots\ldots\ldots\ldots(10),$$

and therefore, by (1),

$$\frac{d\kappa}{dt} = K, \quad \frac{d\kappa'}{dt} = K', \ldots \quad \ldots\ldots\ldots\ldots\ldots(11);$$

which show that the acceleration of κ, under the influence of K, follows simply the law of acceleration of a mass under the influence of a force. Again (for the motions of the solids), let

$$\xi_0 = \xi - \alpha\kappa - \alpha'\kappa' - \&c., \quad \eta_0 = \eta - \beta\kappa - \beta'\kappa' - \&c., \ldots \ldots(12);$$

and let $\mathfrak{D}Q/d\psi$, &c., denote variations of Q on the hypothesis of $\xi_0, \eta_0, \ldots$ each constant.

We have from (5), remembering that $\mathfrak{d}T/d\psi$, &c. denote variations of T, on the hypothesis of $\xi, \eta, \ldots \kappa, \kappa', \ldots$ constant,

$$\frac{\mathfrak{d}T}{d\psi} = \frac{\mathfrak{D}Q}{d\psi} - \frac{dQ}{d\xi}\left(\kappa\frac{d\alpha}{d\psi} + \kappa'\frac{d\alpha'}{d\psi} + \&c.\right)$$
$$- \frac{dQ}{d\eta}\left(\kappa\frac{d\beta}{d\psi} + \kappa'\frac{d\beta'}{d\psi} + \&c.\right) - \&c. + \frac{\mathfrak{d}\mathfrak{Q}}{d\psi},$$

or, by (7)

$$\frac{\mathfrak{d}T}{d\psi} = \frac{\mathfrak{D}Q}{d\psi} - \dot{\psi}\left(\kappa\frac{d\alpha}{d\psi} + \kappa'\frac{d\alpha'}{d\psi} + \&c.\right)$$
$$- \dot{\phi}\left(\kappa\frac{d\beta}{d\psi} + \kappa'\frac{d\beta'}{d\psi} + \&c.\right) - \&c. + \frac{\mathfrak{d}\mathfrak{Q}}{d\psi} \ldots\ldots(13).$$

Hence by (1)

$$\frac{d\xi}{dt} + \frac{\mathfrak{D}Q}{d\psi} - \dot{\psi}\left(\kappa\frac{d\alpha}{d\psi} + \kappa'\frac{d\alpha'}{d\psi} + \&c.\right)$$
$$- \dot{\phi}\left(\kappa\frac{d\beta}{d\psi} + \kappa'\frac{d\beta'}{d\psi} + \&c.\right) - \&c. + \frac{\mathfrak{d}\mathfrak{Q}}{d\psi} = \Psi \ldots\ldots(14).$$

Now, remark that, according to the notation of (12), $\xi_0, \eta_0, \ldots$ are the momentum-components of the solids due to their own motion alone, without cyclic motion of the liquid; and therefore eliminate $\xi, \eta, \ldots$ by (12) from (14). Thus we find

$$\frac{d\xi_0}{dt} + \frac{\mathfrak{D}Q}{d\psi} + \alpha\frac{d\kappa}{dt} + \alpha'\frac{d\kappa'}{dt} + \text{\&c.}$$
$$+\dot{\phi}\left\{\kappa\left(\frac{d\alpha}{d\phi} - \frac{d\beta}{d\psi}\right) + \kappa'\left(\frac{d\alpha'}{d\phi} - \frac{d\beta'}{d\psi}\right) + \text{\&c.}\right\}$$
$$+\dot{\theta}\left\{\kappa\left(\frac{d\alpha}{d\theta} - \frac{d\gamma}{d\psi}\right) + \kappa'\left(\frac{d\alpha'}{d\theta} - \frac{d\gamma'}{d\psi}\right) + \text{\&c.}\right\} + \text{\&c.} = \Psi - \frac{\mathfrak{d}\mathfrak{Q}}{d\psi} \quad \ldots(15),$$

which, with the corresponding equation for ξ_0, &c., and with (11) for κ, κ', &c., are the desired equations of motion.

6. The hypothetical mode of application of K, $K', \ldots$ (§ 1) is impossible, and every other (such as the influence of gravity on a real liquid at different temperatures in different parts) is impossible for our ideal *liquid*, that is to say, a homogeneous incompressible perfect fluid. Hence we have $K = 0$, $K' = 0$, and from (11) conclude that $\kappa, \kappa', \ldots$ are constants. [They are sometimes called the "cyclic constants" (V. M. §§ 62—64).] The equations of motion (15) thus become simply

$$\frac{d\xi_0}{dt} + \frac{\mathfrak{D}Q}{d\psi} + \dot{\phi}\left\{\kappa\left(\frac{d\alpha}{d\phi} - \frac{d\beta}{d\psi}\right) + \kappa'\left(\frac{d\alpha'}{d\phi} - \frac{d\beta'}{d\psi}\right) + \ldots\right\}$$
$$+\dot{\theta}\left\{\kappa\left(\frac{d\alpha}{d\theta} - \frac{d\gamma}{d\psi}\right) + \kappa'\left(\frac{d\alpha'}{d\theta} - \frac{d\gamma'}{d\psi}\right) + \ldots\right\} + \text{\&c.} = \Psi - \frac{\mathfrak{d}\mathfrak{Q}}{d\psi} \quad \ldots(16),$$

with corresponding equations for η_0, ζ_0, and with the following relations from (7), between $\xi_0, \eta_0 \ldots$ and $\dot{\psi}, \dot{\phi}, \ldots$

$$\frac{dQ}{d\xi_0} = \dot{\psi}, \quad \frac{dQ}{d\eta_0} = \dot{\phi}, \quad \frac{dQ}{d\zeta_0} = \dot{\theta}, \text{ \&c.} \quad \ldots\ldots\ldots(17).$$

7. Let

$$\kappa\left(\frac{d\alpha}{d\phi} - \frac{d\beta}{d\psi}\right) + \kappa'\left(\frac{d\alpha'}{d\phi} - \frac{d\beta'}{d\psi}\right) + \text{\&c., be denoted by } \{\phi, \psi\} \quad \ldots(18),$$

so that we have

$$\{\phi, \psi\} = -\{\psi, \phi\} \quad \ldots\ldots\ldots\ldots\ldots\ldots\ldots(19).$$

These quantities $\{\phi, \psi\}$, $\{\theta, \psi\}$, &c., linear functions of the cyclic constants, with coefficients depending on the configuration of the system, are to be generally regarded simply as given functions of the co-ordinates $\psi, \phi, \theta, \ldots$: and the equations of motion are

$$\left.\begin{aligned} \frac{d\xi_0}{dt} + \frac{\mathfrak{D}Q}{d\psi} + \{\phi, \psi\}\dot{\phi} + \{\theta, \psi\}\dot{\theta} + \text{\&c.} &= \Psi - \frac{\mathfrak{d}\mathfrak{Q}}{d\psi} \\ \frac{d\eta_0}{dt} + \frac{\mathfrak{D}Q}{d\phi} + \{\psi, \phi\}\dot{\psi} + \{\theta, \phi\}\dot{\theta} + \text{\&c.} &= \Phi - \frac{\mathfrak{d}\mathfrak{Q}}{d\phi} \end{aligned}\right\} \quad \ldots(20).$$

In these (being of the Hamiltonian form) Q is regarded as a quadratic function of $\xi_0, \eta_0, \zeta_0 \ldots$ with its coefficients functions of ψ, ϕ, θ, &c.; and $\mathfrak{D}$ applied to it indicates variations of these coefficients. If now we eliminate $\xi_0, \eta_0, \zeta_0 \ldots$ from Q by the linear equations, of which (17) is an abbreviated expression, and so have Q expressed as a quadratic function of $\dot\psi, \dot\phi, \dot\theta, \ldots$, with its coefficients functions of ψ, ϕ, θ, &c.; and if we denote by $dQ/d\psi$, $dQ/d\phi$, &c., variations of Q depending on variations of these coefficients; and by $dQ/d\dot\psi$, $dQ/d\dot\phi$, &c., variations of Q depending on variations of $\dot\psi, \dot\phi$, &c., we have [compare Thomson and Tait, § 329 (13) and (15)]

$$\left.\begin{aligned} \xi_0 &= \frac{dQ}{d\dot\psi}, & \eta_0 &= \frac{dQ}{d\dot\phi}, \\ \text{and} \qquad \frac{\mathfrak{D}Q}{d\psi} &= -\frac{dQ}{d\psi}, & \frac{\mathfrak{D}Q}{d\phi} &= -\frac{dQ}{d\phi} \end{aligned}\right\} \quad \ldots\ldots\ldots\ldots(21);$$

and the equations of motion become

$$\left.\begin{aligned} \frac{d}{dt}\frac{dQ}{d\dot\psi} - \frac{dQ}{d\psi} + \{\phi, \psi\}\,\dot\phi + \{\theta, \psi\}\,\dot\theta + \ldots &= \Psi - \frac{\mathfrak{d}\mathfrak{Q}}{d\psi} \\ \frac{d}{dt}\frac{dQ}{d\dot\phi} - \frac{dQ}{d\phi} + \{\psi, \phi\}\,\dot\psi + \{\theta, \phi\}\,\dot\theta + \ldots &= \Phi - \frac{\mathfrak{d}\mathfrak{Q}}{d\phi} \\ \frac{d}{dt}\frac{dQ}{d\dot\theta} - \frac{dQ}{d\theta} + \{\psi, \theta\}\,\dot\psi + \{\phi, \theta\}\,\dot\phi + \ldots &= \Theta - \frac{\mathfrak{d}\mathfrak{Q}}{d\theta} \end{aligned}\right\} \ldots(22).$$

The first members here are of Lagrange's form, with the remarkable addition of the terms involving the velocities simply (in multiplication with the cyclic constants) depending on the cyclic fluid motion. The last terms of the second members contain traces of their Hamiltonian origin in the symbols $\mathfrak{d}/d\psi, \mathfrak{d}/d\phi, \ldots$.

8. As a first application of these equations, let $\dot\psi = 0$, $\dot\phi = 0$, $\dot\theta = 0, \ldots$. This makes $\xi_0 = 0$, $\eta_0 = 0 \ldots$, and therefore also $Q = 0$; and the equations of motion (16) (now equations of equilibrium of the solids under the influence of applied forces Ψ, Φ, &c., balancing the fluid pressure due to the polycyclic motion $\kappa, \kappa', \ldots$), become

$$\Psi = \frac{\mathfrak{d}\mathfrak{Q}}{d\psi}, \qquad \Phi = \frac{\mathfrak{d}\mathfrak{Q}}{d\phi}, \text{ \&c.} \ldots\ldots\ldots\ldots\ldots(23);$$

a result which a direct application of the principle of energy renders obvious (the augmentation of the whole energy produced by an infinitesimal displacement, $\delta\psi$, is $\mathfrak{d}\mathfrak{Q}/d\psi \,.\, \delta\psi$, and $\Psi\delta\psi$ is the work done by the applied forces). It is proved in §§ 724—730 of a volume of collected papers on electricity and magnetism soon to be published, that $\mathfrak{d}\mathfrak{Q}/d\psi$, $\mathfrak{d}\mathfrak{Q}/d\phi$, &c., are the components of the forces experienced by bodies of perfect diamagnetic inductive capacity placed in the magnetic field analogous* to the supposed cyclic irrotational motion. Hence the motive influence of the cyclic motion of the liquid upon the solids in equilibrium is equal and opposite to that of magnetism in the magnetic analogue.

This is proposition II. of the paper *On the Forces experienced by Solids immersed in a Moving Liquid,* which relates to the forces required to keep the movable solids at rest. The present investigation shows Prop. II. of that article to be false. Compare *Reprint*, § 740.

9. Equations (16) for the case of a single perforated movable solid undisturbed by others, agree substantially with equations (6) and (14) of my communication† to the Royal Society of Edinburgh of February 1871. The ξ_0, η_0,... of the present article correspond to the dT/du, dT/dv, &c., of the former; the ξ, η,... mean the same in both. The equations now demonstrated constitute an extension of the theory not readily discovered or proved by that simple consideration of the principle of momentum, and moment of momentum, on which alone was founded the investigation of my former article.

10. Going back to the analytical definition of $\mathfrak{Q}$ in § 3 (5), we see that when none of the movable solids is perforated, this configurational function is equal to the whole kinetic energy (E), which the polycyclic motion would have were there no movable solids, diminished by the energy (W) which would be given up were the liquid, which on this supposition flows through the space of the movable solid or solids, suddenly rigidified and brought to rest. Putting then

$$\mathfrak{Q} = E - W \quad \text{.......................(24)},$$

and remarking that E is independent of the co-ordinates of the movable solids, we may put $-W$ in place of $\mathfrak{Q}$ in the equations

* Proposition I. of article on "The Forces experienced by Solids immersed in a Moving Liquid" (*Proceedings R. S. E.*, February 1870, reprinted in Volume of Electric and Magnetic papers, §§ 733—740).

† See *Proceedings R. S. E.*, Session 1870–71, or reprint in *Philosophical Magazine*, November 1871.

of motion, which, for this slight modification, need not be written out again. W might be directly defined as the whole quantity of work required to remove the movable solids, each to an infinite distance from any other solid having a perforation with circulation through it; and, with this definition, $-W$ may be put for $\mathfrak{Q}$ in the equations of motion without exclusion of cases in which there is circulation through apertures in movable solids.

11. I conclude with a very simple case, the subject of my communication to the Royal Society of last December, in which the result was given without proof. Let there be only one moving body, and it spherical; let the perforated solid or solids be reduced to an infinitely fine immovable rigid curve or group of curves (endless, of course, that is, either finite and closed, or infinite), and let there be no other fixed solid. The rigid curve or curves will be called the *core* or *cores*, as their part is simply that of cores for the cyclic or polycyclic motion. In this case it is convenient to take for ψ, ϕ, θ, the rectangular co-ordinates (x, y, z) of the centre of the movable globe. Then, because the cores, being infinitely fine, offer no obstruction to the motion of the liquid making way for the globe moving through it, we have

$$Q = \tfrac{1}{2}m(\dot{x}^2 + \dot{y}^2 + \dot{z}^2) \quad \text{.................(25)},$$

where m denotes the mass of the globe, together with half that of its bulk of the fluid. Hence

$$\left.\begin{array}{l} \dfrac{dQ}{dx} = 0, \quad \dfrac{dQ}{dy} = 0, \quad \dfrac{dQ}{dz} = 0 \\ \text{and} \qquad \xi_0 \left(= \dfrac{dQ}{d\dot{x}}\right) = m\dot{x}, \quad \eta_0 = m\dot{y}, \quad \zeta_0 = m\dot{z} \end{array}\right\} \text{.........(26)}.$$

A further great simplification occurs, because in the present case $\alpha d\psi + \beta d\phi + \ldots$, or, as we now have it, $\alpha dx + \beta dy + \gamma dz$, is a complete differential*. To prove this, let V be the velocity-potential at any point (a, b, c) due to the motion of the globe, irrespectively of any cyclic motion of the liquid. We have

$$V = \tfrac{1}{2}r^3\left(\dot{x}\frac{d}{dx} + \dot{y}\frac{d}{dy} + \dot{z}\frac{d}{dz}\right)\frac{1}{D},$$

where r denotes the radius of the globe, and

$$D = \{(x-a)^2 + (y-b)^2 + (z-c)^2\}^{\frac{1}{2}}.$$

* Which means that if the globe, after any motion whatever, great or small. comes again to a position in which it has been before, the integral quantity of liquid which this motion has caused to cross any fixed area is zero.

Hence if N denote the component velocity of the liquid at (a, b, c) in any direction (λ, μ, ν) we have

$$N = \left(\dot{x}\frac{d}{dx} + \dot{y}\frac{d}{dy} + \dot{z}\frac{d}{dz}\right) F(x, y, z, a, b, c) \quad \text{......(27)},$$

where $$F(x, y, z, a, b, c) = \tfrac{1}{2}r^3 \left(\lambda\frac{d}{da} + \mu\frac{d}{db} + \nu\frac{d}{dc}\right)\frac{1}{D}.$$

Let now (a, b, c) be any point of the barrier surface Ω (§ 2), and λ, μ, ν, the direction cosines of the normal. By (2) of § 2 we see that the part of $\dot{\chi}$ due to the motion of the globe is $\iint N d\sigma$, or, by (26),

$$\left(\dot{x}\frac{d}{dx} + \dot{y}\frac{d}{dy} + \dot{z}\frac{d}{dz}\right)\iint F(x, y, z, a, b, c)\, d\sigma \quad \text{......(28)}.$$

Hence, putting

$$\iint F(x, y, z, a, b, c)\, d\sigma = U,$$

we see by (8) of § 4, that

$$\alpha = -\frac{dU}{dx}, \qquad \beta = -\frac{dU}{dy}, \qquad \gamma = -\frac{dU}{dz} \quad \text{......(29)}.$$

Hence, with the notation of § 7 (18) for $x, y, \ldots$ instead of $\psi, \phi, \ldots$

$$\{y, z\} = 0, \quad \{z, x\} = 0, \quad \{x, y\} = 0.$$

By this and (25) the equations of motion (22), with (24), become simply

$$m\frac{d^2x}{dt^2} = X + \frac{\partial W}{dx}, \quad m\frac{d^2y}{dt^2} = Y + \frac{\partial W}{dy}, \quad m\frac{d^2z}{dt^2} = Z + \frac{\partial W}{dz} \quad \text{...(30)}.$$

These equations express that the globe moves as a material particle of mass m, with the forces (X, Y, Z) expressly applied to it, would move in a *field of force* having W for potential.

12. The value of W is of course easily found by aid of spherical harmonics, from the velocity potential, P, of the polycyclic motion which would exist were the globe removed, and which we must suppose known: and in working it out (below) it is readily seen that, if, for the hypothetical undisturbed motion, q denote the fluid velocity at the point really occupied by the centre of the rigid globe, we have

$$W = \tfrac{1}{2}\mu q^2 + w \quad \text{......................(31)},$$

where μ denotes one and a half times the volume of the globe, and w denotes the kinetic energy of what we may call the internal motion of the liquid occupying for an instant in the undisturbed motion the space of the rigid globe in the real system. To define w, remark that the harmonic analysis proves the velocity of the centre of inertia of an irrotationally moving liquid globe to be equal to q, the velocity of the liquid at its centre*; and consider the velocity of any part of the liquid sphere, relatively to a rigid body moving with the velocity q. The kinetic energy of this relative motion is what is denoted by w. Remark also that if, by mutual forces between its parts, the liquid globe were suddenly rigidified, the velocity of the whole would be equal to q; and that $\frac{1}{2}mq^2$ is the work given up by the rigidified globe and surrounding liquid when the globe is suddenly brought to rest, being the same as the work required to start the globe with velocity q from rest in a motionless liquid.

Let $P+\psi$ be the velocity potential at (x, y, z) in the actual motion of the liquid when the rigid globe is fixed. Let a be the radius of the globe, r distance of (x, y, z) from its centre, and $\iint d\sigma$ integration over its surface. At any point of the surface of the instantaneous liquid globe, the component velocity perpendicular to the spherical surface in the undisturbed motion is $(dP/dr)_{r=a}$; and hence the impulsive pressure on the spherical surface required to change the velocity potential of the external liquid from P to $P+\psi$, being $-\psi$, undoes an amount of work equal to

$$\iint d\sigma\psi \cdot \frac{1}{2}\frac{dP}{dr},$$

in reducing the normal component from that value to zero. On the other hand, the internal velocity-potential is reduced from P to zero, and the work undone in this process is

$$\iint d\sigma P \cdot \frac{1}{2}\frac{dP}{dr}.$$

Hence $$W = \frac{1}{2}\iint d\sigma\,(P+\psi)\frac{dP}{dr} \quad\text{.................(32).}$$

The condition that with velocity-potential $P+\psi$ there is no flow perpendicular to the spherical surface, gives

$$\left(\frac{dP}{dr}+\frac{d\psi}{dr}\right)_{r=a} = 0 \quad\text{.....................(33).}$$

* This follows immediately from the proposition (Thomson and Tait's *Natural Philosophy*, § 496) that any function V, satisfying Laplace's equation $d^2V/dx^2+d^2V/dy^2+d^2V/dz^2$ throughout a spherical space has for its mean value through this space its value at the centre. For dP/dx satisfies Laplace's equation.

Now let

$$\left.\begin{aligned} P &= P_0 + P_1\frac{r}{a} + \ldots + P_i\left(\frac{r}{a}\right)^i + \&c. \\ \psi &= \Psi_1\left(\frac{a}{r}\right)^2 + \ldots + \Psi_i\left(\frac{a}{r}\right)^{i+1} + \&c. \end{aligned}\right\} \ldots\ldots(34),$$

be the spherical harmonic developments of P and ψ relatively to the centre of the rigid globe as origin; the former necessarily convergent throughout the largest spherical space which can be described from this point as centre without enclosing any part of the core; the latter necessarily convergent throughout space external to the sphere. By (33) we have

$$\Psi_i = \frac{i}{i+1} P_i \ldots\ldots\ldots\ldots\ldots\ldots\ldots\ldots(35).$$

Hence (32) gives

$$W = \frac{1}{2}\iint d\sigma\left(\Sigma\frac{2i+1}{i+1}P_i\right)(\Sigma i P_i),$$

which, by

$$\iint d\sigma P_i P_{i'} = 0,$$

becomes

$$W = \frac{1}{2a}\Sigma\frac{i(2i+1)}{i+1}\iint d\sigma P_i^2 \ldots\ldots\ldots\ldots(36).$$

Now, remarking that a solid spherical harmonic of the first degree may be any linear function of x, y, z, put

$$P_1\frac{r}{a} = Ax + By + Cz \quad \ldots\ldots\ldots\ldots\ldots\ldots(37),$$

which gives

$$q^2 = A^2 + B^2 + C^2,$$

and

$$\frac{1}{a}\iint d\sigma P_1^2 = (A^2 + B^2 + C^2)\cdot\frac{a}{3}\cdot\iint d\sigma = q^2 \times \text{volume of globe} = \frac{2}{3}\mu q^2.$$

Hence by (36)

$$W = \tfrac{1}{2}\mu q^2 + \tfrac{1}{2}\iint d\sigma\left(\frac{2\cdot 5}{3}P_2^2 + \frac{3\cdot 7}{4}P_3^2 + \ldots\right) \ldots(38);$$

and, therefore, by comparison with (31),

$$w = \tfrac{1}{2}\iint d\sigma\left(\frac{2\cdot 5}{3}P_2^2 + \frac{3\cdot 7}{4}P_3^2 + \ldots\right) \ldots\ldots\ldots(39).$$

13. When the radius of the globe is infinitely small,

$$W = \tfrac{1}{2}\mu q^2 \ldots\ldots\ldots\ldots\ldots\ldots\ldots\ldots(40),$$

where μ denotes one and a half times the volume of the globule, and q the undisturbed velocity of the fluid in its neighbourhood. This corresponds to the formula which I gave twenty-five years ago for the force experienced by a small sphere (whether of ferromagnetic or diamagnetic non-crystalline substance) in virtue of the inductive influence which it experiences in a magnetic field*.

14. By taking an infinite straight line for the core a simple but very important example is afforded. In this case, the undisturbed motion of the fluid is in circles having their centres in the core (or axis, as we may now call it), and their planes perpendicular to it. As is well known, the velocity of irrotational revolution round a straight axis is inversely proportional to distance from the axis. Hence the potential function W for the force experienced by an infinitesimal solid sphere in the fluid is inversely as the square of the distance of its centre from the axis, and therefore the force is inversely as the cube of the distance, and is towards the nearest point of the axis. Hence, when the globule moves in a plane perpendicular to the axis, it describes one or other of the forms of Cotesian spirals†. If it be projected obliquely to the axis, the component velocity parallel to the axis will remain constant, and the other component will be unaffected by that one; so that the projection of the globule on the plane perpendicular to the axis will always describe the same Cotesian spiral as would be described were there no motion parallel to the axis. If the globule be left to itself in any position it will commence moving towards the axis as if attracted by a force varying inversely as the cube of the distance. It is remarkable that it traverses at right angles an increasing liquid current without any applied force to prevent it from being (as we might erroneously at first sight expect it to be) carried sideways with the augmented stream. A properly trained dynamical intelligence would at once perceive that the constancy of moment of momentum round the axis requires the globule to move directly towards it.

15. Suppose now the globule to be of the same density as the liquid. If (being infinitely small) it is projected in the

* "On the Forces Experienced by Small Spheres under Magnetic Influence, and some of the Phenomena presented by Diamagnetic Substances" (*Cambridge and Dublin Mathematical Journal*, May 1847); and "Remarks on the Forces experienced by Inductively Magnetised Ferromagnetic or Diamagnetic Non-crystalline Substances" (*Phil. Mag.* October 1850). Reprint of *Papers on Electrostatics and Magnetism*, §§ 634—668; Macmillan, 1872.

† Tait and Steele's *Dynamics of a Particle*, § 149 (15).

direction and with the velocity of the liquid's motion, it will move round the axis in the same circle with the liquid; but this motion would be unstable [and the neglected term w (39) adds to the instability]. Compare Tait and Steele's *Dynamics of a Particle*, § 149 (15), Species IV., case $A = 0$ and AB finite; also limiting variety between Species I. and Species V. The globule will describe the same circle in the opposite direction if projected with the same velocity opposite to that of the fluid. If the globule be projected either in the direction of the liquid's motion or opposite to it, with a velocity less than that of the liquid, it will move along the Cotesian spiral (Species I. of Tait and Steele), from apse to centre in a finite time, with an infinite number of turns. If it be projected in either of those directions with a velocity greater by v than that of the liquid, it will move along the Cotesian spiral (Species V. of Tait and Steele), from apse to asymptote. Its velocity along the asymptote, at an infinite distance from the axis, will be

$$\sqrt{v\left(v+\frac{\kappa}{\pi a}\right)},$$

and the distance of the asymptote from the axis will be

$$a\left(v+\frac{\kappa}{2\pi a}\right)\Big/\sqrt{v\left(v+\frac{\kappa}{\pi a}\right)},$$

where a denotes the distance of the apse from the axis, and $\kappa/2\pi a$ the velocity of the liquid at that distance from the axis. If the globule be projected from any point in the direction of any straight line whose shortest distance from the axis is p, it will be drawn into the vortex or escape from it, according as the component velocity in the plane perpendicular to the axis is less or greater than $\kappa/2\pi p$. It is to be remarked that in every case in which the globule is drawn in to the axis (except the extreme one in which its velocity is infinitely little less than that of the fluid, and its spiral path infinitely nearly perpendicular to the radius vector), the spiral by which it approaches, although it has always an infinite number of convolutions, is of finite length; and therefore, of course, the time taken to reach the axis is finite. Considering, for simplicity, motion in a plane perpendicular to the axis; at any point infinitely distant from the axis, let the globule be projected with a velocity v along a line passing at distance p on either side of the axis. Then if τ denote the velocity of the fluid at distance unity from the axis [which is equal to $\kappa/2\pi$], and if we put

$$n^2 = 1 - \frac{\tau^2}{v^2 p^2} \quad\text{.......................(41)},$$

the polar equation of the path is

$$r = \frac{np}{\cos n\theta} \quad\text{.......................(42)}.$$

Hence the nearest approach to the axis attained by the globule is np, and the whole change of direction which it experiences is $\pi(n^{-1} - 1)$. The case of $n^{-1} = 2{\cdot}3$ is represented in the annexed diagram, copied from Tait and Steele's book [§ 149 (15), Species V.].

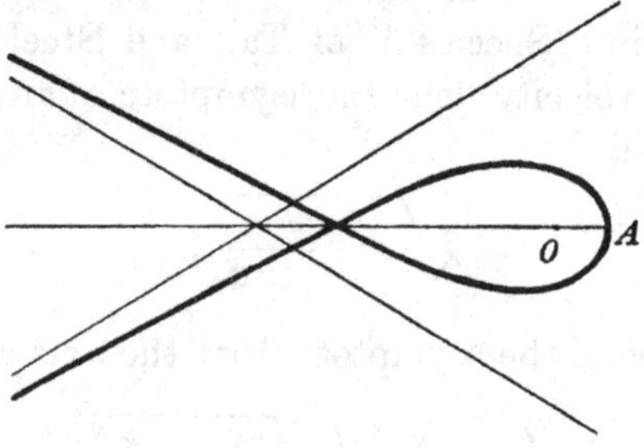

10. Vortex Statics.

[From the *Proceedings of the Royal Society of Edinburgh*, Session 1875–76; reprinted in *Phil. Mag.*, Aug. 1880.]

The subject of this paper is *steady motion* of vortices.

1. Extended definition of "steady motion." The motion of any system of solid, or fluid, or solid and fluid matter is said to be steady when its configuration remains equal and similar, and the velocities of homologous particles equal, however the configuration may move in space, and however distant individual material particles may at one time be from the points homologous to their positions at another time.

2. Examples of steady and not steady motion:—

(1) A rigid body symmetrical round an axis, set to rotate round any axis through its centre of gravity, and left free, performs steady motion. Not so a body having three unequal principal moments of inertia.

(2) A rigid body of any shape, in an infinite homogeneous liquid, rotating uniformly round any, always the same, fixed line, and moving uniformly parallel to this line, is a case of steady motion.

(3) A perforated rigid body in an infinite liquid moving in the manner of example (2), and having cyclic irrotational motion of the liquid through its perforations, is a case of steady motion. To this case belongs the irrotational motion of liquid in the neighbourhood of any rotationally moving portion of fluid of the same shape as the solid, provided the distribution of the rotational motion is such that the shape of the portion endowed with it remains unchanged. The object of the present paper is to investigate general conditions for the fulfilment of this proviso, and to investigate, further, the conditions of stability of distributions of vortex motion satisfying the condition of steadiness.

3. *General Synthetical Condition for Steadiness of Vortex Motion.* The change of the fluid's molecular rotation at any point fixed in space must be the same as if for the rotationally moving portion of the fluid were substituted a solid, with the amount and direction of axis of the fluid's actual molecular rotation inscribed or marked at every point of it, and the whole solid, carrying these inscriptions with it, were compelled to move in some manner answering to the description of example (2). If at any instant the distribution of any molecular rotation* through the fluid and corresponding distribution of fluid-velocity are such as to fulfil this condition, it will be fulfilled through all time.

4. *General Analytical Condition for Steadiness of Vortex Motion.* If, with (§ 24, below) vorticity and "impulse" given, the kinetic energy is a maximum or a minimum, it is obvious that the motion is not only steady, but stable. If, with same conditions, the energy is a maximum-minimum, the motion is clearly steady, but it may be either unstable or stable.

5. The simple circular Helmholtz ring is a case of stable steady motion, with energy maximum-minimum for given vorticity and given impulse. A circular vortex ring, with an inner irrotational annular core, surrounded by a rotationally moving annular shell (or endless tube), with irrotational circulation outside all, is a case of motion which is steady, if the outer and inner contours of the section of the rotational shell are properly shaped, but certainly unstable [if the shell be too thin]†. In this case also the energy is maximum-minimum for circular given vorticity and given impulse.

6. In these examples of steady motion, the "resultant-impulse" (V. M.‡ § 8) is a simple impulsive force, without couple: the corresponding rigid body of example (3) is a toroid; and its motion is purely translational and parallel to the axis of the toroid.

* One of Helmholtz's now well-known fundamental theorems shows that, *from the molecular rotation at every point of an infinite fluid*, the velocity at every point is determinate, being expressed synthetically by the same formulæ as those for finding the "magnetic resultant force" of a pure electromagnet. (Thomson's Reprint of *Papers on Electrostatics and Magnetism.*)

† [The phrase in [ ] is deleted in a copy annotated by Lord Kelvin.]

‡ My first series of papers on Vortex Motion in the *Transactions of the Royal Society of Edinburgh* will be thus referred to henceforth.

We have also exceedingly interesting cases of steady motion in which the impulse is such that, if applied to a rigid body, it would be reducible, according to Poinsot's method, to an impulsive force in a determinate line, *and a couple with this line for axis.* To this category belong certain distributions of vorticity giving longitudinal vibrations, with thickenings and thinnings of the core travelling as waves in one direction or the other round a vortex-ring, which will be investigated in a future communication to the Royal Society. In all such cases the corresponding rigid body of § 2 example (2) has both rotational and translational motion.

7. To find illustrations, suppose, first, the vorticity (defined below, § 24) and the force resultant of the impulse to be (according to the conditions explained below, § 29) such that the cross section is small in comparison with the aperture. Take a ring of flexible wire (a piece of block tin pipe* with its ends soldered together answers well), bend it into an oval form, and then give it a right-handed twist round the long axis of the oval, so that the curve comes to be not in one plane (fig. 1). A properly-shaped twisted ellipse of this kind [a shape perfectly determinate when the vorticity, the force resultant of the impulse, and the rotational moment of the impulse (V. M. § 6), are all given] is the figure of the core in what we may call the first† steady mode of single and simple toroidal vortex motion with rotational moment. To illustrate the second steady mode, commence with a circular ring of flexible wire, and pull it out, at three points 120° from one another, so as to make it into as it were an equilateral triangle with rounded corners. Give now a right-handed twist, round the radius to each corner, to the plane of the curve at and near the corner; and, keeping the character of the twist thus given to the wire, bend it into a certain determinate shape proper for the data of the vortex motion. This is

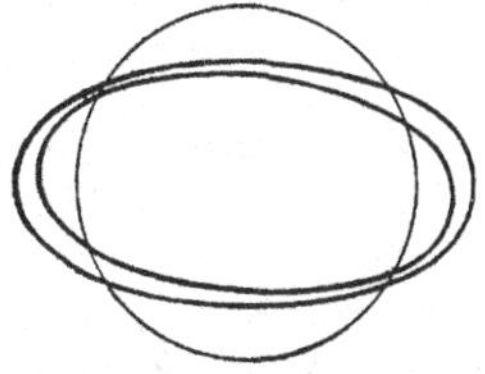

Fig. 1.

* ["Block tin pipe" is substituted here for "very stout lead wire."]

† First or greatest, and second, and third, and higher modes of steady motion to be regarded as analogous to the first, second, third, and higher fundamental modes of an elastic vibrator, or of a stretched cord, or of steady undulatory motion in an endless uniform canal, or in an endless chain of mutually repulsive links.

the shape of the vortex-core in the second steady mode of single and simple toroidal vortex motion with rotational moment. The third is to be similarly arrived at, by twisting the corners of a square having rounded corners; the fourth, by twisting the corners of a regular pentagon having rounded corners; the fifth, by twisting the corners of a hexagon, and so on.

In each of the annexed diagrams of toroidal helices a circle is introduced to guide the judgment as to the relief above and depression below the plane of the diagram which the curve represented in each case must be imagined to have. The circle may be imagined in each case to be the circular axis of a toroidal core on which the helix may be supposed to be wound.

To avoid circumlocution, I have said "give a right-handed twist" in each case. The result in each case, as in fig. 1, illustrates a vortex motion for which the corresponding rigid body describes left-handed helices, by all its particles, round the central axis of the motion. If now, instead of right-handed twists to the plane of the oval, or the corners of the triangle, square, pentagon, &c., we give left-handed twists, as in figs. 2, 3, 4, the result in each case will be a vortex motion for which the corresponding

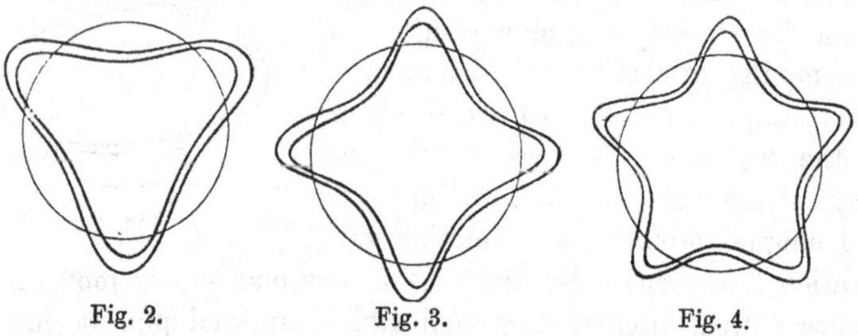

Fig. 2. Fig. 3. Fig. 4.

rigid body describes right-handed helices. It depends, of course, on the relation between the directions of the force resultant and couple resultant of the impulse, with no ambiguity in any case, whether the twists in the forms, and in the lines of motion of the corresponding rigid body, will be right-handed or left-handed.

8. In each of these modes of motion the energy is a maximum-minimum for given force resultant and given couple resultant of impulse. The modes successively described above are successive solutions of the maximum-minimum problem of § 4—a determinate problem with the multiple solutions indicated above, but no other

solution, when the vorticity is given in a single simple ring of the liquid.

9. The problem of steady motion, for the case of a vortex-line with infinitely thin core, bears a close analogy to the following purely geometrical problem:—

Find the curve whose length shall be a minimum with given resultant projectional area, and given resultant areal moment (§ 27 below). This would be identical with the vortex problem if the energy of an infinitely thin vortex-ring of given volume and given cyclic constant were a function simply of its apertural circumference. The geometrical problem clearly has multiple solutions answering precisely to the solutions of the vortex problem.

10. The very high modes of solution are clearly very nearly identical for the two problems (infinitely high modes identical), and are found thus:—

Take the solution derived in the manner explained above, from a regular polygon of N sides, when N is a very great number. It is obvious that either problem must lead to a form of curve like that of a long regular spiral spring of the ordinary kind bent round till its two ends meet, and then having its ends properly cut and joined so as to give a continuous endless helix with axis a circle (instead of the ordinary straight line-axis), and N turns of the spiral round its circular axis. This curve I call a toroidal helix, because it lies on a toroid*, just as the common

* I call a circular toroid a simple ring generated by the revolution of any singly-circumferential closed plane curve round any axis in its plane not cutting it. A "tore," following French usage, is a ring generated by the revolution of a circle round any line in its plane not cutting it. Any simple ring, or any solid with a single hole through it, may be called a toroid; but to deserve this appellation it had better be not very unlike a tore.

The endless closed axis of a toroid is a line through its substance passing somewhat approximately through the centres of gravity of all its cross sections. An apertural circumference of a toroid is any closed line in its surface once round its aperture. An apertural section of a toroid is any section by a plane or curved surface which would cut the toroid into two separate toroids. It must cut the surface of the toroid in just two simple closed curves, one of them completely surrounding the other on the sectional surface: of course it is the space between these curves which is the actual section of the toroidal substance; and the area of the inner one of the two is a section of the aperture.

A section by any surface cutting every apertural circumference, each once and only once, is called a cross section of the toroid. It consists essentially of a simple closed curve.

regular helix lies on a circular cylinder. Let a be the radius of the circle thus formed by the axis of the closed helix; let r denote the radius of the cross section of the ideal toroid on the surface of which the helix lies, supposed small in comparison with a; and let θ denote the inclination of the helix to the normal section of the toroid. We have

$$\tan\theta = \frac{2\pi a}{N\,.\,2\pi r} = \frac{a}{Nr},$$

because $2\pi a/N$ is, as it were, the step of the screw, and $2\pi r$ is the circumference of the cylindrical core on which any short part of it may be approximately supposed to be wound.

Let κ be the cyclic constant, I the given force resultant of the impulse, and μ the given rotational moment. We have (§ 28) approximately

$$I = \kappa\pi a^2, \qquad \mu = \kappa N\pi r^2 a.$$

Hence

$$a = \sqrt{\frac{I}{\kappa\pi}}, \qquad r = \sqrt{\frac{\mu}{N\kappa^{\frac{1}{2}}\pi^{\frac{1}{2}}I^{\frac{1}{2}}}},$$

$$\tan\theta = \sqrt{\frac{I^{\frac{3}{2}}}{N\mu\kappa^{\frac{1}{2}}\pi^{\frac{1}{2}}}}.$$

11. Suppose now, instead of a single thread wound spirally round a toroidal core, we have two separate threads forming, as it were, a "two-threaded screw," and let each thread make a whole number of turns round the toroidal core. The two threads, each endless, will be two helically tortuous rings linked together, and will constitute the core of what will now be a double vortex-ring. The formulæ just now obtained for a single thread would be applicable to each thread, if κ denoted the cyclic constant for the circuit round the two threads, or twice the cyclic constant for either, and N the number of turns of either alone round the toroidal core. But it is more convenient to take N for the number of turns of both threads (so that the number of turns of one thread alone is $\frac{1}{2}N$), and κ the cyclic constant for either thread alone, and thus for very high steady modes of the double vortex-ring,

$$I = 2\kappa\pi a^2, \qquad \mu = \kappa N\pi r^2 a,$$

$$\tan\theta = \sqrt{\frac{(\frac{1}{2}I)^{\frac{3}{2}}}{N\mu\kappa^{\frac{1}{2}}\pi^{\frac{1}{2}}}}.$$

Lower and lower steady modes will correspond to smaller and smaller values of N; but in this case, as in the case of the single vortex-core, the form will be a curve of some ultra-transcendent character, except for very great values of N, or for values of θ infinitely nearly equal to a right angle (this latter limitation leading to the case of infinitely small transverse vibrations).

12. The gravest steady mode of the double vortex-ring corresponds to $N = 2$. This with the single vortex-core gives the case of the twisted ellipse (§ 7). With the double core it gives a system which is most easily understood by taking two plane circular rings of stiff metal linked together. First, place them as nearly coincident as their being linked together permits (fig. 5). Then separate them a little, and incline their planes a little, as shown in the diagram. Then bend each into an unknown shape, determined by the strict solution of the transcendental problem of analysis to which the hydro-kinetic investigation leads for this case.

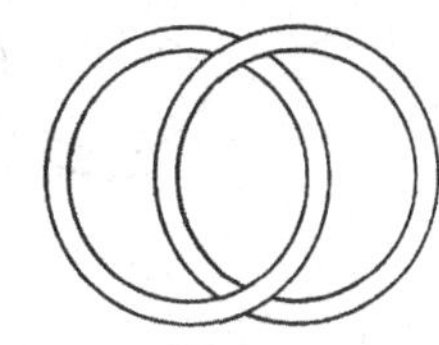

Fig. 5.

13. Go back now to the supposition of § 11, and alter it to this:—

Let each thread make one turn and a half, or any odd number of half turns, round the toroidal core: thus each thread will have an end coincident with an end of the other. Let these coincident ends be united. Thus there will be but one endless thread making an odd number N of turns round the toroidal core. The cases of $N = 3$ and $N = 9$ are represented in the annexed diagrams (figs. 8 and 9)*.

Imagine now a three-threaded toroidal helix, and let N denote the whole number of turns round the toroidal core; we have

$$I = 3\kappa\pi a^2, \qquad \mu = \kappa N \pi r^2 a,$$

$$\tan\theta = \sqrt{\frac{(\frac{1}{3}I)^{\frac{3}{2}}}{N\mu\kappa^{\frac{1}{2}}\pi^{\frac{1}{2}}}}.$$

Suppose now N to be divisible by 3; then the three threads form three separate endless rings linked together. The case of $N=3$ is illustrated by the annexed diagram (fig. 6), which is repeated from the diagram of V. M. § 58. If N be not divisible

* The first of these was given in § 58 of my paper on Vortex Motion. It has since become known far and wide by being seen on the back of the "Unseen Universe."

by 3, the three threads run together into one, as illustrated for the case of $N=14$ in the annexed diagram (fig. 7).

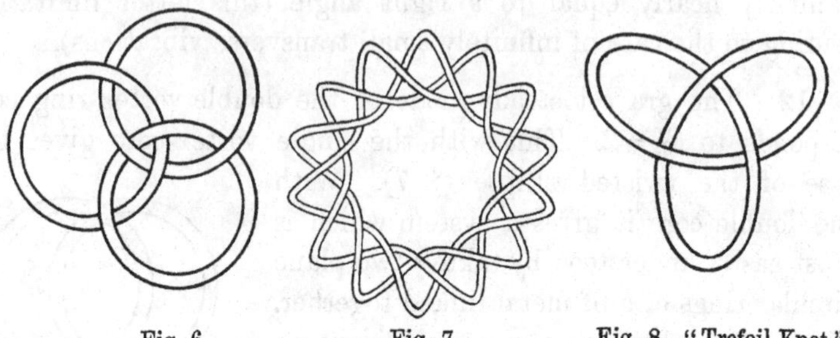

Fig. 6. Fig. 7. Fig. 8. "Trefoil Knot."

14. The irrotational motion of the liquid round the rotational cores in all these cases is such that the fluid-velocity at any point is equal to, and in the same direction as, the resultant magnetic force at the corresponding point in the neighbourhood of a closed galvanic circuit, or galvanic circuits, of the same shape as the core or cores. The setting-forth of this analogy to people familiar, as modern naturalists are, with the distribution of magnetic force in the neighbourhood of an electric circuit, does much to promote a clear understanding of the still somewhat strange fluid-motions with which we are at present occupied.

15. To understand the motion of the liquid in the rotational core itself, take a piece of Indian-rubber gas-pipe stiffened internally with wire in the usual manner, and with it construct any of the forms with which we have been occupied, for instance the symmetrical trefoil knot (fig. 8, § 13), uniting the two ends of the tube carefully, by tying them firmly, by an inch or two of straight cylindrical plug; then turn the tube round and round, round its sinuous axis. The rotational motion of the fluid vortex-core is thus represented. But it must be remembered that the outer form of the core has a motion perpendicular to the plane of the diagram, and a rotation round an axis through the centre of the diagram and perpendicular to the plane, in each of the cases represented by the preceding diagrams. The whole motion of the fluid, rotational and irrotational, is so related in its different parts to

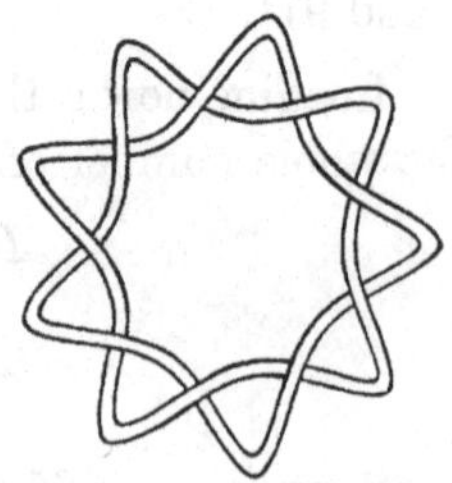

Fig. 9. "Nine-leaved Knot."

one another, and to the translational and rotational motion of the shape of the core, as to be everywhere slipless.

16. Look to the preceding diagrams, and, thinking of what they represent, it is easy to see that there must be a determinate particular shape for each of them which will give steady motion; and I think we may confidently judge that the motion is stable in each, provided only the core is sufficiently thin. It is more easy to judge of the cases in which there are multiple sinuosities by a synthetic view of them (§ 3) than by consideration of the maximum-minimum problem of § 8.

17. It seems probable that the two- or three- or multiple-threaded toroidal helix motions cannot be stable, or even steady, unless I, μ, and N are such as to make the shortest distances between different positions of the core or cores considerable in comparison with the core's diameter. Consider, for example, the simplest case (§ 12, fig. 5) of two simple rings linked together.

18. Go back now to the simple circular Helmholtz ring. It is clear that there must be a shape of absolute maximum energy for given vorticity and given impulse, if we introduce the restriction that the figure is to be a figure of revolution—that is to say, symmetrical round a straight axis. If the given vorticity be given in this determinate shape, the motion will be steady; and there is no other figure of revolution for which it would be steady (it being understood that the impulse has a single force resultant without couple). If the given impulse, divided by the cyclic constant, be very great in comparison with the two-thirds power of the volume of liquid in which the vorticity is given, the figure of steadiness is an exceedingly thin circular ring of large aperture and of approximately circular cross section. This is the case to which chiefly attention is directed by Helmholtz. If, on the other hand, the impulse divided by the cyclic constant be very small compared with the two-thirds power of the volume, the figure becomes like a long oval bored through along its axis of revolution and with the ends of the bore rounded off (or trumpeted) symmetrically, so as to give a figure something like the handle of a child's skipping-rope, but symmetrical on the two sides of the plane through its middle perpendicular to its length. It is certain that, however small the impulse, with given vorticity the figure of steadiness thus indicated is possible, however long in

the direction of the axis and small in diameter perpendicular to the axis and in aperture it may be. I cannot, however, say at present that it is certain that this possible steady motion is stable; for there are figures not of revolution, deviating infinitely little from it, in which, with the same vorticity, there is the same impulse and the same energy, and consideration of the general character of the motion is not reassuring on the point of stability when rigorous demonstration is wanting*.

19. Hitherto I have not indeed succeeded in rigorously demonstrating the stability of the Helmholtz ring in any case. With given vorticity, imagine the ring to be thicker in one place than in another. Imagine the given vorticity, instead of being distributed in a symmetrical circular ring, to be distributed in a ring still with a circular axis, but thinner in one part than in the rest. It is clear that, with the same vorticity and the same impulse, the energy with such a distribution is greater than when the ring is symmetrical. But now let the figure of the cross section of the ring, instead of being approximately circular, be made considerably oval. This will diminish the energy with the same vorticity and the same impulse. Thus from the figure of steadiness we may pass continuously to others with same vorticity, same impulse, and same energy. Thus, we see that the figure of steadiness is, as stated above, a figure of maximum-minimum, and not of absolute maximum, nor of absolute minimum energy. Hence, from the maximum-minimum problem we cannot derive proof of stability.

20. The known phenomena of steam-rings and smoke-rings show us enough of, as it were, the natural history of the subject to convince us beforehand that the steady configuration, with ordinary proportions of diameters of core to diameter of aperture, is stable; and considerations connected with what is rigorously demonstrable in respect to stability of vortex columns (to be given in a later communication to the Royal Society) may lead to a rigorous demonstration of stability for a simple Helmholtz ring, if of thin-enough core in proportion to diameter of aperture. But at present neither natural history nor mathematics gives us perfect assurance of stability when the cross section is considerable in proportion to the area of aperture.

* [Prove steady, W. T., May 10, 1887.]

21. I conclude with a brief statement of general propositions, definitions, and principles used in the preceding abstract, of which some appeared in my series of papers on vortex motion communicated to the Royal Society of Edinburgh in 1867, –68 and –69, and published in the *Transactions* for 1869. The rest will form part of the subject of a continuation of that paper, which I hope to communicate to the Royal Society before the end of the present session.

Any portion of a liquid having vortex motion is called *vortex-core*, or, for brevity, simply "core." Any finite portion of liquid which is all vortex-core, and has contiguous with it over its whole boundary irrotationally moving liquid, is called *a vortex*. A vortex thus defined is essentially a ring of matter. That it must be so was first discovered and published by Helmholtz. Sometimes the word *vortex* is extended to include irrotationally moving liquid circulating round or moving in the neighbourhood of vortex-core; but as different portions of liquid may successively come into the neighbourhood of the core, and pass away again, while the core always remains essentially of the same substance, it is more proper to limit the substantive term *a vortex* as in the definition I have given.

22. *Definition I.* The circulation of a vortex is the circulation [V. M. § 60 (*a*)] in any endless circuit once round its core. Whatever varied configurations a vortex may take, whether on account of its own unsteadiness (§ 1 above), or on account of disturbances by other vortices, or by solids immersed in the liquid, or by the solid boundary of the liquid (if the liquid is not infinite), its "circulation" remains unchanged [V. M. § 59, Prop. (1)]. The circulation of a vortex is sometimes called its *cyclic constant.*

Definition II. An axial line through a fluid moving rotationally, is a line (straight or curved) whose direction at every point coincides with the axis of molecular rotation through that point [V. M. § 59 (2)].

Every axial line in a vortex is essentially a closed curve, [being of course wholly without a vortex]*.

23. *Definition III.* A closed section of a vortex is any section of its core cutting normally the axial lines through every

* [Phrase in [ ] deleted by Lord Kelvin.]

point of it. Divide any closed section of a vortex into smaller areas; the axial lines through the borders of these areas form what are called vortex-tubes. I shall call (after Helmholtz) a vortex-filament any portion of a vortex bounded by a vortex-tube (not necessarily infinitesimal). Of course a complete vortex may be called therefore a vortex-filament; but it is generally convenient to apply this term only to a part of a vortex as just now defined. The boundary of a complete vortex satisfies the definition of a vortex-tube.

A complete vortex-tube is essentially endless. In a vortex-filament infinitely small in all diameters of cross sections "rotation" varies [V. M. § 60 (*e*)] from point to point of the length of the filament, and from time to time, inversely as the area of the cross section. The product of the area of the cross section into the rotation is equal to the circulation or cyclic constant of the filament.

24. Vorticity will be used to designate in a general way the distribution of molecular rotation in the matter of a vortex. Thus, if we imagine a vortex divided into a number of infinitely thin vortex-filaments, the vorticity will be completely given when the volume of each filament and its circulation, or cyclic constant, are given; but the shapes and positions of the filaments must also be given, in order that not only the vorticity, but its distribution, can be regarded as given.

25. The vortex-density at any point of a vortex is the circulation of an infinitesimal filament through this point, divided by the volume of the complete filament. The vortex-density remains always unchanged for the same portion of fluid. By definition it is the same all along any one vortex-filament.

26. Divide a vortex into infinitesimal filaments inversely as their densities, so that their circulations are equal; and let the circulation of each be $1/n$ of unity. Take the projection of all the filaments on one plane. $1/n$ of the sum of the areas of these projections is (V. M. §§ 6, 62) equal to the component impulse of the vortex perpendicular to that plane. Take the projections of the filaments on three planes at right angles to one another, and find the centre of gravity of the areas of these three sets of projections. Find, according to Poinsot's method, the resultant axis, force, and couple of the three forces equal respectively to

$1/n$ of the sums of the areas, and acting in lines through the three centres of gravity perpendicular to the three planes. This will be the resultant axis, the force resultant of the impulse, and the couple resultant of the vortex.

The last of these (that is to say, the couple) is also called the rotational moment of the vortex (V. M. § 6).

27. *Definition IV.* The moment of a plane area round any axis is the product of the area multiplied into the distance from that axis of the perpendicular to its plane through its centre of gravity.

Definition V. The area of the projection of a closed curve on the plane for which the area of projection is a maximum will be called the area of projection of the curve, or simply the area of the curve. The area of the projection on any plane perpendicular to the plane of the resultant area is of course zero.

Definition VI. The resultant axis of a closed curve is a line through the centre of gravity, and perpendicular to the plane of its resultant area. The resultant areal moment of a closed curve is the moment round the resultant axis of the areas of its projections on two planes at right angles to one another, and parallel to this axis. It is understood, of course, that the areas of the projections on these two planes are not evanescent generally, except for the case of a plane curve, and that their zero-values are generally the sums of equal positive and negative portions. Thus their moments are not in general zero.

Thus, according to these definitions, the resultant impulse of a vortex-filament of infinitely small cross section and of unit circulation is equal to the resultant area of its curve. The resultant axis of a vortex is the same as the resultant axis of the curve; and the rotational moment is equal to the resultant areal moment of the curve.

28. Consider for a moment a vortex-filament in an infinite liquid with no disturbing influence of other vortices, or of solids immersed in the liquid. We now see, from the constancy of the impulse (proved generally in V. M. § 19), that the resultant area, and the resultant areal moment of the curve formed by the filament, remain constant however its curve may become contorted; and its resultant axis remains the same line in space.

Hence, whatever motions and contortions the vortex-filament may experience, if it has any motion of translation through space this motion must be on the average along the resultant axis.

29. Consider now the actual vortex made up of an infinite number of infinitely small vortex-filaments. If these be of volumes inversely proportional to their vortex-densities (§ 25), so that their circulations are equal, we now see from the constancy of the impulse that the sum of the resultant areas of all the vortex-filaments remains constant; and so does the sum of their rotational moments: and the resultant areal axis of them all regarded as one system is a fixed line in space. Hence, as in the case of a vortex-filament, the translation, if any, through space is on the average along its resultant axis. All this, of course, is on the supposition that there is no other vortex, and no solid immersed in the liquid, and no bounding surface of the liquid, near enough to produce any sensible influence on the given vortex.

11. On the Precessional Motion of a Liquid [Liquid Gyrostats].

[From *Nature*, Vol. xv. 1877, pp. 297–8; Communicated to Section A of the British Association at Glasgow, September 7, 1876.]

The formulas expressing this motion were laid before the meeting and briefly explained, but the analytical treatment of them was reserved for a more mathematical paper to be communicated to the Section on Saturday. The chief object of the present communication was to illustrate experimentally a conclusion from this theory which had been announced by the author in his opening address to the Section*, to the effect that, if the period of the precession of an oblate spheroidal rigid shell full of liquid is a much greater multiple of the rotational period of the liquid than any diameter of the spheroid is of the difference between the greatest and least diameters, the precessional effect of a given couple acting on the shell is approximately the same as if the whole were a solid rotating with the same rotational velocity. The experiment consisted in showing a liquid gyrostat, in which an oblate spheroid of thin sheet copper filled with water was substituted for the solid fly-wheel of the ordinary gyrostat. In the instrument actually exhibited, the equatorial diameter of the liquid shell exceeded the polar axis by about one-tenth of either.

Supposing the rotational speed to be thirty turns per second, the effect of any motive which, if acting on a rotating solid of the same mass and dimensions, would produce a precession having its period a considerable multiple of $\frac{1}{3}$ of a second, must, according to the theory, produce very approximately the same precession in the thin shell filled with liquid as in the rotating solid. Accordingly the main precessional phenomena of the liquid gyrostat were not noticeably different from those of ordinary solid gyrostats

* *Popular Lectures and Addresses* (Macmillan), vol. ii. pp. 238–272.

which were shown in action for the sake of comparison. It is probable that careful observation without measurement might show very sensible differences between the performances of the liquid and the solid gyrostat in the way of nutational tremors produced by striking the case of the instrument with the fist.

No attempt at measurement either of speeds or forces was included in the communication, and the author merely showed the liquid gyrostat as a rough general illustration, which he hoped might be regarded as an interesting illustration, of that very interesting result of mathematical hydro-kinetics, the quasi-rigidity produced in a frictionless liquid by rotation.

P.S.—Since the communication of this paper to the Association, and the delivery of my opening address which preceded it on the same day, I have received from Prof. Henry No. 240 of the Smithsonian Contributions to Knowledge, of date October, 1871, entitled "Problems of Rotatory Motion presented by the Gyroscope, the Precession of the Equinoxes and the Pendulum," by Brevet Major-Gen. J. G. Barnard, Col. of Engineers, U.S.A., in which I find a dissent, from the portion of my previously-published statements which I had taken the occasion of my address to correct, expressed in the following terms:—

"I do not concur with Sir William Thomson in the opinions quoted in note p. 38, from Thomson and Tait, and expressed in his letter to Mr G. Poulett Scrope (*Nature*, Feb. 1, 1872). So far as regards fluidity, or imperfect rigidity, within an infinitely rigid envelope, I do not think the rate of precession would be affected."

Elsewhere in the same paper Gen. Barnard speaks of "the practical rigidity conferred by rotation." Thus he has anticipated my correction of the statements contained in my paper on the Rigidity of the Earth, so far as regards the effect of interior fluidity on the precessional motion of a perfectly rigid ellipsoidal shell filled with fluid.

I regret to see that the other error of that paper, which I corrected in my opening address, had not been corrected by Gen. Barnard, and that the plausible reasoning which had led me to it had also seemed to him convincing. For myself, I can only say that I took the very earliest opportunity to correct the errors after I found them to be errors, and that I deeply regret any mischief they may have done in the meantime.

Addendum.—Solid and Liquid Gyrostats. The solid gyrostat has been regularly shown for many years in the Natural Philosophy Class of the University of Glasgow as a mechanical illustration of the dynamics of rotating solids, and it has also been exhibited in London and Edinburgh at conversaziones of the Royal Societies

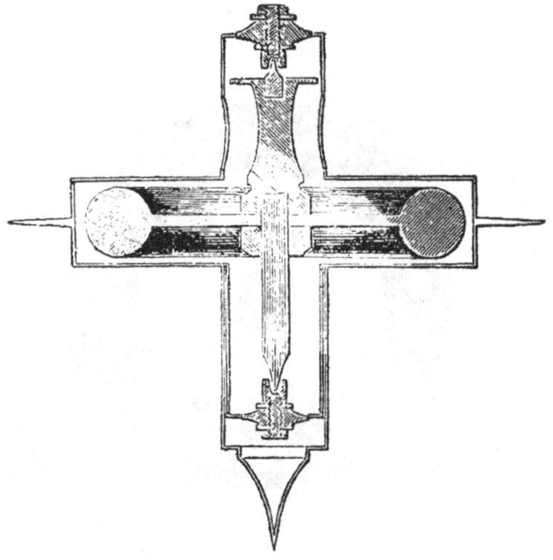

Fig. 1.

and of the Society of Telegraph Engineers, but no account of it has yet been published. The following brief description and drawing may therefore even now be acceptable to readers of *Nature*:—

The solid gyrostat consists essentially of a massive fly-wheel possessing great moment of inertia, pivoted on the two ends of its axis in bearings attached to an outer case which completely incloses it. Fig. 1 represents a section by a plane through the axis of the fly-wheel, and Fig. 2 a section by a plane at right angles to the axis and cutting through the case just above the fly-wheel. The containing case is fitted with a thin projecting edge in the plane of the fly-wheel, which is called the bearing edge. Its boundary forms a regular curvilinear polygon of sixteen sides with its centre at the centre of the fly-wheel. Each side of the polygon is a small arc of a circle of radius greater than the distance of the corners from the centre. The friction of the fly-wheel would, if the bearing-edge were circular, cause the case to roll along on it like a hoop, and it is to prevent this effect that the curved polygonal form described above and represented in the drawing is given to the bearing-edge.

To spin the solid gyrostat a piece of stout cord about forty feet long and a place where a clear run of about sixty feet can be obtained are convenient. The gyrostat having been placed with the

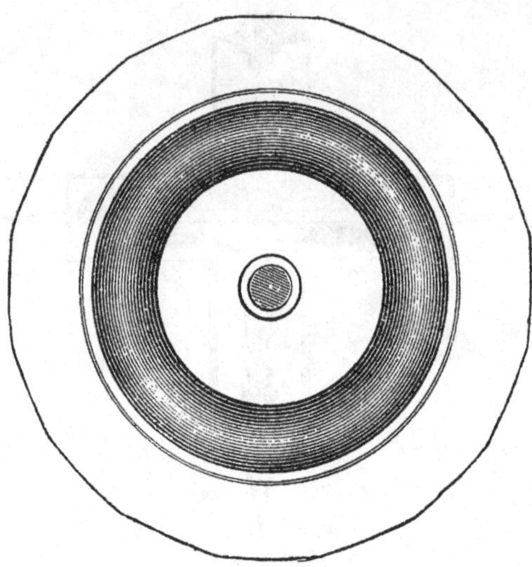

Fig. 2.

axis of its fly-wheel vertical, the cord is passed in through an aperture in the case, two-and-a-half times round the bobbin-shaped part of the shaft, and out again at an aperture on the opposite side. Having taken care that the slack cord is placed clear of all

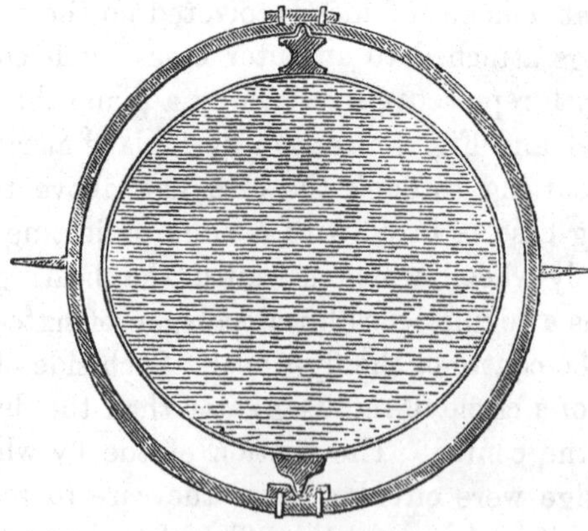

Fig. 3.

obstacles and that it is free from kinks, the operator holds the gyrostat steady so that its case is prevented from turning, while an assistant pulls the cord through by running, at a gradually increasing pace, away from the instrument, while holding the end of the cord in his hand. Sufficient tension is applied to the entering cord to prevent it from slipping round on the shaft. In this way a very great angular velocity is communicated to the fly-wheel, sufficient, indeed, to keep it spinning for upwards of twenty minutes.

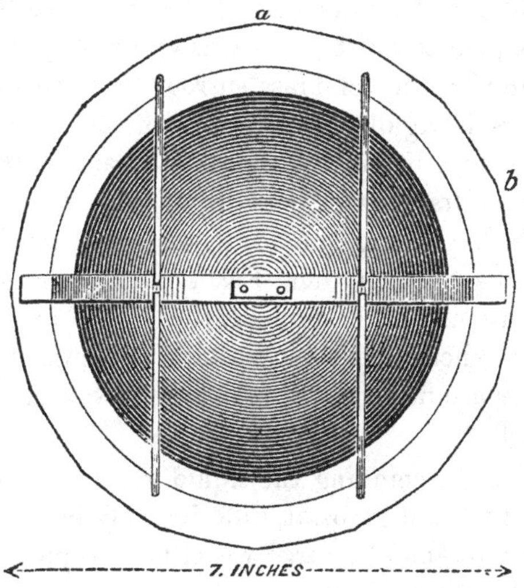

Fig. 4.

If when the gyrostat has been spun it be set on its bearing edge with the centre of gravity exactly over the bearing point, on a smooth horizontal plane such as a piece of plate-glass lying on a table, it will continue apparently stationary and in stable equilibrium. If while it is in this position a couple round a horizontal axis in the plane of the fly-wheel be applied to the case, no deflection of this plane from the vertical is produced, but it rotates slowly round a vertical axis. If a heavy blow with the fist be given to the side of the case, it is met by what seems to the senses the resistance of a very stiff elastic body, and, for a few seconds after the blow, the gyrostat is in a state of violent tremor, which, however, subsides rapidly. As the rotational velocity gradually diminishes, the rapidity of the tremors produced by a blow also diminishes. It is very curious to notice the tottering condition, and slow, seemingly palsied, tremulousness of the gyrostat, when the fly-wheel has nearly ceased to spin.

In the liquid gyrostat the fly-wheel is replaced by an oblate spheroid, made of thin sheet copper, and filled with water. The ellipticity of this shell in the instrument exhibited is $\frac{1}{10}$, that is to say, the equatorial diameter exceeds the polar by that fraction of either. It is pivoted on the two ends of its polar axis in bearings fixed in a circular ring of brass surrounding the spheroid. This circle of brass is rigidly connected with the curved polygonal-bearing edge which lies in the equatorial plane of the instrument, thus forming a frame-work for the support of the axis of the spheroidal shell. In Fig. 3 a section is represented through the axis to show the ellipticity, and Fig. 4 gives a view of the gyrostat as seen from a point in the prolongation of the axis. To prevent accident to the shell when the gyrostat falls down at the end of its spin, cage bars are fitted round it in such a way that no plane can touch the shell.

The method of spinning the liquid gyrostat is similar to that described for the solid gyrostat, differing only in the use of a very much longer cord and of a large wheel for the purpose of pulling it. The cord is first wound on a bobbin, free to rotate round a fixed pin. The end of it is then passed two-and-a-half times round the little pulley shown in the annexed sectional drawing, and thence to a point in the circumference of the large wheel to which it is fixed. An assistant then turns the wheel with gradually increasing velocity, while the frame of the gyrostat is firmly held, and the requisite tension applied to the entering cord to prevent it from slipping round the pulley.

12. Floating Magnets [illustrating Vortex-Systems].

[From *Nature*, Vol. XVIII. 1878, pp. 13, 14.]

The extract from the *American Journal of Science*, describing experiments with floating magnets by Mr Alfred M. Mayer, to illustrate the equilibrium of mutually-repellent molecules each independently attracted towards a fixed centre, which appeared in *Nature*, Vol. XVII. p. 487, must have interested many readers.

It has interested me particularly because the mode of experimenting there described, with a slight modification, gives a perfect mechanical illustration (easily realized with satisfactory enough approximateness) of the kinetic equilibrium of groups of columnar vortices revolving in circles round their common centre of gravity, which formed the subject of a communication I had made to the Royal Society of Edinburgh on the previous Monday. In Mr Mayer's problem the horizontal resultant repulsion between any two of the needles varies according to a complicated function of their mutual distance readily calculable if the distribution of magnetism in each needle were accurately known. Suppose the distributions to be precisely similar in all the bars and in each to be according to the following law:—Let the intensity of magnetisation be rigorously uniform throughout a very large portion, CD, of the whole length of the bar (Fig. 1), and let it vary uniformly from C and D to the two ends A and B. The bar will act as if for its magnetism were substituted ideal magnetic matter*, or polarity, as it may be called, uniformly distributed through the end portions CA and DB; the whole quantity in DB to be equal in amount and opposite in kind to that of CA. For example, suppose true northern polarity in AB and true southern in BD. The lengths of CA and DB need not be equal. Let now $A'C'D'B'$

* Reprint of papers on Electrostatics and Magnetism, § 469 (W. Thomson).

be another bar with an exactly similar distribution of magnetism to that of $ACDB$, and let the two be held parallel to one another. The mutual repulsion will vary inversely as the distance, if the distance be infinitely small in comparison with DB or CA, and if each of these be infinitely small in comparison with CD. If the true south pole S of a powerful bar-magnet be held in a line midway between BA and $B'A'$, at a distance from the ends B and B' infinitely great in comparison with BB', and comparable with the length of each needle, the horizontal component of its effect on each magnet will be a force varying directly as its distance from the central axis. Under these conditions Mr Mayer's experiments will show configurations of equilibrium of two, or three, or four, or any multitude of ideal points in a plane, repelling one another with forces inversely as the mutual distances, and each independently attracted towards a fixed centre with a force varying directly as the distance. This, as I showed in my communication to the Royal Society of Edinburgh, is the configuration of the group of points in which a multitude of straight columnar vortices with infinitely small cores is cut by a plane perpendicular to the columns; the centre of inertia of a group of ideal particles of equal mass placed at these points being the fixed centre in the static analogue.

The consideration of stability referred to by Mr Mayer has occupied me much in the numerical problem, and it is remarkable that the criterion of stability or instability is identical in the static and kinetic problems. In the static problem it is of course that the potential energy of the mutual forces between the particles, together with that of the attraction towards a fixed centre, is less for the configuration of stable equilibrium than for any configuration differing infinitely little from it. The potential energy of the attractive force is a function of distance from the central axis, diminishing as the distance increases, and the statement of the criterion may be conveniently modified to the following:—

For a given value of this function the mutual potential energy of the atoms must be a minimum for stable equilibrium. When, as supposed above, the attractive force varies directly as the distance, its potential energy is

$$C - \tfrac{1}{2}c\Sigma r^2,$$

where C, c, denote constants, and Σr^2 the sum of the squares of the distances of all the particles from the attractive centre. And when the law of force between the particles is the inverse distance, their mutual potential energy is equal to

$$K - k \log (DD'D''...)$$

where K, k, denote constants, and D, D', D'', &c., denote the mutual distances between the particles. Thus the condition of stable equilibrium becomes that the product of the mutual distances between the particles must be a true maximum for a given value of the sum of the squares of their distances from the attractive centre. A first conclusion from this condition must be that the centre of gravity of the particles must be the attractive centre. Now the condition of kinetic equilibrium of a group of vortex columns, that is to say the condition that they may revolve in circles round their common centre of inertia, is, as proved in my communication to the Royal Society of Edinburgh, that the product of their mutual distances must be a maximum or minimum or a maximum-minimum for a given value of the sum of the squares of their distances from the common centre of gravity*; and the condition that this kinetic equilibrium may be stable is that the product be a true minimum for a given value of the sum of the squares of their distances from the centre of inertia. Taking for example a triad of vortices (or of the little magnetic needles of Mr Mayer's problem), it is thus obvious the equilibrium is unstable in the case represented by Fig. 2, and stable in the case represented by Fig. 3. The arrow-heads in Figs. 2 and 3 represent the motions

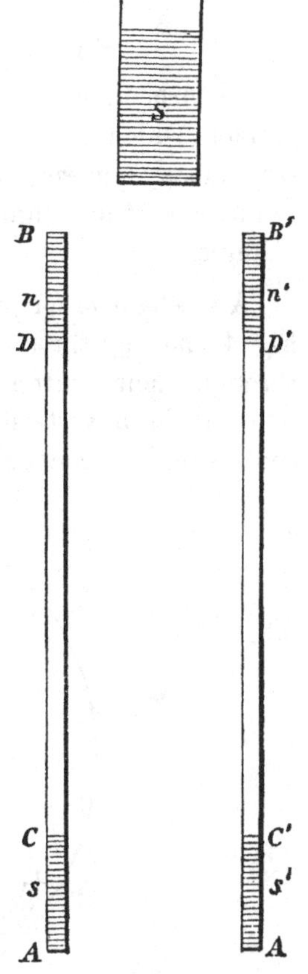

Fig. 1.

* Helmholtz proved that whatever be the complication of motions due to mutual influences among the vortices, their centre of gravity must remain at rest. [Cf. 'Vortex Statics,' *supra*.]

of the vortex columns round their centre of gravity. It must be understood that the core of each column revolves also round its centre of gravity, in the same direction as the group round the common centre of gravity of all, with enormously greater angular velocities.

I have farther considered the problem of oscillations in the

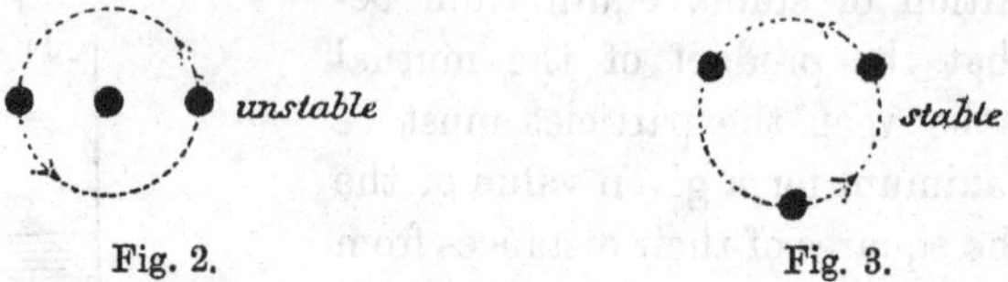

Fig. 2. Fig. 3.

neighbourhood of configuration of stable equilibrium. The general problem which it represents for mathematical analysis has a very easy and simple solution for the case of a triad of equal vortex columns in the neighbourhood of the angles of an equilateral triangle.

A mechanism for producing it kinematically is represented in Fig. 4, showing three circular discs of cardboard pivoted on pins through their centres at the angles of an equilateral triangle rotating in a vertical plane. The plane carrying these three centres may be conveniently made of a circular disc of stiff card-

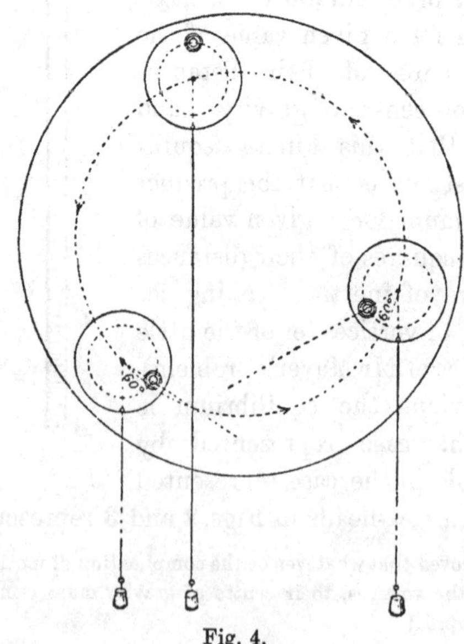

Fig. 4.

board, or of light wood pivoted on a fixed pin through its centre. Each of the small discs or epicycles is prevented from rotation by a fine thread bearing a weight, and attached to a point of its circumference; and on each of them is marked, by a small dark shaded circle, the section of one of the vortex cores in proper position.

The rule for placing the vortices on their epicycles is as follows:—Each vortex keeps a constant distance from its mean position (this being the centre of the epicycle, carrying it in the mechanism); each of the radius vectors drawn from the centres of the epicycles to the centres of the vortices keeps an absolutely fixed direction, while the equilateral triangle of the centres of the epicycles rotates uniformly; and these three fixed directions are inclined to one another at equal angles of 120° measured backwards relatively to the order in which we take the three vortices. It is easily verified that when the distances of the vortices from their mean positions are infinitely small (that is to say, when the triangle of the triad is infinitely nearly equilateral), the product of its three sides remains constant in the movement actually given by the mechanism, and so does the sum of the squares of the distances of its three corners from its centre of gravity. From the stability of the equilateral triangle it follows* that there must be stability with three equal vortices at the corners of an equilateral triangle, and one (whether equal to them or not) at its centre†,  For four equal vortices I have found that the square order, , also is stable. From the stability of the square follows (for vortices or for particles repelling according to inverse distance) the stability of four equals at the corners of the square, and one (whether equal to them or not) at its centre‡,  I have not yet ascertained *mathematically* whether for a

* In the case of vortices or of the static problem when the law of the mutual repulsions is the inverse distance, but not with the law of repulsion with ordinary proportions of linear dimensions and magnetic distributions in Mr Mayer's magnetic arrangement.

† In repetitions of Mr Mayer's experiments, I have always found this configuration unstable, and, for four, only the square stable.

‡ This configuration of the floating magnets I have found stable, but with less wide limits of stability than the pentagon.

pentad of equal vortices there is stability also in the pentagonal arrangement, 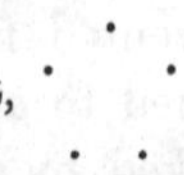 But Mr Mayer's experiment, showing it to be stable for the magnets, is an experimental proof that it must be stable for the vortices; for it is easily proved that if any of the figures is stable with mutual repulsion varying more rapidly (as is the case with the magnets in Mr Mayer's experiment), than according to the inverse distance, *à fortiori*, it must be stable when the force varies inversely as the distance. From the stability of the pentagon I infer (for vortices and for particles repelling according to inverse distance) the stability of the configuration

Mr Mayer's figure*  shows that the hexagonal order was unstable for his six magnets. I had almost convinced myself before seeing the account of his experiments in *Nature*, that the hexagonal order is stable for six equal vortices; and Mr Mayer's last figure shows that with his magnets the hexagonal order is rendered stable by the addition of one in the centre

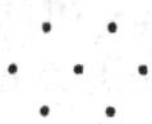

The instability of the hexagon of six magnets shows the simple polygon to be unstable for seven or any other number exceeding six. Thus Mr Mayer's beautiful experiment brings us very near an experimental solution of a problem which has for years been before me unsolved—of vital importance in the theory of vortex atoms—to find the greatest number of bars which a vortex mouse-mill can have.

* I have not found this, nor any other configuration than the pentagon with centre, stable for six floating magnets.

[For further experimental development, see L. Derr, *Proc. American Acad.* May, 1909. For theory, see J. J. Thomson's 'Motion of Vortex Rings,' 1883.]

13. ON GRAVITATIONAL OSCILLATIONS OF ROTATING WATER.

[From the *Proceedings of the Royal Society of Edinburgh*, March 17, 1879: reprinted in *Phil. Mag.* Vol. x, 1880, pp. 97–109.]

THIS is really Laplace's subject in his Dynamical Theory of the Tides; where it is dealt with in its utmost generality except one important restriction—the motion of each particle to be infinitely nearly horizontal, and the velocity to be always equal for all particles in the same vertical. This implies that the greatest depth must be small in comparison with the distance that has to be travelled to find the deviation from levelness of the water-surface altered by a sensible fraction of its maximum amount. In the present short communication I adopt this restriction; and, further, instead of supposing the water to cover the whole or a large part of the surface of a solid spheroid as does Laplace, I take the simpler problem of an area of water so small that the equilibrium-figure of its surface is not sensibly curved. Imagine a basin of water of any shape, and of depth not necessarily uniform, but, at greatest, small in comparison with the least diameter. Let this basin and the water in it rotate round a vertical axis with angular velocity ω so small that the greatest equilibrium-slope due to it may be a small fraction of the radian: in other words, the angular velocity must be small in comparison with $\sqrt{g/\frac{1}{2}A}$, where g denotes gravity, and A the greatest diameter of the basin. The equations of motion are

$$\left.\begin{aligned}\frac{du}{dt} - 2\omega v &= -\frac{1}{\rho}\frac{dp}{dx}\\ \frac{dv}{dt} + 2\omega u &= -\frac{1}{\rho}\frac{dp}{dy}\end{aligned}\right\} \quad\ldots\ldots\ldots\ldots\ldots\ldots(1),$$

where u and v are the component velocities of any point of the fluid in the vertical column through the point (xy), relatively to horizontal axes Ox, Oy revolving with the basin; p the pressure at any point x, y, z of this column; and ρ the uniform density of the liquid. The terms $\omega^2 x$, $\omega^2 y$, which appear in ordinary dynamical equations referred to rotating axes, represent components of centrifugal force, and therefore do not appear in these equations*. Let now D be the mean depth and $D+h$ the actual depth at any time t in the position (xy). The "equation of continuity" is

$$\frac{d(Du)}{dx} + \frac{d(Dv)}{dy} = -\frac{dh}{dt} \quad\text{.....................(2).}$$

Lastly, by the condition that the pressure at the free surface is constant, and that the difference of pressures at any two points in the fluid is equal to $g \times$ difference of levels, we have

$$\left.\begin{aligned} \frac{dp}{dx} &= g\rho\frac{dh}{dx} \\ \frac{dp}{dy} &= g\rho\frac{dh}{dy} \end{aligned}\right\} \quad\text{.........................(3).}$$

Hence for the case of gravitational oscillations (1) becomes

$$\left.\begin{aligned} \frac{du}{dt} - 2\omega v &= -g\frac{dh}{dx} \\ \frac{dv}{dt} + 2\omega u &= -g\frac{dh}{dy} \end{aligned}\right\} \quad\text{.....................(4).}$$

From (1) or (4) we find, by differentiation &c.,

$$\frac{d}{dt}\left(\frac{dv}{dx} - \frac{du}{dy}\right) + 2\omega\left(\frac{du}{dx} + \frac{dv}{dy}\right) = 0 \quad\text{...........(5),}$$

which is the equation of vortex motion in the circumstances.

These equations reduced to polar coordinates, with the following notation,

$$x = r\cos\theta, \qquad y = r\sin\theta,$$
$$u = \zeta\cos\theta - \tau\sin\theta, \qquad v = \zeta\sin\theta + \tau\cos\theta,$$

become

$$\frac{D\zeta}{r} + \frac{d(D\zeta)}{dr} + \frac{d(D\tau)}{rd\theta} = -\frac{dh}{dt} \quad\text{..............(2′),}$$

$$\left.\begin{aligned} \frac{d\zeta}{dt} - 2\omega\tau &= -g\frac{dh}{dr} \\ \frac{d\tau}{dt} + 2\omega\zeta &= -g\frac{dh}{rd\theta} \end{aligned}\right\} \quad\text{..................(4′),}$$

$$\frac{d}{dt}\left(\frac{\tau}{r} + \frac{d\tau}{dr} - \frac{d\zeta}{rd\theta}\right) + 2\omega\left(\frac{\zeta}{r} + \frac{d\zeta}{dr} + \frac{d\tau}{rd\theta}\right) = 0 \quad\text{.....(5′).}$$

[* They only alter the mean level.]

In these cases D may be any function of the co-ordinates. Cases of special interest in connexion with Laplace's tidal equations are had by supposing D to be a function of r alone. For the present, however, we shall suppose D to be constant. Then (2) used in (5) or (2′) in (5′) gives, after integration with respect to t,

$$\frac{dv}{dx} - \frac{du}{dy} = 2\omega \frac{h}{D} \quad \text{.......................(6)},$$

or, in polar coordinates,

$$\frac{\tau}{r} + \frac{d\tau}{dr} - \frac{d\zeta}{rd\theta} = 2\omega \frac{h}{D} \quad \text{..................(6′)}.$$

These equations (6), (6′) are instructive and convenient, though they contain nothing more than is contained in (2) or (2′), and (4) or (4′).

Separating u and v in (4), or ζ and τ in (4′), we find

$$\left.\begin{aligned} \frac{d^2u}{dt^2} + 4\omega^2 u &= -g\left(\frac{d}{dt}\frac{dh}{dx} + 2\omega\frac{dh}{dy}\right) \\ \text{and} \qquad \frac{d^2v}{dt^2} + 4\omega^2 v &= -g\left(2\omega\frac{dh}{dx} - \frac{d}{dt}\frac{dh}{dy}\right) \end{aligned}\right\} \quad \text{................(7)},$$

or, in polar coordinates,

$$\left.\begin{aligned} \frac{d^2\zeta}{dt^2} + 4\omega^2\zeta &= -g\left(\frac{d}{dt}\frac{dh}{dr} + 2\omega\frac{dh}{rd\theta}\right) \\ \frac{d^2\tau}{dt^2} + 4\omega^2\tau &= \quad g\left(2\omega\frac{dh}{dr} - \frac{d}{dt}\frac{dh}{rd\theta}\right) \end{aligned}\right\} \quad \text{...........(7′)},$$

Using in (7) (7′), in (2) (2′), with D constant, or in (6) (6′), we find

$$gD\left(\frac{d^2h}{dx^2} + \frac{d^2h}{dy^2}\right) = \frac{d^2h}{dt^2} + 4\omega^2 h \quad \text{.....................(8)},$$

and

$$gD\left(\frac{d^2h}{dr^2} + \frac{1}{r}\frac{dh}{dr} + \frac{1}{r^2}\frac{d^2h}{rd\theta^2}\right) = \frac{d^2h}{dt^2} + 4\omega^2 h \quad \text{........(8′)}.$$

It is to be remarked that (8) and (8′) are satisfied with u or v substituted for h.

I. Solutions for Rectangular Co-ordinates.

The general type solution of (8) is $h = \epsilon^{\alpha x}\, \epsilon^{\beta y}\, \epsilon^{\gamma t}$, where α, β, γ are connected by the equation

$$\alpha^2 + \beta^2 = \frac{\gamma^2 + 4\omega^2}{gD} \quad \text{.....................(9)}.$$

For waves of oscillations we must have $\gamma = \sigma\sqrt{-1}$, where σ is real.

I *a*. *Nodal Tesseral Oscillations.*

For nodal oscillations of the tesseral type we must have $\theta = m\sqrt{-1}$, $\beta = n\sqrt{-1}$, where m and n are real; and by putting together properly the imaginary constituents we find

$$h = C \frac{\sin}{\cos} \sigma t \frac{\sin}{\cos} mx \frac{\sin}{\cos} ny \quad \ldots\ldots\ldots\ldots\ldots(10),$$

where m, n, σ are connected by the equation

$$m^2 + n^2 = \frac{\sigma^2 - 4\omega^2}{gD} \quad \ldots\ldots\ldots\ldots\ldots\ldots\ldots(11).$$

Finding the corresponding values of u and v, we see what the boundary-conditions must be to allow these tesseral oscillations to exist in a sea of any shape. No bounding-line can be drawn at every part of which the horizontal component velocity perpendicular to it is zero. Therefore to produce or permit oscillations of the simple harmonic type in respect to form, water must be forced in and drawn out alternately all round the boundary, or those parts of it (if not all) for which the horizontal component perpendicular to it is not zero. Hence the oscillations of water in a rotating rectangular trough are not of the simple harmonic type in respect to form, and the problem of finding them remains unsolved.

If $\omega = 0$, we fall on the well-known solution for waves in a non-rotating trough, which are of the simple harmonic type.

I *b*. *Waves or Oscillations in an endless Canal with straight parallel sides.*

For waves in a canal parallel to x, the solution is

$$h = H\epsilon^{-ly} \cos(mx - \sigma t) \ldots\ldots\ldots\ldots\ldots(12);$$

where l, m, σ satisfy the equation

$$m^2 - l^2 = \frac{\sigma^2 - 4\omega^2}{gD} \quad \ldots\ldots\ldots\ldots\ldots\ldots(13),$$

in virtue of (9) or (11).

Using these in (7), we find that v vanishes throughout if we make

$$l = \frac{2\omega m}{\sigma} \ldots\ldots\ldots\ldots\ldots\ldots\ldots\ldots(14);$$

and with this value for l in (12) we find, by (7),

$$u = H\frac{gm}{\sigma}\epsilon^{-ly}\cos(mx - \sigma t) \quad\text{.............(15)};$$

and using (14) and (13) we find

$$m^2 = \frac{\sigma^2}{gD} \quad\text{.........................(16)},$$

from which we infer that the velocity of propagation of waves is the same for the same period as in a fixed canal. Thus the influence of rotation is confined to the effect of the factor $\epsilon^{-2\omega m/\sigma . y}$. Many interesting results follow from the interpretation of this factor with different particular suppositions as to the relation between the period of the oscillation $(2\pi/\sigma)$, the period of the rotation $(2\pi/\omega)$, and the time required to travel at the velocity σ/m across the canal. The more approximately nodal character of the tides on the north coast of the English Channel than on the south or French coast, and of the tides on the west or Irish side of the Irish Channel than on the east or English side, is probably to be accounted for on the principle represented by this factor, taken into account along with frictional resistance, in virtue of which the tides of the English Channel may be roughly represented by more powerful waves travelling from west to east, combined with less powerful waves travelling from east to west, and those of the southern part of the Irish Channel by more powerful waves travelling from south to north combined with less powerful waves travelling from north to south. The problem of standing oscillations in an endless rotating canal is solved by the following equations:—

$$\left.\begin{aligned} h &= H\{\epsilon^{-ly}\cos(mx - \sigma t) - \epsilon^{ly}(\cos mx + \sigma t)\} \\ u &= H\frac{gm}{\sigma}\{\epsilon^{-ly}\cos(mx - \sigma t) + \epsilon^{ly}\cos(mx + \sigma t)\} \\ v &= 0 \end{aligned}\right\} \text{...(17)}.$$

If we give ends to the canal, we fall upon the unsolved problem referred to above of tesseral oscillations. If instead of being rigorously straight we suppose the canal to be circular and endless, provided the breadth of the canal be small in comparison with the radius of the circle, the solution (17) still holds. In this case, if c denote the circumference of the canal, we must have $m = 2i\pi/c$, where i is an integer.

II. Oscillations and Waves in Circular Basin (Polar Coordinates).

Let
$$h = P\cos(i\theta - \sigma t) \quad \text{..................(18)}$$
be the solution for height, where P is a function of r. By (8′) P must satisfy the equation
$$\frac{d^2P}{dr^2} + \frac{1}{r}\frac{dP}{dr} - \frac{i^2P}{r^2} + \frac{\sigma^2 - 4\omega^2}{gD}P = 0 \quad \text{.........(19)};$$
and by (7′) we find
$$\left.\begin{aligned} \zeta &= \frac{g}{\sigma^2 - 4\omega^2}\sin(i\theta - \sigma t)\left(\sigma\frac{dP}{dr} - 2\omega i\frac{P}{r}\right) \\ \tau &= \frac{-g}{\sigma^2 - 4\omega^2}\cos(i\theta - \sigma t)\left(2\omega\frac{dP}{dr} - \sigma i\frac{P}{r}\right) \end{aligned}\right\} \quad \text{......(20).}$$

This is the solution for water in a circular basin, with or without a central circular island. Let a be the radius of the basin; and if there be a central island let a' be its radius. The boundary conditions to be fulfilled are $\zeta = 0$ when $r = a$ and when $r = a'$. The ratio of one to the other of the two constants of integration of (19), and the speed* σ of the oscillation, are the two unknown quantities to be found by these two equations. The ratio of the constants is immediately eliminated; and the result is a transcendental equation for σ. There is no difficulty, only a little labour, in thus finding as many as we please of the fundamental modes, and working out the whole motion of the system for each. The roots of this equation, which are found to be all real by the Fourier-Sturm-Liouville theory, are the speeds of the successive fundamental modes, corresponding to the different circular nodal subdivisions of the i diametral divisions implied by the assumed value of i. Thus, by giving to i the successive values 0, 1, 2, 3, &c., and solving the transcendental equation so found for each, we find all the fundamental modes of vibration of the mass of matter in the supposed circumstances.

* In the last two or three tidal reports of the British Association the word "speed," in reference to a simple harmonic function, has been used to designate the angular velocity of a body moving in a circle in the same period. Thus, if T be the period, $2\pi/T$ is the speed; *vice versa*, if σ be the speed, $2\pi/\sigma$ is the period.

If there is no central island, the solution of (19) which must be taken is that for which P and its differential coefficients are all finite when $r=0$. Hence P is what is called a Bessel's function of the first kind and of order i, and, according to the established notation*, we have

$$P = J_i\left(r\sqrt{\frac{\sigma^2-4\omega^2}{gD}}\right) \quad\text{..................(21).}$$

The solution found above for an endless circular canal is fallen upon by giving a very great value to r. Thus, if we put $2\pi r/i=\lambda$ so that λ may denote wave-length, we have $i/r=2\pi/\lambda$, which will now be the m of former notation. We must now neglect the term $1/r\ dh/dr$ in (19); and thus the differential equation becomes

$$\frac{d^2h}{dr^2}+\left(\frac{\sigma^2-4\omega^2}{gD}-m^2\right)h=0,$$

or

$$\frac{d^2h}{dr^2}-l^2h=0 \quad\text{..................................(22),}$$

where l^2 denotes $m^2-(\sigma^2-4\omega^2)/gD$. A solution of this equation is $h=C\epsilon^{-ly}$, where $y=a-r$; and using this in (20) above, we find $\zeta=g/(\sigma^2-4\omega^2)\ C\sin(mx-\sigma t)(\sigma l-2\omega m)\epsilon^{-ly}$, where $mx=i\theta$. Hence, to make $\zeta=0$ at each boundary, we have $\sigma l=2\omega m$, which makes $\zeta=0$, not only at the boundaries, but throughout the space for which the approximate equation (22) is sufficiently nearly true. And, putting for l^2 its value above, we have

$$4\omega^2m^2=\sigma^2\left(m^2-\frac{\sigma^2-4\omega^2}{gD}\right),$$

whence

$$m^2=\frac{\sigma^2}{gD},$$

which agrees with (16) above.

I hope in a future communication to the Royal Society to go in detail into particular cases, and to give details of the solutions at present indicated, some of which present great interest in relation to tidal theory, and also in relation to the abstract theory of vortex motion. The characteristic differences between cases in which σ is greater than 2ω or less than 2ω are remarkably interesting, and of great importance in respect to the theory of diurnal tides in

* Neumann, *Theorie der Bessel'schen Functionen* (Leipzig, 1867), § 5; and Lommel, *Studien über die Bessel'schen Functionen* (Leipzig, 1868), § 29.

the Mediterranean, or other more or less nearly closed seas in middle latitudes, and of the lunar fortnightly tide of the whole ocean. It is to be remarked that the preceding theory is applicable to waves or vibrations in any narrow lake or portion of the sea covering not more than a few degrees of the earth's surface, if for ω we take the component of the earth's angular velocity round a vertical through the locality—that is to say, $\omega = \gamma \sin l$, where γ denotes the earth's angular velocity, and l the latitude.

14. On the Formation of Coreless Vortices by the Motion of a Solid through an Inviscid Incompressible Fluid.

[From *Proc. Roy. Soc.*, Feb. 3, 1887; *Phil. Mag.*, XXIII. 1887, pp. 255—257.]

Take the simplest case: let the moving solid be a globe, and let the fluid be of infinite extent in all directions. Let its pressure be of any given value, P, at infinite distances from the globe, and let the globe be kept moving with a given constant velocity, V.

If the fluid keeps everywhere in contact with the globe, its velocity relatively to the globe at the equator (which is the place of greatest relative velocity) is $\frac{3}{2}V$. Hence, unless $P > \frac{5}{8}V^2$*, the fluid will not remain in contact with the globe.

Suppose, in the first place, P to have been $> \frac{5}{8}V^2$, and to be suddenly reduced to some constant value $< \frac{5}{8}V^2$. The fluid will be thrown off the globe at a belt of a certain breadth, and a violently disturbed motion will ensue. To describe it, it will be convenient to speak of velocities and motions *relative to the globe*. The fluid must, as indicated by the arrow-heads in fig. 1, flow partly backwards and partly forwards, at the place, I, where it impinges on the globe, after having shot off at a tangent at A. The back-flow along the belt that had been bared must bring to E some fluid in contact with the globe; and the free surface of this fluid must collide with the surface of the fluid leaving the globe at A. It might be thought that the result of this collision is a "vortex-sheet," which, in virtue of its instability, gets drawn out and mixed up indefinitely, and is carried away by the fluid further and further from the globe. A definite amount of kinetic energy would thus be *practically annulled* in a manner which I hope to explain in an early communication to the Royal Society of Edinburgh†.

* The density of the fluid is taken as unity.

† *Infra*, p. 166.

But it is impossible, either in our ideal inviscid incompressible fluid, or in a real fluid such as water or air, to form a vortex-sheet, that is to say, an interface of finite slip, by any natural action. What happens in the case at present under consideration, and in

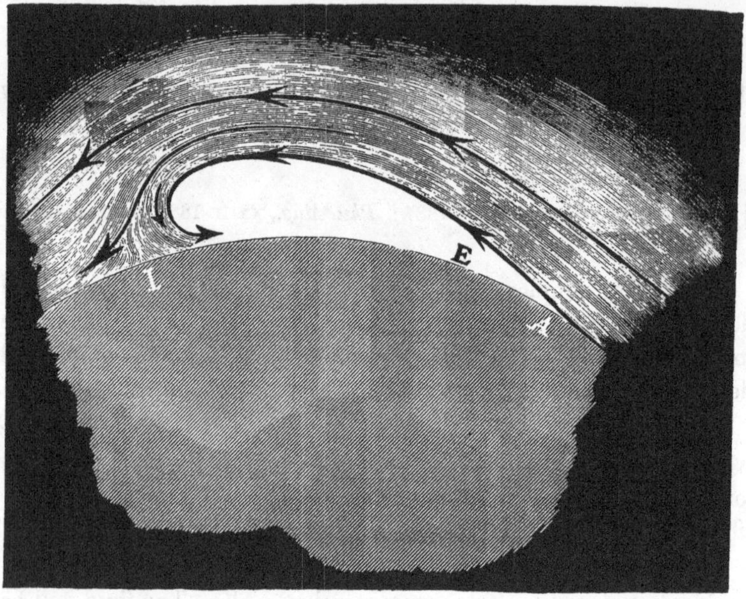

Fig. 1.

every real and imaginable case of two portions of liquid meeting one another (as, for instance, a drop of rain falling directly or obliquely on a horizontal surface of still water), is that continuity and the law of continuous fluid motion become established at the instant of first contact between two points, or between two lines in a class of cases of ideal symmetry to which our present subject belongs.

An inevitable result of the separation of the liquid from the solid, whether our supposed globe or any other figure perfectly symmetrical round an axis, and moving exactly in the line of the axis, is that two circles of the freed liquid surface come into contact and initiate in an instant the enclosure of two rings of vacuum (G and H in fig. 2, which, however, may be enormously far from like the true configuration).

The "circulation" (line-integral of tangential component velocity round any endless curve encircling the ring, as a ring on a ring, or one of two rings linked together) is determinate for each of these vacuum-rings, and remains constant for ever after: unless it divides itself into two or more, or the two first formed unite into one, against which accidents there is no security.

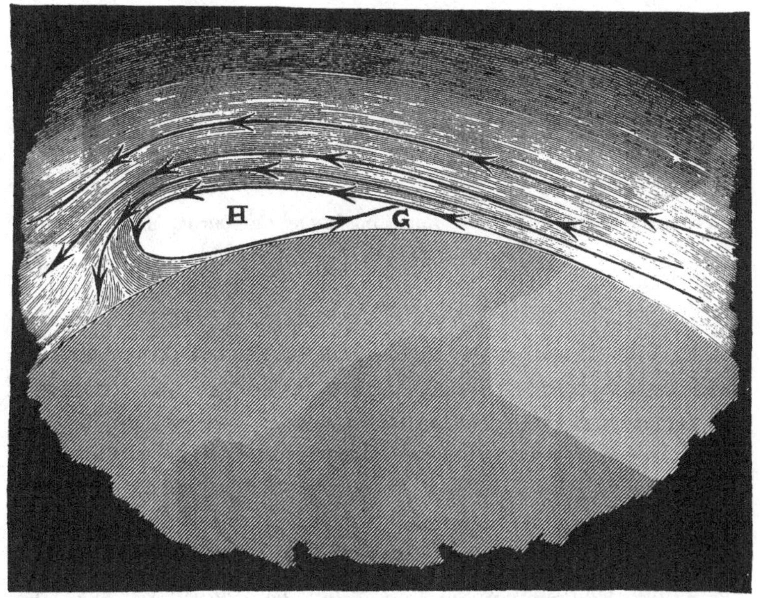

Fig. 2.

It is conceivably possible* that a coreless ring-vortex, with irrotational circulation round its hollow, shall be left oscillating in the neighbourhood of the equator of the globe; *provided* $(\frac{5}{8}V^2 - P)/P$ *be not too great.* If the material of the globe be viscously elastic, the vortex settles to a steady position round the equator, in a shape perfectly symmetrical on the two sides of the equatorial plane; and the whole motion goes on steadily henceforth for ever.

If $(\frac{5}{8}V^2 - P)/P$ exceed a certain limit, I suppose coreless vortices will be successively formed and shed off behind the globe in its motion through the fluid.

* If this conceivable possibility be impossible for a globe, it is certainly possible for some cases of prolate figures of revolution.

15. VIBRATIONS OF A COLUMNAR VORTEX.

[From the *Proceedings of the Royal Society of Edinburgh*, March 1, 1880; *Phil. Mag.*, x. 1880, pp. 155—168.]

THIS is a case of fluid-motion, in which the stream-lines are approximately circles, with their centres in one line (the axis of the vortex) and the velocities approximately constant, and approximately equal at equal distances from the axis. As a preliminary to treating it, it is convenient to express the equations of motion of a homogeneous incompressible inviscid fluid (the description of fluid to which the present investigation is confined) in terms of "columnar coordinates," r, θ, z—that is, coordinates such that $r\cos\theta = x$, $r\sin\theta = y$.

If we call the density unity, and if we denote by $\dot{x}$, $\dot{y}$, $\dot{z}$ the velocity-components of the fluid particle which at time t is passing through the point (x, y, z), and by d/dt, d/dx, d/dy, d/dz differentiations respectively on the supposition of x, y, z constant, t, y, z constant, t, x, z constant, and t, x, y constant, the ordinary equations of motion are

$$\left.\begin{aligned} -\frac{dp}{dx} &= \frac{d\dot{x}}{dt} + \dot{x}\frac{d\dot{x}}{dx} + \dot{y}\frac{d\dot{x}}{dy} + \dot{z}\frac{d\dot{x}}{dz} \\ -\frac{dp}{dy} &= \frac{d\dot{y}}{dt} + \dot{x}\frac{d\dot{y}}{dx} + \dot{y}\frac{d\dot{y}}{dy} + \dot{z}\frac{d\dot{y}}{dz} \\ -\frac{dp}{dz} &= \frac{d\dot{z}}{dt} + \dot{x}\frac{d\dot{z}}{dx} + \dot{y}\frac{d\dot{z}}{dy} + \dot{z}\frac{d\dot{z}}{dz} \end{aligned}\right\} \quad \text{...........(1)},$$

and

$$\frac{d\dot{x}}{dx} + \frac{d\dot{y}}{dy} + \frac{d\dot{z}}{dz} = 0 \quad \text{.....................(2)}.$$

To transform to the columnar coordinates, we have

$$\left.\begin{aligned} x &= r\cos\theta, \quad y = r\sin\theta \\ \dot{x} &= \dot{r}\cos\theta - r\dot{\theta}\sin\theta \\ \dot{y} &= \dot{r}\sin\theta + r\dot{\theta}\cos\theta \\ \frac{d}{dx} &= \cos\theta\frac{d}{dr} - \sin\theta\frac{d}{rd\theta} \\ \frac{d}{dy} &= \sin\theta\frac{d}{dr} + \cos\theta\frac{d}{rd\theta} \end{aligned}\right\}\quad\ldots\ldots\ldots\ldots\ldots\ldots(3).$$

The transformed equations are

$$\left.\begin{aligned} -\frac{dp}{dr} &= \frac{d\dot{r}}{dt} + \dot{r}\frac{d\dot{r}}{dr} - \frac{(r\dot{\theta})^2}{r} + \dot{\theta}\frac{d\dot{r}}{d\theta} + \dot{z}\frac{d\dot{r}}{dz} \\ -\frac{dp}{rd\theta} &= r\frac{d\dot{\theta}}{dt} + \dot{r}\frac{d(r\dot{\theta})}{dr} + \dot{r}\dot{\theta} + \dot{\theta}\frac{d(r\dot{\theta})}{d\theta} + \dot{z}\frac{d(r\dot{\theta})}{dz} \\ -\frac{dp}{dz} &= \frac{d\dot{z}}{dt} + \dot{r}\frac{d\dot{z}}{dr} + \dot{\theta}\frac{d\dot{z}}{d\theta} + \dot{z}\frac{d\dot{z}}{dz} \end{aligned}\right\}\quad\ldots(4),$$

and
$$\frac{d\dot{r}}{dr} + \frac{\dot{r}}{r} + \frac{d(r\dot{\theta})}{rd\theta} + \frac{d\dot{z}}{dz} = 0 \quad\ldots\ldots\ldots\ldots\ldots\ldots(5).$$

Now let the motion be approximately in circles round Oz, with velocity everywhere approximately equal to T, a function of r; and to fulfil these conditions, assume

$$\left.\begin{aligned} \dot{r} &= \rho\cos mz\sin(nt - i\theta), \quad r\dot{\theta} = T + \tau\cos mz\cos(nt - i\theta) \\ \dot{z} &= w\sin mz\sin(nt - i\theta), \quad p = P + \varpi\cos mz\cos(nt - i\theta) \\ \text{with}\quad P &= \int\frac{T^2dr}{r} \end{aligned}\right\}\ldots(6),$$

where ρ, τ, w, and ϖ are functions of r, each infinitely small in comparison with T. Substituting in (4) and (5) and neglecting squares and products of the infinitely small quantities, we find

$$\left.\begin{aligned} -\frac{d\varpi}{dr} &= \left(n - i\frac{T}{r}\right)\rho - 2\frac{T}{r}\tau \\ -\frac{i\varpi}{r} &= -\left(n - i\frac{T}{r}\right)\tau + \left(\frac{T}{r} + \frac{dT}{dr}\right)\rho \\ +m\varpi &= \left(n - i\frac{T}{r}\right)w \end{aligned}\right\}\quad\ldots\ldots\ldots(7),$$

$$\frac{d\rho}{dr} + \frac{\rho}{r} + \frac{i\tau}{r} + mw = 0\ldots\ldots\ldots\ldots\ldots\ldots\ldots(8).$$

Taking (7), eliminating ϖ, and resolving for ρ, τ, we find

$$\left.\begin{aligned}\rho&=\frac{1}{mD}\left(n-i\frac{T}{r}\right)\left\{\left(n-i\frac{T}{r}\right)\frac{dw}{dr}-\frac{i}{r}\left(\frac{T}{r}+\frac{dT}{dr}\right)w\right\}\\ \tau&=\frac{1}{mD}\left\{\left(\frac{T}{r}+\frac{dT}{dr}\right)\left(n-i\frac{T}{r}\right)\frac{dw}{dr}+\frac{i}{r}\left[\frac{T^2}{r^2}-\left(\frac{dT}{dr}\right)^2-\left(n-i\frac{T}{r}\right)^2\right]w\right\}\\ \text{where}\quad D&=\frac{2T}{r}\left(\frac{T}{r}+\frac{dT}{dr}\right)-\left(n-i\frac{T}{r}\right)^2\end{aligned}\right\}\quad\ldots\ldots\ldots(9).$$

For the particular case of $m=0$, or motion in two dimensions (r, θ), it is convenient to put

$$\frac{-w}{m}=\phi\ \ldots\ldots\ldots\ldots\ldots\ldots\ldots\ldots(10).$$

In this case the motion which superimposed on $\dot{r}=0$ and $r\dot{\theta}=T$ gives the disturbed motion is irrotational, and $\phi\sin(nt-i\theta)$ is its velocity-potential. It is also to be remarked that, when m does not vanish, the superimposed motion is irrotational where, if at all, and only where $T=\text{const.}/r$; and that whenever it is irrotational, $\phi\cos mz\sin(nt-i\theta)$, with ϕ as given by (10), is its velocity-potential.

Eliminating ρ and τ from (8) by (9), we have a linear differential equation of the second order for w. The integration of this, and substitution of the result in (9), give w, ρ, and τ in terms of r, and the two arbitrary constants of integration which, with m, n, and i, are to be determined to fulfil whatever surface-conditions, or initial conditions, or conditions of maintenance are prescribed for any particular problem.

Crowds of exceedingly interesting cases present themselves. Taking one of the simplest to begin:—

CASE I.

Let $$T=\omega r\ (\omega\ \text{const.})\ \ldots\ldots\ldots\ldots(11),$$

$$\left.\begin{aligned}&\dot{r}=c\cos mz\sin(nt-i\theta)\ \text{when approximately}\ r=a\\ &\dot{r}=\mathfrak{c}\cos mz\sin(nt-i\theta)\qquad\text{"}\qquad\text{"}\qquad r=\mathfrak{a}\\ &c,\ \mathfrak{c},\ m,\ n,\ a,\ \mathfrak{a}\ \text{being any given quantities and}\ i\\ &\quad\text{any given integer}\end{aligned}\right\}\ldots(12).$$

The condition $T = \omega r$ simplifies (9) to

$$\left.\begin{aligned}\rho &= \frac{(n-i\omega)\left\{(n-i\omega)\dfrac{dw}{dr} - \dfrac{2i\omega}{r}w\right\}}{m\{4\omega^2-(n-i\omega)^2\}}\\ \tau &= \frac{(n-i\omega)\left\{2\omega\dfrac{dw}{dr} - \dfrac{i(n-i\omega)}{r}w\right\}}{m\{4\omega^2-(n-i\omega)^2\}}\end{aligned}\right\}\quad\text{.........(13)};$$

and the elimination of ρ and τ by these from (8) gives

$$\frac{d^2w}{dr^2} + \frac{1}{r}\frac{dw}{dr} - \frac{i^2w}{r^2} + m^2\frac{4\omega^2-(n-i\omega)^2}{(n-i\omega)^2}w = 0 \quad\text{...(14)};$$

or

$$\left.\begin{aligned}&\frac{d^2w}{dr^2} + \frac{1}{r}\frac{dw}{dr} - \frac{i^2w}{r^2} + \nu^2w = 0\\ \text{where}\quad &\nu = m\sqrt{\frac{4\omega^2-(n-i\omega)^2}{(n-i\omega)^2}}\end{aligned}\right\}\quad\text{...........(15)};$$

or

$$\left.\begin{aligned}&\frac{d^2w}{dr^2} + \frac{1}{r}\frac{dw}{dr} - \frac{i^2w}{r^2} - \sigma^2w = 0\\ \text{where}\quad &\sigma = m\sqrt{\frac{(n-i\omega)^2-4\omega^2}{(n-i\omega)^2}}\end{aligned}\right\}\quad\text{...........(16)}.$$

Hence if J_i, $\mathfrak{J}_i$ denote Bessel's functions of order i, and of the first and second kinds* (that is to say, J_i finite or zero for infinitely small values of r, and $\mathfrak{J}_i$ finite or zero for infinitely great values of r), and if I_i and $\mathfrak{I}_i$ denote the corresponding real functions with ν imaginary, we have

$$w = CJ_i(\nu r) + \mathfrak{C}\mathfrak{J}_i(\nu r) \quad\text{..............(17)},$$

or

$$w = CI_i(\sigma r) + \mathfrak{C}\mathfrak{I}_i(\sigma r) \quad\text{..............(18)},$$

where C and $\mathfrak{C}$ denote arbitrary constants, to be determined in the present case by the equations of condition (12). These are equivalent to $\rho = c$ when $r = a$, and $\rho = \mathfrak{c}$ when $r = \mathfrak{a}$, and, when (17) or (18) is used for w in (13), give two simple equations to determine C and $\mathfrak{C}$.

The problem thus solved is the finding of the periodic disturbance in the motion of rotating liquid in a space between two boundaries which are concentric circular cylindric when undis-

* Compare *Proceedings*, March 17, 1879, "Gravitational Oscillations of Rotating Water." Solution II (Case of Circular Basins). *Phil. Mag.*, August 1880, p. 114.

turbed, produced by infinitely small simple harmonic normal motion of these boundaries, distributed over them according to the simple harmonic law in respect to the coordinates z, θ. The most interesting Subcase is had by supposing the inner boundary evanescent ($\mathfrak{a} = 0$), and the liquid continuous throughout the space contained by the outer cylindric boundary of radius a. This, as is easily seen, makes $w = 0$ when $r = 0$, except for the case $i = 1$, and essentially, without exception, requires that $\mathfrak{C}$ be zero. Thus the solution for w becomes

$$w = CJ_i(\nu r) \quad \ldots\ldots\ldots\ldots\ldots\ldots\ldots\ldots(19),$$

or

$$w = CI_i(\sigma r) \ldots\ldots\ldots\ldots\ldots\ldots\ldots\ldots(20);$$

and the condition $\rho = c$ when $r = a$ gives, by (13),

$$C = \frac{\nu^2 c/m}{\nu J_i'(\nu a) - \dfrac{2i\omega}{(n - i\omega)\, a} J_i(\nu a)} \quad \ldots\ldots\ldots\ldots(21),$$

or the corresponding I_i formula.

By summation after the manner of Fourier, we find the solution for any arbitrary distribution of the generative disturbance over the cylindric surface (or over each of the two if we do not confine ourselves to the Subcase), and for any arbitrary periodic function of the time. It is to be remarked that (6) represents an undulation travelling round the cylinder with linear velocity na/i at the surface, or angular velocity n/i throughout. To find the interior effect of a *standing* vibration produced at the surface, we must add to the solution (6), or any sum of solutions of the same type, a solution, or a sum of solutions, in all respects the same, except with $-n$ in place of n.

It is also to be remarked that great enough values of i make ν^2 negative, and therefore ν imaginary; and for such the solutions in terms of σ and the I_i, $\mathfrak{I}_i$ functions must be used.

CASE II. *Hollow Irrotational Vortex in a fixed Cylindric Tube.*

Conditions:—

$$\left.\begin{aligned} &T = \frac{c}{r};\ r = 0 \text{ when } r = a; \\ \text{and}\quad &p = 0 \text{ for the disturbed orbit, } r = \mathfrak{a} + \int \dot{r}_{\mathfrak{a}}\, dt \end{aligned}\right\} \ldots(22),$$

$\mathfrak{a}$ and a being the radii of the hollow cylindric interior, or free boundary, and of the external fixed boundary, and $\dot{r}_{\mathfrak{a}}$ the value of $\dot{r}$ when r is approximately equal to $\mathfrak{a}$. The condition $T = c/r$ simplifies (9) and (8) to

$$\rho = -\frac{1}{m}\frac{dw}{dr}, \text{ and } \tau = \frac{iw}{mr} \quad \text{..............(23)},$$

$$\frac{d^2 w}{dr^2} + \frac{1}{r}\frac{dw}{dr} - \frac{i^2 w}{r^2} - m^2 w = 0 \quad \text{...........(24)};$$

and by (7) we have

$$\varpi = \frac{1}{m}\left(n - \frac{ic}{r^2}\right) w \quad \text{..................(25)}.$$

Hence $$w = CI_i(mr) + \mathfrak{C}\mathfrak{I}_i(mr) \quad \text{...............(26)};$$

and the equation of condition for the fixed boundary (radial velocity zero there) gives

$$CI_i'(ma) + \mathfrak{C}\mathfrak{I}_i'(ma) = 0 \quad \text{..............(27)}.$$

To find the other equation of condition, we must first find an expression for the disturbance from circular figure of the free inner boundary. Let for a moment r, θ be the coordinates of one and the same particle of fluid. We shall have

$$\theta = \int \dot{\theta} dt; \text{ and } r = \int \dot{r} dt + r_0,$$

where r_0 denotes the radius of the "mean circle" of the particle's path.

Hence, to a first approximation,

$$\theta = \frac{ct}{r^2} \quad \text{......................................(28)};$$

and therefore, by (6),

$$\dot{r} = \rho \cos mz \sin\left(n - \frac{ic}{r^2}\right) t;$$

whence $$r = r_0 - \frac{\rho}{n - \frac{ic}{r^2}} \cos mz \cos(nt - i\theta) \quad \text{......(29)}.$$

Hence the equation of the free boundary is

$$r = \mathfrak{a} - \frac{\rho_{(r=\mathfrak{a})}}{n - i\omega} \cos mz \cos(nt - i\theta) \quad \text{......(30)},$$

where $$\omega = \frac{c}{\mathfrak{a}^2} \quad \text{.....................................(31)}.$$

Hence at (r, θ, z) of this surface we have, from $P = \int T^2 dr/r$, of (6) above,

$$P = \frac{T^2}{r}(r - \mathfrak{a})$$

$$= -\frac{c^2}{\mathfrak{a}^3} \frac{\rho_{(r=\mathfrak{a})}}{n - i\omega} \cos mz \cos(nt - i\theta) \text{ ...... } (32).$$

Hence, and by (6), and (26), and (25), and (23), the condition $p = 0$ at the free boundary gives

$$\frac{c^2}{\mathfrak{a}^3}[CI_i'(m\mathfrak{a}) + \mathfrak{C}\mathfrak{I}_i'(m\mathfrak{a})] + \frac{(n - i\omega)^2}{m}[CI_i(m\mathfrak{a}) + \mathfrak{C}\mathfrak{I}_i(m\mathfrak{a})] = 0$$

$$\text{.........}(33).$$

Eliminating $C/\mathfrak{C}$ from this by (27), we get an equation to determine n, by which we find

$$n = \omega(i \pm \sqrt{N}) \text{ ....}(34),$$

where N is an essentially positive numeric.

SUBCASE.

A very interesting Subcase is that of $a = \infty$, which, by (27), makes $C = 0$, and therefore, by (33), gives

$$N = m\mathfrak{a}\frac{-\mathfrak{I}_i'(m\mathfrak{a})}{\mathfrak{I}_i(m\mathfrak{a})} \text{ ....}(35).$$

Whether in Case II or the Subcase, we see that the disturbance consists of an undulation travelling round the cylinder with angular velocity

$$\omega(1 + \sqrt{N}/i) \text{ or } \omega(1 - \sqrt{N}/i),$$

or of two such undulations superimposed on one another, travelling round the cylinder with angular velocities greater than and (algebraically) less than the angular velocity of the mass of the liquid at its free surfaces by equal differences. The propagation of the wave of greater velocity is in the same direction as that in which the liquid revolves; the propagation of the other is in the contrary direction when $N > i^2$ (as it certainly is in some cases).

If the free surface be started in motion with one or other of the two principal angular velocities (34), or linear velocities $\mathfrak{a}\omega(1 \pm \sqrt{N}/i)$, and the liquid be then left to itself, it will perform the simple harmonic undulatory movement represented by (6), (26), (23). But if the free surface be displaced to the corrugated form (30), and then left free either at rest or with any other distribution of normal velocity than either of those, the corrugation will, as it were, split into two sets of waves travelling with the two different velocities $\mathfrak{a}\omega (1 \pm \sqrt{N/i})$.

The case $i=0$ is clearly exceptional, and can present no undulations travelling round the cylinder. It will be considered later.

The case $i=1$ is particularly important and interesting. To evaluate N for it, remark that

$$\left.\begin{aligned} &I_1(mr) = I_0'(mr) \\ \text{and}\quad &\mathfrak{I}_1(mr) = \mathfrak{I}_0'(mr) \end{aligned}\right\} \quad\text{.................(36).}$$

Now the general solution of (24) is

$$\left.\begin{aligned} w = \left(E + D\log\frac{1}{mr}\right)\left(1 + \frac{m^2r^2}{2^2} + \frac{m^4r^4}{2^2\,.\,4^2} + \&c.\right) \\ + D\left(\frac{m^2r^2}{2^2}S_1 + \frac{m^4r^4}{2^2\,.\,4^2}S_2 + \&c.\right) \end{aligned}\right\} \text{...(36*),}$$

where E and D are constants and $S_i = 1^{-1} + 2^{-1} + \ldots i^{-1}$. Hence, according to our notation,

$$I_0(mr) = 1 + \frac{m^2r^2}{2^2} + \frac{m^4r^4}{2^2\,.\,4^2} + \&c. \quad\text{...........(37),}$$

the constant factor being taken so as to make $I_0(0) = 1$.

Stokes* investigated the relation between E and D to make $w=0$ when $r=\infty$, and found it to be

$$\left.\begin{aligned} &E/D = \log 8 + \pi^{-\frac{1}{2}}\Gamma'\tfrac{1}{2} = +2{\cdot}079442 - 1{\cdot}963510 = {\cdot}11593 \\ &\text{or, to 20 places,} \\ &E/D = {\cdot}11593\ 15156\ 58412\ 44881\ldots\ldots\ldots\ldots \end{aligned}\right\} \text{(38).}$$

* "On the Effect of Internal Friction of Fluids on the Motion of Pendulums," equations (93) and (106). (*Camb. Phil. Trans.*, Dec. 1850.)

P.S.—I am informed by Mr J. W. L. Glaisher that Gauss, in section 32 of his "Disquisitiones Generales circa Seriem Infinitam $1 + (\alpha.\beta)/(1.\gamma)\,x + \&c.$" (*Opera*, vol. III. p. 155), gives the value of $-\pi^{-\frac{1}{2}}\Gamma'\frac{1}{2}$, or $-\psi(-\frac{1}{2})$, in his notation, to 23 places as follows:—

1·96351 00260 21423 47944 099.

Thus it appears that the last figure in Stokes's result (106) ought, as in the text, to be 0 instead of 2. In Callet's Tables we find

$\log_e 8 = 2{\cdot}07944\ 15416\ 79835\ 92825;$

and subtracting the former number from this, we have the value of E/D to 20 places given in the text.

Hence, and by convenient assumption for constant factor,

$$\left.\begin{aligned} \mathfrak{I}_0(mr) = \log\frac{1}{mr}\left(1 + \frac{m^2r^2}{2^2} + \frac{m^4r^4}{2^2.4^2} + \&c.\right) \\ + \frac{m^2r^2}{2^2}(S_1 + \cdot 11593) + \frac{m^4r^4}{2^2.4^2}(S_2 + \cdot 11593) + \&c. \end{aligned}\right\} \text{...(39).}$$

It is to be remarked that the series in (36) and (39) are convergent, however great be mr; though for values of mr exceeding 6 or 7 the semiconvergent expressions* will give the values of the functions nearly enough for most practical purposes, with much less arithmetical labour.

From (37) and (39) we find, by differentiation,

$$\left.\begin{aligned} I_1(mr) &= \frac{mr}{2} + \frac{m^3r^3}{2^2.4} + \frac{m^5r^5}{2^2.4^2.6} + \&c. \\ I'_1(mr) &= \frac{1}{2} + \frac{3m^2r^2}{2^2.4} + \frac{5m^4r^4}{2^2.4^2.6} + \&c. \end{aligned}\right\} \text{......(40).}$$

$$\left.\begin{aligned} \mathfrak{I}_1(mr) = \frac{-1}{mr} &+ \frac{mr}{2^2}[-1 + 4(S_1 + \cdot 1159315)] \\ &+ \frac{m^3r^3}{2^2.4^2}[-1 + 4(S_2 + \cdot 1159315)] + \&c. \\ &+ \log\frac{1}{mr}\left(\frac{mr}{2} + \frac{m^3r^3}{2^2.4} + \frac{m^5r^5}{2^2.4^2.6} + \&c.\right) \\ \mathfrak{I}_1'(mr) = \frac{1}{m^2r^2} &+ \frac{1}{2^2}[-3 + 1.2(S_1 + \cdot 1159315)] \\ &+ \frac{m^2r^2}{2^2.4^2}[-7 + 3.4(S_2 + \cdot 1159315)] + \&c. \\ &+ \log\frac{1}{mr}\left(\frac{1}{2} + \frac{3m^2r^2}{2^2.4} + \frac{5m^4r^4}{2^2.4^2.6} + \&c.\right) \end{aligned}\right\} \text{......(41).}$$

For an illustration of the Subcase with $i = 1$, suppose $m\mathfrak{a}$ to be very small. Remarking that $S_1 = 1$, we have

$$N = \frac{-m\mathfrak{a}\mathfrak{I}_1'(m\mathfrak{a})}{\mathfrak{I}_1(m\mathfrak{a})} = \frac{1 + \dfrac{m^2\mathfrak{a}^2}{2}\left[\log\dfrac{1}{m\mathfrak{a}} - \frac{1}{2} + \cdot 1159\right]}{1 - \dfrac{m^2\mathfrak{a}^2}{2}\left[\log\dfrac{1}{m\mathfrak{a}} + \frac{1}{2} + \cdot 1159\right]}$$

$$= 1 + m^2\mathfrak{a}^2\left(\log\frac{1}{m\mathfrak{a}} + \cdot 1159\right) \text{......(42).}$$

* Stokes, *ibid.*

Hence in this case, at all events, $N > i^2$; and the angular velocity of the slow wave, in the reverse direction to that of the liquid's revolution, is

$$-n = \tfrac{1}{2}\omega m^2\mathfrak{a}^2\left(\log\frac{1}{m\mathfrak{a}} + \cdot 1159\right) \quad \text{...........(43).}$$

This is very small in comparison with

$$2\omega + \tfrac{1}{2}\omega m^2\mathfrak{a}^2\left(\log\frac{1}{m\mathfrak{a}} + \cdot 1159\right) \quad \text{...........(44),}$$

the angular velocity of the direct wave; and therefore clearly, if the initial normal velocity of the surface when left free after being displaced from its cylindrical figure of equilibrium be zero or anything small, the amplitude of the quicker direct wave will be very small in proportion to that of the reverse slow one *.

Case III.

A slightly disturbed vortex column in liquid extending through all space between two parallel planes; the undisturbed column consisting of a core of uniform vorticity (that is to say, rotating like a solid), surrounded by irrotationally revolving liquid with no slip at the cylindric interface. Denoting by a the radius of this cylinder, we have

$$\left.\begin{aligned} &T = \omega r \quad \text{when} \quad r < a, \\ \text{and} \qquad &T = \omega\frac{a^2}{r} \quad\ \ ,, \quad\ \ r > a \end{aligned}\right\} \quad \text{.................(45).}$$

Hence (13), (14) hold for $r < a$, and (23), (24) for $r > a$.

Going back to the form of assumption (6), we see that it suits the condition of rigid boundary planes if the axis Oz be perpendicular to them, O in one of them, and the distance between them π/m.

The conditions to be fulfilled at the interface between core and surrounding liquid are that ρ and w must have the same values on the two sides of it: it is easily proved that this implies also equal values of τ on the two sides. The equality of ρ on the two sides of the interface gives, by (13) and (23),

$$\left\{\frac{(i\omega - n)\left[(i\omega - n)\dfrac{dw}{dr} + \dfrac{2i\omega}{r}w\right]}{4\omega^2 - (i\omega - n)^2}\right\}^{\text{internal}}_{r=a} = -\left(\frac{dw}{dr}\right)^{\text{external}}_{r=a} \quad (46);$$

* [Here $m\mathfrak{a}$ is very small, as *supra*, so that the radius of the vortex-core is small compared with the wave-length along its axis.]

and from this and the equality of w on the two sides we have

$$\frac{(i\omega - n)\left[(i\omega - n)\left(\frac{dw}{wdr}\right)^{\text{internal}}_{r=a} + \frac{2i\omega}{a}\right]}{4\omega^2 - (i\omega - n)^2} = -\left(\frac{dw}{wdr}\right)^{\text{external}}_{r=a} \quad (47).$$

The condition that the liquid extends to infinity all round makes $w = 0$, when $r = \infty$. Hence the proper integral of (24) is of the form $\mathfrak{I}_i$: and the condition of undisturbed continuity through the axis shows that the proper integral of (15) is of the form J_i. Hence

$$\left.\begin{aligned} w &= CJ_i(\nu r) \quad \text{for } r < a \\ \text{and} \quad w &= \mathfrak{C}\mathfrak{I}_i(mr) \quad ,, \quad r > a \end{aligned}\right\} \quad \ldots\ldots\ldots\ldots\ldots(48),$$

by which (47) becomes

$$\frac{(i\omega - n)\left[(i\omega - n)\frac{\nu J_i'(\nu a)}{J_i(\nu a)} + \frac{2i\omega}{a}\right]}{4\omega^2 - (i\omega - n)^2} = \frac{-m\mathfrak{I}_i'(ma)}{\mathfrak{I}_i(ma)} \quad \ldots(49);$$

or by (15),

$$\frac{J_i'(q)}{qJ_i(q)} + \frac{i}{q^2\lambda} = \frac{-\mathfrak{I}_i'(ma)}{ma\mathfrak{I}_i(ma)} \quad \ldots\ldots\ldots\ldots\ldots(50);$$

where

$$\lambda = \frac{i\omega - n}{2\omega} \quad \ldots\ldots\ldots\ldots\ldots\ldots\ldots\ldots\ldots(51),$$

and

$$q^2 = m^2a^2\frac{1-\lambda^2}{\lambda^2} \quad \ldots\ldots\ldots\ldots\ldots\ldots\ldots(52).$$

Remarking that $J_i(q)$ is the same for positive and negative values of q, and that it passes from positive through zero to a finite negative maximum, thence through zero to a finite positive maximum, and so on an infinite number of times, while q is increased from 0 to ∞, we see that while λ is increased from -1 to 0, the first member of (50) passes an infinite number of times continuously through all real values from $-\infty$ to $+\infty$, and that it does the same when λ is diminished from $+1$ to 0. Hence (50), regarded as a transcendental equation in λ, has an infinite number of roots between -1 and 0, and an infinite number between 0 and $+1$. And it has no roots except between -1 and $+1$, because its second member is clearly positive, whatever be ma; and its first member is essentially real and negative for all real values of λ except between -1 and $+1$, as we see by remarking that when $\lambda^2 > 1$, $-q^2$ is real and positive, and $-J_i'(q)/qJ_i(q)$ is real and $> i/(-q^2)$; while $i/q^2\lambda$, whether positive or negative, is of less absolute value than $i/(-q^2)$.

Each of the infinite number of values of λ yielded by (50) gives, by (51), (48), and (13), a solution of the problem of finding simple harmonic vibrations of a columnar vortex, with m of any assumed value. All possible simple harmonic vibrations are thus found: and summation, after the manner of Fourier, for different values of m, with different amplitudes and different epochs, gives every possible motion, deviating infinitely little from the undisturbed motion in circular orbits.

The simplest Subcase, that of $i=0$, is curiously interesting. For it (50), (51), (52) give

$$\frac{J_0'(q)}{qJ_0(q)} = \frac{-\mathfrak{I}_0'(ma)}{ma\mathfrak{I}_0(ma)} \quad \text{..................(53)},$$

and

$$n = \frac{2\omega ma}{\sqrt{(m^2a^2+q^2)}} \quad \text{......................(54)}.$$

The successive roots of (53), regarded as a transcendental equation in q, lie between the 1st, 2nd, 3rd,... roots of $J_0(q)=0$, in order of ascending values of q, and the next greater roots of $J_0'(q)=0$, coming nearer and nearer down to the roots of J_0 the greater they are. They are easily calculated by aid of Hansen's Tables of Bessel's functions J_0 and J_1 (which is equal to $-J_0'$) from $q=0$ to $q=20$*. When ma is a small fraction of unity, the second member of (53) is a large number; and even the smallest root exceeds by but a small fraction the first root of $J_0(q)=0$, which, according to Hansen's Table, is 2·4049, or, approximately enough for the present, 2·4. In every case in which q is very large in comparison with ma, whether ma is small or not, (54) gives

$$n = \frac{2\omega ma}{q} \text{ approximately.}$$

Now, going back to (6), we see that the summation of two solutions to constitute waves propagated along the length of the column gives:—

$$\left.\begin{aligned} \dot{r} &= -\rho\sin(nt-mz); & r\dot{\theta} &= T+\tau\cos(nt-mz) \\ \dot{z} &= w\cos(nt-mz); & p &= P+\varpi\cos(nt-mz) \end{aligned}\right\} \text{...(55)}.$$

The velocity of propagation of these waves is n/m. Hence, when q is large in comparison with ma, the velocity of longitudinal waves is $2\omega a/q$, or $2/q$ of the translational velocity of the surface of the core in its circular orbit. This is $1/1·2$, or $\frac{5}{6}$ of the trans-

* Republished in Lömmel's *Bessel'sche Functionen*, Leipzig, 1868.

lational velocity, in the case of *ma* small, and the *mode* corresponding to the smallest root of (53). A full examination of the internal motion of the core, as expressed by (55), (13), (48), (15) is most interesting and instructive. It must form a more developed communication to the Royal Society.

The Subcase of $i=1$, and *ma* very small, is particularly interesting and important. In it we have, by (42), for the second member of (50), approximately,

$$\frac{-\mathfrak{I}_1'(ma)}{ma\mathfrak{I}_1(ma)}=\frac{1}{m^2a^2}\left[1+m^2a^2\left(\log\frac{1}{ma}+\cdot 1159\right)\right] \text{ ...(56).}$$

In this case the smallest root, q, is comparable with ma, and all the others are large in comparison with ma. To find the smallest, remark that when q is very small we have, to a second approximation,

$$\frac{J_1'(q)}{qJ_1(q)}=\frac{1}{q^2}-\frac{1}{4} \text{ ....(57).}$$

Hence (50), with $i=1$, becomes, to a first approximation,

$$\frac{1}{q^2}\left(1+\frac{1}{\lambda}\right)=\frac{1}{m^2a^2} \text{ ....(58).}$$

This and (52), used to find the two unknowns λ and q^2, give

$$\lambda=\tfrac{1}{2}, \text{ and } q^2=3m^2a^2,$$

for a first approximation. Now, with $i=1$, (51) becomes

$$\lambda=\frac{1}{2}\left(1-\frac{n}{\omega}\right),$$

and therefore n/ω is infinitely small. Hence (52) gives for a second approximation,

$$q^2=3m^2a^2\left(1+\frac{8n}{3\omega}\right) \text{ ....(59),}$$

and we have $$\frac{1}{q^2\lambda}=\frac{2}{3}\frac{1}{m^2a^2}\left(1-\frac{5n}{3\omega}\right) \text{ ....(60).}$$

Using now (57), (59), (60), and (56) in (50), we find, to a second approximation,

$$\frac{1}{3m^2a^2}\left(1-\frac{8n}{3\omega}\right)-\frac{1}{4}+\frac{2}{3m^2a^2}\left(1-\frac{5n}{3\omega}\right)$$
$$=\frac{1}{m^2a^2}\left[1+m^2a^2\left(\log\frac{1}{ma}+\cdot 1159\right)\right],$$

whence $$\frac{-n}{\omega}=\frac{1}{2}m^2a^2\left(\log\frac{1}{ma}+\frac{1}{4}+\cdot 1159\right) \text{ ....(61).}$$

Compare this result with (43) above. The fact that, as in (43), $-n$ is positive in (61), shows that in this case also the direction in which the disturbance travels round the cylinder is *retrograde* (or opposite to that of the translation of fluid in the undisturbed vortex); and, as was to be expected, the values of $-n$ are approximately equal in the two cases when ma is small enough; but $-n$ is smaller by a relatively small difference in (43) than in (61), as was also to be expected.

The case of ma small and $i > 1$ has a particularly simple approximate solution for the smallest q-root of the transcendental (50). With any value of i instead of unity we still have (58), as a first approximation for q small. Eliminating q^2/m^2a^2 between this and (52), we still find $\lambda = \frac{1}{2}$; but instead of $n = 0$ by (51), we now have $n = (i-1)\,\omega$. Thus is proved the solution for waves of deformation of sectional figure travelling round a cylindrical vortex, announced thirteen years ago without proof in my first article respecting Vortex Motion*.

* "Vortex Atoms," *Proc. Roy. Soc. Edin.*, Feb. 18, 1867 [*supra*, p. 115].

16. On the Stability of Steady and of Periodic* Fluid Motion.

[From *Phil. Mag.*, xxiii. 1887, pp. 459—464, 529—539, read before *Roy. Soc. Edin.*, April 18, 1887, pp. 172—183, being reprinted from *Nature*, xxiii., Oct. 28, 1880.]

1. The fluid will be taken as incompressible; but the results will generally be applicable to the motion of natural liquids and of air or other gases when the velocity is everywhere small in comparison with the velocity of sound in the particular fluid considered. I shall first suppose the fluid to be inviscid. The

* By steady motion of a system (whether a set of material points, or a rigid body, or a fluid mass, or a set of solids, or portions of fluid, or a system composed of a set of solids or portions of fluid, or of portions of solid and fluid) I mean motion which at any and every time is precisely similar to what it is at one time. By periodic motion I mean motion which is perfectly similar, at all instants of time differing by a certain interval called the period.

Example 1. Every possible adynamic motion of a free rigid body, having two of its principal moments of inertia equal, is steady. So also is that of a solid of revolution filled with irrotational inviscid incompressible fluid.

Example 2. The adynamic motion of a solid of revolution filled with homogeneously rotating inviscid incompressible fluid is essentially periodic, and is steady only in particular cases.

Example 3. The adynamic motion of a free rigid body with three unequal principal moments of inertia is essentially periodic, and is only steady in the particular case of rotation round one or other of the three principal axes; so also, and according to the same law, is the motion of a rigid body having a hollow or hollows filled with irrotational inviscid incompressible fluid, with the three virtual moments of inertia unequal.

Example 4. The adynamic motion of a hollow rigid body filled with rotationally moving fluid is essentially unsteady and non-periodic, except in particular cases. Even in the case of an ellipsoidal hollow and homogeneous molecular rotation the motion is non-periodic. The motion, whether rotational or irrotational, of fluid in an ellipsoidal hollow is fully investigated in a paper under this title published in the *Proceedings of the Royal Society of Edinburgh* for December 7, 1885 [*infra*, p. 193]. Among other results it was proved that the rotation, if initially given homogeneous, remains homogeneous, provided the figure of the hollow be never at any time deformed from being exactly ellipsoidal.

results obtained on this supposition will help in an investigation of effects of viscosity which will follow.

2. I shall suppose the fluid completely enclosed in a containing vessel, which may be either rigid or plastic so that we may at pleasure mould it to any shape, or of natural solid material and therefore viscously elastic (that is to say, returning always to the same shape and size when time is allowed, but resisting all deformations with a force depending on the speed of the change, superimposed upon a force of quasi-perfect elasticity). The whole mass of containing-vessel and fluid will sometimes be considered as absolutely free in space undisturbed by gravity or other force; and sometimes we shall suppose it to be held absolutely fixed. But more frequently we may suppose it to be held by solid supports of real, and therefore viscously elastic, material; so that it will be fixed only in the same sense as a three-legged table resting on the ground is fixed. The fundamental philosophic question, What is fixity? is of paramount importance in our present subject. Directional fixedness is explained in Thomson and Tait's *Natural Philosophy*, 2nd edition, Part I. § 249, and more fully discussed by Prof. James Thomson in a paper "On the Law of Inertia, the Principle of Chronometry, and the Principle of Absolute Clinural Rest and of Absolute Rotation." For our present purpose we shall cut the matter short by assuming our platform, the earth or the floor of our room, to be absolutely fixed in space.

3*. The object of the present communication, so far as it relates to inviscid fluid, is to prove and to illustrate the proof of the three following propositions regarding a mass of fluid given with any rotation in any part of it:—

(I) The energy of the whole motion may be infinitely increased by doing work in a certain systematic manner on the containing-vessel and bringing it ultimately to rest.

(II) If the containing-vessel be simply continuous and be of natural viscously elastic material, the fluid given moving within it will come of itself to rest.

(III) If the containing-vessel be complexly continuous and be of natural viscously elastic material, the fluid will lose energy; not to zero, however, but to a determinate condition of irrotational

* [This § appears, substantially, in *Proc. Roy. Soc. Edin.* XIII. 1885, p. 114, with the title "On Energy in Vortex Motion."]

circulation with a determinate cyclic constant for each circuit through it.

4. To prove § 3 (I) remark, first, that mere distortion of the fluid, by changing the shape of the boundary, can increase the kinetic energy indefinitely. For simplicity, suppose a finite or an infinitely great change of shape of the containing-vessel to be made in an infinitely short time; this will distort the internal fluid precisely as it would have done if the fluid had been given at rest, and thus, by Helmholtz's laws of vortex motion, we can calculate, from the initial state of motion supposed known, the molecular rotation of every part of the fluid, after the change. For example, let the shape of the containing-vessel be altered by homogeneous strain; that is to say, dilated uniformly in one, or in each of two, directions, and contracted uniformly in the other direction or directions, of three at right angles to one another. The liquid will be homogeneously deformed throughout; the axis of molecular rotation in each part will change in direction so as to keep along the changing direction of the same line of fluid particles; and its magnitude will change in inverse simple proportion to the distance between two particles in the line of the axis.

5. But, now, to simplify subsequent operations to the utmost, suppose that anyhow, by quick motion or by slow motion, the containing-vessel be changed to a circular cylinder with perforated diaphragm and two pistons, as shown in fig. 1. In the present circumstances the motion of the liquid may be supposed to have any degree of complexity of molecular rotation throughout. It might chance to have no moment of momentum round the axis of the cylinder, but we shall suppose this not to be the case. If it did chance to be the case (which could be discovered by external tests), a motion of the cylinder, round a diameter, to a fresh position of rest would leave it with moment of momentum of the internal fluid round the axis of the cylinder. Without further preface, however, we shall suppose the cylinder to be given, with the pistons as in fig. 1, containing fluid in an exceedingly irregular state of motion, but with a given moment of momentum M round the axis of the cylinder. The cylinder itself is to be held absolutely fixed, and therefore whatever we do to the pistons we cannot alter the whole moment of momentum of the fluid round the axis of the cylinder.

6. Suppose, now, the piston A to be temporarily fixed in its middle position CC, and the whole containing-vessel of cylinder and pistons to be mounted on a frictionless pivot, so as to be free to turn round AA' the axis of the cylinder. If the vessel be of ideally rigid material, and if its inner surface be an exact figure of revolution, it will, though left free to turn, remain at rest, because the pressure of the fluid on it is everywhere in plane with the axis. But now, instead of being ideally rigid, let the vessel be of natural viscous-elastic solid material. The unsteadiness of the internal fluid motion will cause deformations of the containing-solid with loss of energy, and the result finally approximated to more and more nearly as time advances is necessarily the one determinate condition of minimum energy with the given moment of momentum; which, as is well known and easily proved, is the condition of solid and fluid rotating with equal angular velocity. If the stiffness of the containing-vessel be small enough and its viscosity great enough, it is easily seen that this final condition will be closely approximated to in a very moderate number of times the period of rotation in the final condition. Still we must wait an infinite time before we can find a perfect approximation to this condition reached from our highly complex or irregular initial motion. We shall now, therefore, cut the affair short by simply supposing the fluid to be given rotating with uniform angular velocity, like a solid within the containing-vessel, a true figure of revolution, which we shall now again consider as absolutely rigid, and consisting of cylinder with perforated diaphragm and two movable pistons, as represented in fig. 1.

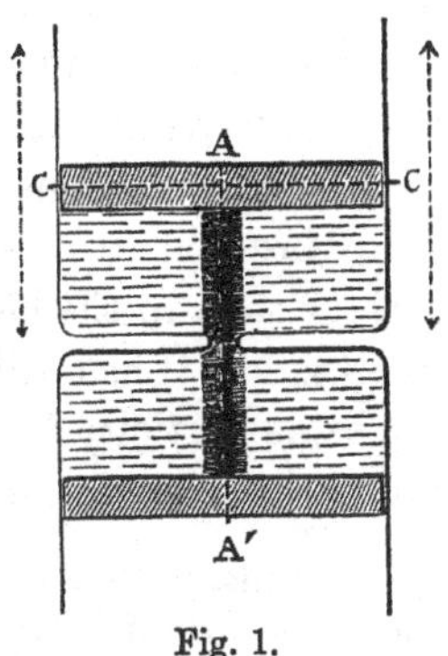

Fig. 1.

7. Give A a sudden pull or push and leave it to itself; it will move a short distance in the direction of the impulse and then spring back*. Keep alternately pulling and pushing it

* The subject of this statement receives an interesting experimental illustration in the following passage, extracted from the *Proceedings* of the Royal Institution of Great Britain for March 4, 1881; being an abstract of a Friday-evening discourse on "Elasticity viewed as possibly a Mode of Motion," and now in the press for republication along with other lectures and addresses in a volume (Vol. I.) of the

always in the direction of its motion. It will not thus be brought into a state of increasing oscillation, but the work done upon it will be spent in augmenting the energy of the fluid motion: so that if, after a great number of to-and-fro motions of the piston with some work done on it during each of them, the piston is once more brought to rest, the energy of the fluid motion will be greater than in the beginning, when it was rotating homogeneously like a solid. It has still exactly the same moment of momentum and the same vorticity* in every part; and the motion is symmetrical round the axis of the cylinder. Hence it is easily seen that the

'Nature Series.' "A little wooden ball, which when thrust down under still water jumped up again in a moment, remained down as if imbedded in jelly when the water was caused to rotate rapidly, and sprang back as if the water had elasticity

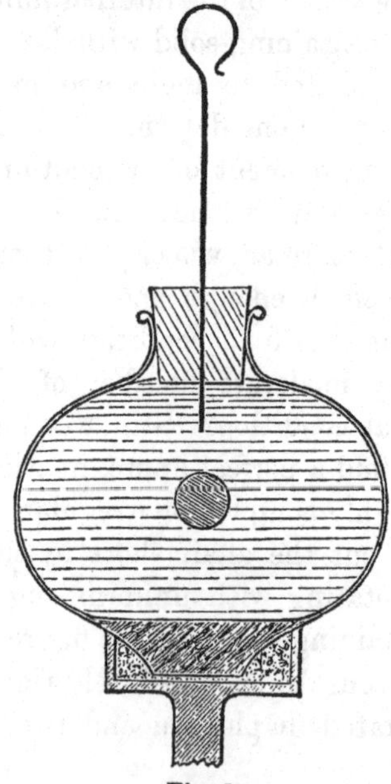

Fig. 2.

like that of jelly when it was struck by a stiff wire pushed down through the centre of the cork by which the glass vessel containing the water was filled."

* The vorticity of an infinitesimal volume dv of fluid is the value of $dv\,.\,\varpi/e$, where ϖ is its molecular rotation, and e the ratio of the distance between two of its particles in the axis of rotation at the time considered, to the distance between the same two particles at a particular time of reference. The amount of the vorticity thus defined for any part of a moving fluid depends on the time of reference chosen. Helmholtz's fundamental theorem of vortex motion proves it to be constant throughout all time for every small portion of an inviscid fluid.

greater energy implies the axial region of the fluid being stretched axially, and so acquiring angular velocity greater than the original angular velocity of the whole fluid mass.

8. The accompanying diagram (fig. 3) represents an easily performed experimental illustration, in which rotating water is churned by quick up-and-down movement of a disc carried on a vertical rod guided to move along the axis of the containing-vessel which is attached to a rotating vertical shaft. The kind of churning motion thus produced is very different from that produced by the perforated diaphragm; but the ultimate result is so far similar, that the statement of § 7 is equally applicable to the two cases. In the experiment, a little air is left under the cork, in the neck of the containing-vessel, to allow something to be seen of the motions of the water. When the vessel has been kept rotating steadily for some time with the churn-disc resting on the bottom, the surface of the water is seen in the paraboloidal form indicated (ideally) by the upper dotted curve (but of course greatly distorted by the refraction of the glass). Now, by finger and thumb applied to the top of the rod, move smartly up and down several times the churn-disk. A hollow vortex (or column of air bounded by water), ending irregularly a little above the disc, is seen to dart down from the neck of the vessel. If, now, the churn-disc is held at rest in any position, the ragged lower end of the air-tube becomes rounded and drawn up, the free surface of the water taking a succession of shapes, like that indicated by the lower dotted curve, until after a few seconds (or about a quarter of a minute) it becomes steady in the paraboloidal shape indicated by the upper dotted curve.

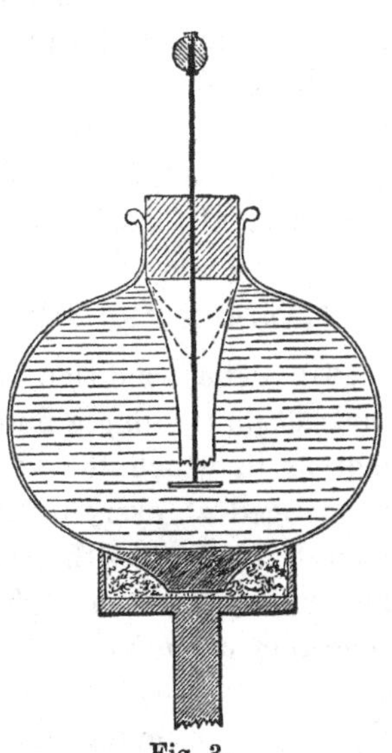
Fig. 3.

9. We have supposed the piston brought to rest after having done work upon the fluid during a vast but finite number of to-and-fro motions. But if left to itself it will not remain at rest;

it will get into a state of irregular oscillation, due to superposition of oscillations of the fluid according to an infinite number of fundamental modes, of the kind investigated in my article "Vibrations of a Columnar Vortex," *Proc. Roy. Soc. Edin.*, March 1, 1880, but not, as there, limited to being infinitesimal! If the motion of the piston be viscously resisted these vibrations will be gradually calmed down; and if time enough is allowed, the whole energy that has been imparted to the liquid by the work done on the pistons will be lost, and it will again be rotating uniformly like a solid, as it was in the beginning.

MAXIMUM AND MINIMUM ENERGY IN VORTEX MOTION*.

10. The condition for steady motion of an incompressible inviscid fluid filling a finite fixed portion of space (that is to say, motion in which the velocity and direction of motion continue unchanged at every point of the space within which the fluid is placed) is that, with given vorticity, the energy is a thorough maximum, or a thorough minimum, or a minimax. The further condition of *stability* is secured, by the consideration of energy alone, for any case of steady motion for which the energy is a thorough maximum or a thorough minimum; because when the boundary is held fixed the energy is of necessity constant. But the mere consideration of energy does not decide the question of stability for any case of steady motion in which the energy is a minimax.

11. It is clear† that, commencing with *any* given motion, the energy may be increased indefinitely by properly-designed operation on the boundary (understood that the primitive boundary is returned to). Hence, with given vorticity, but with no other condition, there is no thorough maximum of energy in any case. There may also, except in the case of irrotational circulation in a multiplexly continuous vessel referred to in § 3 (III) above, be *complete annulment* of the energy by operation on the boundary (with return to the primitive boundary), as we see by the following illustrations:—

* Being a communication read before the British Association, Section A, at the Swansea Meeting, Saturday, August 28, 1880, and published in the Report for that year, p. 473; and in *Nature*, Oct. 28, 1880. Reprinted now with corrections, amendments, and additions.

† See also §§ 3 to 9 above.

(*a*) Two equal, parallel, and oppositely rotating, vortex columns terminated perpendicularly by two fixed parallel planes. By proper operation on the cylindric boundary, they may, in purely two-dimensional motion, be thoroughly and equably mixed in two infinitely thin sheets. In this condition the energy is infinitely small.

(*b*) A single Helmholtz ring, reduced by diminution of its aperture to an infinitely long tube coiled within the enclosure. In this condition the energy is infinitely small.

(*c*) A single vortex column, with two ends on the boundary, bent till its middle infinitely nearly meets the boundary; and further bent and extended till it is broken into two equal and opposite vortex columns, connected, one end of one to one end of the other, by a vanishing vortex ligament infinitely near the boundary; and then further dealt with till these two columns are mixed together to virtual annihilation.

12. To avoid, for the present, the extremely difficult general question illustrated (or suggested) by the consideration of such cases, confine ourselves now to two-dimensional motions in a space bounded by two fixed parallel planes and a closed cylindric, not generally circular cylindric, surface perpendicular to them, subjected to changes of figure (but always truly cylindric and perpendicular to the planes). Also, for simplicity, confine ourselves for the present to vorticity either positive or zero, in every part of the fluid. It is obvious that, with the limitation to two-dimensional motion, the energy cannot be either infinitely small or infinitely great with any given vorticity and given cylindric figure. Hence, under the given conditions, there certainly are at least two stable steady motions—those of absolute maximum and absolute minimum energy. The configuration of absolute maximum energy clearly consists of least vorticity (or zero vorticity, if there be fluid of zero vorticity) next the boundary and greater and greater vorticity inwards. The configuration of absolute minimum energy clearly consists of greatest vorticity next the boundary, and less and less vorticity inwards. If there be any fluid of zero vorticity, all such fluid will be at rest either in one continuous mass, or in isolated portions surrounded by rotationally moving fluid. For illustration, see figs. 4 and 5, where it is seen how, even in so simple a case as that of the

containing-vessel represented in the diagram, there can be an infinite number of stable steady motions, each with maximum (though not greatest maximum) energy; and also an infinite number of stable steady motions of minimum (though not least minimum) energy.

13. That there can be an infinite number of configurations of stable motions, each of them having the energy a thorough minimum (as said in § 12), we see, by considering the case in which the cylindric boundary of the containing-canister consists of two wide portions communicating by a narrow passage, as shown in the drawings. If such a canister be completely filled with rotationally moving fluid of uniform vorticity, the stream-lines must be something like those indicated in fig. 4.

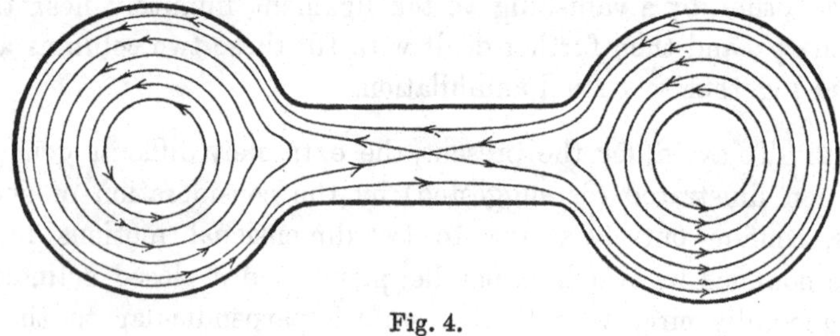

Fig. 4.

Hence, if a not too great portion of the whole fluid is irrotational, it is clear that there may be a *minimum* energy, and therefore a stable configuration of motion, with the whole of this in one of the wide parts of the canister; or the whole in the other; or any proportion in one and the rest in the other.

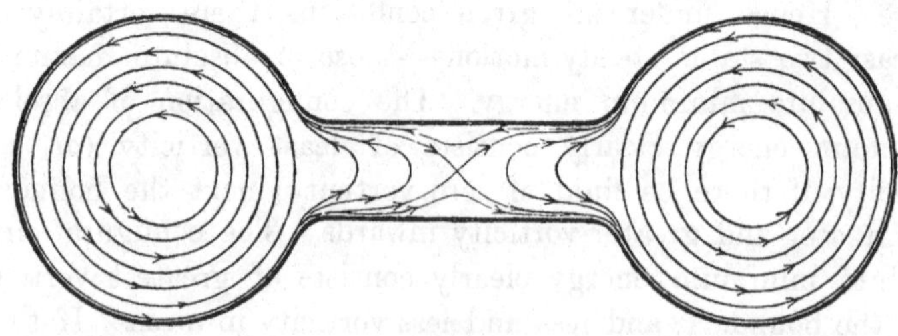

Fig. 5.

Single intersection of stream-lines in rotational motion may be at any angle, as shown in fig. 4. It is essentially at right angles in irrotational motion, as shown in fig. 5, representing the stream-lines of the configuration of *maximum* energy, for which the rotational part of the liquid is in two equal parts, in the middles of two wide parts of the enclosure. There is an infinite number of configurations of maximum energy in which the rotational part of the fluid is unequally distributed between the two wide parts of the enclosure.

14. In every steady motion, when the boundary is circular, the stream-lines are concentric circles and the fluid is distributed in co-axial cylindric layers of equal vorticity. In the stable motion of maximum energy, the vorticity is greatest at the axis of the cylinder, and is less and less outwards to the circumference. In the stable motion of minimum energy the vorticity is smallest at the axis, and greater and greater outwards to the circumference. To express the conditions symbolically, let T be the velocity of the fluid at distance r from the axis (understood that the direction of the motion is perpendicular to the direction of r), and let a be the radius of the boundary. The vorticity at distance r is

$$\frac{1}{2}\left(\frac{T}{r}+\frac{dT}{dr}\right).$$

If the value of this expression diminishes from $r=0$ to $r=a$, the motion is stable, and of maximum energy. If it increases from $r=0$ to $r=a$, the motion is stable and of minimum energy. If it increases and diminishes, or diminishes and increases, as r increases continuously, the motion is unstable*

15. As a simplest Subcase, let the vorticity be uniform through a given portion of the whole fluid, and zero through the remainder. In the stable motion of greatest energy, the portion of fluid having vorticity will be in the shape of a circular cylinder rotating like a solid round its own axis,

* This conclusion I had nearly reached in the year 1875 by rigid mathematical investigation of the vibrations of approximately circular cylindric vortices; but I was anticipated in the publication of it by Lord Rayleigh, who concludes his paper "On the Stability, or Instability, of certain Fluid Motions" (*Proceedings of the London Mathematical Society*, Feb. 12, 1880) with the following statement:—"It may be proved that, if the fluid move between two rigid concentric walls, the motion is stable, provided that in the steady motion the rotation either continually increases or continually decreases in passing outwards from the axis,"—which was unknown to me at the time (August 28, 1880) when I made the communication to Section A of the British Association at Swansea.

coinciding with the axis of the enclosure; and the remainder of the fluid will revolve irrotationally around it, so as to fulfil the condition of no finite slip at the cylindric interface between the rotational and irrotational portions of the fluid. The expression for this motion in symbols is

$$T = \zeta r \quad \text{from } r = 0 \text{ to } r = b,$$

and

$$T = \frac{\zeta b^2}{r} \quad \text{from } r = b \text{ to } r = a.$$

16. In the stable motion of minimum energy the rotational portion of the fluid is in the shape of a cylindric shell, enclosing the irrotational remainder, which in this case is at rest. The symbolical expression for this motion is

$$T = 0, \text{ when } r < \sqrt{(a^2 - b^2)},$$

and

$$T = \zeta\left(r - \frac{a^2 - b^2}{r}\right), \text{ when } r > \sqrt{(a^2 - b^2)}.$$

17. Let now the liquid be given in the configuration (§ 15) of greatest energy, and let the cylindric boundary be a sheet of a real elastic solid, such as sheet-metal with the kind of dereliction from perfectness of elasticity which real elastic solids present; that is to say, let its shape when at rest be a function of the stress applied to it, but let there be a resistance to change of shape depending on the velocity of the change. Let the unstressed shape be truly circular, and let it be capable of slight deformations from the circular figure in cross section, but let it always remain truly cylindrical. Let now the cylindric boundary be slightly deformed and left to itself, but held so as to prevent it from being carried round by the fluid. The central vortex column is set into vibration in such a manner that longer and shorter waves travel round it with less and greater angular velocity*. These waves cause corresponding waves of corrugation to travel round the cylindric bounding sheet, by which energy is consumed, and moment of momentum taken out of the fluid. Let this process go on until a certain quantity M of moment of momentum has been stopped from the fluid, and now let the canister run round freely in space, and, for simplicity, suppose

* See *Proceedings of the Royal Society of Edinburgh* for 1880, or *Philosophical Magazine* for 1880, vol. x. p. 155: "Vibrations of a Columnar Vortex" [*supra*, p. 152].

its material to be devoid of inertia. The whole moment of momentum was initially

$$\pi\zeta b^2(a^2 - \tfrac{1}{2}b^2);$$

it is now

$$\pi\zeta b^2(a^2 - \tfrac{1}{2}b^2) - M,$$

and continues constantly of this amount as long as the boundary is left free in space. The consumption of energy still goes on, and the way in which it goes on is this: the waves of shorter length are indefinitely multiplied and exalted till their crests run out into fine laminæ of liquid, and those of greater length are abated. Thus a certain portion of the irrotationally revolving water becomes mingled with the central vortex column. The process goes on until what may be called a vortex sponge is formed; a mixture homogeneous* on a large scale, but consisting of portions of rotational and irrotational fluid, more and more finely mixed together as time advances. The mixture is altogether analogous to the mixture of the white and yellow of an egg whipped together in the well-known culinary operation. Let b' be the radius of the cylindric vortex sponge, and ζ' its mean molecular rotation, which is the same in all sensibly large parts.

* *Note added May* 13, 1887.—I have had some difficulty in now proving these assertions (§§ 17 and 18) of 1880. Here is proof. Denoting for brevity $1/2\pi$ of the moment of momentum by μ, and $1/2\pi$ of the energy by e, we have

$$\mu = \int_0^a Tr \, . \, r\,dr, \text{ and } e = \tfrac{1}{2}\int_0^a T^2 \, . \, r\,dr.$$

The problem is to make e least possible, subject to the conditions: (1) that μ has a given value; (2) that

$$\tfrac{1}{2}\left(\frac{T}{r} + \frac{dT}{dr}\right) \overline{\overline{<}}\, \zeta, \text{ and } \geqq 0;$$

and (3) that when $r = a$, $T = \zeta b^2/a$; this last condition being the resultant of

$$\int_0^a \tfrac{1}{2}\left(\frac{T}{r} + \frac{dT}{dr}\right) r\,dr = \int_0^b \zeta r\,dr,$$

which expresses that the total vorticity is equal to that of ζ uniform within the radius b. The configurations described in the last three sentences of § 17 and the first three of § 18 clearly solve the problem when

$$M < \tfrac{1}{2}\pi\zeta b^2(a^2 - b^2); \text{ or } \mu > \tfrac{1}{4}\zeta b^2 a^2.$$

The fourth sentence of § 18 solves it when

$$M = \tfrac{1}{2}\pi\zeta b^2(a^2 - b^2); \text{ or } \mu = \tfrac{1}{4}\zeta b^2 a^2.$$

The second paragraph of § 18 solves it when

$$M > \tfrac{1}{2}\pi\zeta b^2(a^2 - b^2); \text{ or } \mu < \tfrac{1}{4}\zeta b^2 a^2.$$

Then, b being as before the radius of the original vortex column, we have

$$T = \zeta' r, \quad \text{from } r = 0 \text{ to } r = b',$$

and
$$T = \zeta' b'^2/r, \text{ from } r = b' \text{ to } r = a;$$

where
$$\zeta' = \zeta b^2/b'^2,$$

and
$$\tfrac{1}{2}b'^2 = \tfrac{1}{2}b^2 + \frac{M}{\pi \zeta b^2}.$$

18. Once more, hold the cylindric case from going round in space, and continue holding it until some more moment of momentum is stopped from the fluid. Then leave it to itself again. The vortex sponge will swell by the mingling with it of an additional portion of irrotational liquid. Continue this process until the sponge occupies the whole enclosure.

After that continue the process further, and the result will be that each time the containing canister is allowed to go round freely in space, the fluid will tend to a condition in which a certain portion of the original vortex core gets filtered into a position next to the boundary (beyond a distance from the axis which we shall denote by c), and the fluid within this space tends to a more and more nearly uniform mixture of vortex with irrotational fluid. This central vortex sponge, on repetition of the process of preventing the canister from going round, and again leaving it free to go round, becomes more and more nearly irrotational fluid, and the outer belt of pure vortex becomes thicker and thicker. The mean resultant motion is now

$$T = \zeta \frac{b^2 + c^2 - a^2}{c^2} r, \text{ for } r < c,$$

$$T = \zeta r - \zeta \frac{a^2 - b^2}{r}, \text{ for } r > c;$$

and the moment of momentum is

$$\tfrac{1}{2}\pi\zeta \{a^2 b^2 - (a^2 - b^2)(a^2 - c^2)\}.$$

The final condition towards which the whole tends is a belt constituted of the original vortex core now next the boundary; and the fluid which originally revolved irrotationally round it now placed at rest within it, being the condition (§ 16 above) of absolute minimum energy. Begin once more with the condition (§ 15 above) of absolute maximum energy, and leave the fluid to itself, whether with the canister free to go round sometimes, or always held fixed, provided only it is ultimately held from going round in space; the ultimate condition is always the same, viz. the condition (§ 16) of absolute minimum energy. The enclosing rotational belt, being the actual substance of the original vortex, is equal in its sectional area to πb^2; and therefore $c^2 = a^2 - b^2$. The moment of momentum is now $\frac{1}{2}\pi\zeta b^4$, being equal to the moment of momentum of the portion of the original configuration consisting of the then central vortex.

19. It is difficult to follow, even in imagination, the very fine—infinitely fine—corrugation and drawing-out of the rotational fluid; and its intermingling with the irrotational fluid; and its ultimate re-separation from the irrotational fluid, which the dynamics of §§ 17, 18 has forced on our consideration. This difficulty is obviated, and we substitute for the "vortex sponge" a much easier (and in some respects more interesting) conception, *vortex spindrift*, if (quite arbitrarily, and merely to help us to understand the minimum-energy-transformation of vortex column into vortex shell) we attribute to the rotational portion of the fluid a Laplacian* mutual attraction between its parts "insensible at sensible distances," and between it and the plane ends of the containing vessel, of such relative amounts as to cause the interface between rotational and irrotational fluid to meet the end planes at right angles. Let the amount of this Laplacian attraction be exceedingly small—so small, for example, that the work required to stretch the surface of the primitive vortex column to a million million times its area is small in comparison with the energy of the given fluid motion. Everything will go on as described in §§ 17, 18 if, instead of "run out into fine laminæ of liquid" (§ 17, line 29) we substitute "*break off into millions of detached fine vortex columns*"; and instead of "sponge" (*passim*) we substitute "*spindrift.*"

20. The solution of minimum energy for given vorticity and given moment of momentum (though clearly not unique, but infinitely multiplex, because magnitudes and orders of breaking-off of the millions of constituent columns of the

* So called to distinguish it from the "Newtonian" attraction, because, I believe, it was Laplace who first thoroughly formulated "attraction insensible at sensible distances," and founded on it a perfect mathematical theory of capillary attraction.

spindrift may be infinitely varied) is fully determinate as to the exact position of each column relatively to the others; and the cloud of spindrift revolves as if its constituent columns were rigidly connected. The viscously elastic containing vessel, each time it is left to itself, as described in §§ 17, 18, flies round with the same angular velocity as the spindrift cloud within; and so the whole motion goes on stably, without loss of energy, until the containing vessel is again stopped or otherwise tampered with.

21. It might be imagined that the Laplacian attraction would cause our slender vortex columns to break into detached drops (as it does in the well-known case of a fine circular jet of water shooting vertically downwards from a circular tube, and would do for a circular column of water given at rest in a region undisturbed by gravity), but it could not, because the energy of the irrotational circulation of the fluid round the vortex column must be infinite before the column could break in any place. The Laplacian attraction might, however, make the *cylindric* form unstable; but we are excluded from all such considerations at present by our limitation (§ 12) to two-dimensional motion.

22. Annul now the Laplacian attraction and return to our purely adynamic system of incompressible fluid acted on only by pressure at its bounding surface, and by mutual pressure between its parts, but by no "applied force" through its interior. For any given moment of momentum between the extreme possible values, $\pi\zeta b^2(a^2-\frac{1}{2}b^2)$ and $\frac{1}{2}\pi\zeta b^4$, there is clearly, besides the §§ 17, 18 solution (minimum energy), another determinate circular solution, viz. the configuration of circular motion of which the energy is greater than that of any other circular motion of same vorticity and same moment of momentum. This solution clearly is found by dividing the vortex into two parts—one a circular central column, and the other a circular cylindric shell lining the containing vessel; the ratio of one part to the other being determined by the condition that the total moment of momentum have the prescribed value. But this solution (as said above, § 14 and footnote) may be proved to be unstable.

I hope to return to this case, among other illustrations of instability of fluid motion—a subject demanding serious consideration *and investigation*, not only by purely scientific coercion, but because of its large practical importance.

23. For the present I conclude with the complete solution, or practical realisation of the solution (only found within the last few days, and after §§ 10—18 of the present article were already in type) of a problem on which I first commenced trials in 1868: *to make the energy an absolute maximum in two-dimensional motion with given moment of momentum and given vorticity in a cylindric canister of given shape.* The solution is, in its terms, essentially unique; "absolute maximum" meaning the greatest of maximums. But the same investigation includes the more extensive problem: To find, of the sets of solutions indicated in § 12, different configurations of the motion having the same moment of momentum. For each of these the energy is a maximum, *but not the greatest maximum,* for the given moment of momentum. The most interesting feature of the practical realisation to which I have now attained is the continuous transition from any one steady or periodic solution, through a series of steady or periodic solutions, to any other steady or periodic solution, produced by a simple mode of operation easily understood, and always under perfect control. The operating instrument is merely a stirrer, a thin round column, or rod, fitted perpendicularly between the two end plates, and movable at pleasure to any position parallel to itself within the enclosure. It is shown, marked *S*, in figs. 6, 7, 8, 9: representing the solution of our problem for the case of a circular enclosure with a small part of its whole volume occupied by vortex fluid, to which exigency of time limits the present communication.

24. Commence with the vortex lining uniformly the enclosing cylinder, and the stirrer in the centre of the still water within the vortex. The velocity of the water in the vortex increases from zero at the inside to $\zeta b^2/a$ at the outside, in contact with the boundary; according to the notation of §§ 15 and 16. Now move the stirrer very slowly from its central position and carry it round with any uniform angular velocity $< \zeta b/a$ and $> \frac{1}{2}\zeta b/a$. A dimple, as shown in fig. 6, will be produced, running round a little in advance of the stirrer, but ultimately falling back to be more and more nearly abreast of it if the stirrer is carried uniformly. If now the stirrer is gradually slowed till the dimple gets again in advance of it as in fig. 6, and is then carried round in a similar relative station, or always a little behind the radius through the middle of the dimple, the angular velocity of the dimple will

decrease gradually and its depth and its concave curvature will increase; till, when the angular velocity is $\frac{1}{2}\zeta b/a$, the dimple reaches the bottom (that is, the enclosing wall) with its concavity a right angle, as shown in fig. 7, and the angular velocity of propagation becomes $\frac{1}{2}\zeta b/a$.

25. The primitively endless vortex belt now becomes divided at the right angle, and the two acquired ends become rounded;

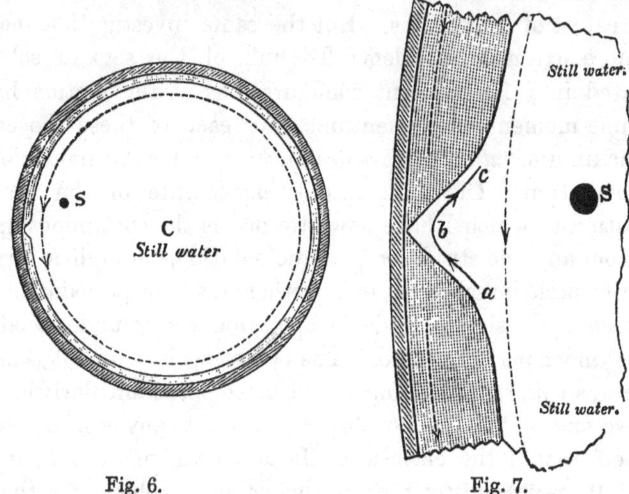

Fig. 6. Fig. 7.

Fig. 6. Dotted circle with arrowheads refers to the velocity of the stirrer and of the dimple, not to the velocity of the fluid.

Fig. 7. Arrowheads in the vortex refer to velocity of fluid. Arrowheads in the irrotational fluid refer to the stirrer and dimple. Arrowheads in *a b c* refer to motion of irrotational fluid relatively to the dimple.

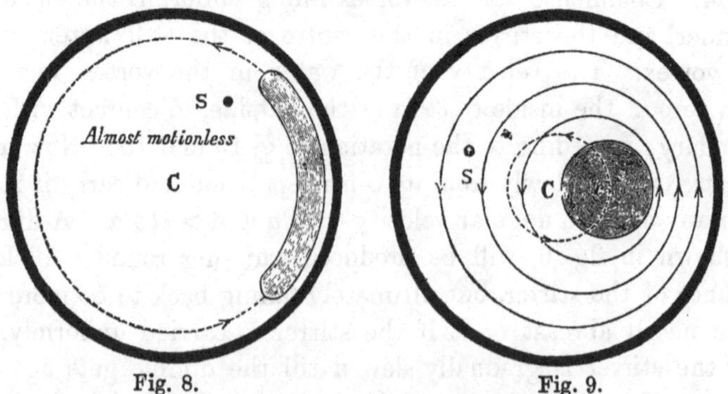

Fig. 8. Fig. 9.

Fig. 8. Arrowheads refer to motion of the stirrer, and of the vortex as a whole.

Fig. 9. Arrowheads on dotted circle refer to orbital motion of *c*, the centre of the vortex. Arrowheads on full fine curves refer to absolute velocity of fluid.

provided the stirrer be carried round always a little rearward, or considerably rearward, of abreast the middle of the gap. Figs. 8 and 9 show the result of continuing the process till ultimately the vortex becomes central and circular (with only the infinitesimal disturbance due to the presence of the stirrer, with which we need not trouble ourselves at present).

26. Suppose, now, at any stage of the process, after the formation of the gap, the stirrer to be carried forward to a station somewhat in advance of abreast the middle of the gap; or somewhat rearward of the rear of the vortex (instead of somewhat in advance of the front as shown in fig. 8). The velocity of propagation will be augmented (*by rearward pull!*), the moment of momentum will be diminished: the vortex train will be elongated till its front reaches round to its rear, each being then sharpened to 45° and brought into absolute contact with the enclosing wall: the front and rear unite in a dimple gradually becoming less; and the process may be continued till we end as we began, with the vortex lining the inside of the wall uniformly, and the stirrer at rest in the middle of the central still-water.

On an Experimental Illustration of Minimum Energy.

[From *Nature*, XXIII., Nov. 18, 1880, pp. 69—70; *British Association Report*, Swansea, 1880, Aug. 30, pp. 491—2.]

THIS illustration consists of a liquid gyrostat of exactly the same construction as that described and represented by the annexed drawing, reprinted from *Nature*, February 1, 1877, pp. 297, 298 *, with the difference that the figure of the shell is prolate instead of oblate. The experiment was in fact conducted with the actual apparatus which was exhibited to the British Association at Glasgow in 1876, altered by the substitution of a shell having its equatorial diameter about $\frac{9}{10}$ of its axial diameter, for the shell with axial diameter $\frac{9}{10}$ of equatorial diameter which was used when the apparatus was shown as a successful gyrostat. The oblate and prolate shells were each of them made from the two hemispheres of sheet copper which plumbers solder together to make their globular floaters. By a little hammering it is easy to alter the hemispheres to the proper shapes to make either the prolate or the oblate figure.

[* *Supra*, p. 129.]

Theory had pointed out that the rotation of a liquid in a rigid shell of oval figure, being a configuration of maximum energy for given vorticity, would be unstable if the containing vessel is left to itself supported on imperfectly elastic supports, although it

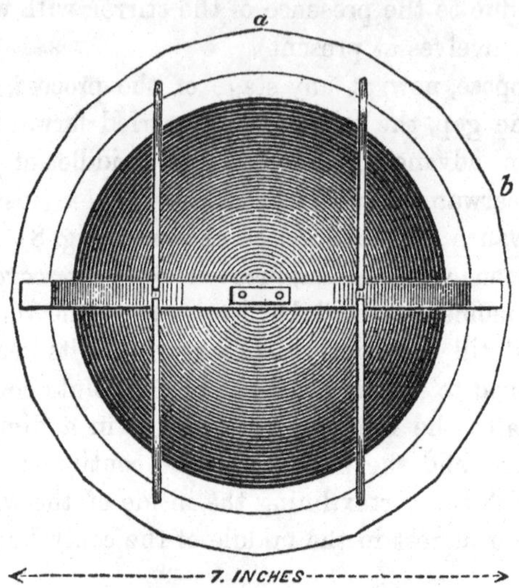

would be stable if the vessel were held absolutely fixed, or borne by perfectly elastic supports, or left to itself in space unacted on by external force; and it was to illustrate this theory that the oval shell was made and filled with water and placed in the apparatus. The result of the first trial was literally startling, although it ought not to have been so, as it was merely a realisation of what had been anticipated by theory. The framework was held as firmly as possible by one person with his two

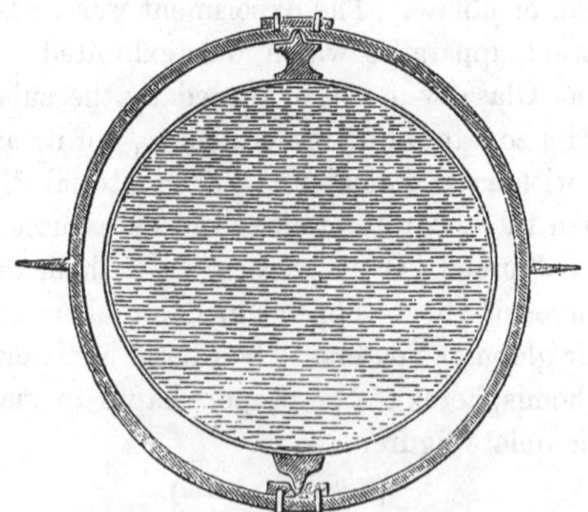

hands, keeping it as steady as he could. The spinning by means of a fine cord*, passing round a small V pulley of ½-inch diameter on the axis of the oval shell and round a large fly-wheel of 3 feet diameter turned at the rate of about one round per second, was continued for several minutes. This in the case of the oblate shell, as was known from previous experiments, would have given amply sufficient rotation to the contained water to cause the apparatus to act with great firmness like a solid gyrostat. In the first experiment with the oval shell, the shell was seen to be rotating with great velocity during the last minute of the spinning; but the moment it was released from the cord, and when, holding the framework in my hands, I commenced carrying it towards the horizontal glass table to test its gyrostatic quality, the framework which I held in my hands gave a violent uncontrollable lurch, and in a few seconds the shell stopped turning. I saw that one of the pivots had become bent over, by yielding of the copper shell in the neighbourhood of the stiff pivot-carrying disk, soldered to it, showing that the liquid had exerted a very strong couple against its containing shell, in a plane through the axis, the effort to resist which by my hands had bent the pivot. The shell was refitted with more strongly attached pivots, and the experiment has been repeated several times. In every case a decided uneasiness of the framework is perceived by the person holding it in his hands during the spinning; and as soon as the cord is cut and the person holding it carries it towards the experimental table, the framework begins, as it were, to wriggle round in his hands, and by the time the framework is placed on the table the rotation is nearly all gone. Its utter failure as a gyrostat is precisely what was expected from the theory, and presents a truly wonderful contrast from what is observed with the apparatus and operations in every respect similar, except in having an oblate instead of a prolate shell to contain the liquid.

* Instead of using a long cord first wound on a bobbin, and finally wound up on the circumference of the large wheel as described in *Nature*, February 1, 1877, p. 297, I have since found it much more convenient to use an endless cord little more than half round the circumference of the large wheel, and less than half round the circumference of the V pulley of the gyrostat; and to keep it tight enough to exert whatever tangential force on the V pulley is desired by the person holding the framework in his hand. After continuing the spinning by turning the fly-wheel for as long a time as is judged proper, the endless cord is cut with a pair of scissors and the gyrostat released.

17. On a Disturbing Infinity in Lord Rayleigh's Solution for Waves in a Plane Vortex Stratum.

[From *Nature*, XXIII., 1880, pp. 45—46; *British Association Report*, Swansea, 1880, pp. 492—3.]

Lord Rayleigh's solution involves a formula equivalent to

$$\frac{d^2v}{dy^2} - \left(m^2 + \frac{\dfrac{d^2T}{dy^2}}{T - \dfrac{n}{m}} \right) v = 0.$$

Where v denotes the maximum value of the y-component of velocity;

„ m denotes a constant such that $2\pi/m$ is the wave-length;

„ T denotes the translational velocity of the vortex-stratum when undisturbed, which is in the x-direction, and is a function of y;

„ n denotes the vibrational speed, or a constant such that $2\pi/n$ is the period.

Now a vortex stratum is stable, if on one side it is bounded by a fixed plane, and if the vorticity (or value of $\frac{1}{2}\,dT/dy$) diminishes as we travel (ideally) from this plane, except in places (if any) where it is constant.

To fulfil this condition, suppose a fixed bounding plane to contain Ox and be perpendicular to Oy; and let dT/dy have its greatest value when $y=0$, and decrease continuously, or by one or more abrupt changes, from this value, to zero at $y=a$ and at all greater values of y.

It is easily proved that the wave-velocity, whatever be the wave-length, is intermediate between the greatest and least values of T. Hence for a certain value of y between 0 and a, the translational velocity is equal to the wave-velocity, or $T = n/m$. Hence for this value of y the second term within the bracket in Lord Rayleigh's formula is infinite unless, for the same value of y, d^2T/dy^2 vanishes.

We evade entirely the consideration of this infinity if we take only the case of a layer of constant vorticity (dT/dy = constant from $y=0$ to $y=a$), as for this case the formula is simply

$$\frac{d^2v}{dy^2} = m^2v\,;$$

but the interpretation of the infinity which occurs in the more comprehensive formula suggests an examination of the stream-lines, by which its interpretation becomes obvious, and which proves that even in the case of constant vorticity the motion has a startlingly peculiar character at the place where the translational velocity is equal to the wave-velocity. This peculiarity is represented by the annexed diagram, which is most easily understood

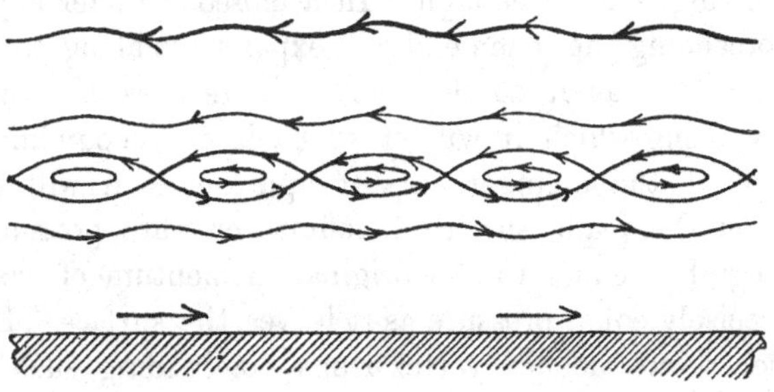

if we imagine the translational velocities at $y=0$ and $y=a$ to be in opposite directions, and of such magnitude that the wave-velocity is zero; so that we have the case of standing waves. For this case the stream-lines are as represented in the annexed diagram, in which the region of translational velocity greater than wave-propagational velocity is separated from the region of translational velocity less than wave-propagational velocity by a cat's-eye border pattern of elliptic whirls.

18. On the Average Pressure due to Impulse of Vortex-Rings on a Solid.

[From *Nature*, Vol. xxiv. 1881, p. 47; read at *Roy. Soc. Edin.*, April 17, 1881.]

When a vortex-ring is approaching a plane large in comparison to the dimensions of the ring, the total pressure over the surface is *nil.* When a ring approaches such a surface it begins to expand, so that if we consider a finite portion of the surface the total pressure upon it due to the ring will have a finite value when the ring is close enough. In a closed cylinder any vortex-ring approaching the plane end will expand out along the surface, losing in speed as it so does, until it reaches the cylindrical boundary, along which it will crawl back, on rebounding, to the other end of the cylinder. As it approaches, it will therefore exert upon the plane surface a definite outward pressure, whose time-integral is equal to the original momentum of the vortex, and a precisely equal pressure as it leaves the surface. Hence, in the case of myriads of vortex-rings bombarding such a plane surface, though no individual vortex-ring leaves the surface immediately after collision, for every vortex-ring that gets entangled in the condensed layer of drawn-out vortex-rings another will get free, so that in the statistics of vortex-impacts the pressure exerted by a gas composed of vortex-atoms is exactly the same as is given by the ordinary kinetic theory, which regards the atoms as hard elastic particles.

19. On the Figures of Equilibrium of a Rotating Mass of Fluid.

[From the *Proceedings of the Royal Society of Edinburgh*, Vol. xi. 1882, p. 610; reprinted in Thomson and Tait's *Natural Philosophy*, Ed. 2, 1883, § 778″.]

(*a*) The oblate spheroid of revolution is proved in Thomson and Tait's *Natural Philosophy* (first edition, § 776, and the Table of § 772) to be stable, if the condition of being an ellipsoid of revolution be imposed. It is obviously not stable for very great eccentricities without this double condition of being both a figure of revolution and ellipsoidal.

(*b*) If the condition of being a figure of revolution is imposed, without the condition of being an ellipsoid, there is, for large enough moment of momentum, an annular figure of equilibrium which is stable, and an ellipsoidal figure which is unstable. It is probable, that for moment of momentum greater than one definite limit and less than another, there is just one *annular* figure of equilibrium, consisting of a *single ring.*

(*c*) For sufficiently large moment of momentum it is certain that the liquid may be in equilibrium in the shape of two, three, four, or more separate rings, with its mass distributed among them in arbitrary portions, all rotating with one angular velocity, like parts of a rigid body. It does not seem probable that the kinetic equilibrium in any such case can be stable.

(*d*) The condition of being a figure of equilibrium being still imposed, the single-ring figure, when annular equilibrium is possible at all, is probably stable. It is certainly stable for very large values of the moment of momentum.

(*e*) On the other hand, let the condition of being ellipsoidal be imposed, but not the condition of being a figure of revolution.

It is proved in Thomson and Tait's *Natural Philosophy*, that whatever be the moment of momentum, there is one, and only one, revolutional figure of equilibrium.

I now* find that,

(1) The equilibrium in the revolutional figure is stable, or unstable, according as $f\left(=\frac{\sqrt{a^2-c^2}}{c}\right)$ is $<$ or $>$ 1·39457.

(2) When the moment of momentum is less than that which makes $f=1{\cdot}39457$ (or eccentricity $=\cdot81266$) for the revolutional figure, this figure is not only stable, but unique.

(3) When the moment of momentum is greater than that which makes $f=1{\cdot}39457$ for the revolutional figure, there is, besides the unstable revolutional figure, the Jacobian figure with three unequal axes, *which is always stable if the condition of being ellipsoidal is imposed.* But, as will be seen in (f) below, the Jacobian figure, without the constraint to ellipsoidal figure, is in some cases certainly unstable, though it seems probable that in other cases it is stable without any constraint.

(f) Referring to Thomson and Tait's *Natural Philosophy*, § 778, and choosing the case of a a great multiple of b, we see obviously that the excess of b above c must in this case be very small in comparison with c. Thus we have a very slender ellipsoid, long in the direction of a, and approximately a prolate figure of revolution relatively to this long a-axis, which, revolving with proper angular velocity round its shortest axis c, is a figure of equilibrium. The motion so constituted, which, without any constraint, is, in virtue of § 778, a configuration of minimum energy or of maximum energy, for the given moment of momentum, is a configuration of *minimum* energy for given moment of momentum, *subject to the condition that the shape is constrainedly an ellipsoid.* From this proposition, which is easily verified, in the light of § 778 of Thomson and Tait's *Natural Philosophy*, it follows that, with the ellipsoidal constraint, the equilibrium is stable. The revolutional ellipsoid of equilibrium, with the same moment of momentum,

* Proof of these results, (1), (2), and (3), will be found in the forthcoming new edition of Thomson and Tait's *Natural Philosophy*, Vol. I. Part II. [The proofs were not these given. The general problem has been analyzed in a comprehensive manner, on the basis of Lord Kelvin's methods, by H. Poincaré, in a classical paper, *Acta Math.*, VII. 1885. Cf. Lamb's *Hydrodynamics*, Ch. XII.]

is a very flat oblate spheroid; for it the energy is a minimax, because clearly it is the smallest energy that a revolutional ellipsoid with the same moment of momentum can have, but it is greater than the energy of the Jacobian figure with the same moment of momentum.

(g) If the condition of being ellipsoidal is removed and the liquid left perfectly free, it is clear that the slender Jacobian ellipsoid of (f) is not stable, because a deviation from ellipsoidal figure in the way of thinning it in the middle and thickening it towards its ends, or of thickening it in the middle and thinning it towards its ends, would with the same moment of momentum give less kinetic energy. With so great a moment of momentum as to give an exceedingly slender Jacobian ellipsoid, it is clear that another possible figure of equilibrium is two detached approximately spherical masses, rotating (as if parts of a solid) round an axis through their centre of inertia, and that this figure is stable. It is also clear that there may be an infinite number of such stable figures, with different proportions of the liquid in the two detached masses. With the same moment of momentum there are also configurations of equilibrium with the liquid in divers proportions in more than two detached approximately spherical masses.

(h) No configuration in more than two detached masses, has secular stability according to the definition of (k) below, and it is doubtful whether any of them, even if undisturbed by viscous influences, could have true kinetic stability; at all events, unless approaching to the case of the three material points proved stable by Gascheau (see Routh's *Rigid Dynamics*, § 475, p. 381).

(i) The transition from the stable kinetic equilibrium of a liquid mass in two equal or unequal portions, so far asunder that each is approximately spherical, but disturbed to slightly prolate figures (found by the well-known investigation of equilibrium tides, given in Thomson and Tait's *Natural Philosophy*, § 804), and to the more and more prolate figures which would result from subtraction of energy without change of moment of momentum, carried so far that the prolate figures, now not even approximately ellipsoidal, cease to be stable, is peculiarly interesting. We have a most interesting gap between the unstable Jacobian ellipsoid when too slender for stability, and the case of smallest moment of momentum consistent with stability in two equal detached portions. The consideration of how to fill up this gap with intermediate figures is a most attractive question, towards answering which I at present offer no contribution.

(*j*) When the energy with given moment of momentum is either a minimum or a maximum, the kinetic equilibrium is clearly stable, if the liquid is perfectly inviscid. It seems probable that it is essentially unstable when the energy is a minimax; but I do not know that this proposition has been ever proved.

(*k*) If there be any viscosity, however slight, in the liquid, or if there be any imperfectly elastic solid, however small, floating on it or sunk within it, the equilibrium in any case of energy either a minimax or a maximum cannot be secularly stable; and the only secularly stable configurations are those in which the energy is a minimum with given moment of momentum. It is not known for certain whether with given moment of momentum there can be more than one secularly stable configuration of equilibrium of a viscous fluid, in one continuous mass, but it seems to me probable that there is only one.

20. On the Motion of a Liquid within an Ellipsoidal Hollow.

[From the *Proceedings of the Royal Society of Edinburgh*, Vol. xiii. 1885, pp. 114, 370—378.]

I have only recently noticed the propositions regarding fluid motion within an ellipsoidal hollow which form the subject of the present communication, and which, though obvious enough and remarkably interesting, do not seem to have been previously discovered.

Preliminary.

I shall use the expression *homogeneous* rotation, or homogeneous molecular rotation, to designate the condition of a fluid in respect to rotation, when throughout it the amounts of its molecular rotation are the same and the axial lines parallel. This designation clearly includes the case of a rotating solid: but it is applicable of course to the more complex case of a fluid, in which irrotational motion is superimposed upon homogeneous rotation as of a solid. To illustrate the complex motion thus signified, consider the following three examples, of which (1) and (2) are included in (3):—

(1) Let a liquid kept in the shape of a figure of revolution, by a rigid containing vessel, be given in a state of homogeneous rotation round the axis of the figure. Let an impulsive rotation round a line perpendicular to this axis be given to the containing vessel. The instantaneous motion of the liquid, at the instant when the impulse is completed, consists of an irrotational motion superimposed on the given homogeneous rotational motion. The molecular rotation of the liquid does not generally remain homo-

geneous after the first instant. But I find it does continue homogeneous, however the containing vessel be moved, provided the shape be ellipsoidal; that is to say (for the present limited case), an ellipsoid of revolution whether prolate or oblate. The possible incident of the containing vessel being brought again to rest in any position after any motion round any succession of diameters perpendicular* to the axis of revolution is of course included.

(2) Given a rigid solid, with a hollow space of any shape not a figure of revolution, within it, full of liquid: solid and liquid all rotating homogeneously. Let the given rotation of the solid be impulsively brought to rest or to any other rotation, whether rotation with changed angular velocity round the same axis, or rotation round another axis. The instantaneous motion of the liquid, at the instant of the completion of the impulse, will be the resultant of the given homogeneous rotation, with an irrotational motion superimposed upon it; this irrotational motion of the liquid being the same as the motion which would be generated from rest, by giving to the solid (whether impulsively or gradually) an angular velocity the same as that which, compounded with the first given angular velocity, produced the second angular velocity to which we supposed the first angular velocity of the solid to be suddenly changed.

In this second example, as in the first, the molecular rotation does not generally continue to be homogeneous in the altered condition in which the solid and liquid do not rotate as if all solid; but it does continue to be homogeneous, if the shape of the hollow is ellipsoidal.

(3) Given a spherical shell full of homogeneously rotating liquid, or a hollow of any shape in a rotating solid full of liquid, rotating homogeneously with the solid. By impulsive pressure at the boundary of the liquid, supposed now to be perfectly yielding, generate any prescribed normal components of velocity in all parts of the boundary. The effect will be to generate throughout the liquid an irrotational motion, the same as would have been generated had the fluid been given at rest. The resultant

* Rotation of the containing vessel round the axis of figure has no effect on the liquid, and need not be included to complicate our considerations.

motion throughout will be the resultant of this irrotational motion, compounded with the given rotational motion. The irrotational motion in the case of the spherical hollow is of course easily calculated by the well-known spherical harmonic analysis for fluid motion. We consider here only the instantaneous motion, which exists at the instant when the impulse is completed. The infinitely more difficult problem of working out the consequences according to any prescribed conditions, as to force, or as to changing shape, for the boundary, we do not follow at present. It will be fully followed up for the case in which the boundary of the liquid is spherical or ellipsoidal to begin with, and is constrained to be always exactly ellipsoidal. It will be proved that in this case the molecular rotation of the fluid remains always homogeneous. We shall see in fact that the geometrical "strain" is essentially homogeneous throughout a liquid contained within a changing ellipsoidal boundary, provided that the motion of the fluid be either wholly irrotational, or be at any one instant homogeneously rotational. The homogeneousness of the geometrical strain being established, it follows from Helmholtz's fundamental principles of vortex motion, that the molecular rotation must continue homogeneous; its magnitude, when there is any stretching or contraction in the axial direction, varying inversely as the length of a line of the substance in this direction, and the axial direction varying so as to keep always along the same substantial line.

If there is the slightest deviation from exactness in the ellipsoidal figure, the homogeneousness of the rotation of the liquid is not maintained, and there is no limit to the amount of deviation from homogeneousness which may supervene in consequence of motions which may be given to the boundary, whether in the way of change of shape, or of motion without change of shape. Confining our attention for the present to motion of the boundary without change of shape, we find it interesting to remark that we may go on indefinitely increasing or indefinitely diminishing the energy of the fluid motion by properly arranged action in the way of moving the containing vessel. To continually increase the energy I believe the following rule may be correct, although I do not yet see a perfect proof of it. Suppose the containing vessel to be given at rest, and the liquid within it to have perfectly homogeneous rotation within the not exactly ellipsoidal hollow, watch it for a little time—it may begin to move or it may not.

If it does not begin to move of itself, give it a very slight motion of rotation round any axis. Generally it will begin to move of itself, but it will not do so if the interior fluid motion fulfils a definite condition of kinetic equilibrium, and therefore if you do not see the containing case beginning to move of itself you must set it in motion. When you see it in motion, act upon it with a couple in any direction to do some positive work upon it, and then suddenly stop it. Left to itself now, it will certainly begin to move of itself. When you see it moving again, again do some work upon it gradually, and stop its motion suddenly. Go on incessantly acting according to this rule. The positive work done gradually will exceed the work undone suddenly each time, or at all events on the aggregate of a large number of times of repetition of the operation. Thus on the whole you will increase the energy of the fluid motion without continually giving kinetic energy to the containing vessel, as might be the case if you continued always to apply a couple in such a direction as to do positive work. Thus by going on long enough operating in the manner described we can present the containing vessel at rest with the liquid moving inside it with any amount of kinetic energy we please.

A simpler rule suffices for diminishing the internal energy. Simply place the containing vessel on flexible imperfectly elastic supports, and leave it to itself, or leave it to itself immersed in a viscous fluid. Watch it for a while till you see it moving; or if you do not see it beginning to move of itself give it a slight motion, then leave it entirely to itself. It will never come to rest unless for an instant, and the internal energy will diminish asymptotically towards zero.

I now proceed to prove the propositions regarding fluid motion in an ellipsoidal hollow referred to above.

Irrotational motion of liquid in a rigid ellipsoidal shell.

Given the motion of the boundary: required the motion of the contained liquid.

Let ϖ, ρ, σ, be the component angular velocities of the shell, and let ϕ be the velocity potential of the corresponding determinate* motion of the internal fluid. The component linear velocities of a point (x, y, z) of the shell are

$$\rho z - \sigma y, \quad \sigma x - \varpi z, \quad \varpi y - \rho x \quad \text{..............(1)},$$

and the component linear velocities of (x, y, z) are

$$\frac{d\phi}{dx}, \quad \frac{d\phi}{dy}, \quad \frac{d\phi}{dz} \quad \text{....................(2)}.$$

If (x, y, z) be any point of the inner surface of the shell, the normal component of velocity (1) must be equal to the normal component of velocity (2); or in symbols

$$\left.\begin{aligned} (\rho z - \sigma y)\frac{px}{a^2} + (\sigma x - \varpi z)\frac{py}{b^2} + (\varpi y - \rho x)\frac{pz}{c^2} & \\ = \frac{d\phi}{dx}\frac{px}{a^2} + \frac{d\phi}{dy}\frac{py}{b^2} + \frac{d\phi}{dz}\frac{pz}{c^2} & \\ \text{where} \quad \frac{x^2}{a^2} + \frac{y^2}{b^2} + \frac{z^2}{c^2} = 1 & \end{aligned}\right\} \text{...(3)};$$

the axes of coordinates being taken as fixed in space and coincident with the axes of the ellipsoid at the instant considered; (a, b, c) being the three semi-axes of the ellipsoid; and p being the perpendicular from the centre to the plane touching the ellipsoid at (x, y, z). To satisfy this, assume

$$\phi = Ayz + Bzx + Cxy \quad \text{.................(4)},$$

and determine A, B, C, to fulfil the first of equations (3). We find that (3) is now satisfied by

$$\phi = \varpi\frac{b^2 - c^2}{b^2 + c^2}yz + \rho\frac{c^2 - a^2}{c^2 + a^2}zx + \sigma\frac{a^2 - b^2}{a^2 + b^2}xy\dagger \text{......(5)}.$$

It is important to remark that this expression for ϕ satisfies the first of equations (3) independently of the second, from which we infer that with the same angular velocity of rotation, the motion of any portion of the contained liquid is independent of the magnitude of the ellipsoidal body, and is determinate from the

* Wm. Thomson, "On the Visviva of a Liquid in Motion," *Camb. and Dub. Math. Journal*, 1849; or Thomson and Tait's *Natural Philosophy*, §§ 312 and 317, example (3).

† This solution is given in Lamb's *Fluid Motion* [ed. I.], § 102. [It had also been given by Ferrers, Beltrami, Bjerknes, and Maxwell. The subsequent analysis of the paths of the particles had been given substantially, probably by Ferrers, in *Cambridge Examination Papers*.]

ratios alone of the three semi-axes. From (4) we find for the velocity components:—

$$\left.\begin{aligned} u &= \rho\,\frac{c^2-a^2}{c^2+a^2}z+\sigma\,\frac{a^2-b^2}{a^2+b^2}y \\ v &= \sigma\,\frac{a^2-b^2}{a^2+b^2}x+\varpi\,\frac{b^2-c^2}{b^2+c^2}z \\ w &= \varpi\,\frac{b^2-c^2}{b^2+c^2}y+\rho\,\frac{c^2-a^2}{c^2+a^2}x \end{aligned}\right\} \quad \ldots\ldots\ldots\ldots\ldots(6),$$

which is the explicit solution of the problem, so far as concerns merely the absolute velocity at any point of the fluid, which is generally considered far enough in the solution of a hydrodynamical problem. But it would be interesting in every case, and it is easy in this case, to complete it up to the determination of the position of every particle of the liquid at any time, and we may therefore go on to do so. Relatively to the axes of the ellipsoid let $(\mathfrak{x}, \mathfrak{y}, \mathfrak{z})$ be the coordinates at time t, of any particular particle $\mathfrak{P}$, of the liquid. The component velocities $(d\mathfrak{x}/dt,\ d\mathfrak{y}/dt,\ d\mathfrak{z}/dt)$ of the particle $\mathfrak{P}$, relatively to the ellipsoid are equal to the differences between the components (u, v, w), of the absolute velocity of $\mathfrak{P}$, and the corresponding components of the absolute velocity of an ideal point (x, y, z) rigidly connected with the ellipsoid, and coincident with $(\mathfrak{x}, \mathfrak{y}, \mathfrak{z})$ at the time t. These last components are

$$\rho\mathfrak{z}-\sigma\mathfrak{y},\quad \sigma\mathfrak{x}-\varpi\mathfrak{z},\quad \varpi\mathfrak{y}-\rho\mathfrak{x} \quad \ldots\ldots\ldots\ldots\ldots(7).$$

Hence, and from (6), at the instant $(\mathfrak{x}, \mathfrak{y}, \mathfrak{z})$ coincide with (x, y, z) we have

$$\left.\begin{aligned} \frac{d\mathfrak{x}}{dt} &= a^2(\gamma\mathfrak{y}-\beta\mathfrak{z}) \\ \frac{d\mathfrak{y}}{dt} &= b^2(\alpha\mathfrak{z}-\gamma\mathfrak{x}) \\ \frac{d\mathfrak{z}}{dt} &= c^2(\beta\mathfrak{x}-\alpha\mathfrak{y}) \\ \text{where}\quad \alpha=\frac{2\varpi}{b^2+c^2},\quad &\beta=\frac{2\rho}{c^2+a^2},\quad \gamma=\frac{2\sigma}{a^2+b^2} \end{aligned}\right\} \quad \ldots\ldots(8).$$

These are linear differential equations of the first order for determining $(\mathfrak{x}, \mathfrak{y}, \mathfrak{z})$ in terms of t. Denoting d/dt by δ, we may write them as follows—

$$\left.\begin{aligned} \frac{\delta}{a^2}\mathfrak{x} - \gamma\mathfrak{y} + \beta\mathfrak{z} &= 0 \\ +\gamma\mathfrak{x} + \frac{\delta}{b^2}\mathfrak{y} - \alpha\mathfrak{z} &= 0 \\ -\beta\mathfrak{x} + \alpha\mathfrak{y} + \frac{\delta}{c^2}\mathfrak{z} &= 0 \end{aligned}\right\} \quad \ldots\ldots\ldots\ldots\ldots\ldots\ldots(9).$$

Operating now on this in the usual manner we find

$$\begin{vmatrix} \frac{\delta}{a^2}, & -\gamma, & \beta \\ \gamma, & \frac{\delta}{b^2}, & -\alpha \\ -\beta, & \alpha, & \frac{\delta}{c^2} \end{vmatrix} = 0 \quad \ldots\ldots\ldots\ldots\ldots(10);$$

whence by expanding the determinant and removing the superfluous factor δ, we have

$$\frac{\delta^2}{a^2b^2c^2} + \frac{\alpha^2}{a^2} + \frac{\beta^2}{b^2} + \frac{\gamma^2}{c^2} = 0 \quad \ldots\ldots\ldots\ldots\ldots\ldots(11),$$

which gives $$\left.\begin{aligned} \delta = \iota\, abc\sqrt{\left(\frac{\alpha^2}{a^2} + \frac{\beta^2}{b^2} + \frac{\gamma^2}{c^2}\right)} \end{aligned}\right\} \quad \ldots\ldots\ldots\ldots(12);$$
where ι denotes $\sqrt{-1}$.

And from the second and third of (9) we have

$$\frac{\mathfrak{x}}{\frac{\delta^2}{b^2c^2} + \alpha^2} = \frac{\mathfrak{y}}{\alpha\beta - \gamma\frac{\delta}{c^2}} = \frac{\mathfrak{z}}{\gamma\alpha + \beta\frac{\delta}{b^2}} \quad \ldots\ldots\ldots\ldots(13),$$

which gives

$$\mathfrak{y} = \mathfrak{x}\,\frac{\alpha\beta - \gamma\frac{\delta}{c^2}}{\frac{\delta^2}{b^2c^2} + \alpha^2}, \quad \text{and} \quad \mathfrak{z} = \mathfrak{x}\,\frac{\gamma\alpha + \beta\frac{\delta}{b^2}}{\frac{\delta^2}{b^2c^2} + \alpha^2} \quad \ldots\ldots\ldots\ldots\ldots(14).$$

In virtue of (12) we may take as the solution for any one of the coordinates, $\mathfrak{x}$ for example, as follows—

$$\left.\begin{aligned} \mathfrak{x} &= A\cos\omega t \\ \text{where}\quad \omega &= 2abc\sqrt{\left[\left(\frac{\varpi/a}{b^2+c^2}\right)^2 + \left(\frac{\rho/b}{c^2+a^2}\right)^2 + \left(\frac{\sigma/c}{a^2+b^2}\right)^2\right]} \end{aligned}\right\} \ldots(15);$$

and from this (14) gives

$$\left.\begin{aligned} \mathfrak{y} &= A\,\frac{\alpha\beta\cos\omega t + \dfrac{\gamma}{c^2}\,\omega\sin\omega t}{\alpha^2 - \dfrac{\omega^2}{b^2c^2}} \\ \mathfrak{z} &= A\,\frac{\gamma\alpha\cos\omega t - \dfrac{\beta}{b^2}\,\omega\sin\omega t}{\alpha^2 - \dfrac{\omega^2}{b^2c^2}} \end{aligned}\right\} \quad \text{............(16).}$$

These equations give explicitly the position of any chosen particle at any time, and of course it would be easy to find from them what the path is; but it is easier to do this from the unintegrated equations (8). Multiplying the first of these by α/a^2, the second by β/b^2, and the third by γ/c^2, and adding, we find

$$\frac{\alpha}{a^2}\frac{d\mathfrak{x}}{dt} + \frac{\beta}{b^2}\frac{d\mathfrak{y}}{dt} + \frac{\gamma}{c^2}\frac{d\mathfrak{z}}{dt} = 0 \quad \text{.....................(17);}$$

which proves that the orbit lies in the plane

$$\frac{\alpha}{a^2}\,\mathfrak{x} + \frac{\beta}{b^2}\,\mathfrak{y} + \frac{\gamma}{c^2}\,\mathfrak{z} = H \quad \text{..................(18),}$$

where H denotes a constant.

Again multiplying the first of equations (8) by $\mathfrak{x}/a^2$, the second by $\mathfrak{y}/b^2$, and the third by $\mathfrak{z}/c^2$ and adding, we find

$$\frac{\mathfrak{x}}{a^2}\frac{d\mathfrak{x}}{dt} + \frac{\mathfrak{y}}{b^2}\frac{d\mathfrak{y}}{dt} + \frac{\mathfrak{z}}{c^2}\frac{d\mathfrak{z}}{dt} = 0 \quad \text{..................(19),}$$

and integrating this we have

$$\frac{\mathfrak{x}^2}{a^2} + \frac{\mathfrak{y}^2}{b^2} + \frac{\mathfrak{z}^2}{c^2} = K \quad \text{......................(20),}$$

where K denotes a constant.

This proves that the orbit lies on the ellipsoid (20); and we conclude that the orbit is the ellipse in which this ellipsoid is cut by the plane (18).

Going back now to the explicit fully integrated solution (15) and (16), we see that a particle of the fluid describes, relatively to the moving solid in which the fluid is contained, the ellipse specified by (18) and (20), according to the law of a single particle describing an ellipse under the influence of a force towards a fixed centre varying in simple proportion to distance from the centre.

Now the period of revolution of the containing shell round its axis of rotation (ϖ, ρ, σ) is $2\pi/\epsilon$ where

$$\epsilon = \sqrt{(\varpi^2 + \rho^2 + \sigma^2)},$$

which is easily seen to be less than $2\pi/\omega$ [the value of ω being given by (15) above]. Hence considering the shell and contained liquid at any instant, and again at the later instant when the shell is again in the same position after a single complete revolution round the axis of its rotation, we see that, relatively to the shell, the liquid will have performed less than a complete period of its retrograde revolution by the difference $(2\pi/\omega - 2\pi/\epsilon)$; or by the fraction $(1 - \omega/\epsilon)$ of the period of the fluid relatively to the shell. In the extreme case of $a = b = c$ (the ellipsoid a sphere), $\omega = \epsilon$ and the retrograde motion of the fluid relatively to the shell is one complete revolution, in the period of the forward revolution of the shell: that is to say, the fluid is perfectly left behind, and remains unmoved while the shell turns. In the other extreme case of any one or any two of the quantities a, b, c being infinitely small, ω is infinitely small: that is to say, the fluid makes an infinitely small fraction of its retrograde revolution during the time of one turn of the shell in the direction which we are calling forward. It must not from this be inferred that the fluid moves very nearly, as if solid, with the shell. On the contrary, it experiences large distortion even in the first complete turn of the shell, and largely increasing to a maximum in the course of the first quarter period of the liquid relatively to the shell.

21. On the Stability and Small Oscillation of a Perfect Liquid full of Nearly Straight Coreless Vortices.

[Extract from a Letter to Professor G. F. FitzGerald. From the *Proceedings of the Royal Irish Academy*; read November 30, 1889.]

"I HAVE quite confirmed one thing I was going to write to you (in continuation with my letter of October 26), viz. that rotational vortex cores must be absolutely discarded, and we must have nothing but irrotational revolution and vacuous cores. So not to speak of my little piece of coreless vortex work ('Vibrations of a Columnar Vortex,' *Proc. R.S.E.*, March 1, 1880), Hicks' Paper, 'On the Steady Motion and small Vibrations of a Hollow Vortex,' *Transactions Roy. Society*, 1884), will be the beginning of the Vortex Theory of ether and matter, if it is ever to be a theory. Steady motion, with crossing lines of vortex column, is impossible with rotational cores, but is possible with vacuous cores and purely irrotational circulations around them. The accompanying diagram (fig. 1) helps to explain by an illustration. It shows the shape of an infinitely long cylindrical vacuous vortex column as disturbed by a rigid tore*, held fixed in a plane perpendicular to the axis of the column, and having irrotational circulation through itself. The column represents vacuum; the space on each side, liquid; and the two black circles section of the tore. The curves representing the boundary of the vortex are calculated to give uniform resultant fluid velocity over the whole surface of the hollow core. This velocity is the resultant of the velocities due to the circulation around the vacuous core, and to circulation through the tore. The former is rigorously in inverse proportion to distance from the axis of the vacuous column. The latter is approximately parallel to this axis, and in inverse proportion to the cube of the distance

* Or circular ring of circular cross section, like an anchor ring.

from the circular axis of the tore. The equation of the curve is easily written down. It is calculated for the case in which the velocity at the centre of the tore (were there no vacuity) due to circulation through the tore, is equal to $\sqrt{8}/3$ of the velocity at the boundary of the vacuous column at great enough distances on either side to be undisturbed by the circulation through the tore. This makes the maximum diameter of the vacuous core three times the undisturbed diameter. If the velocity-component, due to the disturbance, is small in comparison with the surface-velocity of the vortex column, the swelling will, of course, be but a small fraction of the radius of the undisturbed column. Try to get a corresponding problem of steady motion *with rotational core,* and you will see why I now abjure rotational motion, and definitively adopt vacuum for all cores.

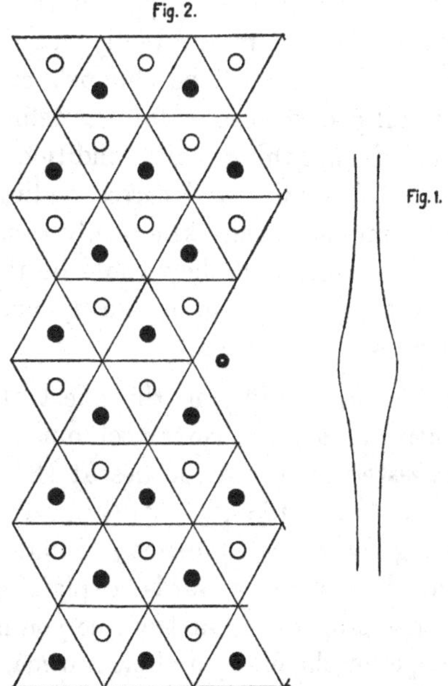

"Now, consider a uniform distribution throughout space, of vacuous vortex columns; represented by section perpendicular to the length in fig. 2; rings and discs, each representing vacuum, but with opposite circulations around them. But, instead of the proportions of the diagram, let the distance from each column to its three nearest neighbours be enormously great in comparison with the diameters of the columns. I think I can now prove that this arrangement of vortices is stable; but its quasi-rigidity relatively to two-dimensional motion, without change of volume of the cores, is exceedingly small, and the corresponding laminar wave-motion exceedingly sluggish in comparison with the tensile quasi-rigidity, and corresponding wave-velocity, which we should find by considering laminar motion in planes parallel to the plane of the diagram.

"Now, imagine a very great number of planes in all directions—as many within an angle of 1° of any one plane, as within 1° of any other. And let there be a distribution of straight vortex vacuous cores (as represented in fig. 2), perpendicular to every one of these planes. The cores being thin enough, they may be placed along straight lines, no one of which intersects any other. The mutual influence of the vortices will produce disturbances from the straight lines in which we supposed them given, and slight swellings and deviations from exactly circular figure, in their cross sections; and there will be sluggish motions of the cores, unless they are all placed so as to fulfil a definite condition giving steady motion. Even if this definite condition is not exactly fulfilled, the tensile quasi-rigidity, and corresponding velocity of laminar waves of the medium, thus kinetically constituted, will certainly differ but little from what they would be if the vortices were arranged so as to give absolutely steady motion for the equilibrium condition of the medium.

"I have been anxiously considering the effect of free vortex rings with vacuous cores among the vortex columns of this tensile vortex ether, as suggested for cored vortices at the end of your communication of April 26, 1889, to *Nature*. It will be an exceedingly interesting dynamical question; though it seems to promise at present but little towards explaining universal gravitation or any other property of matter; so you may imagine I do not see much hope for chemistry and electro-magnetism."

22. TOWARDS THE EFFICIENCY OF SAILS, WINDMILLS, SCREW-PROPELLERS IN WATER AND AIR, AND AEROPLANES.

[From *Nature*, Vol. L. 1894, p. 425.]

THE discussion of this day week, on flying machines, in the British Association was not, for want of time, carried so far as to prove from the numerical results of observation put before the meeting by Mr Maxim, that the resistance of the air against a thin stiff plane caused to move at sixty miles an hour through it, in a direction inclined to the plane at a slope of about one in eight, was found to be about fifty-three times as great as the estimate given by the old "theoretical" (!) formula, and something like five or ten times* that calculated from a formula written on the black-board by Lord Rayleigh, as from a previous communication to the British Association at its Glasgow meeting in 1876.

I had always felt that there was no validity, even for rough or probable estimates, in any of the "theoretical" investigations hitherto published: but how wildly they all fall short of the truth I did not know until I have had opportunity in the last few days, *procul negotiis,* to examine some of the observational results which Maxim gave us in the introduction to his paper. On the other hand, I have never doubted but that the true theory was to be found in what I was taught conversationally by William Froude twenty years ago, which, though I do not know of its having been anywhere published hitherto, is clearly and tersely expressed in the following sentence which I quote from a type-written copy, kindly given me by Mr Maxim, of his paper of last week:—

"The advantages arising from driving the aeroplanes on to new air, the inertia of which has not been disturbed, is clearly shown in these experiments."

* [Substitute 4·8 times; see footnote, p. 219 *infra.*]

Founding on this principle, I have at last, I believe, succeeded in calculating, with some approach to accuracy, the force required to keep a long, narrow, rectangular plane moving through the air with a given constant velocity, V, in a direction perpendicular to its length, l, and inclined at any small angle, i, to its breadth, a. In a paper, which I hope to be able to communicate to the *Philosophical Magazine* in time for publication in its next October number, I intend to give the investigation, including consideration of "skin-resistance" and proof that it is of comparatively small importance when i is not much more than 1/10, or 1/20, of a radian, and the "plane" is of some practically smooth, real, solid material. In the meantime, here is the result, with skin-resistance neglected:—The resultant force (perpendicular, therefore, to the plane) is $\frac{1}{2}\pi V^2 \sin i \cos i la$ *; which is $\pi \cos i/\sin i$ times (or for the case of $\sin i = 1/32$, one hundred times), the old miscalled "theoretical" result.

* [In *Nature* the numerical factor was 2 instead of $\frac{1}{2}$.]

23. On the Resistance of a Fluid to a Plane kept moving uniformly in a direction inclined to it at a small angle.

[From the *Philosophical Magazine*, Vol. xxxviii. 1894, pp. 409—413.]

1. Let q be the velocity; i its inclination to the plane; and u, v its components in and perpendicular to the plane. We have

$$u = q \cos i, \quad v = q \sin i.$$

2. Suppose now the moving body to be not an ideal infinitely thin plane, but a disc of finite thickness very small in comparison with its least diameter, and having its edges everywhere smoothly rounded. If the fluid is inviscid and incompressible, and the boundary containing it perfectly unyielding, the motion produced in the fluid from rest, by any motion given to the disc, is determinately the unique motion of which the energy is less than that of any other motion possible to the fluid with the given motion of the disc. We suppose the disc to be *very* thin, and therefore the profile-curvature at every point of its edge to be *very* great: there is no limit to the thinness at which the proposition could cease to be true; so it still holds in the ideal case of an infinitely thin disc, when the fluid and its boundary fulfil the ideal conditions of the enunciation.

3. But in nature every fluid has some degree of viscous resistance to change of shape; and any viscosity however small (even with ideally perfect incompressibility of the fluid and unyieldingness of the boundary) would prevent the infinitely great velocities at the edge of the disc which the unique minimum-energy solution gives when the disc is infinitely thin; and would originate so great a disturbance in the motion of the fluid that the resistance to the motion of the disc would probably be very nearly the same whatever the actual value of the viscosity, if not too great in comparison with the velocity of the disc multiplied

by the least radius of curvature of the boundary of its area. No approach, however, has hitherto been made towards a complete mathematical solution of any case of this problem, or indeed of the motion of a body of any shape through a viscous fluid, except when, as in Stokes's original solutions for the globe and circular cylinder, the motion is so slow that its configuration is the same as it would be if it were infinitely slow, and when therefore* the velocity of the fluid at every point is equal to, and in the same direction as, the infinitesimal static displacement of an elastic solid when a rigid body imbedded in it is held in a position infinitesimally displaced from its position of equilibrium, in the manner translationally and rotationally corresponding to the translational and rotational velocity given to the rigid body in the fluid.

4. It has occurred to me, guided by the teaching of William Froude regarding the continued communication of momentum to a fluid by the application of force to keep a solid moving with uniform translational velocity through it, that an approximate determination of the resistance, which is the subject of the present communication, may probably be found by the following method, with result expressed in § 9, which I venture to give as a *guess*, and not as a satisfactory mathematical investigation.

5. Considering a disc of finite thickness, however small, moving in an inviscid incompressible liquid within an unyielding boundary, and, for a moment, thinking only of the u-component of the motion, according to the notation of § 1, let E and E' denote the front and the rear parts of the edge, respectively. Imagine now instead of the real motion of the unvarying solid disc through the fluid, that the disc grows all over E, by rigidification and accretion of the fluid in front of it, and melts away from E' by liquefaction of the solid. In an infinitesimal time δt, the extent of the accretion in front of E will be $u\delta t$. Now if the v-component of the motion of the disc is maintained without diminution during this accretion, a force, F, equal to $(I'-I)/\delta t$, must be applied from without, perpendicular to the disc; I denoting the impulsive force which would be required to give the v-component velocity to the unaugmented disc, and I' that required to give the same velocity to the augmented disc. The point of application of the force

* The equations for the *steady* infinitely slow motion of a viscous fluid are identical with those for the equilibrium of an elastic solid. See *Mathematical and Physical Papers* (Sir W. Thomson), Vol. III. Art. cxix. §§ 17, 18.

$(I'-I)/\delta t$ must be that of the resultant of impulses I' and $-I$, applied at the hydraulic centres of inertia* of the augmented disc and the unaugmented disc respectively.

6. Sudden cessation of the rigidity by liquefaction of any portion (finite or infinitely small) of matter of the disc at E' requires no instantaneous application of force, to prevent change of the v-motion of the residual solid. The continued gradual liquefaction which we are supposing performed, leaves a Helmholtz "vortex sheet" of finite slip growing out in the liquid, behind E', the evolutions and contortions of which are not easily followed in imagination. This sheet is in the form of a pocket of which the lip remains always attached to the solid disc. The space enclosed between it and the disc is filled by the liquid which was solid. It grows always longer and longer by gain of liquid from the melting solid at E' in front of it, and probably also by its rear extending farther and farther, far away in the wake of the disc.

7. Suppose now that, after having been performed during a certain time T, the ideal processes of §§ 5, 6 are discontinued, and the resulting solid disc, equal and similar to the original disc, but carried in the u-direction through a space equal to uT, is left with simply its v-motion through the fluid maintained. The pocket of liquefied solid will be left farther and farther behind the disc. Its mouth, still always stopped by the solid, will shrink from its original area which was the whole of E'; and will become always smaller and smaller, but not infinitely small in any finite time. The neck of the pocket in the wake of the disc will become narrower and narrower, and the whole pocket will be drawn out longer and longer behind; but, through all time, the fluid which was solid will remain separated by a surface of finite slip, or Helmholtz "vortex sheet," from the surrounding fluid, except over the ever diminishing area of the disc, which stops the mouth of the pocket. The motion of the fluid is irrotational outside the pocket, and rotational within it. To keep the solid disc moving with its v-motion constant, and with no other motion whether rotational or translational, it is necessary to apply force to it. But this force becomes less and less, and approximates to zero, as the vortex-

* I call the "hydraulic centre of inertia" of a massless rigid disc immersed in liquid the point at which it must be struck perpendicularly by an impulse, to give it a simple translational motion.

trail becomes finer and finer; and the motion of the fluid in the neighbourhood of the disc approximates more and more nearly to perfect agreement with the unique irrotational motion due to v-motion of the solid through the fluid.

8. So far we have, in §§ 5, 6, 7, been on sure ground, and every statement is rigorously true, not only for a "disc" of any shape of boundary and of any thickness however small, but also for a solid of any shape, dealt with according to § 5, provided only that the fluid is inviscid and incompressible, and its boundary unyielding. My hypothesis, or "guess" (§ 4), which forms the subject of the present paper, is that default from infinitely perfect fulfilment of all these three conditions would, for an infinitely thin disc kept moving with uniform translational motion (u, v, § 1), require the continued application to it of force determined in magnitude and position by § 5; *provided v be very small in comparison with u.*

9. The result is worked out with great ease for the case of a rectangular disc of which the length, l, is very great in comparison with the breadth, a. For this case, by the well-known hydrokinetics of an ellipsoid or elliptic cylinder moving translationally in an inviscid incompressible fluid of unit density, we have

$$I = \tfrac{1}{4}\pi a^2 lv;$$

and, still using the notation of § 5,

$$I' = \tfrac{1}{4}\pi (a + u\delta t)^2 lv.$$

Hence

$$F = \tfrac{1}{2}\pi aluv;$$

and the distance of the point of application of this force from the middle line of the rectangle is $\tfrac{1}{4}a$.

Comparison of this hypothetical result, with observation, in respect both to the magnitude of the force and its point of application, will, I hope, form the subject of a future communication.

24. On the Motion of a Heterogeneous Liquid, commencing from Rest with a given Motion of its Boundary.

[From the *Proceedings of the Royal Society of Edinburgh*, Vol. xxi. 1896, pp. 119—122.]

I use the word "liquid" for brevity to denote an incompressible fluid, viscid or inviscid, but inviscid unless the contrary is expressly stated. A finite portion of liquid, viscid or inviscid, being given at rest, within a bounding vessel of any shape, whether simply or multiply continuous; let any motion be *suddenly* produced in some part of the boundary, or throughout the boundary, subject only to the enforced condition of unchanging volume. Every particle of the liquid will instantaneously commence moving with the determinate velocity and in the determinate direction, such that the kinetic energy of the whole is less than that of any other motion which the liquid could have with the given motion of its boundary*. This proposition is also true for an incompressible elastic solid, manifestly (and for the ideal "ether" of *Proc. R.S.E.*, March 7 1890; and Art. xcix. vol. iii. of my *Collected Mathematical and Physical Papers*). The truth of the proposition for the case of a viscous liquid is very important in practical hydraulics. As an example of its application to inviscid and viscous fluid and to elastic solid, consider an elastic jelly standing in an open rigid mould, and equal bulks of water and of an inviscid

* *Cambridge and Dublin Mathematical Journal*, Feb. 1849. This is only a particular case of a general kinetic theorem for any material system whatever, communicated to the Royal Society, Edinburgh, April 6, 1863, without proof (*Proceedings*, 1862–63, p. 114), and proved in Thomson and Tait's *Natural Philosophy*, § 317, with several examples. Mutual forces between the containing vessel and the liquid or elastic solid, such as are called into play by viscosity, elasticity, hesivity (or resistance to sliding between solid and solid), cannot modify the conclusion, and do not enter into the equations used in the demonstration.

liquid in two vessels equal and similar to it. Give equal sudden motions to the three containing vessels: the instantaneous motions of the three contained substances will be the same. Take, as a particular case, a figure of revolution with its axis vertical for the containing vessel; and let the given motion be rotation round this axis suddenly commenced and afterwards maintained with uniform angular velocity. The initial kinetic energy will be zero for each of the three substances. The inviscid liquid will remain for ever at rest; the water will acquire motion according to the Fourier law of diffusion of which we know something for this case by observation of the result of giving an approximately uniform angular motion round the vertical axis to a cup of tea initially at rest. The jelly will acquire laminar wave motion proceeding inwards from the boundary. But in the present communication we confine our attention to the case of inviscid liquid.

The now well-known solution* of the minimum problem thus presented, when the bounding surface is simply continuous, is, simply:—that the initial motion of the liquid is irrotational. That the *initial* motion *must be irrotational*† is indeed obvious, when we consider that the impulsive pressure by which any portion of the liquid is set in motion is everywhere perpendicular to the interface between it and the contiguous matter around it, and therefore the initial moment of momentum round any diameter of every spherical portion, large or small, is zero. But that irrotationality of the motion of every spherical portion of the liquid suffices to determine the motion within a simply continuous boundary having any stated motion, is not obvious without mathematical investigation.

Whether the boundary is simply continuous, or multiply continuous, irrotationality suffices to determine the motion produced, as we now suppose it to be produced, from rest by a given motion of the boundary.

Now in a homogeneous liquid acted on by no bodily force, or only by such force (gravity, for example) as could not move it when its boundary is fixed, the motion started from rest by any movement of the boundary remains always irrotational, as we

* Thomson and Tait's *Natural Philosophy*, § 312.

† That is to say, motion such that the moment of momentum of every spherical portion, large or small, is zero round every diameter.

know from elementary hydrokinetics. Hence, if at any time the boundary is suddenly or gradually brought to rest, the motion of every particle of the liquid is brought to rest at the same instant. But it is not so with a heterogeneous liquid. Of the following conclusions Nos. (1), (2), (3) need no proof. To prove No. (4), remark that as long as there is any motion of the heterogeneous liquid within the imperfectly elastic vessel the liquid must be losing energy; and the energy cannot become infinitely small with any finite spherical portion of the liquid homogeneous.

(1) The initial motion of a heterogeneous liquid is irrotational only at the first instant after being *quite suddenly* started from rest by motion of its boundary. Whatever motion be subsequently given to the boundary the motion of the liquid is never again irrotational. Hence

(2) If the boundary be suddenly brought to rest at any time, the liquid, unless homogeneous throughout, is not thereby brought to rest; and it would go on for ever with undiminished energy if the liquid were perfectly inviscid and the boundary absolutely fixed. The ultimate condition of the liquid, if there is no *positive* surface tension in the interfaces between heterogeneous portions, is an infinitely fine mixture of the heterogeneous parts*. And if there were no gravity or other bodily force acting on the liquid, the density would ultimately become uniform throughout. Take, for example, a corked bottle half full of water or other liquid with air above it given at rest. Move the bottle and bring it to rest again: the liquid will remain shaking for some time. An ordinary non-scientific person will scarcely thank us for this result of our mathematical theory. But, when we tell him that if air and the liquid were both perfectly fluid (that is to say perfectly free from viscosity), the well-known shaking of the liquid surface would, after a little time, give rise to spherules tossed up from the main body of the liquid; and that the shaking of the liquid, left to itself in the bottle supposed perfectly rigid, will end in spindrift of spherules which would be infinitely fine if the capillary tension of the interface between liquid and air were infinitely small, he

* *Popular Lectures and Addresses*, by Lord Kelvin, vol. I. pp. 19, 20, and 53, 54. See also *Philosophical Magazine*, 1887, second half-year: "On the formation of coreless vortices by the motion of a solid through an inviscid incompressible fluid"; "On the stability of steady and of periodic fluid motion"; "On maximum and minimum energy in vortex motion." [*Supra*, pp. 149, 172, and *infra*

may be incredulous unless he tends to have faith in all assertions made in the name of science.

(3) If the boundary is an enclosing vessel of any real material (and therefore neither perfectly rigid nor perfectly elastic), and if it is laid on a table and left to itself, under the influence of gravity, the liquid, supposed perfectly inviscid, will lose energy continually by generation of heat in the containing vessel, and will come asymptotically to rest in the configuration of stable equilibrium with surfaces of equal density horizontal and increasing density downwards.

(4) With other conditions as in (3), but no gravity, the ultimate configuration of rest will be infinitely fine mixture (probably, I think, of equal density throughout). Consider, for example, two homogeneous liquids of different densities filling the closed vessel, or a single homogeneous liquid not filling it. As an illustration, take a bottle half full of water, and shake it violently. Observe how you get the whole bottle full of a mixture of fine bubbles of air, nearly homogeneous throughout. Think what the result would be if there were no gravity, and if the water and air were inviscid and the bottle shaken as gently as you please; and if there were perfect vacuum in place of the air; or, if for air were substituted any liquid of density different from that of water.

25. ON THE DOCTRINE OF DISCONTINUITY OF FLUID MOTION, IN CONNECTION WITH THE RESISTANCE AGAINST A SOLID MOVING THROUGH A FLUID *

[From *Nature*, Vol. L. 1894, pp. 524, 549, 573, 597.]

I.

§ 1. THE doctrine that "discontinuity," that is to say finite difference of velocity on two sides of a surface in a fluid, would be produced if an inviscid incompressible fluid were caused to flow past a sharp edge of a rigid solid *with no vacant space between fluid and solid* was, I believe, first given by Stokes in 1847 †.

It is inconsistent with the now well-known dynamical theorem that an incompressible inviscid fluid initially at rest, and set in motion by pressure applied to its boundary, acquires *the* unique distribution of motion throughout its mass, of which the kinetic energy is less than that of any other motion of the fluid with the same motion of its boundary.

§ 2. The reason assigned for the formation of a surface of finite slip between fluid and fluid was the infinitely great velocity of the fluid *at* the edge, and the corresponding negative-infinite pressure, implied by the unique solution, *unless the fluid is allowed to separate itself from contact with the solid.* This an inviscid incompressible fluid certainly would do, unless the pressure of the fluid were infinitely great everywhere except at the edge. In nature the tendency to very great negative pressure arising from

* [These communications formed the subject of a prolonged playful controversy between Lord Kelvin and his intimate friend Sir George Stokes, in a series of letters which have been preserved.]

† *Collected Papers*, Vol. I. pp. 310, 311.

greatness of velocity of a fluid flowing round a corner is always obviated by each one of three defalcations from our ideal:—

(I) Viscosity of the fluid, preventing the exceeding greatness of the velocity.

(II) Compressibility of the fluid.

(III) Yieldingness of the outer boundary of the fluid.

§ 3. Defalcation (I) is in many practical cases largely operative when air is the fluid; but (II) is also largely operative in some very interesting cases, such as the *whistling* of a strong wind blowing round a sharp corner or through a chink; the blowing against the sharp edge in the embouchure of an organ-pipe, and in the mouthpiece of a flageolet or of a small "whistle"; and the blowing across the end of a tube or a hole in the side of a tube, to cause a key or a flute to sound.

§ 4. Defalcation (III) is largely operative, and (II) but little, in many practical cases of most common occurrence in the flow of water. It is probable that much of the foam seen near the sides and in the wake of a screw steamer going at a high speed through glassy-calm water, is due to "vacuum" behind edges and roughnesses causing dissolved air to be extracted from the water. A stiff circular disc of 10-inch diameter, and 1/10 of an inch thick in its middle, shaped truly to the figure of an oblate ellipsoid of revolution, would cause a vacuum* to be formed all round its edge, if moved at even so small a velocity as 1 foot per second under water of any depth less than 63 feet; if water were inviscid: and at greater depths the motion would, on the same supposition, be wholly continuous, with no vacuum, and would be exactly in accordance with the unique minimum energy solution†.

* Single word to denote space vacated by water.

† From the elementary hydrokinetics of the motion of an ellipsoid through an inviscid incompressible fluid, originated by Green, who first gave the solution for the case of translational motion of the ellipsoid, we know that, if θ denotes the angle between the normal to the surface at any point and the axis of an oblate ellipsoid of revolution, of which the equatorial and polar semiaxes are a, b, the velocity of the fluid flowing over this point of the surface is

$$\frac{(a^2-b^2)\,V\sin\theta}{b\left\{\dfrac{a^2}{\sqrt{(a^2-b^2)}}\sin^{-1}\dfrac{\sqrt{(a^2-b^2)}}{a}-b\right\}},$$

if the velocity of the fluid at great distances from the solid is V, and in parallel lines, and the solid is held fixed in the fluid, with its axis parallel to these lines.

While the velocity of the fluid across the equator is 63·7 feet per second, the velocity across each of the two parallel circles whose radii are 4·218 inches (the radius of the equator being 5 inches) is only 1 foot per second.

§ 5. The exceedingly rapid change of shape of the fluid flowing across the equatorial zone between these circles, with velocity at the surface augmenting from 1 foot per second to 63·7 feet per second in advancing over a distance of less than ·85 of an inch of the surface from one of the small circles to the equator, and diminishing again from 63·7 to 1 from the equator to the other parallel, in a small fraction of a second of time would, if the fluid is water or any other real liquid, give rise, through viscosity, to forces greatly diminishing the maximum velocity, and causing, through fluid pressure, the motion of the water to differ greatly from that of the minimum-energy solution; not only near the equator, or in its wake, or over the rear side of the disc, but over all the front side also, though no doubt much more on the rear side and in the wake, than on the front side and in the fluid before it.

The viscosity would also, at less depths than 63 feet, have great effect in keeping down the maximum velocity; and it is possible that even at 10 or 20 feet a greater velocity than 1 foot per second might be required to make vacuum round the equator of our disc of 10 inches diameter and the 1/2000 of an inch radius of curvature which its elliptic meridional section gives it. But it seems quite certain that there must be much forming of vacuum, and consequent extraction of air and rising of bubbles to the surface, from the somewhat sharp corners and roughnesses, of iron, in the hull of an ordinary iron sailing ship or steamer, going through the water at twelve knots (that is, 20 ft. per second).

II.

§ 6. In every case in which vacuum is formed at an edge of a solid moving in an inviscid incompressible fluid, under pressure constant at all infinitely great distances from the solid, a succession

Taking $a=100b$ in this formula, we reduce it to $200/\pi.(V\sin\theta)$ approximately within 1 per cent.; and taking $\sin\theta=1$, and $V=1$ foot per second, we find 63·7 feet per second for the velocity across the equator. Hence the gravitational head corresponding to the "negative-pressure" is $(63{\cdot}7^2-1^2)/64{\cdot}4$, or very approximately 63 feet, which proves the statement in the text.

of finite individual vortices is sent from the edge into the liquid, *and the motion is essentially unsteady.* Each individual vortex has a finite endless vacuum for its core instead of the rotationally moving ring of fluid of the Helmholtz vortex ring. But it should be noticed that it would not be rings of vacuum, but bubbles, that would in many cases be first detached from the solid; that by the tumultuous collapse of bubbles they become rings; and that the case in which the collapse of a bubble, in our ideal fluid, could be completed to an annulment of volume, is of necessity infinitely rare; and that the case in which, when a bubble becomes a ring by the meeting of two points of collapsing boundary, there is exactly no circulation through the aperture, is infinitely rare*.

§ 7. In the case of our circular disc, it would be circular vortex rings that, if the water were inviscid, would be shed off from its edge when the depth is less than 63 feet. If the depth is very little less than 63 feet these rings would be exceedingly fine, and would follow one another at exceedingly short intervals of time. Thus quite close to the edge there would be something somewhat like to Stokes' "rift," but with a rapid succession of vacuum rollers, as it were, and *no slipping* between the portions of the fluid on its two sides.

§ 8. At greater depths than 63 feet, if the water had absolutely no viscosity†, the motion would be continuous and irrotational, as described in § 4, text and foot-note: but any degree of viscosity, however slight, would, if the edge were infinitely sharp (instead of having a radius of curvature of 1/2000 of an inch, as has our supposed disc), give rise to a state of motion in its neighbourhood somewhat like to Stokes' rift‡, "a

* The whole subject of the motion of an incompressible inviscid fluid, with vacuum on the other side of the whole or any part of its boundary, is of surpassing interest. Consider for example an open fixed basin, with water poured into it and left not quite at rest, under the influence of gravity swinging slightly from side to side; let us suppose for example the water perfectly inviscid, and vapourless, but, may be, either cohesional or cohesionless; there being perfect vacuum over all its free surface. Very soon it will certainly throw up a drop somewhere: and before very long it will become covered with spin-drift and will thus illustrate Maxwell's important allegation (which I believe to be true though it has been much doubted), that any conservative system must "sooner or later" pass through every possible configuration.

† Viscosity is resistance to change of shape in proportion to the speed of the change.

‡ Stokes, *Mathematical and Physical Papers*, Vol. I. p. 310.

surface of discontinuity extending some way into the fluid," but with the difference that there is no slip of fluid on fluid. A trail of rotationally moving liquid, a Helmholtz' "vortex sheet" of exceedingly small thickness, is thus left in the wake of the circular edge; which, while becoming thicker as its gets farther from the edge, becomes rolled up in a wildly tumultuous manner, giving the appearance of an irregular crowd of detached circular ring-vortices. This crowd follows the disc at an ever diminishing speed and widens outward farther and farther, and· inward encroaches more and more on the comparatively undisturbed middle of the wake, as it is left farther and farther behind the disc.

§ 9. Whether as in § 7 for an ideal inviscid incompressible fluid, or as in § 8 for a natural liquid such as water, the "wake," that is to say, the fluid on the rear side of the plane of the disc, as far as it is sensibly affected by the motion of the disc, must be as described in the last sentence of § 8. The rear of the wake is always moving forwards, that is to say, following the disc; but at a continually diminishing speed. Hence, if the disc has been set in motion from rest some finite time, T, ago, the whole wake must be included between the plane from which the disc started, and the plane in which the disc is now, at the time when we are thinking of it. These two planes are at the finite distance, VT, asunder. In other words the wake extends to some distance less than VT, rearwards from the disc.

§ 10. The shedding off of vortex rings from the edge of the disc, to follow in its wake at less speed than its own, essentially gives a contribution to negative pressure on the rear side of the disc equal to $d/dt\,.\,\Sigma\kappa$; where* $\Sigma\kappa$ denotes the sum of the circulations of all the coreless ring-vortices, or of all the rotationally moving liquid, which have or has left the edge since the beginning of the motion. This, with the commonly assumed velocities of the fluid on the two sides of the rigid plane, seems insufficient to account for the excess of observed pressure above that calculated for a long blade by Lord Rayleigh's formula† referred to in my letter to *Nature*, "Towards the Efficiency of Sails, &c.," and leaves some correction to be made on those assumed velocities. But the

* [*Cf.* § 17 *infra.*]

† In lines 9 and 10 of the printed letter (*Nature*, Aug. 30, 1894, p. 426), for "something like five or ten," substitute "4·8." I unfortunately had not Lord Rayleigh's formula by me at the time the letter was written.

working out of this interesting piece of mathematical hydrokinetics must be deferred for a continuation of the present article in which supposed discontinuity of fluid motion, extending far and wide, as taught by many writers in many scientific papers and text-books since Stokes' infinitesimal rift started it in 1847, will be considered.

III.

§ 11. The accompanying diagram (fig. 1) illustrates the application of the doctrine in question, to a disc kept moving

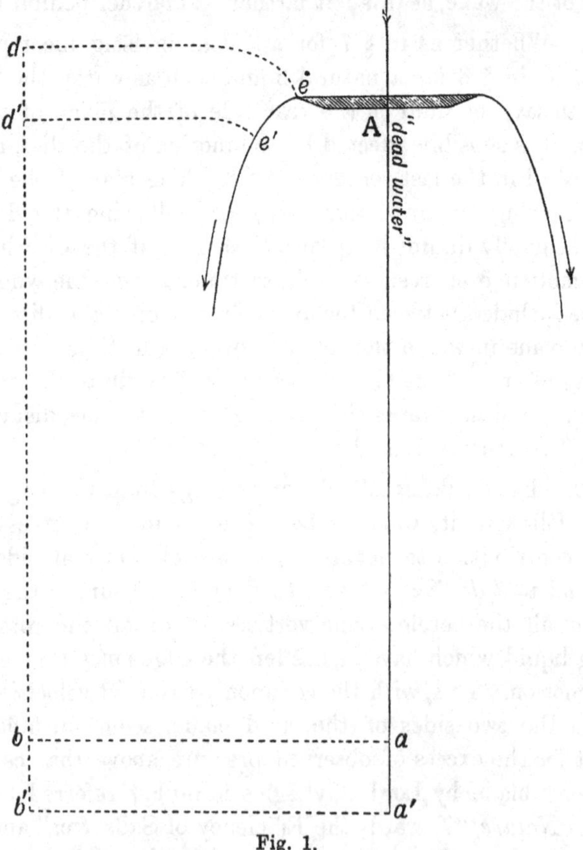

Fig. 1.

through water or air with a constant velocity, V, perpendicular to its own plane. The assumption to which I object as being inconsistent with hydrodynamics, and very far from any approximation to the truth for an inviscid incompressible fluid in any circumstances, and utterly at variance with observation of discs or blades (as oar blades) caused to move through water, is, that starting from the edge as represented by the two continuous curves in the diagram, and extending indefinitely rearwards, there is a "surface of discontinuity" on the outside of which the water flows, relatively to the disc, with velocity V, and on the inside of which there is a rearless mass of "dead water"* following close after the disc.

§ 12. The supposed constancy of the velocity on the outside of the supposed surface of discontinuity entails for the inside a constant pressure, and therefore quiescence relatively to the disc, and rearlessness of the "dead water." How could such a state of motion be produced? and what is it in respect to rear? are questions which I may suggest to the teachers of the doctrine; but which happily, not going in for an examination in hydrokinetics, I need not try to answer.

§ 13. But now, supposing the motion of the disc to have been started some finite time, t, ago, and considering the consequent necessity (§ 9) for finiteness of its wake, let ab, bd be lines sufficiently far behind the rear, and beyond one side, of the disturbed water, to pass only through water not sensibly disturbed. We thus have a real finite case of motion to deal with, instead of the inexplicably infinite one of § 11. Let us try if it is possible that for some finite distance from the edge, and from the disc on each side, the motion could be even approximately, if not rigorously, that described in § 11, and indicated by the diagram.

§ 14. Let v be the velocity at any point in the axis, Aa, at distance y from the disc, rearwards. Draw ed perpendicular to the stream lines of the fluid, relatively to the disc supposed at rest.

The "flow"† in the line ed is 0;
„ „ „ db „ $V \times db$;
„ „ „ ba „ 0;
„ „ „ aA „ $-\int_0^{Aa} v\,dy$;
„ „ „ Ae „ 0, by hypothesis.

* This is a technical expression of practical hydraulics, adopted by the English teachers of the doctrine of finite slip between two parts of a homogeneous fluid, to designate water at rest relatively to the disc.

† "Vortex Motion" (Thomson), *Trans. R. S. E.* 1869.

Hence for the "circulation" in the closed polygon *edbaAe*, we have

$$V \times db - \int_0^{Aa} v\,dy.$$

Similarly, for the circulation in the same circuit* at a time later by any interval, τ, when the line *ba* has moved to the position $b'a'$, and *ed* to $e'd'$, we have

$$V \times db - \int_0^{Aa} v'\,dy,$$

where v' denotes, for the later time, $t+\tau$, the velocity in *Aa*, at distance y from *A*. Hence the circulation in *edbaAe*† gains in time τ an amount equal to

$$-\int_0^{Aa} (v'-v)\,dy;$$

which is the same as

$$-\int_0^{\infty} (v'-v)\,dy.$$

This, by the general theorem of "circulation,"‡ must be equal to the gain of circulation in time τ, of all the vortex-sheet in its growth from the edge according to the statement of § 11. Hence, with the notation of § 10,

$$(\Sigma\kappa)' - \Sigma\kappa = -\int_0^{\infty} (v'-v)\,dy.$$

§ 15. Remarking now that the fluid has only continuous irrotational motion through a finite space all round each of the lines *ed*, *db*, *ba*, *aA*; and all round *Ae* except the space occupied by the disc and the fluid beyond its front side, we have, for the velocity-potential of this motion, relatively to the disc,

$$Vy + \phi(x, y, z, t),$$

where ϕ denotes the velocity-potential of the motion relative to the infinitely distant fluid all round: and we have along *Aa*

$$v = V + \frac{d}{dy}\phi(0, y, 0, t).$$

With this the equation of § 14 becomes

$$(\Sigma\kappa)' - \Sigma\kappa = \{\phi(0, 0, 0, t+\tau)\} - \{\phi(0, 0, 0, t)\}.$$

* Remark that the circulation in $abb'a'$ is zero, and therefore the circulation in $edb'a'Ae$ is equal to that in *edbaAe*.

† [This requires that there is a smooth stream line from the neighbourhood of e to e', which is outside all the vortex motion; for only then is the circulation in $edd'e'$ null.]

‡ "Vortex Motion," *Trans. R. S. E.* 1869.

Hence, by taking τ infinitely small,

$$\frac{d}{dt}\Sigma\kappa = \frac{d}{dt}\phi(0, 0, 0, t).$$

§ 16. Now in the time from t to $t+\tau$, there has been, according to the supposition stated in § 11, a growth of vortex-sheet from e, at the rate $\frac{1}{2}V$, being the mean between the velocities of the fluid on its two sides*, and the circulation, per length l of the sheet thus growing is lV. Hence the vortex-circulation of the growing sheet augments, in time τ, by $\frac{1}{2}V\tau \times V$: and therefore, by § 15,

$$\frac{d}{dt}\phi(0, 0, 0, t) = \tfrac{1}{2}V^2.$$

§ 17. Now, if Π denotes the pressure of the fluid at great distances, where its velocity relative to the disc is V, and p the pressure at any point of the rear side of the disc, being the same as the pressure at A, we have, by elementary hydrokinetics,

$$p = \Pi + \tfrac{1}{2}V^2 - \frac{d}{dt}\phi(0, 0, 0, t),$$

because the velocity of the fluid at every point of the rear side of the disc is zero according to the assumption of "dead water." Hence, by § 16,

$$p = \Pi,$$

which, being the same as the pressure on the rear side given by the unmitigated assumption of an endless ever broadening wake of "dead water," proves that our substitution (§ 13) of a finite configuration of motion conceivably possible as the consequence of setting the disc in motion at some finite time, t, ago, instead of the inconceivable configuration described in § 11, does not alter the pressure on the rear side of the disc.

§ 18. Hence were the motion of the fluid for some finite distance from the disc, on both its sides, the same, or very approximately the same, as that described in § 11, the force that must be applied to keep it moving uniformly would be the same, or very approximately the same, as that calculated by Lord Rayleigh from the motion of the fluid supposed to be wholly as described in § 11.

* Helmholtz, *Wissenschaftliche Abhandlungen*, Vol. I. foot of p. 151.

§ 19. But what reason have we for supposing the velocity of the fluid at the edge, on the front side of the disc, to be exactly or even approximately equal to the undisturbed velocity, V, of the fluid at great distances from the disc? None that I can see. It seems to me indeed probable that it is in reality much greater than V, when we consider that, with inviscid incompressible fluid in an unyielding outer boundary, the velocity, in the case considered in § 4, is equal to V at even so far from the edge as ·85 of an inch, and increases from V to $63{\cdot}7 \times V$ between that distance from the edge, and the edge with its 1/2000 of an inch radius of curvature.

§ 20. And what of the "dead water" in contact with the whole rear side of the disc which the doctrine of discontinuity assumes? Look at the reality and you will see the water in the rear exceedingly lively everywhere except at the very centre of the disc. You will see it eddying round from the edge and returning outwards very close along the rear surface, often I believe with much greater velocity than V, but with no steadiness; on the contrary, with a turbulent unsteadiness utterly unlike the steady regular motion generally assumed in the doctrine of discontinuity.

§ 21. We may I think safely conclude that on the front side the opposing pressure is less than that calculated by Rayleigh. That this diminution of resistance is partially compensated or is over-compensated by diminution of pressure on the rear, is more than we are able to say from theory alone, in a problem of motion so complex and so far beyond our powers of calculation: but we are entitled to say so, I believe, by experiment. Rayleigh's investigation of the resistance experienced by an infinitely thin rigid plane blade bounded by two parallel straight edges, when caused to move through an inviscid incompressible fluid, with constant velocity, V, in a direction perpendicular to the edges and inclined at an angle i to the plane, gives a force cutting the plane perpendicularly at a distance from its middle equal to

$$\frac{3 \cos i}{4(4 + \pi \sin i)}$$

of its breadth, and gives for the amount of this force in gravitation measure,

$$\frac{2\pi \sin i}{4 + \pi \sin i} PA,$$

where A denotes the area of one side of the blade, and P the weight of a column of the fluid of unit cross-sectional area, and of height equal to the height from which a body must fall to acquire a velocity equal to V.

§ 22. The assumption (§ 11) on which this investigation is founded admits no velocity of fluid motion, relatively to the disc, greater anywhere than V. It gives velocity reaching this value only at the edges of the blade; and at the supposed surface of discontinuity; and in the fluid at infinite distances all round except in the infinitely broad wake of "dead water" where the velocity is zero. It makes the pressure equal to Π all through the "dead water," and makes it increase through the moving fluid, from Π at an infinite distance, and at the "surface of discontinuity" to a maximum value $\Pi + P$ attained at the water-shed line of the disc. If the fluid is air, and if V be even so great as 120 feet per second (1/10 of the velocity of sound)* P would be only 7/1000 of Π. The corresponding augmentation of density could cause no very serious change of the motion from that assumed: and therefore in Rayleigh's investigation air may be regarded as an incompressible fluid if the velocity of the disc is anything less than 120 feet per second.

We may therefore test his formula for the resistance, by comparison with results of careful experiments made by Dines† on the resistance of air to discs and blades moved through it at velocities of from 40 to 70 statute miles per hour (59 to 103 feet per second).

§ 23. Dines finds for normal incidence the resistance against a foot-square plate, moving through air at m British statute miles per hour, to be equal to $\cdot 0029m^2$ of a pound weight.

This, if we take the specific gravity of the air as 1/800, gives according to our notation of § 21,

$$2\cdot 157 \times PA$$

as the resistance to a square plate of area A. At the foot of p. 255 (*Proc. R. S.* June 1890) Dines says that he finds the resistance to a long narrow blade to be more than 20 per cent. greater than to a square plate. For a blade we may therefore take

$$2\cdot 59 \times PA$$

* Or $\frac{1}{10} \times \sqrt{1\cdot 4 \times gH}$, where H is "the height of the homogeneous atmosphere."

† *Proc. R. S.* June 1890.

as the resistance according to Dines' experiments. This is 2·94 times the resistance calculated from Rayleigh's formula (§ 21 above), which is

$$88 \times PA,$$

for normal incidence.

§ 24. For incidences more and more oblique, the discrepancy is greater and greater. Thus, from curves given by Dines (p. 256) showing his own and Rayleigh's results, I find the normal resistance to a blade moved through air in a direction inclined 30° to its plane, to be 3·52 times that given by Rayleigh's formula. And by drawing a tangent to Dines' curve at the point in which it cuts the line of zero pressure, I find that, for very small values of i, it gives

$$6{\cdot}39 \times \sin i \times PA.$$

This is rather more than four times the value of the force given by Rayleigh's formula for very small values of i, which is

$$\tfrac{1}{2}\pi \sin i \,.\, PA.$$

It is somewhat more than double that given by my conjectural formula (*Nature*, August 30, p. 426; and *Phil. Mag.* October 1894) for very small values of i, which is

$$\pi \sin i \cos i \,.\, PA.$$

My formula is, however, merely conjectural; and I was inclined to think that it may considerably under-estimate the force*. That it does so to some degree is perhaps made probable by its somewhat close agreement with Dines; because the blade in his experiments was 3 inches broad and $\frac{3}{8}$ of an inch thick in the middle with edges "feathered off." An infinitely thin blade would probably have shown greater resistances, at all angles, and especially at those of small inclination to the wind.

* [The numbers above have here been doubled, in addition to other slight modification. This throws out the approximate agreement which was found by Lord Kelvin.

In reply to a request for information regarding the results of recent investigation at the National Physical Laboratory, Dr T. E. Stanton writes (Oct. 7, 1909) as follows, and sends the diagram annexed. "The value of Dines' coefficient quoted by Lord Kelvin agrees remarkably well with our results here when account is taken of the variation in the size of the plate, as you will see from the enclosed diagram. I find that the increase in total resistance per unit area with size is entirely due to the increased suction effect at the back of the plate as the dimensions increase.

"As regards the pressures on long narrow plates and on inclined plates, our results are in practical agreement with those of Dines; and on our investigating the

IV.

§ 25. Another decisive demonstration that the doctrine of discontinuity is very far from an approximation to the truth, is afforded, in an exceedingly interesting and instructive manner, by Dines' observations of the pressures on the two sides of a disc held at right angles to a relative wind of 60 statute miles per hour (88 ft. per sec.), produced by carrying it round at the end of the revolving arm of his machine. The observations were described in a communication to the Royal Meteorological Society in May 1890. In his paper of June of the same year, in the *Royal Society Proceedings* already referred to, he states the results, which are, that at the middle of the front side an augmentation of pressure, and at the middle of the rear side a diminution of pressure, measured respectively by 1·82 and ·89 inches of water, were found. These correspond to heads of air, of density 1/800 of that of water, equal respectively to $121\frac{1}{3}$ and $59\frac{1}{3}$ feet. The former is in almost exact accordance with rigorous mathematical theory for an inviscid incompressible fluid; which gives $88^2/64{\cdot}4$, or $120\frac{1}{4}$ feet for the

distribution of pressure we found that the excess in the total resistance over that given by Lord Rayleigh's formula was, as in the case of normal impingement, due to the suction effect of the eddies on the leeward side."

Air-resistance of square plates.

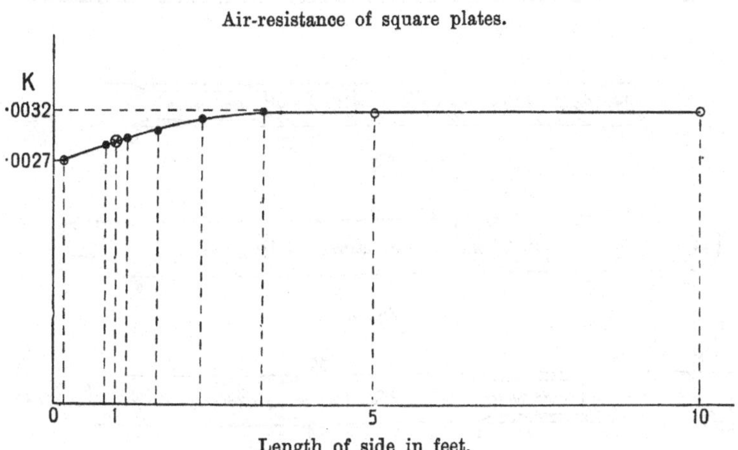

Length of side in feet.

● Means of observations made by M. Eiffel on falling plates.
⊕ " " " at N.P.L. in a current of air.
○ " " " " in the wind.
⊗ " " " by Mr Dines on whirling table.]

depth corresponding to the pressure at the water-shed point or points, of a solid of any shape moving through it at the rate of 88 feet per second. The latter shows that there is a "suction" at the centre of the rear side very nearly equal to half the augmentation of pressure on the front; instead of there being neither suction nor augmented pressure as taught in the doctrine of discontinuity!

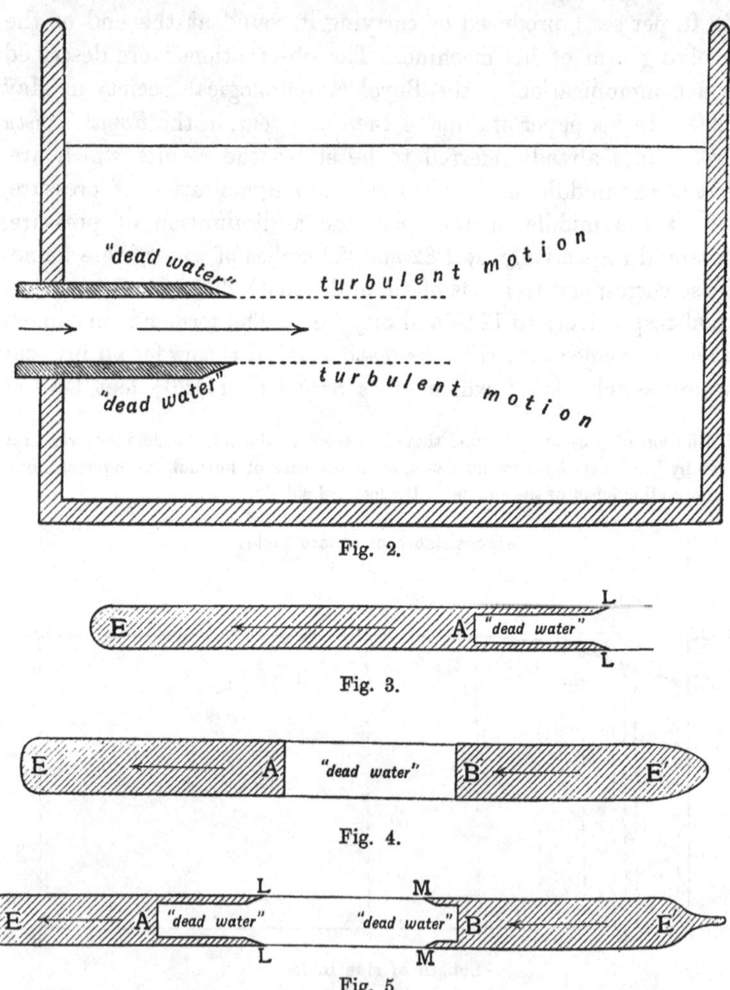

Fig. 2.

Fig. 3.

Fig. 4.

Fig. 5.

§ 26. The accompanying diagrams (2, 3, 4, 5) represent several illustrations of the doctrine of discontinuity in the motion of an inviscid fluid, less attractive to writers on mathematical hydrokinetics than that represented in Fig. 1 (whether as its stands, or varied to suit oblique incidence), because each is instantly soluble without mathematical analysis, and they do not, like it in the two-dimensional case, constitute illustrations of the beautiful mathematical method for finding surfaces of constant fluid velocity in prolongation of given surfaces along which the velocity is not constant, originated by Helmholtz*, developed in a mathematically most interesting manner by Kirchhoff†, and validly applied to the theory of the "vena contracta" by Rayleigh‡.

§ 27. A cylindric jet (not necessarily of circular cross-section) issuing from a tube with *sharp edge*, into a *very large* volume of fluid of the same density as that of the jet, is represented in Fig. 2. This case was carefully considered by Helmholtz§, both for the ideal inviscid incompressible fluid and for real water or real air. He gave good reason for believing that, with real water or real air, and at distances from the mouth as great as several times the diameter of the tube (or the least diameter, if it is not of circular cross-section) the surrounding fluid is nearly at rest, and the jet is but little disturbed from the kind of motion it had in passing out of the tube: and therefore that the efflux is nearly the same as, other circumstances the same, it would be if the atmosphere into which the jet is discharged were inertia-less. This conclusion, which is of great importance in practical hydraulics, has been confirmed by careful experiments made eight years ago in the physical laboratory of the University of Glasgow by two young officers of the American Navy, Mr Capps and the late Mr Hewes. I believe it has been tested and confirmed by other experimenters.

§ 28. The very simplest application of the doctrine of discontinuity to the theory of the resistance of fluids to solids moving through them, is represented in Fig. 3, and the result is no resistance at all! Surely this case, requiring no calculation, might have been a warning of the extreme wrongness of the doctrine in connection with resistance of fluids against solids moving through them. The nullity of the resistance in the case represented by Fig. 3 according to the assumption of a wake of "dead water"

* "Wissenschaftliche Abhandlungen," Vol. I. pp. 153–156.

† "Vorlesungen über Mathematische Physik," Vol. XXI.

‡ "Notes on Hydrodynamics," *Phil. Mag.* 1876, second half-year.

§ "Wiss. Abh.," Vol. I. pp. 152–153.

having the same pressure, Π, as the distant and near water flowing uniformly in parallel lines, follows immediately from an easily proved theorem which I stated in the combined meeting of Sections A and G [Brit. Assoc.], in Oxford last August, to the effect that the longitudinal component of the pressure on each of the ends, E, E′, in Figs. 3, 4, 5, whatever their shapes, and whether "bow" or "stern," provided only that it ends tangentially in a cylindric "mid-body" long in comparison with the greatest transverse diameter of the solid, is equal to ΠA, where A is the area of the cross-section of the cylindric part of the solid.

§ 29. Figs. 4 and 5 represent two varieties of a case wholly free from the inconceivable endlessness of Fig. 1, and carefully chosen as thoroughly defensible by holders of the doctrine of discontinuity if it has any defensibility at all. I venture to leave it with them for their consideration.

THEORY OF THE TIDES*.

26. ON AN ALLEGED ERROR IN LAPLACE'S THEORY OF THE TIDES.

[From the *Philosophical Magazine*, Vol. L. 1875, pp. 227—242.]

1. AIRY, in his article on Tides and Waves in the *Encyclopedia Metropolitana,* gives a version of Laplace's theory of the tides which has the undoubtedly great merit of being freed from certain inappropriate applications of "Laplace's coefficients," or, as they are now more commonly called, "Spherical harmonics," by which its illustrious author attempted, not quite successfully, to design a method for taking into account the alteration of gravity due to the tidal disturbance of the surface of the sea in the solution of his dynamical equations.

It is the repetition of this attempt for each of the "three species of oscillations" (*Mécanique Céleste,* Liv. IV. arts. 5, 7, 9), and the preparation for it [in the course of working out the fundamental differential equations (art. 3)] by the assertion that "a, b, c, a' are to be rational functions of μ and $\sqrt{1-\mu^2}$," that throws a cloud over nearly the whole chapter (Liv. IV. chap. i) of

[* The practical side of tidal theory was undertaken by a Committee of the British Association, which published various reports (*B.A. Reports*, 1868, 70, 71, 72, 76, 78) on the practical methods of harmonic analysis of tides and the results for various parts of the ocean. One of the longer of these (*Brit. Assoc. Report*, 1872, pp. 355–395) was "drawn up by Mr E. Roberts under the direction of the Committee": the next (*Brit. Assoc. Report*, 1876, pp. 275–307) was "drawn up by Sir W. Thomson." Subsequently the direction of this work was taken over mainly by Sir George Darwin; cf. Thomson & Tait's *Nat. Phil.* ed. 2 and Sir George Darwin's *Collected Scientific Papers*, Vol. II. (which include further *B.A. Reports*, 1881, etc.) and his book on *The Tides*.

Since the papers here reprinted drew attention afresh to the Laplacian theory, it has been very much improved and developed by various writers, including Darwin, Poincaré, Lamb, and in particular by S. S. Hough in two memoirs in *Phil. Trans.* Vol. 189 (1897), p. 201, and Vol. 191, A (1898), p. 139: cf. Lamb's *Hydrodynamics*, ed. 2, or 3.

In *Popular Lectures and Addresses*, vol. III. Navigation, 1891, a British Association lecture (Southampton, 1882) on "The Tides" is reprinted, pp. 139—190 with Appendices A, B, C, D, E, of which B, C are papers (*Brit. Assoc.*, Dublin, 1878) on the "Influence of the Straits of Dover on the Tides in the British Channel and the North Sea," and "On the Tides of the Southern Hemisphere and of the Mediterranean," the latter in conjunction with Capt. Evans, while D is a "Sketch of Proposed Plan of Procedure in Tidal Observation and Analysis" from *Brit. Assoc. Report*, Norwich, 1868.]

the *Mécanique Céleste* devoted to the dynamical theory of the tides, and fully justifies the following statement with which Airy ["Tides and Waves," art. (66)] introduces his own version of Laplace's theory:—

"It would be useless to offer this theory in the same shape in which Laplace has given it; for the part of the *Mécanique Céleste* which contains the Theory of Tides is perhaps on the whole more obscure than any other part of the same extent in that work. We shall give the theory in a form equivalent to Laplace's, and indeed so nearly related to it that a person familiar with the latter will perceive the parallelism of the successive steps. The results at which we shall arrive are the same as those of Laplace."

2. The only good thing lost in Airy's treatise through the omission of the spherical harmonic analysis is Laplace's complete solution for the case of no rotation and equal depth of the sea all round the earth. When the earth's rotation is taken into account, or when the sea is of unequal depth, the differential equation to be solved takes a form altogether unsuited for the introduction of spherical harmonics*; and Airy's investigation is substantially the same as Laplace's, except in the judicious omission of the unsuccessful attempts referred to above.

3. In giving Laplace's solution for the semi-diurnal tide with the change of gravity due to the change of figure of the water not taken into account, Airy points out what he believed to be an error, so serious that, after correcting it, it was "needless to observe that Laplace's numerical calculations of the heights of tides in certain latitudes, and his inferences as to the latitude where there is no tide &c., fall to the ground." When I first read Airy's treatise ten years ago on board the 'Great Eastern,' I could not assent to his correction of Laplace, but, on the contrary, satisfied myself that Laplace was quite right. Not having the *Mécanique Céleste* at hand, I set the subject aside for a time, intending to return to it for the second volume of 'Thomson and Tait's *Natural Philosophy*,' with the first volume of which I was then occupied.

4. My attention has recently been recalled to it by reading in a volume of 'Tidal Researches,' constituting an appendix contributed by W. Ferrel to the '*United States Coast-Survey*

[* Yet S. S. Hough subsequently succeeded in introducing harmonic analysis appropriately. G. H. D.]

Report for 1874,' the following passage referring to Laplace's solution for the semidiurnal tides:—

"The results show that the form which the surface of the sea assumes in this case differs very much from that of a prolate spheroid with its longer axis in the direction of the disturbing body, and that for certain depths of the ocean the tides at the equator are inverted, low water taking place under the attracting body. For great depths, however, the tides were found to be direct in all latitudes; and even in the cases in which they are inverted at the equator they were found to be direct toward the poles, and consequently there is a latitude in such cases where there are no tides. Laplace, however, failed to interpret correctly in this case his own solution, so that the numerical results which he has given for different assumed depths of the ocean are erroneous; but still the general results just stated are readily seen from the solution. Having failed to see the indeterminate character of the problem, he adopted a singular and unwarranted principle for determining the value of a constant which is entirely arbitrary in the case of no friction, but which vanishes in the case of friction, however small. This oversight of Laplace and the indeterminateness of this constant were subsequently pointed out by Airy." The "singular and unwarranted principle" thus referred to is in fact an exquisitely subtle method by which it seems Laplace had determined a constant which is not arbitrary in any case, and which cannot be more than infinitesimally modified by infinitesimal friction. Ferrel further extends to Laplace's integration for the "diurnal tide" the objection of indeterminateness which Airy had raised only against his integration for the semidiurnal; and he follows Airy in an integration (not given by Laplace) for the "long-period tide," in which a false appearance of determinateness (strangely inconsistent with the indeterminateness asserted of the solutions for the semidiurnal and the diurnal) is produced by the inadvertent omission of a constant*, the true value of which is to be determined by a proper application of Laplace's method. With these results before me, I cannot wait two or three years more for the second volume of 'Thomson and Tait's *Natural Philosophy*' to defend Laplace's process, but must speak out on the subject

* See an article in the next (October) Number of the *Phil. Mag.*, entitled "Note on the Oscillations of the First Species in Laplace's Theory of the Tides." [*Infra.*]

without delay; and therefore I offer the present article for publication in the Philosophical Magazine, regretting that I did not do so ten years earlier.

5. To Airy's statement of his case against Laplace, quoted in full below, I premise, by way of explanation:—

I. The tide-generating force for the case in question (the "semidiurnal tide") is such that the equilibrium tide-height is represented by the formula

$$H\sin^2\theta\cos 2\phi, \text{ or } Hx^2\cos 2\phi \quad \text{..............(1)},$$

where H is a constant, θ is the colatitude of the place, or x the cosine of the latitude, and ϕ the hour-angle of the disturbing body, which may be conveniently supposed to consist of moon and anti-moon (two halves of the moon's mass) placed opposite to one another at distances equal to the moon's mean distance from the earth in a line kept always in a fixed direction through the earth's centre and in the plane of the equator.

II. Instead of $H\sin^2\theta$ or Hx^2 in the equilibrium formula, put a, so that it becomes $a\cos 2\phi$. This expresses the actual tide-height if a be a function of the latitude fulfilling over the whole surface the differential equation (*Méc. Cél.* Liv. IV. art. 10)

$$(1-x^2)\,x^2\frac{d^2a}{dx^2}-x\frac{da}{dx}-2\left(4-x^2-\frac{2mr}{\gamma}\,x^4\right)a=-8Hx^2 \text{...(2)},$$

where m denotes the ratio of centrifugal force to gravity at the earth's equator (its value being actually about $\frac{1}{289}$), r the earth's radius, and γ the depth of the sea, in the present case assumed to be uniform all over the earth's surface, and but a small fraction of the radius.

III. Remark that the period of the disturbance thus investigated is rigorously half the period of the earth's rotation—that is to say, half the sidereal day. This supposition is no doubt quite a close enough approximation for the solar semi-diurnal tide; but it is certainly not practically close enough for the lunar semi-diurnal tide, its period exceeding, as it does, the half-sidereal day by about $\frac{1}{28}$ of its value.

IV. Remark also that if the earth's rotation is infinitely slow, m is infinitely small, and the differential equation is satisfied by

$$a=Hx^2;$$

that is to say, agreement with the equilibrium tide.

V. Lastly, Laplace and Airy assume

$$a = K_2 x^2 + K_4 x^4 + \ldots + K_{2k} x^{2k} + K_{2k+2} x^{2k+2} + \&c. \quad \ldots(3)$$

as a solution of the differential equation (2). Then, by equating the coefficient of x^2 in the left-hand member to $-8H$, its coefficient in the right, and by equating to zero the coefficient of x^{2k+4} for all values of k from 0 to ∞, they find

$$-2.4K_2 = -8H \quad \ldots\ldots\ldots\ldots\ldots\ldots(4)$$

and $$2k(2k+6)K_{2k+4} - 2k(2k+3)K_{2k+2} + \frac{4mr}{\gamma}K_{2k} = 0 \quad \ldots(5)$$

for all positive integral values of k [the case of $k=0$ justifies the omission of K_0 in (3)]. The first of these equations of condition gives $K_2 = H$. The second, if for brevity we put $mr/\gamma = e$, gives,

$$K_{2k+4} = \frac{2k+3}{2k+6}K_{2k+2} - \frac{e}{k(k+3)}K_{2k} \quad \ldots\ldots\ldots\ldots(6),$$

and so determines successively K_6, K_8, $K_{10}, \ldots$ &c., all in terms of K_2, K_4. Thus the differential equation (2) is satisfied by (3) with K_2 given by (4), K_4 arbitrary, and the other coefficients given by (6).

6. On this and Laplace's process for completing the solution Airy [art. (111)] remarks:—

"The indeterminateness of K_4 is a circumstance that admits of very easy interpretation. It is one of the arbitrary constants in a complete solution of the equation. It shows that we may give to K_4 any value that we please, even if $H^* = 0$; and then, provided that we accompany our arbitrary K_4 with the corresponding values of K_6, K_8, &c., we shall have a series which expresses a value of a† that will satisfy the equation *when there is no external disturbing force whatever*, and which therefore may be added, multiplied by any number, to the expression determined as corresponding to a given force. In the next section we shall find several instances exactly similar to this. Yet this obvious view of the interpretation of this circumstance appears to have escaped Laplace, and he has actually persuaded himself to adopt

* G in Airy's notation, $3L/4r^3g$ in Laplace's.

† ga in Airy's notation, aa in Laplace's.

the following process. Putting the general equation among the coefficients into the form

$$\frac{K_{2k+2}}{K_{2k}} = \frac{2bm/l}{(2k^2+3k)-(2k^2+6k)\,K_{2k+4}/K_{2k+2}},$$

he has unwarrantably conceived that this must apply when $k=1$ for the determination of K_4; and thus applying the same equation to each quotient of terms which occurs in the denominator of the fraction, he finds

$$\frac{K_4}{K_2} = \frac{2bm/l}{2\,.\,1^2+3\,.\,1-}\;\frac{(2\,.\,1^2+6\,.\,1)\,2bm/l}{2\,.\,2^2+3\,.\,2-}\;\frac{(2\,.\,2^2+6\,.\,2)\,2bm/l}{2\,.\,3^2+3\,.\,3-}\;\&c.$$

in an infinite continued fraction. And upon this he founds some numerical calculations adapted to different suppositions of the depth of the sea. We state, as a thing upon which no person after examination can have any doubt, that this operation is entirely unfounded."

7. A careful examination at the time when I first read this led me to the opposite conclusion, and showed me that Laplace was perfectly right. If K_{2k+2}/K_{2k} vanishes when k is infinitely great, then K_4/K_2 cannot but be equal to the continued fraction. What, then, must be the case if K_4 has any other value than that determined by Laplace? K_{2k+2}/K_{2k} cannot then converge to zero for greater and greater values of k. But unless K_{2k+2}/K_{2k} is infinitely small when k is infinitely great, the second term of the second member of (6) is infinitely small in comparison with the first, and therefore ultimately

$$K_{2k+4} = \frac{2k+3}{2k+6}\,K_{2k+2},$$

or

$$K_{2k+2} = \frac{2k+1}{2k+4}\,K_{2k}$$

$$= \left(1-\frac{3}{2k}\right)K_{2k},$$

when k is infinitely great. Now this is precisely the degree of ultimate convergence of the coefficients of x^{2k}, x^{2k+2}, &c. in the expansion of $(1-x^2)^{\frac{1}{2}}$. Hence, when x is infinitely nearly equal to unity, a is finite, and so also is $\sqrt{(1-x^2)}\,da/dx$, or $da/d\theta$. Now clearly at the equator (or when $x=1$) we must have $da/d\theta=0$, because of the symmetry of the disturbance in the northern and southern hemispheres in the case proposed for solution by Laplace and Airy. Hence in this case K_{2k+2}/K_{2k} must converge to zero, and therefore K_4 must have the value given to it by Laplace *.

8. Look now to the degree of convergence obtained by Laplace's evaluation of K_4 and verify that it secures $da/d\theta = 0$ when $\theta = \frac{1}{2}\pi$. Put

$$K_{2k+2}/K_{2k} = R_k \text{ ....(7).}$$

By this we have

$$K_{2k+4} = R_k \,.\, R_{k+1} \,.\, K_{2k};$$

and (6) resolved for R_k gives

$$R_k = \frac{e}{k(k+3)} \Big/ \left(\frac{2k+3}{2k+6} - R_{k+1}\right) \text{ ....(8).}$$

Hence, unless the ratios converge to unity, (8) gives

$$R_k = \frac{2e}{k(2k+3)} \text{ when } k \text{ is great ....(9).}$$

Now Laplace's determination of K_4 by his continued fraction implies the determination of the ratios by taking $R_{k+1} = 0$ for some very great value of k and calculating

$$R_k, \quad R_{k-1}, \quad R_{k-2}$$

by successive applications of (8) with $k-1, k-2, \ldots$ substituted for k. Hence it gives to the series (3) a degree of convergency (approximately the same as that of the expansion of $\epsilon^{x\sqrt{e}} + \epsilon^{-x\sqrt{e}}$ in powers of x, and) such that da/dx, d^2a/dx^2, d^3a/dx^3,... &c. are all finite for every finite value of x. Hence $da/d\theta$, being equal to $\sqrt{(1-x^2)}\, da/dx$, is zero when $x = 1$.

9. Thus it appears that Laplace's process simply determines K_4 to fulfil the condition that $da/d\theta = 0$ at the equator. And the assumed form of solution (3) has the requisite convergency to zero when $x = 0$, for the poles. Laplace's result is therefore the solution of the determinate problem of finding the tidal motion in an ocean covering the whole earth continuously from pole to pole. *Whatever other motion the sea could have in virtue of any initial disturbance, cannot, except for certain critical depths, have the same period as that of the assumed tide-generating force.*

10. If the sea be precisely of such a depth that some one of the possible free vibrations in which the height of the surface at any instant is expressible by the formula $v \cos 2\psi$, where ψ

* [Sir G. Airy's remarks in reply, *Phil. Mag.* Oct. 1875, referred specially to this point.]

denotes longitude, and v some function of the latitude having the same value for equal north and south latitudes, has its period equal to that of the tide-generating influence, it is easily seen that the solution of the differential equation (2) gives an infinitely great value to a. It is only when the depth has one of these critical values that the arbitrary solutions introduced by Airy and adopted by Ferrel are applicable to an ocean covering the whole earth continuously.

11. Yet Laplace himself fell into the same error of imagining that the general integration of the differential equation (2), with the proper arbitrary constants, includes oscillations depending on the primitive state of the sea, as the following passage (Liv. IV. chap. I. art. 4) shows:—

"L'intégration de l'équation (4)* dans le cas général où n n'est pas nul, et où la mer a une profondeur variable, surpasse les forces de l'analyse; mais pour déterminer les oscillations de l'océan, il n'est pas nécessaire de l'intégrer généralement; il suffit d'y satisfaire; car il est clair que la partie des oscillations qui dépend de l'état primitif de la mer, a dû bientôt disparaître par les résistances de tout genre que les eaux de la mer éprouvent dans leurs mouvemens; en sorte que sans l'action du soleil et de la lune, la mer serait depuis longtemps parvenue à un état permanent d'équilibre: l'action de ces deux astres l'en écarte sans cesse, et il nous suffit de connaître les oscillations qui en dépendent."

Laplace, however, did not suffer himself to be led into wrong action by this misconception, and he seems to have entirely forgotten it when he goes direct to the right result, without note or comment, by the truly "singular" process referred to above.

12. On the other hand, Airy, after having, in the passage quoted in § 6 above, allowed the same misconception to fatally influence his practical dealing with the solution, closes with a perfectly correct statement which is sufficient to show the groundlessness of his objection to Laplace's result, and the untenability of what he substitutes for it. This passage has not only the merit of inconsistency with the article which precedes it, but it also constitutes a very decided advance in the theory beyond

* A general equation, of which equation (2) of our numbering above is a particular case.

any thing that Laplace either did or suggested, and for both reasons I am glad to quote it. [Airy, "Tides and Waves," art. (113).] "If, using the more complete values of a that we have just found, we proceed to form the values of a''', b, and u, we find that u* will contain a series of terms multiplied by the indeterminate K_4. We may determine K_4 so that, for a given value of θ, u shall $=0$; that is to say, so that, in a given latitude, the water shall have no north-and-south motion. We might therefore suppose an east-and-west barrier (following a parallel of latitude) to be erected in the sea, and the investigation would still apply. Thus, then, we have a complete solution for a sea which is bounded by a shore whose course is east and west."

13. Now in fact Laplace's process by the continued fraction is a particular case of the determination of K_4 thus suggested by Airy, though one for which Airy's method fails through non-convergence; that is to say, the case in which the proposed east-and-west barrier coincides with the equator. For as we have seen (§ 8), Laplace's determination makes $da/d\theta=0$ when $\theta=\frac{1}{2}\pi$, and therefore makes the north-and-south motion zero at the equator, as is obvious from symmetry, or as we see from the general expression [Laplace, Liv. IV. art. 3; or Airy, arts. (85), (95)]

$$u=\frac{1}{4m\sin^2\theta}\left(\frac{da}{d\theta}+\frac{2\cos\theta}{\sin\theta}a-4H\sin\theta\cos\theta\right)\cos 2\phi \quad \ldots(10)$$

for the southward component of the displacement of the water by the semidiurnal tide.

14. By § 7 above we see that Laplace's solution (3), with K_4 left arbitrary, is convergent for all values of $x<1$. Therefore it is continuously convergent for all values of $\theta<\frac{1}{2}\pi$. Hence Airy's article (113), with the formulæ which he gives in his article (112), or equations (3) and (4), (5), and (10) of §§ 5 and 13 above, constitute a complete and convergent numerical solution of the problem of finding the semi-diurnal tide in a polar basin, or ocean continuous and equally deep from either pole to a shore lying along any circle of latitude on the near side of the equator. Laplace's result, as we have seen, does the same for a hemispherical sea from pole to equator. But for a sea extending from either pole to a coast coinciding with a circle of latitude beyond

* This denotes the meridional component of the displacement of the water in any part of the sea.

the equator another form of solution (still, however, with but one arbitrary constant) must be sought*, because Laplace's form [(3) of § 5 above], ceasing to converge when x (or the sine of the polar distance θ) increases up to unity, fails to provide for the continuous variation of θ from zero to any value exceeding $\frac{1}{2}\pi$. And, further, the method suggested by Airy, when extended to a complete solution of the differential equation with its two arbitrary constants†, completely solves the problem of finding the semidiurnal tide in a zonal sea of equal depth between coasts coinciding with any two parallels of latitude.

15. Returning to Laplace's solution for the whole earth covered with water, we find in the *Mécanique Céleste* the numerical results referred to by Airy (but not quoted, because of the supposed error in the process by which they were obtained). They are of exceedingly great interest (when we know them to be correct); and, in the circumstances, I may be permitted to quote them here. They are obtained by working out numerically the process indicated in §§ 5 and 6 above, for three different depths of the sea, 1/2890, 1/722·5, 1/361·25 of the earth's radius. The values of e, or mr/γ corresponding to these depths, are 10, 2·5, 1·25 respectively; and Laplace finds for the solution [(3) § 5] in the three cases as follows:—

$$\begin{aligned}(e=10),\ a=H\{&1{\cdot}0000\,.\,x^2+20{\cdot}1862\,.\,x^4+10{\cdot}1164\,.\,x^6\\ &-13{\cdot}1047\,.\,x^8-15{\cdot}4488\,.\,x^{10}-7{\cdot}4581\,.\,x^{12}\\ &-2{\cdot}1975\,.\,x^{14}-0{\cdot}4501\,.\,x^{16}-0{\cdot}0687\,.\,x^{18}\\ &-0{\cdot}0082\,.\,x^{20}-0{\cdot}0008\,.\,x^{22}-0{\cdot}0001\,.\,x^{24}\};\end{aligned}$$

$$\begin{aligned}(e=2{\cdot}5),\ a=H\{&1{\cdot}0000\,.\,x^2+6{\cdot}1960\,.\,x^4+3{\cdot}2474\,.\,x^6\\ &+0{\cdot}7238\,.\,x^8+0{\cdot}0919\,.\,x^{10}+0{\cdot}0076\,.\,x^{12}\\ &+0{\cdot}0004\,.\,x^{14}\};\end{aligned}$$

$$\begin{aligned}(e=1{\cdot}25),\ a=H\{&1{\cdot}0000\,.\,x^2+0{\cdot}7504\,.\,x^4+0{\cdot}1566\,.\,x^6\\ &+0{\cdot}01574\,.\,x^8+0{\cdot}0009\,.\,x^{10}\}.\end{aligned}$$

* It is to be found by using Laplace's first differential equation [the one from which he derives (2) of § 5 above by putting $1-\mu^2=x^2$ (Liv. IV. art 10)],

$$(1-\mu^2)^2\frac{d^2a}{d\mu^2}-2\left[3+\mu^2-2e\left(1-\mu^2\right)^2\right]a=-8H\left(1-\mu^2\right),$$

and satisfying it by the assumption

$$a=A_0+A_1\mu+A_2\mu^2+\&c.;$$

which, however, is a complete solution with two arbitrary constants, to be reduced to one by the proper condition to make $u=0$ at *one* pole (say, when $\mu=+1$).

† The general solution indicated in the preceding footnote suffices for this purpose.

By putting $x = 0$ in each case we find $a = 0$, showing that there is no rise and fall at the poles. Putting $x = 1$, we find in the three cases,

$$a = -\ 7{\cdot}434 \,.\, H \ldots \text{(depth 1/2890 of radius)},$$
$$a = \ 11{\cdot}267 \,.\, H \ldots (\ \text{,,}\ \ 1/722{\cdot}5\ \ \text{,,}\ \ ),$$
$$a = \ 1{\cdot}924 \,.\, H \ldots (\ \text{,,}\ \ 1/361{\cdot}25\ \ \text{,,}\ \ ).$$

The negative sign in the first case shows that the tide is "inverted" at the equator; or there is low water when the disturbing body is on the meridian, and high water when it is rising or setting. For small values of x (that is to say, for polar regions) the sign is positive, and therefore the tides are direct for this, as clearly for every other depth (because in every case the first term is $+ Hx^2$). In the particular case in question (depth 1/2890), as we see from the formula given above, the value of a increases from zero to a positive maximum, and then decreases to the negative value stated above as x is increased from 0 to 1; and the intermediate value of x which makes it 0 is roughly ·95, or the cosine of 18°. Hence Laplace concludes the tides are inverted in the whole zone between the parallels of 18° north and south latitude, while throughout the regions north and south of these latitudes the tides are direct. The formulæ given above for the second and third of the depths chosen by Laplace show that in these cases the tides are everywhere direct and increase continuously from poles to equator.

The results of the summation for the equatorial tide in the three cases given above are very interesting as showing how much greater it is in each case than H (the equilibrium height). Upon this Laplace remarks that for still greater depths the value of a diminishes; but this diminution has a limit, namely the equilibrium value, which it soon approximately reaches. To find what is meant by "soon" ("bientôt") take the case of $e = \frac{1}{4}$, or depth 1/72·25 of the radius. For a rough approximation to R_3 take $R_4 = 0$, and use formula (9) § 8 with $k = 3$. Thus we have

$$R_3 = 1/54.$$

Then by successive applications of (8) with $k = 2$, and $k = 1$, we find

$$R_2 = {\cdot}0367, \text{ and } R_1 = {\cdot}104.$$

Hence in this case we have roughly

$$a = H\,(x^2 + {\cdot}104 \,.\, x^4 + {\cdot}104 \,.\, {\cdot}0367 \,.\, x^6 + {\cdot}104 \,.\, {\cdot}0367 \,.\, {\cdot}0185 x^8)$$
$$= H\,(x^2 + {\cdot}104 \,.\, x^4 + {\cdot}00382 \,.\, x^6 + {\cdot}000071 \,.\, x^8),$$

which shows that when the depth is about a seventieth of the radius, the actual amount of equatorial tide exceeds the equilibrium amount by nearly eleven per cent.

16. From the first and second of Laplace's numerical formulæ for a given above (§ 15), we may infer that when e is increased from 2·5 continuously to 10, the value of a for any value of x must increase continuously to $+\infty$, then suddenly become $-\infty$, and increase continuously from that till it has the value given by the formula for $e=10$. When e has a value exceeding by however small a difference the value which makes $a=\pm\infty$, the value of a for very small values of x is positive, and diminishes through 0 to very large negative values as x is increased to 1; that is to say, there are nodes coinciding with two very small circles of latitude, one round each pole, direct tides within these circles, and very great inverted tides round the rest of the earth. As e is increased continuously from this first critical value, the nodal circles expand until (as seen above) when $e=10$ they coincide approximately with 18° North and South latitude. From the greatness of the coefficient of x^4 in Laplace's result for this case we may judge that e cannot be increased much above 10 without reaching a second critical value, for which the coefficient of x^4, after increasing to $+\infty$, suddenly becomes $-\infty$. It is probable that the nodal circles do not get much nearer the equator than 18° North and South before this critical value is reached. When e is increased above it, a second pair of nodal circles commence at the two poles, spreading outwards and getting nearer to the former pair of nodal circles, which themselves are getting nearer and nearer to the equator. Then there are direct tides in the equatorial belt, inverted tides in the zones between the nodal parallels of latitude in each hemisphere, and direct tides in the north and south polar areas beyond. This is the state of things for any value of e greater than the second critical value just considered and less than a third. When e is increased through this third critical value, a third pair of nodal circles grows out from the poles; and there are inverted tides at the equator, direct tides in the zone between the nodal circles of the first and second pair, inverted tides in the zones between the second and third nodal circles of each hemisphere, and still, as in every case, direct tides in the areas round the poles; a fourth critical value of e introduces a fourth pair of nodal circles, and so on.

17. The critical values of e which we have just been considering are of course those corresponding to depths for which free vibrations of the several types described are symperiodic with the disturbing force; and the free oscillations without disturbing force are in these cases expressed by the formula (3) of § 5, with $K_2 = 0$ —that is to say, by

$$a = K_4 x^4 + K_6 x^6 + K_8 x^8 + \&c.,$$

where K_4, K_6, &c. are to be found by giving an arbitrary value to any one of them, and determining the ratios R_1, R_2, &c. by successive applications of Laplace's formula,

$$R_k = \frac{e}{k(k+3)} \Big/ \left(\frac{2k+3}{2k+6} - R_{k+1}\right) \quad \text{...[(8) of § 8]},$$

with diminishing values of k, commencing with a value corresponding to the highest ratio to be used in calculating coefficients in the series. If we thus find $R_2 = \frac{5}{8}$, the next application of the formula gives $R_1 = \infty$, which is the test that the value of e used in the calculation corresponds to a depth for which the period of one of the free oscillations is exactly half the earth's period of rotation.

18. The calculation of the ratios R_1, R_2, R_3 is an exceedingly curious and interesting subject of pure mathematics or arithmetic. First, remark the rapid extinction of the error resulting from taking 0 or any other than its true value for R_{k+1} in the first application of the formula (8). Supposing k to be so large that $e/k(k+3)$ is a small fraction, we know that this is somewhat approximately the value of R_k, and that $e/(k+1)(k+4)$ is still more approximately the value of R_{k+1}. Hence we see at once how small the error is if we take 0 instead of R_{k+1}. If we take $\pm\infty$ for R_{k+1}, the formula gives 0 for R_k, and then rapid convergence to the true values of R_{k-1}, R_{k-2}, &c. If we take R_{k+1} exactly equal to $\frac{2k+3}{2k+6}$, we get $R_k = \infty$, $R_{k-1} = 0$, and then rapid convergence to the true values for R_{k-2}, &c. But if we take for R_{k+1} a value less than $\frac{2k+3}{2k+6}$ by a certain very small difference, we find for R_k a value less than $\frac{2k+1}{2k+4}$ by a corresponding very small difference, and then for R_{k-1} a value less than $\frac{2k-1}{2k+2}$ by a

corresponding small difference, and so on. Any value of R_{k+1}, except precisely the one particular value last indicated, will, provided k be large enough, lead to the desired values of the lower ratios after the two, three, four or more successive applications of the formula required to dissipate the effects of the initial error. It is a curious and instructive arithmetical exercise to calculate R_k, R_{k-1}, and so on down to R_1; and then by successive reverse applications of the formula to calculate $R_2, R_3, \ldots R_{k-1}, R_k$. If the calculation has been rigorous, of course the initial value of R_k will be that found at the end of the process; but if the calculation has been approximate (say, with always the same number of significant figures retained in each step), the value found for R_k will be not the initial value, but $\frac{2k+1}{2k+4}$, or, more approximately, $\frac{2k+1}{2k+4} - \frac{e}{(k-1)(k+2)}$. And if we choose for R_1 any other value than precisely that obtained by an infinitely accurate application of Laplace's process, then work up by successive reverse applications of the formula, we find for R_k a value approximately equal to $\frac{2k+1}{2k+4}$*. Laplace has not warned us of this; on the contrary, his instructions, literally followed, would lead us simply to calculate K_4 by his continued fraction and then to calculate K_6, K_8, &c. successively from K_2 and K_4 by successive applications of the

* Compare with the calculation of the formula

$$r_i = \frac{1}{2a - r_{i-1}},$$

where a denotes any numerical quantity >1. Take any value at random for r_0, and calculate $r_1, r_2, r_3 \ldots$ by successive applications of the formula. For larger and larger values of i, r_i will be found more and more nearly equal to the smaller root of the equation

$$x^2 - 2ax + 1 = 0.$$

Now calculate backwards to r_0 by the reversed formula

$$r_{i-1} = 2a - 1/r_i;$$

and instead of finding the initial value of r_0 again, the result (unless the calculation has been rigorously accurate in every step) will be approximately the *greater* root of the quadratic, or approximately equal to $1/r_i$. Thus, for example, take the equation

$$x^2 - 6x + 1 = 0,$$

of which the roots are

$$x = \cdot 171573 \text{ and } x = 5\cdot 828427.$$

To find successive approximations to the smaller root, take

formula (6) of § 5. This, except with infinitely rigorous arithmetic, will bring out for very large values of k not the true rapidly diminishing values of the coefficients K_{2k}, K_{2k+2}, &c., but sluggishly converging values corresponding to the ratio $R_k = \frac{2k+1}{2k+4}$. But this dissipation of accuracy is avoided, and at the same time the labour of the process is much diminished, by using for the ratios the values already found for them in the successive steps in the calculation of the continued fraction for R_1.

19. The law of variation of R_1, R_2, &c., considered as functions of e (§ 8), is of fundamental importance. Some of the remarkable characteristics which it presents have been already noticed (§§ 15—17). Remark now that as e (which is essentially positive in the actual problem) is increased from 0 to $+\infty$, each of the ratios $R_1, R_2, \ldots R_i, R_{i+1}, \ldots$ increases from zero, each one more rapidly than the next in ascending order, until R_1 becomes $+\infty$, and suddenly changes to $-\infty$, and again goes on increasing till it again reaches $+\infty$ and suddenly $-\infty$, and so on. But before R_1 becomes ∞ the second time, R_2 becomes $+\infty, -\infty$, and again increases towards $+\infty$. The same holds for each of the other ratios; that is to say, as e increases continuously, each one of the

$r_0 = 0,$	$r'_0 = 6 - \frac{1}{r'_1} = 5{\cdot}8284,$
$r_1 = \frac{1}{6 - r_0} = {\cdot}1667,$	$r'_1 = 6 - \frac{1}{r'_2} = 5{\cdot}8284,$
$r_2 = \frac{1}{6 - r_1} = {\cdot}1714,$	$r'_2 = 6 - \frac{1}{r'_3} = 5{\cdot}8277,$
$r_3 = \frac{1}{6 - r_2} = {\cdot}1716,$	$r'_3 = 6 - \frac{1}{r'_4} = 5{\cdot}803,$
$r_4 = \frac{1}{6 - r_3} = {\cdot}1716,$	$r'_4 = 6 - \frac{1}{r'_5} = 5{\cdot}067,$
$r_5 = \frac{1}{6 - r_4} = {\cdot}1716,$	$r'_5 = 6 - \frac{1}{r'_6} = 1{\cdot}072,$
$r_6 = \frac{1}{6 - r_5} = {\cdot}1716,$	$r'_6 = 6 - \frac{1}{r'_7} = {\cdot}2029,$
$r_7 = \frac{1}{6 - r_6} = {\cdot}1716,$	$r'_7 = 6 - \frac{1}{r_8} = {\cdot}1725,$
$r_8 = \frac{1}{6 - r_7} = {\cdot}1716,$	$r'_8 = {\cdot}1716.$

If the arithmetic at each step had been rigorous, we should have found $r_7' = r_7$, $r_6' = r_6$, and so on! Instead of coming back on the value 0 assumed for r_0, we find $r_0' = 5{\cdot}8284$, the greater root of the equation!

ratios is always increasing except when its value reaches $+\infty$ and passes suddenly to $-\infty$. The order in which the values of the different ratios pass through ∞ is a subject of great interest and importance which requires careful examination. I hope to return to it, and meantime only remark that the formula (8) for calculating R_i from R_{i+1} shows:—

(1) That no two consecutive ratios can be simultaneously negative.

(2) While R_{i+1} increases from $-\infty$ to 0, R_i increases from 0 to a value somewhat (but very slightly) greater than $e/i(i+3)$, and goes on increasing till it reaches ∞, when $R_{i+1}=\frac{2i+3}{2(i+3)}$.

(3) When R_{i+1} is $>\frac{2i+3}{2(i+3)}$, and therefore when $R_{i+1}>1$, R_i is negative.

From (1) it follows that in the series of coefficients

$$K_2,\quad K_4,\quad K_6,\ldots$$

there cannot be two consecutive changes of sign. From (2), (3) it follows that each coefficient is less in absolute value than its predecessor if of the same sign, except when the predecessor is of opposite sign to the coefficient preceding *it*; and of two coefficients immediately following a change of sign, the second *may* be less than the first, but if so, only by a very small proportion of the value of either; but through nearly the whole range of values of e for which there is a change of sign from, say, K_i to K_{i+1}, K_{i+2} is $>K_{i+1}$ in absolute value. (For illustration of this see Laplace's series above, for his case of $e=10$, for which he gives

$$K_2=1,\quad K_4=20{\cdot}1862,\quad K_6=10{\cdot}1164,\quad K_8=-13{\cdot}1047,$$
$$K_{10}=-15{\cdot}4488,\quad K_{12}=-7{\cdot}4581,\quad K_{14}=-2{\cdot}1975\ldots \text{\&c.})$$

20. Laplace's brilliant invention which forms the subject of this article is capable of great extension, as I hope to show in a future communication. I have not hitherto found any trace of it in treatises on differential equations; but I can scarcely think it probable that in some form or other it is not known to mathematicians who have occupied themselves with this subject. Known or unknown, it is of exceeding value and beauty as a purely mathematical method. As to Laplace's Dynamical Theory of Tides in general, I have much pleasure in concluding with

a warmly appreciative statement by Airy which I find in his "Tides and Waves," art. (117).

"If, now, putting from our thoughts the details of the investigation, we consider its general plan and objects, we must allow it to be one of the most splendid works of the greatest mathematician of the past age. To appreciate this, the reader must consider, first, the boldness of the writer who, having a clear understanding of the gross imperfection of the methods of his predecessors, had also the courage deliberately to take up the problem on grounds fundamentally correct (however it might be limited by suppositions afterwards introduced); secondly, the general difficulty of treating the motions of fluids; thirdly, the peculiar difficulty of treating the motions when the fluids cover an area which is not plane, but convex; and, fourthly, the sagacity of perceiving that it was necessary to consider the earth as a revolving body, and the skill of correctly introducing this consideration. The last point alone, in our opinion, gives a greater claim for reputation than the boasted explanation of the long inequality of Jupiter and Saturn."

27. NOTE ON THE "OSCILLATIONS OF THE FIRST SPECIES" IN LAPLACE'S THEORY OF THE TIDES.

[From the *Philosophical Magazine*, L. 1875, pp. 279—284.]

LAPLACE'S "Oscillations of the First Species" are simple harmonic oscillations, in which the surface of the water is always a figure of revolution round the axis of rotation. The "tide-generating influence" in this case is such that the equilibrium tide-height would be $\Theta \cos \sigma t$ at time t, σ denoting a constant (called the "speed" in the British-Association Tidal Committee's Report for 1871), and Θ a function of the latitude. Θ being supposed known, the problem consists in finding a' a function of the latitude such that $(\Theta + a') \cos \sigma t$ is the actual tide-height at time t, and, for the case of the sea equally deep everywhere, it is to be solved by finding the proper solution of the differential equation

$$\frac{d}{d\mu}\left(\frac{1-\mu^2}{\mu^2-f^2}\frac{da'}{d\mu}\right) - 4ea' = 4e\Theta \quad \text{..............(1)};$$

where μ denotes the sine of the latitude, and e and f are constants defined by the equations

$$f = \frac{\sigma}{2n}, \quad e = \frac{n^2r^2}{g\gamma} = \frac{mr}{\gamma};$$

r being the earth's radius,

g the force of gravity at its surface,

m the ratio of gravity to equatorial centrifugal force, being equal to 1/289,

n the angular velocity of the earth's rotation,

and γ the depth of the sea, supposed small in comparison with r—not greater, say, than $r/50$.

The quickest of the "Oscillations of the First Species" is the lunar fortnightly (declinational); and for it σ is about 1/14 of n, which makes $f=1/28$. Even for this, and more decidedly for the lunar monthly (elliptic) and solar semi-annual (declinational) and annual (elliptic), a good approximation to the result might be obtained by taking $\sigma=0$. Laplace does not enter on the integration of the equation, but contents himself by pointing out that an infinitesimal degree of friction will, when $\sigma=0$, cause the actual tide-height to be the same as the equilibrium tide-height, and that even for the lunar fortnightly the actual height must be sensibly the same as the equilibrium height if there is enough of friction to reduce in a fortnight a free oscillation to a small fraction of its original amount. The result of *any* tide-generating influence of sufficiently long period would obviously be more and more nearly in exact agreement with the equilibrium theory the longer the period, were it not for the earth's rotation. But, because of the earth's rotation, a long-period tide does not approximate to agreement with the equilibrium tide if the water be perfectly frictionless; and the solution of the beautiful "vortex problem" thus presented is what is aimed at by Airy* and Ferrel† in their integration of the preceding equation for the case $\sigma=0$, in which it is reduced to the comparatively simple form‡

$$\frac{d}{d\mu}\left(\frac{1-\mu^2}{\mu^2}\frac{da'}{d\mu}\right)-4ea'=4e\Theta \quad \ldots\ldots\ldots\ldots\ldots(2)$$

* "Tides and Waves" (*Encyclopædia Metropolitana*), art. (97).

† "Tidal Researches" (Appendix to *United States Coast-Survey Report*, 1874), § 151.

‡ [Note added, Bristol, September 2, 1875.]—Without this simplification, the equation (1) is susceptible of nearly as simple a solution as with it. Assume

$$\frac{1}{\mu^2-f^2}\frac{da'}{d\mu}=\Sigma K_i\mu^i.$$

This gives
$$a'=\Sigma\frac{\mu^i}{i}(K_{i-3}-f^2K_{i-1})$$

and
$$\frac{d}{d\mu}\left(\frac{1-\mu^2}{\mu^2-f^2}\frac{da'}{d\mu}\right)=\Sigma\mu^i(i+1)(K_{i+1}-K_{i-1});$$

so that, to determine the coefficients K_i, we have the equation of condition

$$(i+1)K_{i+1}-\left(i+1-\frac{4ef^2}{i}\right)K_{i-1}-\frac{4e}{i}K_{i-3}=4e\Theta_i,$$

if
$$\Theta=\Sigma\Theta_i\mu^i.$$

This is a particular case of an almost equally simple solution of Laplace's general equation of the Tides, which has been communicated to the British Association at its meeting now concluded, and will be published [*infra*, p. 254] also in the November Number of the *Philosophical Magazine*.]

[which substantially agrees with Airy's equation of art. (97) (with $q=0$, to make the depth constant as we now suppose it), and with Ferrel's § 151, equation (288); but is simpler in form, partly through the use of Laplace's notation μ for $\cos\theta$]. For each of the "long-period tides" in the actual case of the earth under the influence of the sun and moon, the function Θ is given by the formula

$$\Theta = H(1-3\mu^2) \quad\text{.......................(3)},$$

where H denotes the equilibrium value of the tide-height at the equator. Airy, with this value of Θ, finds an integral of the differential equation by assuming

$$a' = B_2\mu^2 + B_4\mu^4 + \ldots + B_i\mu^i + \&c.,$$

and determining the coefficients so as to satisfy it. But this assumption errs in making the tide-height at the equator equal to the equilibrium height. The correct assumption for the particular problem proposed (or for any case in which Θ involves only even powers of μ) is

$$a' = B_0 + B_2\mu^2 + B_4\mu^4 + \ldots + B_i\mu^i + \&c.;$$

but the more general assumption,

$$a' = B_0 + B_1\mu + B_2\mu^2 + \ldots + B_i\mu^i + \&c. \quad\text{............(4)},$$

is as easily dealt with (and includes oscillations in which the equator is a line of nodes). With it we have

$$\frac{d}{d\mu}\left(\frac{1-\mu^2}{\mu^2}\frac{da'}{d\mu}\right) - 4ea' = \sum_{i=-3}^{i=\infty}\mu^i\{(i+4)(i+1)B_{i+4} - (i+2)(i+1)B_{i+2} - 4eB_i\},$$

which is to be equated to $4e\Theta$. Thus, for the case of

$$\Theta = H(1-3\mu^2),$$

we find, by putting $i=-2$, $i=0$, $i=2$, &c.:—

$$\left.\begin{array}{l} 2\,.\,(-1)\,.\,B_2 = 0 \\ 4\,.\,1\,.\,B_4 - 2\,.\,1\,.\,B_2 - 4eB_0 = 4eH \\ 6\,.\,3\,.\,B_6 - 4\,.\,3\,.\,B_4 - 4eB_2 = -12eH \end{array}\right\} \quad\text{...........(5)},$$

and $\quad (i+4)(i+1)B_{i+4} - (i+2)(i+1)B_{i+2} - 4eB_i = 0$ ...(6)

for all even positive values of i except 0 and 2.

The first of equations (5) gives $B_2 = 0$; and with this the second and third give

$$B_4 = e(B_0 + H), \quad B_6 = \tfrac{2}{3}eB_0 \quad \ldots\ldots\ldots\ldots\ldots(7);$$

and if in (6) we put successively $i = 4$, $i = 6$, $i = 8, \ldots$ and use in this order the equations so found, we can calculate successively by means of them, B_8, B_{10}, $B_{12}, \ldots$ each in terms of B_0; and we thus have a solution of (2), with one arbitrary constant, B_0, which may be written thus,

$$a' = H \,.\, f(\mu, e) + B_0 \,.\, F(\mu, e) \quad \ldots\ldots\ldots\ldots\ldots(8),$$

where $f(\mu, e)$ denotes the function of μ and e expressed by the series (4), with the coefficients calculated for the case $B_0 = 0$ and $H = 1$; and $F(\mu, e)$ the function similarly found by taking $H = 0$ and $B_0 = 1$.

The constant B_0, as Airy has pointed out ["Tides and Waves," art. (113)] with reference to a corresponding question in the solution for semi-diurnal tides, may be assigned so as to make the north and south component motion of the water zero in a given latitude. In the present case (that is, the case of symmetry round the axis of rotation) we have [Airy, art. (95), or Laplace, Liv. IV. chap. i. art. (3)]

$$\text{northward displacement of water} = \frac{\cos \sigma t}{4m\sqrt{(1 - \mu^2)}} \frac{da'}{d\mu} \quad \ldots(9);$$

and therefore to make the north and south motion zero we must have

$$da'/d\mu = 0 \quad \ldots\ldots\ldots\ldots\ldots\ldots\ldots\ldots(10);$$

whence, by (8),

$$\frac{B_0}{H} = -\frac{df(\mu, e)}{d\mu} \Big/ \frac{dF(\mu, e)}{d\mu} \quad \ldots\ldots\ldots\ldots\ldots(11).$$

If, then, we find B_0 by this equation for any given value of μ, we have a solution of the determinate problem of finding the motion of the water under the given tide-generating influence when, instead of covering the whole earth, the sea covers only an equatorial belt between two equal circular polar islands.

The solution thus obtained is in a series essentially convergent, except in the extreme case of the polar islands vanishing. For, taking the equations (6) in the order indicated above, and so calculating B_{i+4} successively from smaller to greater values of i by the formula

$$B_{i+4} = \frac{i+2}{i+4} B_{i+2} + \frac{4eB_i}{(i+4)(i+1)} \quad \text{............(12)},$$

we inevitably find for greater and greater values of i,

$$\frac{B_{i+4}}{B_{i+2}} = \frac{i+2}{i+4}, \text{ more and more nearly the greater is } i \text{...(13)},$$

and this whatever value, zero or other, we give to B_0 (unless we give it *precisely* the value found by Laplace's method below, and then perform each step of the calculation with infinite accuracy). Hence, whatever be the value of e, the series expressing the solution converges for every value of $\mu < 1$. Thus the solution is thoroughly satisfactory for the supposed case of two equal polar islands of any finite magnitude. But the ultimate convergence is shown by (13) to be the same as that of the series

$$\frac{\mu^2}{1} + \frac{\mu^4}{2} + \dots \frac{\mu^{2i}}{i} + \dots,$$

which is equal to $\qquad \log \dfrac{1}{1-\mu^2}$

Hence, when $\mu = 1$, the series for a' becomes infinitely great; and *à fortiori* it gives an infinitely great value for $da'/d\mu$, unless it has been calculated for *precisely* the particular value of B_0 sought. Hence equation (11) fails to determine this value. Thus the solution fails for the very case for which it was sought, the case proposed originally by Laplace, and taken by Airy and Ferrel as the subject of their investigation—that is, the case of the whole earth covered with water. Here Laplace's brilliant process, referred to in an article in the preceding Number of the *Philosophical Magazine*, comes to our aid marvellously.

Let
$$\frac{B_{i+2}}{B_i} = \frac{-1}{N_i} \quad \text{........................(14)}.$$

We have, by (6),

$$N_i = \frac{1}{4e}\left\{(i+2)(i+1) + \frac{(i+4)(i+1)}{N_{i+2}}\right\} \quad \text{......(15)}.$$

From this equation applied to any moderately great even value of i (greater or less great according to the degree of approximation required), taking $N_{i+2} = \infty$, calculate N_i, and then, by successive applications N_{i-2}, N_{i-4}, ... N_6, N_4 successively. Equations (7), with

$$B_4 = -N_4 B_6 \quad \text{.......................(16)},$$

then give

$$B_0 = -\frac{3H}{3+2N_4}, \quad B_4 = e\frac{2N_4H}{3+2N_4}, \quad B_6 = -e\frac{2H}{3+2N_4} \quad \text{...(17)}.$$

Thus, finally, the solution is

$$a' = \frac{2eH}{3+2N_4}\left\{-\frac{3}{2e} + N_4\mu^4 - \mu^6 + \frac{\mu^8}{N_6} - \frac{\mu^{10}}{N_6 . N_8} + \frac{\mu^{12}}{N_6 . N_8 . N_{10}} - \&c.\right\} \quad \text{...(18)},$$

where N_4, N_6, N_8, ... are functions of e determined by (15).

28. General Integration of Laplace's Differential Equation of the Tides.

[From the *Philosophical Magazine*, L. 1875, pp. 388—402.]

1. Laplace considers the ocean as a rotating mass of frictionless incompressible liquid covering a rotating rigid spheroid to a depth everywhere infinitely small in proportion to the radius, and investigates its oscillations under the influence of periodic disturbing forces, with the limitation that the rise and fall is nowhere more than an infinitely small fraction of the depth, the condition that the mean angular velocity of every part of the liquid is the same as that of the solid, and the assumption that the distance from summit to summit of the disturbed water-surface is nowhere less than a large multiple of the depth. This last assumption is, though not explicitly stated by Laplace, implied in, and is virtually equivalent to, his assumptions (*Mécanique Céleste*, Livre I. No. 36) that the vertical motion of the water is small in comparison with its horizontal motion, and that the horizontal motion is sensibly the same for all depths.

2. Let now h be the elevation of the water-surface above mean level, and ξ and $\eta \sin\theta$ the southward and eastward horizontal component displacements of the water at time t, and at the place whose north latitude is $\frac{1}{2}\pi - \theta$ (or north-polar distance θ) and east longitude ψ. The "equation of continuity" [*Méc. Cél.* Liv. I. No. 36, or Airy, "Tides and Waves" (*Encyclopædia Metropolitana*), art. (72)] is

$$\frac{d(\gamma\xi)}{d\theta} + \frac{\gamma\xi\cos\theta}{\sin\theta} + \frac{d(\gamma\eta)}{d\psi} + h = 0 \quad\text{..............(1)},$$

or

$$\frac{d(\gamma\xi\sin\theta)}{\sin\theta\, d\theta} + \frac{d(\gamma\eta)}{d\psi} + h = 0 \quad\text{............(1) } bis,$$

where γ denotes the ratio of the depth of the sea to the earth's radius. And the dynamical equations [*Méc. Cél.* Liv. I. No. 36 (M), Airy (87)], are

$$\left.\begin{aligned} \frac{d^2\xi}{dt^2} - 2n\sin\theta\cos\theta\frac{d\eta}{dt} &= -\frac{g}{r^2}\frac{d(h-e)}{d\theta} \\ \sin^2\theta\frac{d^2\eta}{dt^2} + 2n\sin\theta\cos\theta\frac{d\xi}{dt} &= -\frac{g}{r^2}\frac{d(h-e)}{d\psi} \end{aligned}\right\} \quad \text{......(2)},$$

where r denotes the earth's radius, n the angular velocity of its rotation, g the force of gravity at its surface, and e the "equilibrium tide-height" at time t, and co-latitude and longitude θ and ψ; that is to say [Thomson and Tait's *Natural Philosophy*, § 805], the height at which the water would stand above the mean level if it were so placed at rest relatively to the rotating solid that it would remain at rest if the disturbing force were kept constantly what it is in reality at time t.

3. Laplace remarks that the general integration of these equations presents great difficulties; and he confines himself to a very extensive case, that in which γ is a function of latitude simply, and is the same in all longitudes. In this case the complete integration is to be effected by assuming

$$\left.\begin{aligned} h &= Hr\cos(\sigma t + s\psi) \\ \xi &= a\cos(\sigma t + s\psi) \\ \eta &= b\sin(\sigma t + s\psi) \end{aligned}\right\} \quad \text{....................(3)},$$

provided the disturbing force is such that

$$e = Er\cos(\sigma t + s\psi) \quad \text{....................(4)},$$

where H, a, b, E are functions of the latitude, of which E is given, and H, a, b are to be found by integration of the equations. With this assumption (1) *bis* and (2) give

$$\frac{d(\gamma a\sin\theta)}{\sin\theta d\theta} + s\gamma b + H = 0 \quad \text{..................(5)},$$

$$\left.\begin{aligned} \sigma^2 a + 2n\sigma\sin\theta\cos\theta . b &= \frac{g}{r}\frac{d(H-E)}{d\theta} \\ \sigma^2 b + 2n\sigma\frac{\cos\theta}{\sin\theta}a &= -\frac{g}{r}\frac{s(H-E)}{\sin^2\theta} \end{aligned}\right\} \quad \text{...........(6)},$$

Putting, in these, $\qquad H - E = u$ ....(7), we find

$$\left.\begin{aligned} a &= \frac{g}{r}\,\frac{\dfrac{du}{d\theta} + \dfrac{2ns}{\sigma}\dfrac{\cos\theta}{\sin\theta}u}{\sigma^2 - 4n^2\cos^2\theta} \\ b &= -\frac{g}{r}\,\frac{\dfrac{2n}{\sigma}\dfrac{\cos\theta}{\sin\theta}\dfrac{du}{d\theta} + \dfrac{su}{\sin^2\theta}}{\sigma^2 - 4n^2\cos^2\theta} \end{aligned}\right\} \ldots\ldots\ldots\ldots\ldots(8),$$

and then, eliminating a, b, H from (5), by (7) and (8),

$$\frac{g}{r}\left\{\frac{d}{\sin\theta\, d\theta}\,\frac{\gamma\left(\sin\theta\dfrac{du}{d\theta} + s\dfrac{2n}{\sigma}\cos\theta\, u\right)}{\sigma^2 - 4n^2\cos^2\theta} - \frac{s\gamma\left(\dfrac{2n}{\sigma}\dfrac{\cos\theta}{\sin\theta}\dfrac{du}{d\theta} + \dfrac{su}{\sin^2\theta}\right)}{\sigma^2 - 4n^2\cos^2\theta}\right\} + u = -E \ldots(9).$$

This is Laplace's differential equation of the tides [*Mécanique Céleste*, Liv. IV. No. 3, equation (4); or Airy, "Tides and Waves," *Encyclopædia Metropolitana*, art. (95)]. It is a linear differential equation of the second order, the complete integration of which gives u, and thence, by (8), a and b, in terms of θ, with two arbitrary constants to be determined so as to fulfil proper terminal conditions (§§ 11–17, below). It is essentially in the form in which Airy gave it, being that in which it comes direct from the formulæ preceding it in the investigation. It originally appeared in the *Mécanique Céleste*, masked somewhat by the addition and subtraction of a certain term which gives it a different form, not seeming at first sight better or simpler; but this as it were capricious modification suggests the following very substantial simplification.

4. Put $(\sin\theta)^{\frac{2ns}{\sigma}} u = \phi$, and $(\sin\theta)^{\frac{2ns}{\sigma}} E = \Phi$.........(10); then we have

$$\left.\begin{aligned} a &= \frac{g}{r}(\sin\theta)^{-\frac{2ns}{\sigma}+1}\,\frac{\dfrac{d\phi}{\sin\theta\, d\theta}}{\sigma^2 - 4n^2\cos^2\theta} \\ b &= -\frac{g}{r}(\sin\theta)^{-\frac{2ns}{\sigma}}\left\{\frac{\dfrac{2n}{\sigma}\cos\theta\dfrac{d\phi}{\sin\theta\, d\theta}}{\sigma^2 - 4n^2\cos^2\theta} + \frac{s\phi}{\sigma^2\sin^2\theta}\right\} \end{aligned}\right\} \ldots(11).$$

If (with Laplace) we put $\cos\theta = \mu$, and for brevity

$$n^2 r/g = m, \text{ and } \sigma/2n = f \quad \ldots\ldots\ldots\ldots\ldots(12),$$

these equations (11) become

$$\left.\begin{aligned} a &= -\frac{1}{4m}(\sin\theta)^{-\frac{s}{f}+1}\frac{\dfrac{d\phi}{d\mu}}{f^2-\mu^2} \\ b &= \frac{1}{4m}(\sin\theta)^{-\frac{s}{f}}\left\{\frac{\dfrac{1}{f}\mu\dfrac{d\phi}{d\mu}}{f^2-\mu^2} - \frac{s\phi}{f^2\sin^2\theta}\right\} \end{aligned}\right\} \quad \ldots\ldots(13).$$

Using these instead of (8) in the process by which (9) was found above, and multiplying the resulting equation by $4m(\sin\theta)^{s/f+2}$, we find

$$(1-\mu^2)^2\frac{d}{d\mu}\frac{\gamma}{f^2-\mu^2}\frac{d\phi}{d\mu} + 2\left(\frac{s}{f}-1\right)\frac{\gamma\mu(1-\mu^2)}{f^2-\mu^2}\frac{d\phi}{d\mu}$$

$$+\left[-\frac{s^2}{f^2}\gamma + 4m(1-\mu^2)\right]\phi = -4m(1-\mu^2)\Phi \ldots(14).$$

5. To integrate this, take first the case of $\Phi = 0$ (free oscillations), and assume

$$\frac{1}{f^2-\mu^2}\frac{d\phi}{d\mu} = K_0 + K_1\mu + K_2\mu^2 + \ldots + K_i\mu^i + \&c. \quad \ldots\ldots(15).$$

This gives

$$\phi = C + \mu f^2 K_0 + \frac{\mu^2}{2}f^2K_1 + \frac{\mu^3}{3}(f^2K_2 - K_0) + \ldots$$

$$+ \frac{\mu^i}{i}(f^2K_{i-1} - K_{i-3}) + \&c. \ldots(16),$$

where C denotes a constant of integration. Now let ϖ denote a symbol of operation such that

$$\varpi K_i = K_{i-1}, \text{ or generally } \varpi F(i) = F(i-1) \quad \ldots(17),$$

$F(i)$ being any function of i. By aid of this notation we may write (16) short thus,

$$\phi = \Sigma\mu^i\frac{1}{i}(f^2 - \varpi^2)\varpi K_i \quad \ldots\ldots\ldots\ldots\ldots\ldots(18);$$

understanding that, when $i = 0$,

$$\frac{f^2K_{i-1} - K_{i-3}}{i} = C \quad \ldots\ldots\ldots\ldots\ldots\ldots\ldots(19),$$

and that $\qquad K_i = 0$ for all negative values of i ....(20).

Let now $\gamma(\varpi)$ denote a symbol of operation obtained by putting ϖ for μ in γ (which, be it remembered, is a function of μ). Then from (18) we have

$$\left[-\frac{s^2}{f^2}\gamma+4m(1-\mu^2)\right]\phi$$

$$=\Sigma\mu^i\left[-\frac{s^2}{f^2}\gamma(\varpi)+4m(1-\varpi^2)\right]\frac{1}{i}(f^2-\varpi^2)\varpi K_i \text{...(21).}$$

Going back to (15) we have

$$\gamma\,.\,\frac{\mu(1-\mu^2)}{f^2-\mu^2}\frac{d\phi}{d\mu}=\Sigma\mu^i\gamma(\varpi)\,\varpi(1-\varpi^2)K_i \text{ ....(22),}$$

$$\frac{d}{d\mu}\left(\frac{\gamma}{f^2-\mu^2}\frac{d\phi}{d\mu}\right)=\Sigma\mu^i(i+1)\,\varpi^{-1}\gamma(\varpi)K_i \text{ ....(23),}$$

and

$$(1-\mu^2)^2\frac{d}{d\mu}\left(\frac{\gamma}{f^2-\mu^2}\frac{d\phi}{d\mu}\right)$$

$$=\Sigma\mu^i(1-\varpi^2)^2(i+1)\,\varpi^{-1}\gamma(\varpi)K_i$$

$$=\Sigma\mu^i(1-2\varpi^2+\varpi^4)(i+1)\,\varpi^{-1}\gamma(\varpi)K_i$$

$$=\Sigma\mu^i[i+1-2(i-1)\varpi^2+(i-3)\varpi^4]\,\varpi^{-1}\gamma(\varpi)K_i$$

$$=\Sigma\mu^i[i(1-\varpi^2)^2+1+2\varpi^2-3\varpi^4]\,\varpi^{-1}\gamma(\varpi)K_i$$

$$=\Sigma\mu^i[i(1-\varpi^2)+1+3\varpi^2](1-\varpi^2)\,\varpi^{-1}\gamma(\varpi)K_i$$

$$=\Sigma\mu^i[i(1-\varpi^2)+1+3\varpi^2](1-\varpi^2)\gamma(\varpi)K_{i+1} \text{ ...(24).}$$

Lastly, using (24), (22), and (21) in (14), and equating to zero the coefficient of μ^i, we have

$$\left\{[i(1-\varpi^2)+1+\left(\frac{2s}{f}+1\right)\varpi^2](1-\varpi^2)\gamma(\varpi)\right.$$

$$\left.+\left[-\frac{s^2}{f^2}\gamma(\varpi)+4m(1-\varpi^2)\right]\frac{1}{i}(f^2-\varpi^2)\varpi^2\right\}K_{i+1}=0 \text{ ...(25).}$$

By giving i successively in this formula all integral values from $-\infty$ to 0 and $+\infty$, and attending to (19) and (20), we have a succession of equations which successively determine K_1, K_2, K_3, &c. in terms of the arbitraries C and K_0; and using the values found in (16) we have the complete solution sought.

6. Laplace takes $\gamma=l(1-q\mu^2)$,

where l and q are constants; so that the bottom and the undisturbed free surface of the water may be both elliptic spheroids of revolution. With this or any other rational integral function of μ for γ, there is no difficulty in developing (§ 7 below) the first member of (25), and working out a practical solution of the problem. Laplace's most interesting and instructive results, however, are confined to the case of an ocean of uniform depth (for which in his notation $q=0$, or $\gamma=$ constant). Taking this case first, putting

$$4m/\gamma = \alpha \quad \text{.........................(26)},$$

and expanding the first member of (25), we have

$$(i+1)K_{i+1} + \left(-2i + \frac{2s}{f} - \frac{s^2 - \alpha f^2}{i}\right)K_{i-1}$$
$$+\left[i - \frac{2s}{f} - 1 + \frac{s^2 - \alpha f^2}{if^2} - \frac{\alpha f^2}{i-2}\right]K_{i-3} + \frac{\alpha}{i-2}K_{i-5} = 0 \quad \text{...(27)}.$$

For all negative values of i up to -2 this equation is an identity in virtue of (20); and for $i=-1$ it becomes

$$0K_0 = 0 \quad \text{.........................(28)},$$

and leaves K_0 arbitrary. For $i=0$ it becomes, in virtue of (19) and (20),

$$K_1 - \frac{s^2 - \alpha f^2}{f^2}C = 0 \quad \text{...................(29)};$$

for $i=1$, in virtue of (20),

$$2K_2 + \left(-2 + \frac{2s}{f} - s^2 + \alpha f^2\right)K_0 = 0 \quad \text{...........(30)};$$

and for $i=2$, in virtue of (19),

$$3K_3 + \left(-4 + \frac{2s}{f} - \frac{s^2 - \alpha f^2}{2}\right)K_1 - \alpha C = 0 \quad \text{......(31)}.$$

Then directly, for $i=3$, $i=4$, etc.

$$\left.\begin{aligned} &4K_4 + \left(-6 + \frac{2s}{f} - \frac{s^2 - \alpha f^2}{3}\right)K_2 \\ &\qquad + \left(2 - \frac{2s}{f} + \frac{s^2 - \alpha f^2}{3f^2} - \alpha f^2\right)K_0 = 0 \\ &5K_5 + \left(-8 + \frac{2s}{f} - \frac{s^2 - \alpha f^2}{4}\right)K_3 \\ &\qquad + \left(3 - \frac{2s}{f} + \frac{s^2 - \alpha f^2}{4f^2} - \frac{\alpha f^2}{2}\right)K_1 = 0 \end{aligned}\right\} \quad \text{......(32)},$$

$$\left.\begin{aligned}
&6K_6+\left(-10+\frac{2s}{f}-\frac{s^2-af^2}{5}\right)K_4\\
&\qquad+\left(4-\frac{2s}{f}+\frac{s^2-af^2}{5f^2}-\frac{af^2}{3}\right)K_2+\frac{a}{3}K_0=0\\
&7K_7+\left(-12+\frac{2s}{f}-\frac{s^2-af^2}{6}\right)K_5\\
&\qquad+\left(5-\frac{2s}{f}+\frac{s^2-af^2}{6f^2}-\frac{af^2}{4}\right)K_3+\frac{a}{4}K_1=0\\
&8K_8+\left(-14+\frac{2s}{f}-\frac{s^2-af^2}{7}\right)K_6\\
&\qquad+\left(6-\frac{2s}{f}+\frac{s^2-af^2}{7f^2}-\frac{af^2}{5}\right)K_4+\frac{a}{5}K_2=0
\end{aligned}\right\}\text{......(33)},$$

and so on.

Of these equations, (29), (31), the second of (32), the second of (33), and every second equation thenceforward, determine successively K_1, K_3, K_5, K_7, and so forth, all in terms of the arbitrary C; and (30), the first of (32) the first and third of (33), determine successively K_2, K_4, K_6, &c. in terms of the arbitrary K_0.

7. Returning now to the more general supposition of the depth varying with the latitude, we may assume, without practically restricting the problem further,

$$\gamma=\gamma_0+\gamma_1\mu+\gamma_2\mu^2+\ldots+\gamma_n\mu^n \text{ ....(34)},$$

$\gamma_0, \gamma_1, \ldots \gamma_n$ being given constants. This makes

$$\gamma(\varpi)K_i=\gamma_0K_i+\gamma_1K_{i-1}+\gamma_2K_{i-2}+\ldots+\gamma_nK_{i-n}\text{...(35)}.$$

Using this in (25) and proceeding precisely as in § 6, we find K_1, K_2, K_3, K_4, K_5, &c., each in terms of two arbitraries C and K_0—unless γ contains only even powers of μ, in which case, as in that of uniform depth (§ 6), we find $K_1, K_3, K_5, \ldots$ in terms of one arbitrary C alone, and K_2, K_4, $K_6\ldots$ in terms of the other arbitrary K_0 alone. The first two of the equations by which this is done, those namely which correspond to (29) and (30), being found by putting $i=0$ and $i=1$ in (25), are

$$K_1+\frac{\gamma_1}{\gamma_0}K_0-\left(\frac{s^2}{f^2}-\frac{4m}{\gamma_0}\right)C=0 \text{ ....(36)}$$

and

$$2K_2+2\frac{\gamma_1}{\gamma_0}K_1+\left(-2+\frac{2s}{f}-s^2+\frac{4m}{\gamma_0}f^2+\frac{2\gamma_2}{\gamma_0}\right)K_0-\frac{s^2}{f^2}\frac{\gamma_1}{\gamma_0}C=0 \text{ ....(37)}.$$

8. In §§ 5, 6, 7 we supposed $\Phi=0$, and so made, for the time, "free oscillations" our subject. Now suppose Φ to be any given function of μ. For the actual problem of tides of any species, it is a rational integral function of μ, or of μ and $\sqrt{(1-\mu^2)}$, if we neglect the influence produced by the change of attraction of the water due to its change of figure. A proper way of taking into account this influence by successive approximations will be explained later. Meantime, without losing generality, I assume

$$\Phi=\Phi_0+\Phi_1\mu+\Phi_2\mu^2+\ldots+\Phi_i\mu^i+\&c. \quad \ldots\ldots(38),$$

where Φ_0, Φ_1, Φ_2, &c. are given constants, either finite in number, or of such magnitudes as to render the series convergent for values of μ within the limits used in each particular case. With this for Φ, the second member of (14) becomes

$$-4m\Sigma\mu^i(\Phi_i-\Phi_{i-2}) \quad \ldots\ldots\ldots\ldots\ldots\ldots(39),$$

and instead of (25) we have

$$\left\{\left[i(1-\varpi^2)+1+\left(\frac{2s}{f}+1\right)\varpi^2\right](1-\varpi^2)\gamma(\varpi)\right.$$
$$\left.+\left[-\frac{s^2}{f^2}\gamma(\varpi)+4m(1-\varpi^2)\right]\frac{1}{i}(f^2-\varpi^2)\varpi^2\right\}K_{i+1}=-4m(\Phi_i-\Phi_{i-2})$$
$$\ldots\ldots\ldots(40).$$

The proper modification, according to this formula, must be made in (27), and in each of the particular equations (29), (30), (31), (32), (33), (36), (37) when required.

9. Before considering the conditions which may be fulfilled by proper determination of the two arbitrary constants C and K_0, it is convenient to investigate the convergency of the series (16) which we have found for the complete solution. For this purpose put (40) [including (25) as the case for which $\Phi=0$] into the form

$$(1-\varpi^2)^2\gamma(\varpi)K_{i+1}=\frac{1}{i}\left\{-\left[1+\left(\frac{2s}{f}+1\right)\varpi^2\right](1-\varpi^2)\gamma(\varpi)K_{i+1}\right.$$
$$\left.+\left[\frac{s^2}{f^2}\gamma(\varpi)-4m(1-\varpi^2)\right]\frac{1}{i}(f^2K_{i-1}-K_{i-3})-4m(\Phi_i-\Phi_{i-2})\right\}$$
$$\ldots\ldots\ldots(41).$$

In a certain very important class of cases, of which the first example known to mathematicians is that so splendidly and successfully treated by Laplace in the process defended and controverted in the two preceding Numbers of this Magazine, terms of the second member of this equation are, for infinitely great values of i, comparable in magnitude with terms of the first member, through $\frac{K_{i-1}}{K_{i+1}}$ or $\frac{K_i}{K_{i+2}}$, being infinitely great of the order i^2. These cases can only occur when γ is either constant or expressed in (34) by the first two terms, $\gamma_0 + \gamma_1\mu$. Reserving them for consideration later, we see by (41) that, except in those special cases, K_i must for very great values of i fulfil, more and more nearly the greater is i, the equation

$$(1 - \varpi^2)^2 \gamma(\varpi) K_{i+1} = 0 \quad \text{.................(42).}$$

Calling κ_i the complete and rigorous solution of this equation in finite differences, we have

$$\kappa_i = l + l'i + (l'' + l'''i)(-1)^i + \frac{k}{\rho^i} + \frac{k'}{\rho'^i} + \text{\&c.} \quad \text{...(43),}$$

where ρ, ρ', &c. denote the roots of the equation $\gamma = 0$, and l, l', l'', l''', k, k', &c. constants. Hence for great values of i, K_i must be approximately equal to (43) with some particular values for the constants l, l', &c. But for very great values of i all the terms of (43) except one leading term, or [because of the equal roots of $(1 - \varpi^2)^2 = 0$] one leading pair of terms, vanish in comparison with this term or pair of terms. Hence we must have, for very great values of i,

$$\left.\begin{array}{ll} & K_i = l + l''(-1)^i, \text{ or } K_i = [l' + l'''(-1)^i]\, i \\ \text{or} & K_i = \dfrac{k}{\rho^i}, \text{ or } K_i = \dfrac{k'}{\rho'^i}, \text{ and so on} \end{array}\right\} \quad \text{...(44).}$$

Thus we see that if each of the roots ρ, ρ', &c. is greater than unity, the series (15) and (16) are necessarily convergent for all values of μ from $\mu = -1$ to $\mu = +1$, and they are divergent for values of μ beyond these limits unless conditions proper to make $l = 0$, $l' = 0$, $l'' = 0$, $l''' = 0$ are fulfilled. But if one or more of the roots ρ, ρ', &c. is less than unity, and ρ the absolutely least of them all, then unconditionally the series (15) and (16) are necessarily convergent for all values of μ from $-\rho$ to $+\rho$, and they are divergent for all values of μ beyond these limits unless a condition proper to make $k = 0$ is fulfilled*. When $\gamma = 0$ has imaginary

* Mr W. H. L. Russell, as I am informed by himself and Professor Cayley, has given, in perfectly general terms, this criterion for the convergency of the series in ascending powers of x for the integral of

$$\phi(x)\frac{d^2u}{dx^2} + \psi(x)\frac{du}{dx} + \chi(x) = 0,$$

in a paper communicated to the Royal Society, of which certain extracts have been published in the *Proceedings* for 1870, 1871, 1872.

roots, as $\alpha \pm \beta\sqrt{-1}$, the absolute magnitude of either of the pair is to be reckoned as $(\alpha^2+\beta^2)^{\frac{1}{2}}$, and with this understanding the same statement as to convergency and divergency holds as for real roots. But there is this distinction in the circumstances of the loss of convergency in the two cases, of transition through a real root and through the absolute value of a pair of imaginary roots. In the latter case there is no discontinuity when μ is continuously increased through the critical value $\sqrt{(\alpha^2+\beta^2)}$; in the former, ϕ and its differential coefficients become infinite and imaginary, as μ is increased continuously up to and beyond any real root of $\gamma=0$. The interpretation of the circumstances when imaginary roots of $\gamma=0$ influence the solution is an exceedingly interesting subject, to which I hope to return in a future communication. The remainder of the present article must be confined to the case of $\gamma=0$ having two real roots, each less than unity.

10. Let ρ be any real root of $\gamma=0$, and put $\mu=z+\rho$. Then, for infinitely small values of z, the differential equation (14) becomes

$$a\frac{d}{dz}\left(z\frac{d\phi}{dz}\right)+bz\frac{d\phi}{dz}+(c+\mathrm{d}z)\phi=e+fz\ldots\ldots\ldots(45),$$

where a, b, c, d, e, f denote constants. The complete solution of this approximate equation may be found by assuming

$$\left.\begin{aligned}\phi=\log z\,(H_0+H_1z+H_2z^2+\&c.)\\ +K_0+K_1z+K_2z^2+\&c.\end{aligned}\right\}\ldots\ldots\ldots\ldots(46),$$

and determining H_1, H_2, &c. in terms of H_0, arbitrary, by equating coefficients of $\log z$, $z\log z$, $z^2\log z$, &c. to zero, and lastly determining K_1, K_2, K_3 in terms of K_0 and H_0, each arbitrary, and H_1, H_2, H_3, &c. previously found. This shows the kind of discontinuity which any complete solution of the exact equation (14) necessarily presents when the value of μ passes through a real root of $\gamma=0$, and how this discontinuity is averted by an assignment of the two constants of integration in the rigorous solution proper to make $H_0=0$ in the approximate solution (46).

11. Return now to the question (§ 9) of assigning the two constants of integration so as to fulfil any proper physical conditions of our problem. First, to work out the general solution

in ascending powers of μ, use (40), and calculate K_1, K_2, &c. successively with C and K_0 arbitrary. Thus we find

$$K_i = C\alpha_i + K_0\beta_i + \Phi_0\lambda_i^{(0)} + \Phi_1\lambda_i^{(1)} + \Phi_2\lambda_i^{(2)} + \&c. \ldots (47),$$

where α_i, β_i, $\lambda_i^{(0)}$, $\lambda_i^{(1)}$, $\lambda_i^{(2)}$, &c. are numbers calculated by the process, supposing f, s, m, γ_0, γ_1, γ_2, &c. to have had any particular numerical values assigned to them, and Φ_0, Φ_1, Φ_2,... to denote given heights. Or if before we begin the arithmetical process particular values are assigned to Φ_0, Φ_1, Φ_2, &c. so that we may put

$$\Phi_0 = n_0 L, \quad \Phi_1 = n_1 L, \quad \Phi_2 = n_2 L, \&c. \ldots\ldots\ldots (48),$$

L denoting a given line, and n_0, n_1, n_2, &c. given numerical quantities, the result of the process of calculation of K_i from (40) will take the form

$$K_i = \alpha_i C + \beta_i K_0 + \lambda_i L \ldots\ldots\ldots\ldots\ldots\ldots (49),$$

where α_i, β_i, λ_i are calculated numbers. Then we have, by (15) and (16),

$$\left.\begin{aligned} \frac{1}{f^2-\mu^2}\frac{d\phi}{d\mu} &= \alpha(\mu).C + \beta(\mu).K_0 + \lambda(\mu).L \\ \text{where} \qquad \alpha(\mu) &= \alpha_0 + \alpha_1\mu + \alpha_2\mu^2 + \&c. \\ \beta(\mu) &= \beta_0 + \beta_1\mu + \beta_2\mu^2 + \&c. \\ \lambda(\mu) &= \lambda_0 + \lambda_1\mu + \lambda_2\mu^2 + \&c. \end{aligned}\right\} \ldots\ldots (50),$$

$$\text{and} \qquad \phi = \bar{\alpha}(\mu).C + \bar{\beta}(\mu).K_0 + \bar{\lambda}(\mu).L$$

$$\left.\begin{aligned} \text{where } \bar{\alpha}(\mu) &= 1 + f^2\alpha_0\mu + \tfrac{1}{2}f^2\alpha_1\mu^2 + \tfrac{1}{3}(f^2\alpha_2 - \alpha_0)\mu^3 + \&c. \\ \bar{\beta}(\mu) &= \qquad f^2\beta_0\mu + \tfrac{1}{2}f^2\beta_1\mu^2 + \tfrac{1}{3}(f^2\beta_2 - \beta_0)\mu^3 + \&c. \\ \bar{\lambda}(\mu) &= \qquad f^2\lambda_0\mu + \tfrac{1}{2}f^2\lambda_1\mu^2 + \tfrac{1}{3}(f^2\lambda_2 - \lambda_0)\mu^3 + \&c. \end{aligned}\right\} (51).$$

It remains to determine the constants of integration so as to fulfil prescribed conditions rendering the problem determinate. This we shall actually do for two typical cases:—first, the sea bounded north and south by two vertical cliffs; secondly, by two sloping beaches with gradual deepening from each to a single maximum depth along an intermediate parallel of latitude.

12. First, let the ocean be a belt of water between vertical cliffs in two given latitudes, either both in the same hemisphere or one north and the other south. The conditions of this case are that there is no north and south motion of the water at either

of the bounding parallels of latitude; and they are to be fulfilled [§ 4 (13)], by putting

$$d\phi/d\mu = 0 \quad\text{.........................(52)}$$

for each of the terminal values of μ (that is to say, the sines of the bounding latitudes). If each of these is less in absolute value than the least root of $\gamma = 0$, each of the series in (50) and (51) is convergent through the whole range of values of μ corresponding to the supposed ocean.

Calling, then, μ', μ'' the sines of the two bounding latitudes (to be reckoned negative for south latitude if either or both be south), we have, by using (52), in (50),

$$\left.\begin{array}{ll} & \alpha(\mu').C + \beta(\mu').K_0 + \lambda(\mu').L = 0 \\ \text{and} & \alpha(\mu'').C + \beta(\mu'').K_0 + \lambda(\mu'').L = 0 \end{array}\right\} \text{.........(53)},$$

which give

$$\left.\begin{aligned} C &= -\frac{\beta(\mu'').\lambda(\mu') - \beta(\mu').\lambda(\mu'')}{\beta(\mu'').\alpha(\mu') - \beta(\mu').\alpha(\mu'')} L \\ K_0 &= \frac{\alpha(\mu'').\lambda(\mu') - \alpha(\mu').\lambda(\mu'')}{\beta(\mu'').\alpha(\mu') - \beta(\mu').\alpha(\mu'')} L \end{aligned}\right\} \text{......(54)}.$$

With these values for C and K_0, (50) and (51) give $\dfrac{1}{f^2-\mu^2}\dfrac{d\phi}{d\mu}$, and ϕ for every value of μ through the range of the supposed ocean; and then the following formulæ [which it is convenient to recall from (13), (7), (3), (10), (38), and (48) above] give h the height of the free surface, ξ the southward displacement, and $\eta\sqrt{(1-\mu^2)}$ the eastward displacement of the water at time t, latitude $\sin^{-1}\mu$, and longitude ψ:

$$\left.\begin{aligned} h &= \left\{\frac{\phi}{(1-\mu^2)^{ns/\sigma}} + E\right\}\cos(\sigma t + s\psi) \\ \xi &= -\frac{1}{4m(1-\mu^2)^{ns/\sigma-\frac{1}{2}}}\cdot\frac{1}{f^2-\mu^2}\frac{d\phi}{d\mu}\cdot\cos(\sigma t + s\psi) \\ \eta &= \frac{1}{4m(1-\mu^2)^{ns/\sigma}}\cdot\left\{\frac{\mu}{f}\cdot\frac{1}{f^2-\mu^2}\frac{d\phi}{d\mu} - \frac{s}{f^2(1-\mu^2)}\phi\right\}\sin(\sigma t + s\psi) \end{aligned}\right\} \text{.........(55)},$$

where f denotes $2n/\sigma$, and m/r the ratio of equatorial centrifugal force to gravity.

This fully determined solution expresses the motion of the supposed zonal ocean due to a disturbing influence, of which the equilibrium tide-height is $E\cos(\sigma t+s\psi)$, E being expressed by the formula

$$E=\frac{L}{(1-\mu^2)^{ns/\sigma}}(n_0+n_1\mu+n_2\mu^2+\&c.) \quad \text{......(56)},$$

where n_0, n_1, n_2, &c. are any given numbers.

13. If $L=0$, equation (54) gives, except in a certain critical case to be considered presently, $C=0$ and $K_0=0$, and therefore the solution expresses determinately that there is no motion; that is to say, there cannot be any "free oscillation" of the assumed type and period, § 3 (3)*, except in the critical case alluded to. This critical case is the case in which the denominator of the expressions for C and K_0 vanishes, or

$$\frac{\beta(\mu'')}{\alpha(\mu'')}=\frac{\beta(\mu')}{\alpha(\mu')} \quad \text{......................(57)}.$$

Then (54) gives infinite values to C and K_0 unless L is zero; and if L is zero, (53) gives

$$\frac{C}{K_0}=-\frac{\beta(\mu'')}{\alpha(\mu'')}=-\frac{\beta(\mu')}{\alpha(\mu')} \quad \text{..................(58)},$$

thus determining the ratio of C to K_0 but leaving the magnitude of either indeterminate.

14. The problem of finding all the fundamental modes of free oscillation of our supposed zonal sea is solved by giving to s the values of 0, 1, 2, &c., and for each value of s treating (57) as a transcendental equation for the determination of σ. After the manner of Fourier, and Sturm and Liouville, it may be proved that this transcendental equation cannot have imaginary roots and has necessarily an infinite number of real roots more and more nearly equi-different when taken in order of magnitude from the smallest positive to larger and larger positive, or from the smallest negative to larger and larger negative. In the case of $s=0$ the positive and negative roots are equal, unequal in all other cases ($s=1$, $s=2$, &c.).

15. For the convergency of the series in (50) and (51) it is necessary and sufficient (§ 9) that there be no root, real or imaginary, of $\gamma=0$ whose absolute magnitude is less than that

* This equation defines perfectly the configuration of the assumed motion, and specifies also that its period is $2\pi/\sigma$, or its "speed" σ.

of the absolutely greater of the two quantities μ' and μ''. But it is only when, with algebraic signs taken into account, there is a real root actually between μ' and μ'' (that is to say, when γ becomes zero for some value of μ on the direct range from μ' to μ'') that any of the six functions $\alpha(\mu)$, $\beta(\mu)$, $\lambda(\mu)$, $\bar{\alpha}(\mu)$, $\bar{\beta}(\mu)$, $\bar{\lambda}(\mu)$ used in the processes (53), (54), (57), (58), and in the final solution (51), (50), (55), is discontinuous. Why some or all of these functions should be discontinuous in this case is obvious: the sea's depth being zero along any parallel of latitude limits the physical problem to the side on which the depth is positive, or (case of equal roots of $\gamma=0$) separates the problem into two independent ones, to find the motions of the water on the two sides of a reef just "awash." An imaginary root of $\gamma=0$ having its absolute magnitude R between μ' and μ'', or a real root of contrary sign to the absolutely greater of μ' and μ'', and of absolute magnitude R between them, renders the series for $\alpha(\mu)$, $\beta(\mu)$, &c. in ascending powers of μ divergent for the portion of our range of latitude which lies beyond $\pm\sin^{-1}R$. Still the solution of the problem is fully given by (55) in terms of six functions $\alpha(\mu)$, $\beta(\mu)$, &c., *each continuous throughout the range*, but calculable, by the series in ascending powers of μ set forth in our preceding formulæ, only for the part of the range of latitude which lies between $-\sin^{-1}R$ and $+\sin^{-1}R$. The mode of dealing with the case of imaginary roots so as to obtain convenient formulæ for the numerical calculation of $\alpha(\mu)$ &c. is an interesting and important subject to which I hope to return. Being (§ 9) at present limited to the case of real roots, it is enough to remark that in this case for each of the six functions $\alpha(\mu)$, $\beta(\mu)$, &c. a series continuously convergent throughout the range from μ' to μ'' may be found thus:—Let ρ and ρ' be consecutive real roots of $\gamma=0$, and let ρ, μ', μ'', ρ' be in order of algebraic magnitude. Let a be any quantity such that algebraically

$$a>\tfrac{1}{2}(\mu''+\rho) \text{ and } a<\tfrac{1}{2}(\mu'+\rho') \qquad (59).$$

Then, putting

$$\mu=z+a \qquad (60)$$

in § 4 (14), and working precisely as in § 5, but with z instead of μ in the second member of (15) and the proper corresponding modification of (16) &c., we obtain a solution in ascending powers of z, or $\mu-a$, which is necessarily convergent throughout the range of our problem. The degree of convergence of the series so found for each of the six functions,

$$\alpha(z+a),\quad \beta(z+a),\quad \lambda(z+a),$$
$$\bar{\alpha}(z+a),\quad \bar{\beta}(z+a),\quad \bar{\lambda}(z+a),$$

is, for any value of z, the same as that of the geometrical series

$$1 \pm \frac{z}{c} + \left(\frac{z}{c}\right)^2 \pm \text{\&c.},$$

where c is the less of the two quantities $a-\rho$, $\rho'-a$.

16. For our second proposed case (§ 11) let $\rho_{\prime}, \rho, \rho', \rho''$ be four consecutive roots of $(1-\mu^2)\gamma=0$; let ρ, ρ' be each between -1 and $+1$; and let γ be positive for values of μ between ρ and ρ'. Required, to determine tides, and the free oscillations, of the zone of water corresponding to these intermediate values of ρ. Take any quantity, a, between $\rho_{\prime}$ and ρ'', such that $\rho \sim a$ and $\rho' \sim a$ are each less than the less of the two differences $a-\rho_{\prime}$, $\rho''-a$. Put $\mu=z+a$, and solve in ascending powers of z, as in § 15. Let α_i, β_i, λ_i be the coefficients of z^i in the series thus found for $\alpha(\mu)$, $\beta(\mu)$, $\lambda(\mu)$ in formulæ corresponding to (50), but with z for μ in the second members, so that we have

$$\frac{1}{f^2-\mu^2}\frac{d\phi}{d\mu} = (\Sigma\alpha_i z^i)\,C + (\Sigma\beta_i z^i)\,K_0 + (\Sigma\lambda_i z^i)\,L \quad \text{...(61).}$$

Let now p, q be two values of i, and put

$$\left.\begin{aligned} \alpha_p C + \beta_p K_0 + \lambda_p L = 0 \\ \alpha_q C + \beta_q K_0 + \lambda_q L = 0 \end{aligned}\right\} \quad \text{..............(62).}$$

If p, q, and $p-q$* be each infinitely great, the values of C and K_0 determined by these equations and used in (61) and (55), give the tides due to the tide-generating influence

$$\Phi = L\,(n_0 + n_1\mu + n_2\mu^2 + \text{\&c.}).$$

The periods of the fundamental free oscillations of the supposed zone of water are determined by finding σ so as to make

$$\alpha_p/\beta_p = \alpha_q/\beta_q \quad \text{.......................(63),}$$

and the oscillations are then expressed by taking

$$K_0 = -\frac{\alpha_p}{\beta_p}\,C = -\frac{\alpha_q}{\beta_q}\,C \quad \text{.....................(64)}$$

* Except in the case of $\rho-a=-(\rho'-a)$, when we must take $p-q=1$, or any odd integer; but $p-q=1$ is best in this case.

in (61) and (55). By giving moderately great values to p, q, and $p-q$, the rigorous solution may be satisfactorily approximated to by easy, methodical, and not very laborious arithmetic. The proof is obvious from § 9 (44).

17. Corresponding investigations to find solutions for the case of water over one or both poles must be reserved for a future article. They involve highly interesting extensions of Laplace's admirable process referred to in § 9 of the present article and in several places of the last two Numbers of the *Philosophical Magazine*.

[Sir George Darwin has by request kindly supplied the following note on these papers on the tides.

The procedure of Laplace in his discussion of tidal oscillations is now universally accepted as correct, and the paper (No. 26) on "an alleged error in Laplace's Theory" is acknowledged to have finally settled the controversy raised by the strictures of Airy and Ferrel.

With regard to the paper (No. 27) on the "oscillations of the first species," now commonly called tides of long period, I pointed out (*Proc. Roy. Soc.* Vol. 41 (1886), p. 337, or Vol. I of collected papers) that Laplace's argument was unconvincing, namely that friction was adequate to cause these tides to conform to the equilibrium theory. Following Lord Kelvin I found numerical solutions according to Laplace's method; other solutions have been found by Lamb (*Hydrodynamics*, § 216), and by Hough (see references on p. 231). It appeared from these solutions that on an ocean-covered planet the equilibrium theory might be widely in error.

Accepting however Laplace's argument as to friction, I evaluated (Thomson and Tait's *Nat. Phil.* § 840', or my collected papers, Vol. I) the elastic yielding of the solid earth as derived from observation of the oceanic tides of long period. Dr W. Schweydar (*Beitr. z. Geophysik*, Vol. 9 (1907), p. 41), using far more extensive data arrived at a closely similar result. Since the rigidity of the earth derived in this way agrees admirably with that found from observations with the horizontal pendulum, we may feel confident that Laplace was in fact right in supposing these oceanic tides to conform to the equilibrium law. The result cannot however be explained by friction; and at length Lord Rayleigh (*Phil. Mag.* Vol. V (1903), p. 136) showed that land barriers, as on the earth, would annul those modes of fluid motion which, in the case of the ocean-covered planet, cause so wide a divergence from the equilibrium law.

It would appear then that Laplace was correct in fact, as regards the earth, but wrong in his reasoning.

Further references on the subject will be found in Vol. VII of the *Encyklopädie der Mathematischen Wissenschaften*, Art. "Bewegung der Hydrosphäre."

The paper (No. 28) on "the general integration" of Laplace's tidal equation now possesses less interest than the two preceding ones, since its subject is to a large extent covered by the two memoirs of Hough (see p. 231), who also succeeded in introducing the effects of the mutual attraction of the water.]

WAVES ON WATER.

29. On Stationary Waves in Flowing Water.

[From the *Philosophical Magazine*, XXII. October 1886, pp. 353—357; having been read before Section A of the British Association, Birmingham, Sept. 7, 1886.]

Part I.

This subject includes the beautiful wave-group produced by a ship propelled uniformly through previously still water, but the present communication* is limited to two dimensional motion.

Imagine frictionless water flowing in uniform regime through an infinitely long canal with vertical sides; and bottom horizontal except where modified by transverse ridges or hollows, or slopes between portions of horizontal bottom at different levels. Included among such inequalities we may suppose bars above the bottom, fixed perpendicularly between the sides. Let these inequalities be all within a finite portion, AB, of the length, and let f denote the difference of levels of the bottom on the two sides of this portion, positive if the bottom beyond A is higher than the bottom beyond B.

Now, let the water be given at an infinite, or very great, distance beyond A, perpetually flowing towards A with any prescribed constant velocity u, and filling up the canal to a prescribed constant depth a. It is required to find the motion of the water towards A, through AB, and beyond B as disturbed by the inequalities between A and B. This problem is essentially

* I have since found, in a sufficiently practical form, the solution for the wave-group produced by the ship, which I hope to communicate to the *Philosophical Magazine* for publication in the November number.

determinate; and it has only one solution if we confine it to cases in which the vertical component of the water's velocity is everywhere small in comparison with the velocity acquired by a falling body falling from a height equal to half the depth. Let b be the mean depth, and v the mean horizontal velocity at very great distances beyond B; and (to have w to denote wave-energy) let w be such that

$$(\tfrac{1}{2}v^2 + \tfrac{1}{2}gb)\,b + w \;\ldots\ldots\ldots\ldots\ldots\ldots\ldots(1)$$

is the whole energy, kinetic and potential, per unit of the canal's breadth, and per unit of its length. In cases in which the water flows away unruffled at great distances from B, w is zero. But, in general, the surface is ruffled, and the water flows "*steadily*" between the plane bottom and a corrugated free surface, as in the well-known appearance of water flowing in a mill-lead, or Highland burn, or in the clear rivulet on the east side of Trumpington Street, Cambridge, or in the race of Portland, or Islay overfalls. The train of diminishing waves which we see in the wake of each little irregularity of the bottom would, of course, extend to infinity if the stream were infinitely long, and the water absolutely inviscid (frictionless); and a single inequality, or group of inequalities, in any part AB of the stream would give rise to corrugation in the whole of the flow after passing the inequalities, more and more nearly uniform, and with ridges and hollows more and more nearly perpendicular to the sides of the canal, the farther we are from the last of the inequalities. Observation, with a little common sense of the mathematical kind, shows that at a distance of two or three wave-lengths from the last of the irregularities if the breadth of the canal is small in comparison with the wave-length, or at a distance of nine or ten breadths of the canal if the breadth is large in comparison with the wave-length, the condition of uniform corrugations with straight ridges perpendicular to the sides of the canal, would be fairly well approximated to; even though the irregularity were a single projection or hollow in the middle of the stream. But the subject of the present communication is simpler, as it is limited to two-dimensional motion; and our inequalities are bars, or ridges, or hollows, perpendicular to the sides of the canal. Thus, in our present case, we see that the condition of ultimate uniformity of the standing waves in the wake of the irregularities is closely approximated to at a distance of two or three wave-lengths from the last of the inequalities.

Let SA, SB denote two fixed vertical sections of the canal at infinitely great distances beyond A and beyond B; and p, q the mean fluid pressures in these planes. It will simplify considerations and formulas if we take SB at a node (or place where the depth is equal to b, the mean depth), and we therefore take it so; although this is not necessary for the following kinematical and dynamical statements:—

I. The volumes of fluid crossing SA and SB in the same or equal times are equal; or, in symbols,

$$au = bv = M \quad\text{..........................(2)},$$

where M denotes the volume of water passing per unit of time.

II. The excess (positive or negative) of the work done by p on any volume of the water entering across SA, above the work done by q on an equal volume of the water passing away across SB is equal to the excess of the energy, potential and kinetic, of the water passing away above that of the water entering. Hence, and by (1), taking the volume of water unity, we have

$$p - q = \tfrac{1}{2}(v^2 + gb) + \frac{w}{b} - [\tfrac{1}{2}u^2 + g(f + \tfrac{1}{2}a)] \quad\text{......(3)}.$$

Now calling the pressure at the free surface zero, we have

$$p = \tfrac{1}{2}ga\text{; and } q = \tfrac{1}{2}gb - \frac{w'}{b} \quad\text{..............(4)};$$

w' denoting a quantity depending on wave-disturbance. Hence, and by (2),

$$\tfrac{1}{2}M^2\frac{a^2 - b^2}{a^2b^2} - g(a - b + f) + \frac{w - w'}{b} = 0 \quad\text{.........(5)}.$$

Now, put $$\frac{\tfrac{1}{2}(a+b)}{a^2b^2} = \frac{1}{D^3}\text{; and } M = VD \quad\text{..............(6)}.$$

Thus D will denote a mean depth (intermediate between a and b and approximately equal to their arithmetic mean, when their difference is small in comparison with either); and V will denote a corresponding mean velocity of flow (intermediate between u and v, and approximately equal to their arithmetic mean, when their difference is small in comparison with either).

With this notation, (5) gives

$$b - a = \left(f - \frac{w - w'}{gb}\right)\Big/\left(1 - \frac{V^2}{gD}\right) \quad\text{...........(7)}.$$

If $b-a$ were exactly equal to f, and if there were no beruffllement of the water beyond B, the mean level of the water would be the same in the entering and leaving water at great distances on the two sides of AB; but this is not generally the case, and there is a (positive or negative) rise of level, given by the formula

$$y = b - a - f = \left(\frac{V^2}{gD} f - \frac{w - w'}{gb}\right) \Big/ \left(1 - \frac{V^2}{gD}\right) \quad \text{......(8).}$$

Consider now the case of no corrugation (that is to say, of plane free surface and uniform flow) at great distances beyond B. We have $w - w' = 0$; and therefore

$$y = b - a - f = \left(\frac{V^2}{gD} f\right) \Big/ \left(1 - \frac{V^2}{gD}\right) \quad \text{...........(9);}$$

or, with V^2 replaced by M^2/D^2,

$$y = b - a - f = \left(\frac{M^2}{gD^3} f\right) \Big/ \left(1 - \frac{M^2}{gD^3}\right) \quad \text{.........(10),}$$

where, as above, $$D^3 = \frac{a^2 b^2}{\frac{1}{2}(a+b)} \quad \text{.........................(11).}$$

The elimination of b and D between these three equations gives y as a function of f. It is clear that the change of level of the bottom may be sufficiently gradual to obviate any of the corrugational effect; and when this is the case, the equation of the free surface will be found from y in terms of f; f being a given function of the horizontal coordinate, x.

If f is everywhere small in comparison with a, D is approximately constant [much more approximately equal to $\frac{1}{2}(a+b)$], and y is approximately in constant proportion to f.

When the flow is so gentle that V is small in comparison with $\sqrt{gD}$, $\frac{M^2}{gD^3}$ is a small proper fraction, and y is approximately equal to this fraction of f.

Generally, in every case when $V < \sqrt{gD}$ the upper surface of the water rises when the bottom falls, and the water falls when the bottom rises.

On the other hand, when $V > \sqrt{gD}$, the water surface rises convex over every projection of the bottom, and falls concave over hollows of the bottom; and the rise and fall of the water are each greater in amount than the rise and fall of the bottom; so that the water is deeper over elevations of the bottom, and is shallower over depressions of the bottom.

Returning now to the subject of standing waves (or corrugations of the surface) of frictionless water flowing over a horizontal bottom of a canal with vertical sides, I shall not at present enter on the mathematical analysis by which the effect of a given set of inequalities within a limited space AB of the canal's length, in producing such corrugations in the water after passing such inequalities, can be calculated, provided the slopes of the inequalities and of the surface corrugations are everywhere very small fractions of a radian. I hope before long to communicate a paper to the *Philosophical Magazine* on this subject for publication. I shall only just now make the following remarks:—

1. Any set of inequalities large or small must in general give rise to stationary corrugations large or small, but perfectly stationary, however large, short of the limit that would produce infinite convex curvature (according to Stokes's theory an obtuse angle of 120°) at any transverse line of the water surface.

2. But in particular cases the water flowing away from the inequalities may be perfectly smooth and horizontal. This is obvious because of the following reasons:—

(i) If water is flowing over a plane bottom with infinitesimal corrugations, an inequality which could produce such corrugations may be placed on the bottom so as either to double those previously existing corrugations of the surface or to annul them.

(ii) The wave-length (that is to say the length from crest to crest) is a determinate function of the mean depth of the water and of the height of the corrugations above the bottom, and of the volume of water flowing per unit of time. This function is determined graphically in Stokes's theory of finite waves. It is independent of the height, and is given by the well-known formula when the height is infinitesimal.

(iii) From No. (ii) it follows that, as it is always possible to diminish the height of the corrugations by properly adjusted obstacles in the bottom, it is always possible to annul them.

3. The fundamental principle in this mode of considering the subject is that whatever disturbance there may be in a perpetually sustained stream, the motion becomes ultimately steady, all

agitations being carried away down stream. The explanation of this will be more fully developed in Part III., to be published in December.

In Part II., to be published in the November number of the Magazine, the integral horizontal component of fluid pressure on any number of inequalities in the bottom, or bars, will be found from consideration of the work done in generating stationary waves, and the obvious application to the work done by wave-making in towing a boat through a canal will be considered. The definite investigation of the wave-making effect when the inequalities in the bottom are geometrically defined, to which I have just now referred, will follow; and I hope to include in Part II., or at all events in Part III. to be published in December, a complete investigation, illustrated by drawings, of the beautiful pattern of waves produced by a ship propelled uniformly through calm deep water.

[From the *Philosophical Magazine*, XXII. November 1886, pp. 445—452.]

PART II.

To find, as promised in Part I., the sum of horizontal pressures on an inequality of the bottom, or on a bar, or on a series of inequalities or bars, consider the horizontal components of momentum of different portions of the water in the following manner. Because the motion is steady, the momentum of the matter at any instant within any fixed volume of space S remains constant; and therefore the rate of delivery of momentum from S by water flowing out on one side above gain of momentum by water flowing into S on the other side must be equal to the total amount of horizontal force acting on the water which at any instant is within S; the direction of this force being that of the flow when the momentum of the leaving water exceeds that of the entering water. Now let S be the space bounded by the bottom, the free surface of the water, and four vertical planes, two of them, called A, A_0, perpendicular to the stream, and two of them parallel to the stream and at unit distance from one another. Let $\mathfrak{P}PB$, and $\mathfrak{P}_0B_0$ be vertical lines on the two transverse ends A and A_0 of the space S; $\mathfrak{P}$, $\mathfrak{P}_0$ being points of the surface, and B, B_0 points of the bottom. Let

$$\mathfrak{P}B = D \text{ and } \mathfrak{P}P = y,$$

and let u be the horizontal component velocity at P. The rate of delivery of momentum (per unit of time understood) from S by water flowing across A is equal to

$$\int_0^D u^2 dy \text{ ....(1)};$$

and the excess of delivery of momentum from S across A above receipt of momentum across A_0 is equal to

$$\int_0^D u^2 dy - \left\{\int_0^D u^2 dy\right\}_0 \text{ ....(2)}.$$

When this is positive, the water between A_0 and A must experience, on the whole, a pressure in the direction from A_0 towards A, made up of difference of fluid-pressures on the end sections A_0 and A, and pressures upon the water by fixed inequalities, if there are any, between A_0 and A. Hence if X, X_0 denote the integral fluid-pressures on the ideal planes A, A_0, and F the sum of horizontal pressures of the inequalities on the fluid, regarded as positive when the direction of the total is from A towards A_0, (2) must be equal to

$$X_0 - X - F \text{ ....(3)}.$$

Hence we have

$$F = \left\{X + \int_0^D u^2 dy\right\}_0 - \left(X + \int_0^D u^2 dy\right) \text{..........(4)}.$$

Now the fluid-pressure at P is equal to $gy + \frac{1}{2}(\mathfrak{q}^2 - q^2)$, by the elementary formula for pressure in steady motion (the pressure at the free surface being taken as zero), $\mathfrak{q}$ and q denoting the velocity of the fluid at $\mathfrak{P}$ and P respectively.

Hence

$$X = \int_0^D [gy + \tfrac{1}{2}(\mathfrak{q}^2 - q^2)]\, dy = \tfrac{1}{2}(gD + \mathfrak{q}^2) D - \tfrac{1}{2}\int_0^D q^2 dy \text{...(5)}.$$

Hence

$$X + \int_0^D u^2 dy = \tfrac{1}{2}(gD + \mathfrak{q}^2) D + \tfrac{1}{2}\int_0^D (u^2 - v^2)\, dy \text{..........(6)},$$

if v be the vertical component velocity at P.

This and the corresponding expression relatively to A_0, give, by (3), the sum of horizontal pressures on all inequalities between A_0 and A, when the problem of the fluid motion in the circumstances is so far solved as to give D, $\mathfrak{q}$, and $u^2 - v^2$ for each of the end sections A_0, A.

Suppose, now, A_0 to be so far on the up-stream side of the inequalities that the motion of the water across it is sensibly uniform and horizontal, with velocity which we shall denote by U_0; so that, for A_0, (6) becomes

$$\left\{X + \int_0^D u^2 dy\right\}_0 = \tfrac{1}{2}gD_0^2 + U_0^2 D_0 \ldots\ldots\ldots\ldots\ldots(7).$$

Hence, and by (6) and (4),

$$F = \tfrac{1}{2}g\,(D_0^2 - D^2) + U_0^2 D_0 - \tfrac{1}{2}\mathfrak{q}^2 D - \tfrac{1}{2}\int_0^D (u^2 - v^2)\,dy \ldots(8).$$

Now, by the law of velocity at the free surface in steady motion, we have

$$\tfrac{1}{2}\mathfrak{q}^2 = \tfrac{1}{2}U_0^2 + g\,(D_0 - D) \ldots\ldots\ldots\ldots\ldots\ldots(9);$$

because, the points B_0, B of the bottom being on the same level, $D_0 - D$ is the difference of levels between the surface-points $\mathfrak{P}_0$ and $\mathfrak{P}$. Hence (8) becomes

$$F = \tfrac{1}{2}g\,(D_0 - D)^2 + U_0^2\,(D_0 - D) - \tfrac{1}{2}\,(U^2 - U_0^2)\,D$$
$$+ \tfrac{1}{2}\int_0^D (v^2 + U^2 - u^2)\,dy \ldots(10),$$

where U denotes a constant which may have any value. It is convenient to make it the mean horizontal component velocity across $\mathfrak{P}B$: we therefore take

$$U = \frac{1}{D}\int_0^D u\,dy \ldots\ldots\ldots\ldots\ldots\ldots\ldots(11):$$

and, because the quantities flowing in across A_0 and out across A are equal, as the motion is steady, we have

$$UD = U_0 D_0 \ldots\ldots\ldots\ldots\ldots\ldots\ldots\ldots(12).$$

Using this to eliminate U_0 from (10), we find

$$F = \tfrac{1}{2}\left(g - \frac{U^2 D}{D_0^2}\right)(D_0 - D)^2 + \tfrac{1}{2}\int_0^D (v^2 + U^2 - u^2)\,dy \ldots(13).$$

To evaluate $D_0 - D$ when we know enough about the motion, and to see how its value is related to other characteristic quantities, let us look back to (9), and in it take

$$\mathfrak{q}^2 = U^2 + \mathfrak{v}^2 \ldots\ldots\ldots\ldots\ldots\ldots\ldots\ldots(14).$$

Thus, if $\mathfrak{P}$ be chosen at a point of the water-surface where the horizontal component velocity is rigorously or approximately equal to U, then $\mathfrak{v}$ is rigorously or approximately the vertical component velocity at $\mathfrak{P}$. Using now (14) in (9), with UD/D_0 for U_0, we find

$$D_0 - D = \tfrac{1}{2}\mathfrak{v}^2 \Big/ \left\{ g - \frac{\frac{1}{2}(D_0 + D)\, U^2}{D_0^2} \right\} \quad \text{.........(15)};$$

which, used in (13), gives

$$F = \frac{\mathfrak{v}^4}{8}\left(g - \frac{U^2 D}{D_0^2}\right) \Big/ \left[g - \frac{\frac{1}{2}(D_0 + D)\, U^2}{D_0^2} \right]^2 + \tfrac{1}{2}\int_0^D (v^2 + U^2 - u^2)\, dy \quad \text{........(16)}.$$

Hence, when the change of level, $D_0 - D$, is but small, in comparison with D or D_0, we have

$$F \doteqdot \tfrac{1}{8}\mathfrak{v}^4 \Big/ \left(g - \frac{U^2}{D} \right) + \tfrac{1}{2}\int_0^D (v^2 + U^2 - u^2)\, dy \quad \text{......(17)},$$

where $\doteqdot$ denotes approximate equality. Going back to (16), let $\mathfrak{P}$ be so chosen on the water-surface that

$$\int_0^D u^2 dy = U^2 D \quad \text{........................(18)},$$

which it is clear we can do, because at a crest the first member is less than the second, and at a hollow greater. When the motion is infinitely nearly simple harmonic (the stream-lines curves of sines), the position of $\mathfrak{P}$ thus chosen will be exactly the middle between crest and hollow. When the motion is anything, however great, up to Stokes's highest possible wave, the chosen place of $\mathfrak{P}$ is a less or more rough approximation to the mid-level point of a wave: it is always rigorously determinate. For brevity we shall call it, that is to say a point defined by (18), a nodal point. Thus, when $\mathfrak{P}$ is taken as a nodal point, (16) becomes simplified to

$$F = \frac{\mathfrak{v}^4}{8}\left(g - \frac{U^2 D}{D_0^2}\right) \Big/ \left[g - \frac{\frac{1}{2}(D_0 + D)\, U^2}{D_0^2} \right]^2 + \tfrac{1}{2}\int_0^D v^2 dy \quad \text{...(19)}.$$

This expression is rigorous. In it $\mathfrak{v}$, which is given rigorously by (14), is approximately (not rigorously) equal to the vertical component velocity at $\mathfrak{P}$: and if we suppose D given, D_0 is found by (15), which is a cubic equation in D_0, most easily solved by successive approximations according to the process obviously indicated by the form in which the equation appears in (15). (As a first approximation take D for D_0 in the second member and so on.)

To work out the formula (19) for the case of infinitesimal displacement, we may take $\mathfrak{P}$ at a great enough distance from inequalities to let the surface in its neighbourhood be sensibly a curve of sines, and the motion simple harmonic. The investigation is facilitated by also taking $\mathfrak{P}$ at a node, as in the diagrams. If we take

$$\mathfrak{h} = h \sin mx \text{ ....}(20)$$

as the equation of the free surface, the known solution for simple harmonic waves in water of depth D gives,

$$\left.\begin{aligned} u &= U\left\{1 - mh\,\frac{\epsilon^{m(D-y)} + \epsilon^{-m(D-y)}}{\epsilon^{mD} - \epsilon^{-mD}} \sin mx\right\} \\ v &= \qquad Umh\,\frac{\epsilon^{m(D-y)} - \epsilon^{-m(D-y)}}{\epsilon^{mD} - \epsilon^{-mD}} \cos mx, \\ \text{where}\quad U &= \sqrt{\left\{\frac{g}{m}\,\frac{\epsilon^{mD} - \epsilon^{-mD}}{\epsilon^{mD} + \epsilon^{-mD}}\right\}} \end{aligned}\right\} \text{ ......}(21).$$

Hence, where $x = 0$, as in the nodal section $\mathfrak{P}PB$,

$$u = U, \text{ and } v = Umh\,\frac{\epsilon^{m(D-y)} - \epsilon^{-m(D-y)}}{\epsilon^{mD} - \epsilon^{-mD}} \text{ ......}(22);$$

also

$$\int_0^D v^2 dy = \tfrac{1}{2}U^2 mh^2 \frac{\epsilon^{2mD} - \epsilon^{-2mD} - 4mD}{(\epsilon^{mD} - \epsilon^{-mD})^2} \text{ ......}(23)$$

$$= \tfrac{1}{2}gh^2\left\{1 - \frac{4mD}{\epsilon^{2mD} - \epsilon^{-2mD}}\right\} \text{..............}(24).$$

Now going back to (19) we see that when U approaches the critical velocity

$$\sqrt{gD\frac{D_0^2}{\frac{1}{2}(D_0 + D)\,D}},$$

the first term might become important, even though the corrugations at a great distance down-stream from the inequalities were infinitesimal. Reserving considerations of this case, and supposing for the present U to be considerably smaller than the critical value, we may neglect the first term in comparison with the second, remembering that in fact quantities comparable with the first term are neglected in the approximation (24) to the value of the second; and we have, as our final approximate result,

$$F = \tfrac{1}{4}gh^2\left(1 - \frac{4mD}{\epsilon^{2mD} - \epsilon^{-2mD}}\right) \text{ ....}(25).$$

There is no difficulty in understanding the permanent steadiness of the motion which we have now been considering: to any finite distance, however great, on either the up-stream or down-stream side of the inequalities, if the water in the finite space considered is given in this state of motion, and if water is admitted on the one side and carried away on the other side conformably. But it is very interesting and instructive to consider the initiation of such a state of things from an antecedent condition of uniform flow over a plane bottom. Suppose, as the primary condition, an inequality, whether elevation or depression, to exist in the bottom, but to be carried along with the water, so that the flow of the water is everywhere uniform and in parallel lines. If the inequality is an elevation above the bottom, our supposition is that the whole projecting piece, moving with the water, slips along the bottom. If the inequality be a depression in the bottom, the more awkward supposition must be made of a plasticity of the bottom, and the form of the inequality carried along, while the bottom is kept rigidly plane before and after this depression.

Suppose, now, the inequality is gradually or suddenly brought to rest, what will be the resulting motion of the water? The question is identical with that of finding the motion of water in a canal, when by an external force, such as that of a towing-rope, a boat is gradually or suddenly set in motion through it; or, rather, it would be identical if the boat were a beam filling the whole breadth across the canal, so that the motion of the water shall be purely two-dimensional. I hope in a later article (Part III. or Part IV. of the present series) to investigate the formation of the procession of standing waves in the wake of the obstacle, and its gradual extension farther and farther down-stream from the obstacle, the motion having become sensibly steady in its neighbourhood, and becoming so to greater and greater distances down-stream by the completion of the growth of fresh waves. The disturbance sent up-stream from the initiating irregularity must also be considered. Equation (15) shows that whether the irregularity be an elevation, as in our first diagram (fig. 1), or a depression, as in fig. 2, a rising of level must travel up-stream, at a velocity relatively to the water which we know must be $\sqrt{gD_0'}$, where D_0' is intermediate between D_0 and the smaller depth, which we shall call D', in the undisturbed stream above. But however gradually the initiating irregularity may have been instituted, this travelling of an elevation up-stream must develop a bore; because the velocity of propagation is, as it were, different in different parts of the slope, being $\sqrt{gD'}$ at the commencement of the slope, and ranging from this, through $\sqrt{gD_0'}$, to $\sqrt{gD_0}$ as the depth rises from D' to D_0; so that, as it were, the brow of the plateau in its advance up-stream overtakes the talus, till the slope becomes too steep for our approximation. The inevitable bore and "broken" water (inevitable without viscidity of the water, or some surface-action preventing the excessive steepness) would modify affairs down-stream in a manner which it is difficult to imagine. It becomes, therefore, interesting to see how it may be avoided, whether by surface-action, or by giving some viscosity

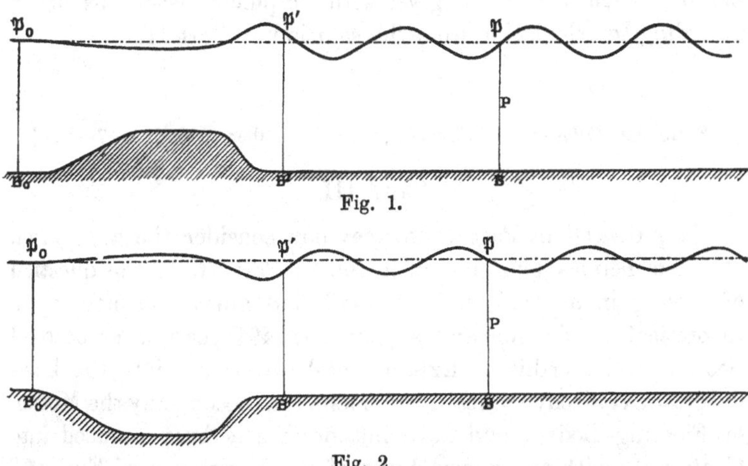

Fig. 1.

Fig. 2.

to the water. It is more interesting to do this by surface-action, and to allow the water to be perfectly inviscid, so that our standing waves down-stream may be perfectly unimpaired. And we may do it very simply by covering the free surface all over (up-stream and down-stream) with an infinitely thin viscously elastic flexible membrane, stiffened transversely (after the manner of the sail of a Chinese junk) by rigid massless bars with ends travelling up and down in vertical guides on the sides of the canal. If we suppose the motion of these ends to be resisted by forces proportional to their velocities, and the membrane to exercise (positive or negative) contractile tensional force in simple proportion to the velocity of the change of its length in each infinitely small part; we have a mechanical arrangement by which is realized the mathematical condition of a surface normal pressure varying according to normal component velocity of the otherwise free surface, and in simple proportion to this normal velocity when the slope is infinitesimal. By making the viscous forces sufficiently great, we may make the progress of the rise of level up-stream as gradual as we please, and perfectly avoid the bore. We may also make the progress of the procession of stationary waves down-stream as slow as we please. The form of the water-surface over the inequality or inequalities, and to any distance from them, both up-stream and down-stream, is not ultimately affected at all by the viscous covering; and it becomes, as time advances, more and more nearly that of the mathematical solution for steady motion, which I hope to give, with graphic illustrations drawn according to calculation from the solution, in Part III.

[From the *Philosophical Magazine*, XXII. December 1886, pp. 517—530.]

PART III.

As promised in Part I., we may now consider the application of the principles developed in it and in Part II., to the question of towing in a canal, and we shall find almost surprisingly a theoretical verification and explanation, 49½ years after date, of Scott Russell's brilliant "Experimental Researches into the Laws of certain Hydrodynamical Phenomena that accompany the Motion of Floating Bodies, and have not previously been reduced into Conformity with the known Laws of the Resistance of Fluids*," which had led to the Scottish system of "fly-boat," carrying passengers on the Glasgow and Ardrossan Canal and between Edinburgh and Glasgow on the Forth and Clyde Canal, at speeds of from 8 to 12 or 13 miles an hour† by a horse, or a pair of horses, galloping along the bank. The practical method originated from the accident of a spirited horse, whose duty it was to drag a boat along at a slow speed (I suppose a walking speed), taking fright and running off, drawing the boat after him, and so discovering that when the speed exceeded $\sqrt{gD}$ the resistance was

* By John Scott Russell, Esq., M.A., F.R.S.E. Read before the Royal Society of Edinburgh on April 4, 1837, and published in the *Transactions* in 1840.

† One mile an hour is English and American reckoning of velocity, which, when not at sea, signifies 1·60933 kilometres per hour, or ·44704 metre per second.

less than at lower speeds. Mr Scott Russell's description of the incident, and of how Mr Houston took advantage for his Company of his horse's discovery, is so interesting that I quote it *in extenso*:—
"Canal navigation furnishes at once the most interesting illustrations of the interference of the wave, and most important opportunities for the application of its principles to an improved system of practice. It is to the diminished anterior section of displacement, produced by raising a vessel with a sudden impulse to the summit of the progressive wave, that a very great improvement recently introduced into canal transport owes its existence. As far as I am able to learn, the isolated fact was discovered accidentally on the Glasgow and Ardrossan Canal of small dimensions. A spirited horse in the boat of William Houston, Esq., one of the proprietors of the works, took fright and ran off, dragging the boat with it, and it was then observed, to Mr Houston's astonishment, that the foaming stern surge which used to devastate the banks had ceased, and the vessel was carried on through water comparatively smooth, with a resistance very greatly diminished. Mr Houston had the tact to perceive the mercantile value of this fact to the Canal Company with which he was connected, and devoted himself to introducing on that canal vessels moving with this high velocity. The result of this improvement was so valuable in a mercantile point of view, as to bring, from the conveyance of passengers at a high velocity, a large increase of revenue to the Canal Proprietors. The passengers and luggage are conveyed in light boats, about sixty feet long and six feet wide, made of thin sheet iron and drawn by a pair of horses. The boat starts at a slow velocity behind the wave, and at a given signal it is by a sudden jerk of the horses drawn up on the top of the wave, where it moves with diminished resistance, at the rate of 7, 8, or 9 miles an hour*."

The "diminished anterior section of displacement produced by raising a vessel with a sudden impulse to the summit of the progressive wave" is no doubt a correct observation of an essential feature of the phenomenon; but it is the annulment of "the foaming stern surge which [at the lower speeds] used to devastate the banks" that gives the direct explanation of the diminished resistance. It is in fact easy to see that when the motion is steady, no waves

* *Trans. Roy. Soc. Edin.* vol. XIV. (1840), p. 79.

can be left astern of a boat towed through a canal at a speed greater than $\sqrt{gD}$, the velocity of an infinitely long wave in the canal; and therefore (the water being supposed inviscid) the resistance to towage must be *nil* when the velocity exceeds $\sqrt{gD}$. This holds true also obviously for towage in an infinite expanse of open water of depth D over a plane bottom.

The formula (25) of Part II. for the whole horizontal component force upon an inequality or succession of inequalities on the bottom, allows us to calculate the resistance on a boat of any dimensions and any shape provided we know the height of the regular waves which follow it steadily at its own speed in the canal, at a sufficiently great distance behind it to be sensibly uniform across the breadth of the canal, according to the principle explained on page 273 of Part I. The principle, upon which the values of $\mathfrak{H}$ [the h of formula (25), Part II.] may be calculated are partly given in the remainder of the present article, and will be more fully developed in Part IV.

To find the steady motion of water flowing in a rectangular channel over a bottom with geometrically specified inequalities, it is convenient, after the manner of Fourier, to first solve the problem for the case in which the profile of the bottom is a curve of sines deviating infinitesimally from a horizontal plane.

For convenience, take OX along the mean level of the bottom, positive in the direction of U the mean velocity of the stream; and OY vertical, positive upwards. Let

$$h = H \cos mx \quad\text{..........................(1)}$$

be the equation of the bottom; and

$$y - D = \mathfrak{h} = \mathfrak{H} \cos mx \quad\text{...................(2)}$$

be the equation of the free surface, $\mathfrak{h}$ being height above its mean level. Let ϕ be the velocity potential; u, v the velocity components; and p the pressure at any point (x, y) of the water at time t: so that we have

$$u = \frac{d\phi}{dx} \text{ and } v = \frac{d\phi}{dy} \quad\text{...................(3)},$$

and

$$p = C - gy - \tfrac{1}{2}(u^2 + v^2) \quad\text{..............(4)}.$$

Now the deviation from uniform horizontal velocity is infinitesimal, and therefore v and $u - U$ are infinitely small. Hence (4) gives

$$p = C - gy - \tfrac{1}{2}U^2 - U(u - U) \quad\text{..............(5)}.$$

ϕ must be a solution of the equation of continuity

$$\frac{d^2\phi}{dx^2}+\frac{d^2\phi}{dy^2}=0,$$

and the proper one for our present case clearly is

$$\phi = Ux + \sin mx\,(K\epsilon^{my} + K'\epsilon^{-my}) \quad \text{.............(6)},$$

where, because the motion is steady, K and K' are constants. This, in virtue of (3), gives

$$u - U = m\cos mx\,(K\epsilon^{my} + K'\epsilon^{-my}) \quad \text{.............(7)};$$

$$v = m\sin mx\,(K\epsilon^{my} - K'\epsilon^{-my}) \quad \text{.............(8)}.$$

Hence, as the values of y at the bottom and at the surface are infinitely nearly 0 and D respectively, we find respectively for the vertical component velocity at the bottom and at the surface,

$$m\sin mx\,(K - K'), \text{ and } m\sin mx\,(K\epsilon^{mD} - K'\epsilon^{-mD}).$$

Hence, to make the bottom-stream-lines and surface-stream-lines agree respectively with the assumed forms (1) and (2), we clearly have

$$m\,(K - K') = -mHU \quad \text{.....................(9)},$$

and

$$m\,(K\epsilon^{mD} - K'\epsilon^{-mD}) = -m\mathfrak{H}U \quad \text{.............(10)};$$

whence

$$\left.\begin{aligned} K &= -U\frac{\mathfrak{H} - H\epsilon^{-mD}}{\epsilon^{mD} - \epsilon^{-mD}} \\ K' &= -U\frac{\mathfrak{H} - H\epsilon^{mD}}{\epsilon^{mD} - \epsilon^{-mD}} \end{aligned}\right\} \quad \text{..................(11)}.$$

Now at the free surface the pressure is constant, and hence, by (5), we have

$$-gy - U(u - U) = \text{constant} \quad \text{...............(12)}:$$

from which, by (2), (7), and (11), we find

$$0 = -g\mathfrak{H} + mU^2\frac{\mathfrak{H}(\epsilon^{mD} + \epsilon^{-mD}) - 2H}{\epsilon^{mD} - \epsilon^{-mD}},$$

whence

$$\mathfrak{H} = \frac{2H}{\epsilon^{mD} + \epsilon^{-mD} - \dfrac{g}{mU^2}(\epsilon^{mD} - \epsilon^{-mD})} \quad \text{.............(13)},$$

which is the solution of our problem, for the case of the bottom a simple harmonic curve.

Suppose now the equation of the bottom to be

$$h = (\kappa\cos mx + \kappa^2\cos 2mx + \kappa^3\cos 3mx + \&c.)\,mA/\pi \text{...(14)};$$

the equation of the surface, found by superposition of solutions given by (13), allowable because the motion deviates infinitely little from horizontal uniform motion throughout the water, is

$$y - D = \mathfrak{h} = \sum_{i=1}^{i=\infty} \frac{2\kappa^i \cos imx \,.\, mA/\pi}{\epsilon^{imD} + \epsilon^{-imD} - \dfrac{g}{imU^2}(\epsilon^{imD} - \epsilon^{-imD})} \quad \text{...(15)}.$$

To interpret the equation (14) by which the bottom is defined, remark that, by the well-known summation of its second member, it is equivalent to

$$h = \frac{\frac{1}{2}mA/\pi \,.\, (1-\kappa^2)}{1 - 2\kappa\cos mx + \kappa^2} - \tfrac{1}{2}mA/\pi = \frac{mA/\pi \,.\, \kappa(\cos mx - \kappa)}{1 - 2\kappa\cos mx + \kappa^2} \quad \text{...(16)}.$$

The series (14) is convergent for all values of κ less than unity*. According to the method of Fourier, Cauchy, and Poisson, the extreme case of κ infinitely little less than unity will be made the foundation of our practical solutions. By (14) we see that

$$\int_{-\pi/m}^{\pi/m} dx h = 0 \quad \text{........................(17)};$$

and hence by the first of equations (16) we see that

$$\int_{-\pi/m}^{\pi/m} dx \frac{\frac{1}{2}mA/\pi \,.\, (1-\kappa^2)}{1 - 2\kappa\cos mx + \kappa^2} = A \quad \text{..............(18)}.$$

Now when κ is infinitely little short of unity the factor of dx in the first member of (18) is zero for all values of x differing finitely from zero or $2i\pi/m$ (i being an integer); and it is infinitely great when $x = 0$ or $2i\pi/m$. Hence we infer from (17) and (18) that a vertical longitudinal section of the bottom presents a regular row of similar elevations and depressions above and below its mean level; the elevations being confined to very small spaces on the two sides of each of the points $x = 0$ and $x = 2i\pi/m$, and the profile-area of each elevation being A. The depths of the depressions below the average level in the intermediate spaces between the elevations, are of course extremely small because of the exceeding shortness of the spaces over which are the elevations. For our complete analytical solution, not only must A be infinitely small, but the steepness of the slope up to the summit of h must everywhere be an infinitely small fraction of a radian; and of course therefore the infinitesimal lowering of the bottom between the

[* The value of h is plotted in § 43 of "Deep Water Ship Waves," *infra.*]

ridges, which the adoption of a mean bottom-level for our datum line has necessarily introduced, may be left out of account in our dynamical problem.

If the slope of the ridge is not an infinitely small fraction of a radian our solution will still hold, provided its height is very small in comparison with the depth of the water over it. But the effective potency of the ridge would then not be its profile-area A, but something much greater; of which the amount would be found by taking a stream-line over it, far enough above it to have nowhere more than an infinitesimal slope, and finding the profile-area of such a stream-line above its own average level considered as the virtual bottom. With these explanations we shall speak of a ridge for brevity instead of an "irregularity" or "obstacle," and call its profile-area A, simply the "magnitude of the ridge"; this being, as we see by (15), the measure of its potency in disturbing the surface. When instead of a ridge we have a hollow, A is negative; and when convenient we may, of course, call a hollow a negative ridge.

It is clear that (15) converges, and does not depend for its convergence on κ being less than unity; so that in it we may take κ absolutely equal to unity, and we shall do so accordingly.

To find now the effect of a single ridge, remark that if l be the length from ridge to ridge,

$$m = 2\pi/l \qquad \text{.........................(19).}$$

After the manner of Fourier now suppose l infinitely large; which makes m infinitely small; and put

$$im = q \text{ and } m = dq \qquad \text{..................(20);}$$

then with $\kappa = 1$, (15) becomes

$$\mathfrak{h} = \int_0^\infty dq \frac{2A/\pi \, . \cos qx}{\epsilon^{qD} + \epsilon^{-qD} - \dfrac{1}{qb}(\epsilon^{qD} - \epsilon^{-qD})} \qquad \text{.........(21);}$$

where

$$b = U^2/g \qquad \text{.........................(22).}$$

Equation (21) will be shortened, and for some interpretations simplified, by making $qD = \sigma$, when it becomes

$$\mathfrak{h} = \int_0^\infty d\sigma \frac{2A/D\pi \, . \cos(\sigma x/D)}{\epsilon^{\sigma} + \epsilon^{-\sigma} - \dfrac{D}{b\sigma}(\epsilon^{\sigma} - \epsilon^{-\sigma})} \qquad \text{..............(23).}$$

The definite integral (21) or (23) seemed rather intractable, and the quadratures required to evaluate it, for many and wide-spread enough values of x to show the shape of the surface for any one particular value of D/b, would be very laborious. But I had found a method of evaluating it from the periodic solution for an endless succession of equidistant equal ridges (15), wholly analogous to analytical deductions from corresponding solutions for cases of thermal conduction and of signalling through submarine cables, to be found in vol. II. pp. 49 and 56 of my collected *Mathematical and Physical Papers*; and, towards applying this method to a particular case of the disturbance due to a single ridge, I had fully worked out the periodic solution for the case represented by the diagram of curves (fig. 3, p. 295), when I found a direct and complete analytical solution for the single-ridge problem in a form exceedingly convenient for arithmetical computation, except for the case of x equal to zero, or from zero to a quarter or a half of the depth. The previous method happily gives the solution for small values of x, and indeed for values up to two or three times the depth, by very rapidly converging series, and thus between the two methods we have a remarkably satisfactory solution of the whole problem.

Before explaining the curves and their relation to the problem of the single ridge, I shall give the new direct solution of this problem. It is founded on a well-known analytical method of Cauchy's, of which examples are given in the Eighteenth note (p. 284) to his Memoir on the Theory of Waves*.

First, bring the denominator of (23) to the form of the product of an infinite number of quadratic factors, as follows:—Let

$$W = \tfrac{1}{2}\left(1-\frac{D}{b}\right)^{-1}\left\{\epsilon^{\sigma}+\epsilon^{-\sigma}-\frac{D}{b\sigma}\left(\epsilon^{\sigma}-\epsilon^{-\sigma}\right)\right\}\text{......(24).}$$

Expanding in powers of σ, we have

$$W = 1+\left(1-\frac{D}{b}\right)^{-1}\left\{\frac{1}{1\,.\,2}\left(1-\frac{1}{3}\frac{D}{b}\right)\sigma^{2}\right.$$
$$\left.+\frac{1}{1\,.\,2\,.\,3\,.\,4}\left(1-\frac{1}{5}\frac{D}{b}\right)\sigma^{4}+\text{\&c.}\right\}\text{...(25).}$$

Hence, when b is greater than D, W is positive for all real values of σ. But when b has any positive value less than D, W (which

* *Memoires de l'Académie Royale de l'Institut de France, savans étrangers,* tome I. (1827).

is always positive for small values of σ^2) is negative for large values of σ^2; and therefore at least one positive value of σ^2 makes W zero. We shall see presently thåt only one positive value of σ^2 does so. We shall see that all the zeros of W when b is greater than D and all but one when b is less than D, correspond to real negative values of σ^2. This indeed is obvious if for σ^2 we put $-\theta^2$, which gives

$$W = \left(1 - \frac{D}{b}\right)^{-1}\left(\cos\theta - \frac{D}{b}\frac{\sin\theta}{\theta}\right) \text{............(26)};$$

and which shows that the zeros of W are given by the roots of the well-known transcendental equation

$$\frac{\tan\theta}{\theta} = \frac{b}{D} \text{ ....(27)}.$$

When b is greater than D this equation has all its roots real, and in the first, third, fifth, &c. quadrants. When b is less than D the root in the first quadrant is lost, and in its stead we clearly have a pure imaginary; while the roots in the third, fifth, &c. quadrants remain real. Let θ_1, θ_2, θ_3, &c. be the roots of the first, third, fifth, &c. quadrants. As the first term of equation (25) is unity, we have

$$\left.\begin{aligned} W &= \left(1 - \frac{\theta^2}{\theta_1^2}\right)\left(1 - \frac{\theta^2}{\theta_2^2}\right)\left(1 - \frac{\theta^2}{\theta_3^2}\right) \text{\&c.} \\ \text{or} \quad W &= \left(1 + \frac{\sigma^2}{\theta_1^2}\right)\left(1 + \frac{\sigma^2}{\theta_2^2}\right)\left(1 + \frac{\sigma^2}{\theta_3^2}\right) \text{\&c.} \end{aligned}\right\} \text{.........(28)};$$

where θ_2^2, θ_3^2, &c. are real positive numerics, while θ_1^2 is real positive or real negative according as b is greater than D or less than D.

Resolving now the reciprocal of W into partial fractions, we find

$$\frac{1}{W} = \frac{N_1}{1 + \dfrac{\sigma^2}{\theta_1^2}} + \frac{N_2}{1 + \dfrac{\sigma^2}{\theta_2^2}} + \frac{N_3}{1 + \dfrac{\sigma^2}{\theta_3^2}} + \text{\&c. ....(29)};$$

where

$$\begin{aligned} N_i = \frac{-1}{\theta_i^2\left[\dfrac{dW}{d(\theta^2)}\right]_i} = \frac{-2}{\theta_i\left(\dfrac{dW}{d\theta}\right)_i} &= \frac{2(1 - D/b)\cos\theta_i}{D/b - \cos^2\theta_i} \\ &= \frac{2(1 - D/b)\sin\theta_i}{\theta_i(1 - b/D\,.\cos^2\theta_i)} \text{.........(30)}. \end{aligned}$$

For $i=1$ and $D>b$, θ_i is, as we have seen, imaginary (its square real negative), and for this case the formula (30) may be conveniently written

$$N_1 = -\frac{(D/b-1)(\epsilon^{\sigma_1}+\epsilon^{-\sigma_1})}{D/b-\frac{1}{2}-\frac{1}{4}(\epsilon^{2\sigma_1}+\epsilon^{-2\sigma_1})} \text{ ....(31)};$$

and the equation for finding σ_1 is

$$\epsilon^{\sigma_1}+\epsilon^{-\sigma_1}-D/b\sigma_1 \,.\, (\epsilon^{\sigma_1}-\epsilon^{-\sigma_1})=0 \text{ ....(32)},$$

an equation which has one, and only one, real root when $D>b$ and no real root when $D<b$.

When b/D is given, it is easy to find, as the case may be, σ_1 of (32) or θ_1 the first-quadrant root of (27), by arithmetical trial and error; and the successive roots θ_2, θ_3, &c. more and more easily, by the solution of (27). It is to be remarked that, whatever be the value of b/D, these roots approach more and more nearly to the superior limits of the quadrants in which they lie: thus if we put

$$\theta_i = (i-\tfrac{1}{2})\pi - \alpha_i \text{ ....(33)},$$

we have

$$N_i\theta_i = (-1)^{i+1}\, 2\, \frac{(1-D/b)\sin\alpha_i}{D/b-\sin^2\alpha_i}[(i-\tfrac{1}{2})\pi-\alpha_i]$$

$$= (-1)^{i+1}\, 2\, \frac{(1-D/b)\cos\alpha_i}{1-b/D \,.\, \sin^2\alpha_i} \text{ ....(34)};$$

and
$$\sin\alpha_i[(i-\tfrac{1}{2})\pi-\alpha_i] = D/b \,.\, \cos\alpha_i \text{...........(35)};$$

or, as is convenient for approximation when i is very large,

$$\alpha_i[(i-\tfrac{1}{2})\pi-\alpha_i] = D/b \,.\, \alpha_i/\tan\alpha_i \text{ .... (36)},$$

which shows that as i is increased to infinity, the value of α_i approaches asymptotically to $D/[b(i-\frac{1}{2})\pi]$. Hence when i is very large, the second member of (36) becomes approximately $D/b \,.\, (1-\frac{1}{3}\alpha_i^2)$; and the equation becomes

$$(1-\tfrac{1}{3}D/b)\,\alpha_i^2 - (i-\tfrac{1}{2})\,\pi\alpha_i = -D/b \text{ ....(37)};$$

a quadratic, of which the smaller root when D is less than $3b$, and the positive root when D is greater than $3b$, is the required value of α_i.

Going back now to (23) and modifying it by (24) and (29), we have

$$\mathfrak{h} = \frac{A/D\pi}{1-D/b} \,.\, \Sigma N_i \int_0^\infty d\sigma \frac{\cos x\sigma/D}{1+\sigma^2/\theta_i^2} \text{ ....(38)};$$

or, according to the well-known evaluation (attributed by Cauchy to Laplace) of the definite integral indicated,

$$\mathfrak{h} = \frac{\frac{1}{2}A/D}{1-D/b} \cdot \Sigma \theta_i N_i \epsilon^{-\frac{\theta_i x}{D}} \quad \text{..............(39);}$$

or with θ_i, N_i eliminated by (33) and (34),

$$\mathfrak{h} = \frac{A}{D} \cdot \Sigma \frac{(-1)^{i+1} \cos \alpha_i}{1 - b/D \,.\, \sin^2 \alpha_i} \epsilon^{-\frac{[(i-\frac{1}{2})\pi - \alpha_i] x}{D}} \quad \text{.........(40),}$$

where $\alpha_1, \alpha_2, \ldots \alpha_i$ denote all the positive roots of (35).

This series converges with exceeding rapidity when x is any thing greater than D, and with very convenient rapidity for calculation when x is even as small as a tenth of D. When $x = 0$, the convergence has [finally] the same order as that of $1 - e + e^2 -$ &c., when $e = 1$; and we find the sum by taking as remainder half the term after the last term included. The true value of the sum is intermediate between the values which we obtain by this rule for a certain number of terms, and then for one term more. When it is desired to obtain the result with considerable accuracy, a large number of terms would be required; and it will no doubt be preferable to use my first method as indicated above.

It remains to deal with the first term for the case $D > b$, which makes it imaginary in the form (39), but real in the form (38) with $-\sigma_1^2$ substituted for θ_i^2. For this case we have, by the well-known definite integral, first, I believe, evaluated by Cauchy,

$$\mathfrak{h}_1 = \frac{\frac{1}{2}A/D}{1-D/b} \cdot \sigma_1 N_1 \sin \frac{\sigma_1 x}{D} \quad \text{..............(41);}$$

where σ_1 and N_1 are given by (32) and (31)*.

It is to be remarked that, inasmuch as (38) has the same value for equal positive and negative values of x, the evaluations expressed in (39) and (41) are essentially discontinuous at $x = 0$; and when x is negative, $-x$ must be substituted for x in the second member of the formulas. I hope in Part IV. to give numerical illustrations; but with or without numerical illustrations,

* [Here and elsewhere in the integrals the "principal value" of Cauchy is adopted. This simply neglects the infinite amplitudes in the integrand, which arise from synchronism with free vibrations; in nature such very large amplitudes are always depressed by frictional agencies, and when the friction is slight the range of this depression is narrow, confined to the very near neighbourhood of the free period, so that their actual contribution is negligible, and the "principal value" is thus practically justified.]

the analytical formula (39), with (41) for its first term and the sign of x changed throughout when x is negative, is particularly interesting as a discontinuous expression for a curve passing continuously from one to the other of the two curves

$$\left.\begin{aligned} y &= \frac{\frac{1}{2}A/D}{1-D/b}\cdot\sigma_1 N_1 \sin\frac{\sigma_1 x}{D} \text{ for large positive values of } x \\ \text{and}\quad y &= -\frac{\frac{1}{2}A/D}{1-D/b}\,\sigma_1 N_1 \sin\frac{\sigma_1 x}{D} \text{ for large negative values of } x \end{aligned}\right\} \qquad \text{.........(42).}$$

For the case of $b > D$ every term of (39) is real, and (remembering that the sign of x is changed when x is negative) we see that it makes $\mathfrak{y}$ equal for equal positive and negative values of x, and diminish asymptotically to zero as x becomes greater and greater in either direction. It expresses unambiguously the solution (clearly unique when $b > D$) of the problem of steady motion of water in a uniform rectangular canal interrupted only by a single ridge of magnitude A across the bottom. This is the case of velocity of flow greater than that acquired by a body in falling through a height equal to half the depth.

It is otherwise in respect to uniqueness of the solution when the velocity of flow is less than that acquired by a body in falling through a height equal to half the depth ($b < D$). For this case the formulas (39) and (41) express a particular solution of the problem of steady motion through a rectangular canal, when regularity of the canal is only interrupted by the single ridge of magnitude A. But we clearly have an infinite number of solutions of this problem; because in still water in a canal of depth D we can have free waves of any velocity from zero to $\sqrt{gD}$, which is the velocity of an infinitely long wave in water of depth D. In our flowing water then superimpose upon the solution, (39) (41), any wave-motion of arbitrary magnitude, and arbitrarily chosen position for one of the zeros, with wave-length such that the velocity of wave-propagation is U, and the direction of motion such as to cause the progression of the wave to be up-stream. The wave-motion thus instituted constitutes a set of free stationary waves, and the superposition of this upon the case of motion represented by our symmetrical solution constitutes the general solution of the problem of single-ridge steady motion. To find the arbitrary addition which we must thus make to our symmetrical solution to find the general solution, put (13) into the following form:

$$\frac{2H}{\mathfrak{H}} = \epsilon^{mD} + \epsilon^{-mD} - \frac{g}{mU^2}(\epsilon^{mD} - \epsilon^{-mD}) \;\ldots\ldots\ldots(43).$$

This shows that if $H = 0$, $\mathfrak{H}$ may have any value (that is to say, we may have stationary waves of any magnitude over a plane bottom) if

$$\epsilon^{mD} + \epsilon^{-mD} - \frac{g}{mU^2}(\epsilon^{mD} - \epsilon^{-mD}) = 0 \;\ldots\ldots\ldots(44).$$

This is in fact the well-known equation to find the velocity U relatively to the water, of periodic waves of wave-length $2\pi/m$ in a canal of depth D. For us at present equation (44) is to be looked upon as a transcendental equation for determining the wave-length corresponding to U a given velocity of progress; and it has, as we have seen, only one real root when $U < \sqrt{gD}$; but no real root when $U > \sqrt{gD}$. Putting now in (43) $U^2 = gb$, and comparing with (32), we see that $mD = \sigma_1$; and going back to equation (2) above we see that

$$\mathfrak{H}\cos\frac{\sigma_1(x-a)}{D} \;\ldots\ldots\ldots\ldots\ldots\ldots\ldots(45);$$

where $\mathfrak{H}$ and a are arbitrary constants, is the addition which we must make to (39) to give the general solution for the case $b < D$. Putting together this and (39) and (41), we accordingly have for the general solution of the single-ridge steady-motion problem, for the case of $U < \sqrt{gD}$,

$$\mathfrak{h} = C\cos\frac{\sigma_1 x}{D} + \left(C' + \frac{\frac{1}{2}A/D}{1-D/b}\cdot\sigma_1 N_1\right)\sin\frac{\sigma_1 x}{D} + \frac{\frac{1}{2}A/D}{1-D/b}\cdot\sum_2^\infty \theta_i N_i \epsilon^{-\frac{\theta_i x}{D}}$$

when x is positive, and

$$\mathfrak{h} = C\cos\frac{\sigma_1 x}{D} + \left(C' - \frac{\frac{1}{2}A/D}{1-D/b}\cdot\sigma_1 N_1\right)\sin\frac{\sigma_1 x}{D} + \frac{\frac{1}{2}A/D}{1-D/b}\cdot\sum_2^\infty \theta_i N_i \epsilon^{\frac{\theta_i x}{D}}$$

when x is negative

$$\ldots\ldots\ldots(46);$$

where C and C' denote arbitrary constants, and A is the profile-sectional area of the ridge on the bottom.

The motion represented by this solution, with any values of C and C', is steady and stable throughout any finite length of the canal on each side of the ridge, provided the water is introduced at one end of the portion considered and taken away at the other conformably. If the canal extends to infinity in both directions, and if the water throughout be given in the state of motion corresponding to the solution (46); the motion throughout any finite distance on each side of the ridge will continue for an infinite time conformable to (46). The water, if given at rest, might be started into this state of motion in the following manner:—First displace its surface to the shape represented by equation (46), and apply a rigid corrugated lid to keep it exactly in this shape, so that it is now enclosed as it were in a rectangular tube with one side corrugated, two sides plane, and the fourth side (the bottom) plane, except at the place of the ridge. Next by means of a piston set the water gradually in motion in this tube. To begin with, the pressure on the lid will, in virtue of gravity, be non-uniform; less at the high parts and greater at the low parts. If too great a velocity be given to the water by the piston the pressure will, in virtue of fluid motion, be greater at the high parts and less at the low parts. If the average velocity be made exactly U, the pressure will be uniform over the lid, which may then be annulled; thus the liquid is left moving steadily under the surface represented by equation (46) as free surface. But it is only in virtue of this motion being given to the fluid throughout an infinite length of the canal on each side of the ridge, that the motion can remain steady on each side of the ridge conformable to (46), except for the particular case of this general solution, corresponding to

$$C = 0 \quad \text{and} \quad C' = \frac{\frac{1}{2}A/D}{1-D/b} \cdot \sigma_1 N_1 \quad \text{...........(47)},$$

which reduces (46) to

$$\mathfrak{h} = \frac{A/D}{1-D/b}\left(\sigma_1 N_1 \sin\frac{\sigma_1 x}{D} + \tfrac{1}{2}\sum_2^\infty \theta_i N_i \epsilon^{-\frac{\theta_i x}{D}}\right) \text{ when } x \text{ is positive}$$

$$\text{and } \mathfrak{h} = \frac{\frac{1}{2}A/D}{1-D/b}\sum_2^\infty \theta_i N_i \epsilon^{\frac{\theta_i x}{D}} \qquad \text{when } x \text{ is negative} \quad \text{........(48)};$$

this being the practical solution for the case of water flowing from the side of x negative over the single ridge and towards the side of x positive. It is the mathematical realization, for the case of a single ridge, of the circumstances described in Part I. above (*ante*, pp. 274—5), and is the mathematical solution promised in the last sentence of Part II. The demonstration that this is the practically unique solution for inviscid water flowing in a canal with a single ridge, and the explanation of how any other state of motion, such, for example, as that represented by (46) with any value of C and C', but given to the water throughout only a finite distance on each side of the ridge, settles into the permanent steady motion represented by (48), must be reserved for Part IV., which I hope will appear in the January number.

Meantime the accompanying diagram represents by two curves two cases of the solution (46) for the particular value 2·456 for D/b; that is to say, for velocity $=$ ·6381 of the critical velocity $\sqrt{gD}$. The faint curve represents the solution (46) with $C=0$ and $C'=0$. The heavy curve represents the practical solution (48). These curves were drawn from calculations of a periodic solution, according to the first of the two methods indicated above, before I had found the analytical solution (39) by which the desired result could have been arrived at with much less labour. The faint curve was drawn first by direct calculation from the periodic solution: the letters $\frac{1}{2}l$, $\frac{1}{4}l$, $-\frac{1}{4}l$, $-\frac{1}{2}l$, show, on the two sides of one ridge, quarters of the distance from ridge to ridge in the periodic solution, one of the ridges being in the middle of the diagram. The heavy curve is found by adding to the ordinates of the faint curve the ordinates of a curve of sines, found by trial to as nearly as possible annul on the one side, and to double on the other side, the ordinates of the original curve. How nearly perfect was the annulment on the one side and the doubling on the other is illustrated by the small-scale diagram annexed (fig. 3), which has been drawn by the

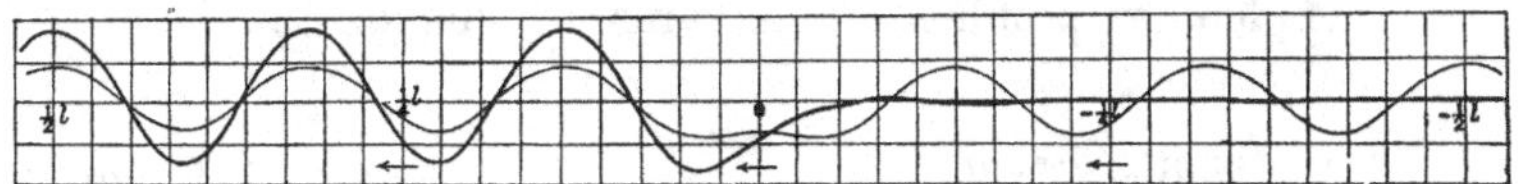

Fig. 3.

engraver from a [ten] times larger copy. How nearly perfect the annulment and the doubling ought to be at any particular distance from a single ridge is now easily calculated from the second line of equation (48), and will be actually calculated for the case of these curves, and probably also for some other cases for numerical illustrations, which I hope to give in Part IV.

* [An extension of the present investigation to the effect of an inequality of any form in the bed of the stream is given by V. Ekman, *Archiv för Matematik, Astronomi ock Physik*, Band 3, No. 2, 1906.]

PART IV. STATIONARY WAVES ON THE SURFACE PRODUCED BY EQUIDISTANT RIDGES ON THE BOTTOM.

[From the *Philosophical Magazine*, Vol. XXIII. *January* 1887, pp. 52—57.]

THE most obvious way of solving this problem is by the use of periodic functions, which we have been so well taught by Fourier in his *Mathematical Theory of Heat*; and in this way it was solved in Part III. (formulas 1 to 15); the solution being (15) Part III. with

$$\kappa = 1, \quad m = 2\pi/a \text{ ....(1)};$$

where a denotes the distance from ridge to ridge. Thus, reproducing (15) Part III. with the notation modified to shorten it in form and to suit it for numerical computation, we have

$$\mathfrak{h} = \sum_{i=1}^{i=\infty} \frac{4A/a \,.\, \cos i\psi}{e^i + e^{-i} - Mi^{-1}(e^i - e^{-i})} \text{ ....(2)};$$

where $\mathfrak{h}$ denotes height above mean level of the water at distance x from the point over one of the ridges;
A denotes profile-sectional area of one of the ridges;
ψ denotes $2\pi x/a$;
e denotes $\epsilon^{2\pi D/a}$;
M denotes the g/mU^2 of Part III. (6) to (18) or $a/2\pi b$;
b denotes U^2/g;
and D denotes the depth ...(3).

Thus, in (2) we have an expression for the surface-effect of an endless succession of equidistant ridges on the bottom. We shall see presently that if the succession of ridges is finite, the result expressed by (2) will not be approximated to by increasing the number of ridges. The difference in the effect of a million equidistant ridges from that of a million and one equidistant ridges, in respect to the corrugations on the surface of the fluid over any part of the series, may be as great as the difference between the effects of a thousand and of a thousand and one, or between the effects of ten and of eleven: and the absolute effect of four, or six, or eight, may be sensibly the same as, or may be greater than, or may be less than, the effect of a million, in respect to the condition of the surface over the space between the two middle ridges. The awkwardness of the consideration of infinity for our present case is beautifully done away with, after the manner of Fourier, by substituting for an "infinite canal" an "endless* canal," or a canal forming a complete circuit†: a circular canal as we may imagine it to be, although it might be curved, of any form, provided only that, whether it be circular or not circular, the radius of curvature at any point is very great compared with the breadth of the canal. This condition is all that is necessary to allow the motion of the water in every part of the canal to be so nearly two-dimensional, that our formulas for two-dimensional motion in a straight canal shall be practically applicable to the water in the curved canal.

Now let there be any integral number n of equidistant ridges in the circuit, and let a be the distance from ridge to ridge. Superposition by simple addition of solutions of the formula (2) gives, for the surface effect,

$$\mathfrak{h} = \sum_{i=1}^{i=\infty} \frac{4A/a \,.\, \sum_{j=0}^{j=n-1} \cos i\left(\psi + \frac{2j\pi}{n}\right)}{e^{i} + e^{-i} - Mi^{-1}(e^{i} - e^{-i})} \quad\ldots\ldots\ldots\ldots(4).$$

* It is curious that the word "endless" should in common usage, and especially in technology, have so different a meaning from "infinite." Thus every one understands what is meant by an "endless cord." An "infinite cord" means, in common language, an infinitely long cord—a cord which has no limit to the greatness of its length.

† A curious piece of illogical usage in mathematical language, according to which an enclosing curve is called a "closed curve," must henceforth be absolutely avoided. It has already led to endless trouble in electrical nomenclature, according to which, in common language, an electric circuit is said to be closed when a current *can pass through it*, and to be open when a current *cannot pass through it.* I believe all, or almost all, English writers on electrical subjects have been guilty of this absurdity. I doubt whether any one of them would say a road round a park is open when a gate on it is closed, and is closed when every gate on it is open.

The consideration of cases of different values of n, even or odd, leads to interesting illustrations both of mathematical principles and of practical results in dynamics; but for the present I confine myself to the case of $n=1$, for which (4) becomes identical with (2).

Remark, now, that if $M(e^i - e^{-i})/(e^i + e^{-i})$ [practically constant for large values of i] is an integer, the denominator of (2) vanishes for the case of i equal to this integer. This is the case in which the length of the circuit of the canal is an integral number of times the wave-length of free waves in water of depth D. The interpretation is obvious, and is interesting both in itself and in its relation to corresponding problems in many branches of physical science.

Meantime remark only that, when the value of

$$M(e^i - e^{-i})/(e^i + e^{-i})$$

approaches very nearly to any integer j, the chief term of (2) is that for which $i=j$, and all the other terms are relatively very small. Thus the chief effect is forced stationary waves of wave-length a/j. Thus, if we consider different velocities of flow approaching more and more nearly to the velocity which makes $M(e^i - e^{-i})/(e^i + e^{-i})$ an integer, the magnitude of the forced stationary waves is greater and greater for the same magnitude of ridge, but the motion is still perfectly determinate. Suppose, now, we make the ridge smaller and smaller, so that the wave-height of the stationary wave may have any moderate value; as the velocity approaches more and more nearly to that which makes $M(e^i - e^{-i})/(e^i + e^{-i})$ an integer, the magnitude of the ridge must be smaller and smaller, and in the limit must be zero. Thus, with no ridge at all, we may have stationary waves of any given moderate value, in the limiting case,—that in which the velocity of the flow equals the velocity of a wave of wave-length a/j.

But now let us consider the case of $M(e^i - e^{-i})/(e^i + e^{-i})$ as far as possible from being an integer; that is to say,

$$M(e^i - e^{-i})/(e^i + e^{-i}) = j + \tfrac{1}{2} \quad \ldots\ldots\ldots\ldots\ldots(5),$$

where j is an integer. For all values of i less than $j+1$ the denominator of (2) is clearly negative, with decreasing absolute values up to $i=j$; and for all values of i greater than j it is positive, with increasing values from $i=j+1$ to $i=\infty$. Thus the absolute magnitudes of the coefficients of $\cos i\psi$ in the successive terms of the series from the beginning are negative, with increasing absolute values up to $i=j$; and after that positive, with decreasing values converging ultimately according to the ratio e^{-1}. Remembering that $e=\epsilon^{2\pi D/a}$, we see that the convergence is sluggish when a, the distance from ridge to ridge (or the length of the circuit in the case of an endless canal with one ridge only), is very large in comparison with the depth; but that when a is less than the depth, or not more than five or ten times the depth (an exceedingly interesting class of cases), the convergence is very rapid.

We shall find presently, however, another solution still more convergent, much more convergent indeed for the greater part of the configuration, whatever be the ratio of D to a; a solution which is highly convergent in every case except for values of x considerably smaller than the depth. The calculation for these small values of x is necessary to give the shape of the water-surface at distances on each side of the vertical through the ridge small in comparison with the depth: for this purpose, and for this purpose only, is the solution (2) indispensable. For investigating all other parts of the configuration the new solution is much more convenient, and involves, on the whole, very much less of arithmetical labour. It is found by summation from the solution of the single-ridge problem given in Part III. (40), (41), as follows.

Let the whole number of ridges be $j+j'+1$, and let it be required to find the shape of the surface between the verticals through ridges numbers $j+1$ and $j+2$. Take the origin of the coordinate x in the vertical through number $j+1$ ridge, and let number $j+2$ be on the positive side of it. The solution will be found by adding to the solution (40) Part III., j solutions differing from (40) only in having respectively $x+a$, $x+2a$, ..., $x+ja$ substituted for x; and j' solutions each the same as (40) Part III., but having $-x+a, -x+2a, \ldots, -x+j'a$ substituted for x. Thus, denoting by S the sum of the effects of the $j+j'+1$ single-ridges, we find

$$S=\sum_{i=1}^{i=\infty} C_i\left\{\frac{(1-f_i^{j+1})f_i^{x/a}+(1-f_i^{j'})f_i^{1-x/a}}{1-f_i}\right\} \quad \text{......(6)};$$

where

$$\left.\begin{array}{l}
C_i \text{ denotes } \dfrac{A}{D}\cdot\dfrac{(-1)^{i+1}\cos\alpha_i}{1-b/D\,.\sin^2\alpha_i};\\
f_i \text{ denotes } \epsilon^{-\frac{[(i-\frac{1}{2})\pi-\alpha_i]\,a}{D}} \text{ or } \epsilon^{-\frac{\theta_i a}{D}};\\
\alpha_i \text{ denotes } (i-\tfrac{1}{2})\pi-\theta_i; \text{ or the numeric between zero and } \pi/2 \text{ which satisfies the equation } [(i-\tfrac{1}{2})\pi-\alpha_i]\tan\alpha_i - D/b = 0;\\
D \text{ denotes the depth};\\
b \text{ denotes } U^2/g;\\
U \text{ denotes the velocity of the flow};\\
a \text{ denotes the distance from ridge to ridge};\\
A \text{ denotes the profile-sectional area of one of the ridges};\\
S \text{ denotes, for the horizontal coordinate } x, \text{ the height of the water above the mean level of places infinitely distant either upstream or downstream from the ridges}
\end{array}\right\}\ldots(7).$$

Take first the case of $b > D$. In this case, as we have already remarked in Part III., $\alpha_1, \alpha_2, \ldots, \alpha_i$ are all real; and therefore $f_1, f_2, \ldots, f_i$ are each real and less than unity. Hence in this case the j series and the j' series, of which the sums appear in (6), are each convergent, and if we take $j = \infty$ and $j' = \infty$, (6) becomes

$$S = \sum_{i=1}^{i=\infty} C_i \frac{f_i^{x/a} + f_i^{1-x/a}}{1-f_i} \quad \ldots\ldots\ldots\ldots\ldots\ldots(8).$$

We have now the same expression for S whichever of the ridges be chosen for the origin of x; and the value for $x = a$ is equal to the value for $x = 0$. The water-disturbance is therefore equal and similar in all the spaces from ridge to ridge, and the solution (8), from $x = 0$ to $x = a$, expresses within the period the height of the water above a certain level; not now, as in (2), the mean level throughout the period, but a level at a height $\int_0^a S\,.\,dx/a$ above the mean level. Now, by integration of (8), we find

$$\frac{1}{a}\int_0^a S dx = \sum_{i=1}^{i=\infty} \frac{2C_i}{\log(1/f_i)} \quad \ldots\ldots\ldots\ldots\ldots(9).$$

To evaluate the series forming the second member of this expression, remark that by (7) above and (34) Part III., we have

$$\frac{2C_i}{\log(1/f_i)} = A/a \frac{(-1)^{i+1}\cos\alpha_i}{\theta_i(1-b/D.\sin^2\alpha_i)} = \frac{A/a.N_i}{1-D/b} \quad \text{.....(10).}$$

Now by putting $\sigma = 0$ in Part III. (29) and (24), we find

$$\Sigma N_i = 1 - D/b \quad \text{.......................(11).}$$

Hence, and by (10), (9) becomes

$$\frac{1}{a}\int_0^a S\,dx = A/a \quad \text{......................(12).}$$

Denoting now, as before, by $\mathfrak{h}$ the height above mean level from ridge to ridge, we find from (8),

$$\mathfrak{h} = \sum_{i=1}^{i=\infty} C_i \frac{f_i^{x/a} + f_i^{1-x/a}}{1-f_i} - A/a \quad \text{.............(13).}$$

The comparison between this and (2) above, two different expressions for the same quantity (with, for simplicity, $D=1$), leads to the following remarkable theorem of pure analysis,

$$\sum_{=1}^{i=\infty} \frac{4/a.\cos i\dfrac{2\pi x}{a}}{e^i + e^{-i} - \dfrac{1}{i}\cdot\dfrac{a}{2\pi b}(e^i - e^{-i})}$$

$$= \frac{1}{2}\sum_{i=1}^{i=\infty} \frac{(-1)^{i+1}\cos\alpha_i}{1-b\sin^2\alpha_i} \frac{\epsilon^{-\theta_i x} + \epsilon^{-\theta_i(a-x)}}{1-\epsilon^{-\theta_i a}} - \frac{1}{a} \quad \text{....(14);}$$

where

a denotes any real positive numeric;
b denotes any numeric > 1;
e denotes $\epsilon^{2\pi/a}$;
α_i denotes the numeric between zero and $\pi/2$ which satisfies the equation
$$[(i-\tfrac{1}{2})\pi - \alpha_i]\tan\alpha_i - 1/b = 0;$$
θ_i denotes $(i-\tfrac{1}{2})\pi - \alpha_i$;
x denotes any real positive numeric $< a$

...(15).

The theorem (14) is easily verified by taking $\int_0^a dx.\cos j\dfrac{2\pi x}{a}$ of both members. The first member of the result is obviously $2\Big/\left[e^j + e^{-j} - \dfrac{1}{j}\cdot\dfrac{a}{2\pi b}(e^j - e^{-j})\right]$ The second member, modified by

(34), (29), and (24) of Part III., is found to have the same value. For the particular case of $j=0$ (that is to say, the mere integral $\int_0^a dx$ of each member), the equality is proved by (12).

For the most interesting cases of our physical problem, the solution (13) converges with great rapidity, except for small values of x; and for these the form of the surface is more easily calculated by (2). Numerical illustrations and the working out of the solution corresponding to (13) for the case of $b < D$ are reserved for Part V., which, I am sorry to say, must be set aside for some time. I hope it will appear in the April or May number, and that it, or Part VI., will contain practical illustrations, such as the stationary waves produced by a deeper place, or a less deep place, extending over a considerable length of the stream, which is very easily worked out from our solution (40) (48) Part III., for the effect of a single infinitesimal ridge. I hope to pass next to the effect of surface disturbance, with interesting applications to the question of the towage of a boat in a canal, and the beautiful practical discoveries of Mr Houston and Mr Scott Russell referred to at the commencement of Part III. If I succeed in carrying out my intention, this series of Articles on Stationary Waves will end with the investigation of the wave-group produced by a ship moving through the water with uniform velocity, promised at the commencement of Part I.; and suggestions for extension in the direction towards the theory of the effect of the wind in generating waves at sea.

30. On the Waves Produced by a Single Impulse in Water of Any Depth, or in a Dispersive Medium.

[From the *Philosophical Magazine*, Vol. XXIII. March 1887, pp. 252—255; having been read before the Royal Society, 3rd February, 1887, *Proceedings*, Vol. XLII. p. 80.]

FOR brevity and simplicity consider only the case of *two-dimensional motion.*

All that it is necessary to know of the medium is the relation between the wave-velocity and the wave-length of an endless procession of periodic waves. The result of our work will show us that the velocity of progress of a zero, or maximum, or minimum, in any part of a varying group of waves is equal to the velocity of progress of periodic waves of wave-length equal to a certain length, which may be defined as the wave-length in the neighbourhood of the particular point looked to in the group (a length which will generally be intermediate between the distances from the point considered to its next-neighbour corresponding points on the preceding and following waves).

Let $f(m)$ denote the velocity of propagation corresponding to wave-length $2\pi/m$. The Fourier-Cauchy-Poisson synthesis gives

$$u = \frac{1}{2\pi}\int_0^\infty dm \cos m\,[x - t\,f(m)] \quad\text{............(1)}$$

for the effect at place and time (x, t) of an infinitely intense disturbance at place and time (0, 0). The principle of interference, as set forth by Prof. Stokes and Lord Rayleigh in their theory of group-velocity and wave-velocity, suggests the following treatment for this integral:—

When $x - tf(m)$ is very large, the parts of the integral (1) which lie on the two sides of a small range, $\mu - \alpha$ to $\mu + \alpha$, vanish by annulling interference; μ being a value, or the value, of m which makes

$$\frac{d}{dm}\{m[x - tf(m)]\} = 0 \quad\text{.................(2);}$$

so that we have

$$x = t\{f(\mu) + \mu f'(\mu)\} = Vt \quad\text{.................(3),}$$

where

$$V = f(\mu) + \mu f'(\mu)^* \quad\text{.....................(4);}$$

and we have by Taylor's theorem for $m - \mu$ very small,

$$m[x - tf(m)] = \mu[x - tf(\mu)] - t[\mu f''(\mu) + 2f'(\mu)]\tfrac{1}{2}(m-\mu)^2 \text{...(5);}$$

or, modifying by (3),

$$m[x - tf(m)] = t\{\mu^2 f'(\mu) + [-\mu f''(\mu) - 2f'(\mu)]\tfrac{1}{2}(m-\mu)^2\} \text{...(6).}$$

Put now

$$m - \mu = \frac{\sigma\sqrt{2}}{t^{\frac{1}{2}}[-\mu f''(\mu) - 2f'(\mu)]^{\frac{1}{2}}} \quad\text{............(7),}$$

and using the result in (1), we find

$$u = \frac{\sqrt{2}\int_{-\infty}^{\infty} d\sigma \cos[t\mu^2 f'(\mu) + \sigma^2]}{2\pi t^{\frac{1}{2}}[-\mu f''(\mu) - 2f'(\mu)]^{\frac{1}{2}}} \quad\text{...........(8);}$$

the limits of the integral being here $-\infty$ to ∞, because the denominator of (7) is so infinitely great that, though $\pm\alpha$, the arbitrary limits of $m - \mu$, are infinitely small, α multiplied by it is infinitely great†.

Now we have

$$\int_{-\infty}^{\infty} d\sigma \cos^2\sigma = \int_{-\infty}^{\infty} d\sigma \sin^2\sigma = \sqrt{\frac{\pi}{2}} \quad\text{........(9).}$$

Hence (8) becomes

$$u = \frac{\cos[t\mu^2 f'(\mu)] - \sin[t\mu^2 f'(\mu)]}{2\pi^{\frac{1}{2}} t^{\frac{1}{2}}[-\mu f''(\mu) - 2f'(\mu)]^{\frac{1}{2}}} = \frac{\sqrt{2}\cos[t\mu^2 f'(\mu) + \frac{1}{4}\pi]}{2\pi^{\frac{1}{2}} t^{\frac{1}{2}}[-\mu f''(\mu) - 2f'(\mu)]^{\frac{1}{2}}}$$

.........(10).

* This is the group-velocity according to Lord Rayleigh's generalization of Prof. Stokes's original result. [For further extension on the lines of the present paper, see 'Baltimore Lectures,' App. C, pp. 528—531, and paper on 'Deep-Sea Ship Waves' reprinted *infra* §§ 80 *seq.*; also H. Lamb, *Hydrodynamics*, 3rd ed., 1906, § 253, T. H. Havelock, *Proc. Roy. Soc.* Aug. 1908, pp. 398—430, and G. Green, *Proc. R. S. E:* vol. XXIX, July, 1909.]

† [Mr Green points out that this condition of very great denominator is not needed, greatness of t sufficing by itself to justify the infinite limits: cf. equation (14) *infra*.]

To prove the law of wave-length and wave-velocity for any point of the group, remark that, by (3),

$$t\mu^2 f'(\mu) = \mu\,[x - tf(\mu)],$$

and therefore the numerator of (10) is equal to $\sqrt{2}\cos\theta$, where

$$\theta = \mu\,[x - tf(\mu)] + \tfrac{1}{4}\pi \quad\text{................}(10'),$$

and by (2) and (3),

$$d/d\mu\,\{\mu\,[x - tf(\mu)]\} = 0\,;$$

by which we see that

$$d\theta/dx = \mu,\ \text{ and }\ d\theta/dt = -\,\mu f(\mu) \quad\text{.........}(10''),$$

which proves the proposition.

Example (1). As a first example take deep-sea waves; we have

$$f(m) = \sqrt{\frac{g}{m}} \quad\text{..........................}(11),$$

which reduces (4), (3), and (10) to

$$V = \tfrac{1}{2}\sqrt{\frac{g}{\mu}} \quad\text{........................}(12),$$

and

$$x = \tfrac{1}{2}\sqrt{\frac{g}{\mu}}\,.\,t \quad\text{.....................}(13),$$

$$u = \frac{g^{\frac{1}{2}}}{2^{\frac{3}{2}}\pi^{\frac{1}{2}}}\frac{t}{x^{\frac{3}{2}}}\left(\cos\frac{gt^2}{4x} + \sin\frac{gt^2}{4x}\right) = \frac{g^{\frac{1}{2}}t}{2^{\frac{3}{2}}\pi^{\frac{1}{2}}x^{\frac{3}{2}}}\cos\left(\frac{gt^2}{4x} - \frac{\pi}{4}\right)\text{...}(14);$$

which is Cauchy and Poisson's result for places where x is very great in comparison with the wave-length $2\pi/\mu$; that is to say, for place and time such that $gt^2/4x$ is very large.

Example (2). Waves in water of depth D,

$$f(m) = \sqrt{\left\{\frac{g}{m}\frac{1 - \epsilon^{-2mD}}{1 + \epsilon^{-2mD}}\right\}} \quad\text{..............}(15).$$

Example (3). Light in a dispersive medium.

Example (4). Capillary gravitational waves,

$$f(m) = \sqrt{\left(\frac{g}{m} + Tm\right)} \quad\text{.....................}(16).$$

Example (5). Capillary waves,

$$f(m) = \sqrt{(Tm)} \quad\text{.......... ....}(17).$$

Example (6). Waves of flexure running along a uniform elastic rod

$$f(m) = m\sqrt{\frac{B}{w}} \quad \text{.......................(18),}$$

where B denotes the flexural rigidity and w the mass per unit of length.

These last three examples have been taken by Lord Rayleigh as applications of his generalization of the theory of group-velocity; and he has pointed out, in his 'Standing Waves in Running Water" (London Mathematical Society, December 13, 1883), the important peculiarity of example (4) in respect to the critical wave-length which gives minimum wave-velocity, and therefore group-velocity equal to wave-velocity. The working out of our present problem for this case, or any case in which there are either minimums or maximums, or both maximums and minimums, of wave-velocity, is particularly interesting, but time does not permit its being included in the present communication.

For examples (5) and (6) the denominator of (10) is imaginary; and the proper modification, from (7) forwards, gives for these and such cases, instead of (10), the following:—

$$u = \frac{\cos\left[t\mu^2 f'(\mu)\right] + \sin\left[t\mu^2 f'(\mu)\right]}{2\pi^{\frac{1}{2}} t^{\frac{1}{2}} \left[\mu f''(\mu) + 2f'(\mu)\right]^{\frac{1}{2}}} \quad \text{...........(19).}$$

The result is easily written down for each of the two last cases [Examples (5) and (6)].

31. ON THE FRONT AND REAR OF A FREE PROCESSION OF WAVES IN DEEP WATER.

[*Proc. Roy. Soc. Edin.* Jan. 7, 1887; *Phil. Mag.* Vol. XXIII. February 1887, pp. 113—120.]

[Replaced later by a Paper on different lines in *Phil. Mag.*, Oct. 1904: also substantially included in a Paper in *Phil. Mag.*, Jan. 1907, §§ 127—158: both reprinted *infra.*]

"Not to be printed because in my *R. S. E.* paper of Feb. 1, and its successor now in hand, the whole substance of it, with promised extensions, is given in much better, and more easily read, form. K. (Mentone, March 30, 1904)."

32. ON SHIP WAVES.

[A Lecture in connexion with the Institution of Mechanical Engineers' Conference at Edinburgh, Aug. 3, 1887.

Reprinted from *Proc. Inst. Mech. Eng.* 1887, in *Popular Lectures and Addresses*, Vol. III. pp. 450—500, some of the illustrations being omitted.]

33. On the Propagation of Laminar Motion through a Turbulently moving Inviscid Liquid.

[From *Brit. Assoc. Report*, 1887, pp. 486—495; *Phil. Mag.* Vol. XXIV. Oct. 1887, pp. 342—353.]

1. In endeavouring to investigate turbulent motion of water between two fixed planes, for a promised communication to Section A of the British Association at its Meeting in Manchester, I have found something seemingly towards a solution (many times tried for within the last twenty years) of the problem to construct, by giving vortex motion to an incompressible inviscid fluid, a medium which shall transmit waves of laminar motion as the luminiferous ether transmits waves of light *.

2. Let the fluid be unbounded on all sides, and let u, v, w be the velocity-components, and p the pressure at (x, y, z, t). We have

$$\frac{du}{dx}+\frac{dv}{dy}+\frac{dw}{dz}=0 \quad \text{.....................(1)},$$

$$\frac{du}{dt}=-\left(u\frac{du}{dx}+v\frac{du}{dy}+w\frac{du}{dz}+\frac{dp}{dx}\right) \quad \text{...........(2)},$$

$$\frac{dv}{dt}=-\left(u\frac{dv}{dx}+v\frac{dv}{dy}+w\frac{dv}{dz}+\frac{dp}{dy}\right) \quad \text{...........(3)},$$

$$\frac{dw}{dt}=-\left(u\frac{dw}{dx}+v\frac{dw}{dy}+w\frac{dw}{dz}+\frac{dp}{dz}\right) \quad \text{...........(4)}.$$

From (2), (3), (4) we find, taking (1) into account,

$$-\nabla^2 p=\left(\frac{du}{dx}\right)^2+\left(\frac{dv}{dy}\right)^2+\left(\frac{dw}{dz}\right)^2+2\left(\frac{dv}{dz}\frac{dw}{dy}+\frac{dw}{dx}\frac{du}{dz}+\frac{du}{dy}\frac{dv}{dx}\right) \quad \text{.........(5)}.$$

* [Cf. G. F. FitzGerald, *Nature*, May 9, 1889, *Proc. Roy. Dub. Soc.* 1899, and *B. A. Report*, 1899; or in *Scientific Papers*, 1902, pp. 254, 472, 484. See also *snpra*, p. 202.]

3. The velocity-components u, v, w may have any values whatever through all space, subject only to (1). Hence, on Fourier's principles, we have, as a perfectly comprehensive expression for the motion at any instant,

$$u = \overset{e\,f\,g}{\underset{m\,n\,q}{\Sigma\Sigma\Sigma\Sigma\Sigma\Sigma}}\, \alpha^{(e,f,g)}_{(m,n,q)} \sin(mx+e)\cos(ny+f)\cos(qz+g) \quad \ldots(6),$$

$$v = \overset{e\,f\,g}{\underset{m\,n\,q}{\Sigma\Sigma\Sigma\Sigma\Sigma\Sigma}}\, \beta^{(e,f,g)}_{(m,n,q)} \cos(mx+e)\sin(ny+f)\cos(qz+g) \quad \ldots(7),$$

$$w = \overset{e\,f\,g}{\underset{m\,n\,q}{\Sigma\Sigma\Sigma\Sigma\Sigma\Sigma}}\, \gamma^{(e,f,g)}_{(m,n,q)} \cos(mx+e)\cos(ny+f)\sin(qz+g) \quad \ldots(8),$$

where $\alpha^{(e,f,g)}_{(m,n,q)}$, $\beta^{(e,f,g)}_{(m,n,q)}$, $\gamma^{(e,f,g)}_{(m,n,q)}$ are any three velocities satisfying the equation

$$0 = m\alpha^{(e,f,g)}_{(m,n,q)} + n\beta^{(e,f,g)}_{(m,n,q)} + q\gamma^{(e,f,g)}_{(m,n,q)} \quad \ldots\ldots\ldots\ldots(9);$$

and $\Sigma\Sigma\Sigma\Sigma\Sigma\Sigma$ summation (or integration) for different values of m, n, q, e, f, g. The summations for e, f, g may, without loss of generality, be each confined to two values: $e=0$, and $e=\frac{1}{2}\pi$; $f=0$, and $f=\frac{1}{2}\pi$; $g=0$, and $g=\frac{1}{2}\pi$. We shall admit large values, and infinite values of m^{-1}, n^{-1}, q^{-1}, under certain conditions [§ 4 (10), (11), (12), and § 15 below], but otherwise we shall suppose the greatest value of each of them to be of some moderate, or exceedingly small, linear magnitude. This is an essential of the averagings to which we now proceed.

4. Let xav, xzav, xyzav denote space-averages, linear, surface, and solid, through infinitely great spaces, defined and illustrated by examples, each worked out from (6), (7), (8), as follows, L denoting an infinitely great length, or a very great multiple of whichever of m^{-1}, n^{-1}, q^{-1} may be concerned:—

$$\text{xav}\, u = \frac{1}{2L}\int_{-L}^{L} dx\, u = \overset{f\,g}{\underset{n\,q}{\Sigma\Sigma\Sigma\Sigma}}\, \alpha^{(\frac{1}{2}\pi,f,g)}_{(0,n,q)} \cos(ny+f)\cos(qz+g) \quad \ldots\ldots\ldots(10),$$

$$\text{xzav}\, u = \left(\frac{1}{2L}\right)^2 \int_{-L}^{L}\int_{-L}^{L} dz\,dx\, u = \overset{f}{\underset{n}{\Sigma\Sigma}}\, \alpha^{(\frac{1}{2}\pi,f,0)}_{(0,n,0)} \cos(ny+f) \ldots(11),$$

$$\text{xyzav}\, u = \left(\frac{1}{2L}\right)^3 \int_{-L}^{L}\int_{-L}^{L}\int_{-L}^{L} dz\,dy\,dx\, u = \alpha^{(\frac{1}{2}\pi,0,0)}_{(0,0,0)} \quad \ldots\ldots\ldots\ldots(12),$$

$$\text{xav}\, u^2 = \tfrac{1}{2}\overset{e\,f\,g}{\underset{m\,n\,q}{\Sigma\Sigma\Sigma\Sigma\Sigma\Sigma}}\, [\alpha^{(e,f,g)}_{(m,n,q)}]^2 \cos^2(ny+f)\cos^2(qz+g) \quad \ldots(13);$$

this with the exceptions that

in the case of $m=0$, $e=0$, we take 0 in place of $\frac{1}{2}$,
and in the case of $m=0$, $e=\frac{1}{2}\pi$ „ 1 „ „ $\frac{1}{2}$.

$$\text{xzav}\, u^2 = \tfrac{1}{4}\sum_m^e\sum\sum_n^f\sum\sum_q^g\sum [\alpha^{(e,f,g)}_{(m,n,q)}]^2 \cos^2(ny+f) \quad\ldots\ldots\ldots\ldots\ldots(14),$$

$$\text{xzav}\, uv = \tfrac{1}{4} \sum_m\sum\sum_n^f\sum\sum_q^g [\alpha^{(\frac{1}{2}\pi,f,g)}_{(m,n,q)}\beta^{(0,f,g)}_{(m,n,q)}$$

$$-\alpha^{(0,f,g)}_{(m,n,q)}\beta^{(\frac{1}{2}\pi,f,g)}_{(m,n,q)}]\cos(ny+f)\sin(ny+f) \quad\ldots\ldots\ldots(15);$$

with the exceptions for (14) that

in the case of $m=0$ and $e=0$
and in the case of $q=0$ and $g=\frac{1}{2}\pi$ } we take 0 instead of $\frac{1}{4}$;

in the case of $m=0$ and $e=\frac{1}{2}\pi$
and in the case of $q=0$ and $g=0$ } „ $\frac{1}{2}$ „ „ $\frac{1}{4}$;

in the case of $m=0$, $e=\frac{1}{2}\pi$, $n=0$, $f=\frac{1}{2}\pi$ „ 1 „ „ $\frac{1}{4}$;

and analogous exceptions for (15).

$$\text{xyzav}\, u^2 = \tfrac{1}{8}\sum_m^e\sum\sum_n^f\sum\sum_q^g\sum \left[\alpha^{(e,f,g)}_{(m,n,q)}\right]^2 \ldots\ldots\ldots\ldots\ldots\ldots(16),$$

with exceptions for zeros of m and q, analogous to those of (14).

5. As a last example of averagings for the present, take xyzav of (5). Thus we find

$$\left.\begin{aligned} -\text{xyzav}\,\nabla^2 p &= \tfrac{1}{8}\sum_m^e\sum\sum_n^f\sum\sum_q^g\sum \left\{m\alpha^{(e,f,g)}_{(m,n,q)} + n\beta^{(e,f,g)}_{(m,n,q)} + q\gamma^{(e,f,g)}_{(m,n,q)}\right\}^2 \\ &= 0 \text{ by } (9) \end{aligned}\right\} \ldots\ldots\ldots(17).$$

The interpretation is obvious.

6. Remark, as a general property of this kind of averaging,

$$\text{xav}\,\frac{dQ}{dx} = 0 \ldots\ldots\ldots\ldots\ldots\ldots\ldots\ldots(18),$$

if Q be any quantity which is finite for infinitely great values of x.

7. Suppose now the motion to be homogeneously distributed through all space. This implies that the centres of inertia of all great volumes of the fluid have equal parallel motions, if any motions at all. Conveniently, therefore, we take our reference lines OX,

OY, OZ, as fixed relatively to the centres of inertia of three (and therefore of all) centres of inertia of large volumes; in other words, we assume no translatory motion of the fluid as a whole. This makes zero of every large average of u and of v and of w; and, in passing, we may remark, with reference to our notation of § 3, that it makes, as we see by (10), (11), (12),

$$0 = \alpha_{(0,\, n,\, q)} = \alpha_{(m,\, 0,\, q)} = \alpha_{(m,\, n,\, 0)} = \beta_{(0,\, n,\, q)} \dots = \gamma_{(m,\, n,\, 0)} \dots (19).$$

Without for the present, however, encumbering ourselves with the Fourier-expression and notation of § 3, we may write, as the general expression for nullity of translational movement in large volumes,

$$0 = \text{ave}\, u = \text{ave}\, v = \text{ave}\, w \quad \dots\dots\dots\dots\dots (20);$$

where ave denotes the average through any great length of straight or curved line, or area of plane or curved surface, or through any great volume of space.

8. In terms of this generalized notation of averages, homogeneousness implies

$$\text{ave}\, u^2 = U^2, \qquad \text{ave}\, v^2 = V^2, \qquad \text{ave}\, w^2 = W^2 \quad \dots(21),$$

$$\text{ave}\, vw = BC, \qquad \text{ave}\, wu = CA, \qquad \text{ave}\, uv = AB \quad \dots(22);$$

where U, V, W, A, B, C are six velocities independent of the positions of the spaces in which the averages are taken. These equations are, however, infinitely short of implying, though implied by, homogeneousness.

9. Suppose now the distribution of motion to be isotropic. This implies, but is infinitely more than is implied by, the following equations in terms of the notation of § 8, with further notation, R, to denote what we shall call THE AVERAGE VELOCITY of the turbulent motion:—

$$U^2 = V^2 = W^2 = \tfrac{1}{3} R^2 \quad \dots\dots\dots\dots\dots\dots (23),$$

$$0 = A = B = C \dots\dots\dots\dots\dots\dots\dots\dots\dots (24).$$

10. Large questions now present themselves as to transformations which the distribution of turbulent motion will experience in an infinite liquid left to itself with any distribution given to it initially. If the initial distribution be homogeneous through all large volumes of space, except a certain large finite space, S, through which there is initially either no motion, or turbulent motion homogeneous or not, but not homogeneous with the motion through the surrounding space, will the fluid which at any time is within S acquire more and more nearly as time

advances the same homogeneous distribution of motion as that of the surrounding space, till ultimately the motion is homogeneous throughout?

11. If the answer were *yes*, could it be that this equalization would come to pass through smaller and smaller spaces as time advances? In other words, would any given distribution, homogeneous on a large enough scale, become more and more *fine-grained* as time advances? Probably *yes* for some initial distributions; probably *no* for others. Probably *yes* for vortex motion given continuously through all of one large portion of the fluid, while all the rest is irrotational.

12. Probably *no* for the initial motion given in the shape of equal and similar Helmholtz rings, of proportions suitable for individual stability, and each of overall diameter considerably smaller than the average distance from nearest neighbours. Probably also *no*, though the rings be of very different volumes and vorticities. But probably *yes** if the diameters of the rings, or of many of them, be not small in comparison with distances from neighbours, or if the individual rings, each an endless slender filament, be entangled or nearly entangled among one another.

13. Again a question: If the initial distribution be *homogeneous and æolotropic*, will it become more and more nearly isotropic as time advances, and *ultimately quite isotropic*? Probably *yes*†, for any random initial distribution, whether of continuous rotationally-moving fluid or of separate finite vortex rings. Possibly *no* for some symmetrical initial distribution of vortex rings, conceivably stable.

14. If the initial distribution be homogeneous and isotropic (and therefore utterly *random* in respect to direction), will it remain so? Certainly *yes*. I proceed to investigate a mathematical formula, deducible from the answer, which will be of use to us later (§ 18). By (22) and (24) we have

$$\text{xzav}\, uv = 0, \text{ for all values of } t \quad \text{..........(25)}.$$

But by (2) and (3) we find

$$\frac{d}{dt}(\text{xzav}\, uv) = -\,\text{xzav}\left\{u\frac{d(uv)}{dx} + v\frac{d(uv)}{dy} + w\frac{d(uv)}{dz} + v\frac{dp}{dx} + u\frac{dp}{dy}\right\} \quad \text{........(26)}.$$

* [No? W. T., Netherhall, Aug. 10, 1889.] See p. 202 *supra*.

† [No? Because in fact such æolotropy as that of § 20 is merely translational motion of liquid and vortices. W.T.]

Hence

$$0 = \text{xzav}\left\{u\frac{d(uv)}{dx} + v\frac{d(uv)}{dy} + w\frac{d(uv)}{dz} + v\frac{dp}{dx} + u\frac{dp}{dy}\right\} \text{...(27)}.$$

This equation in fact holds for every random case of motion satisfying (30) below, because positive and negative values of u, v, w are all equally probable, and therefore the value of the second member of (27) is doubled by adding to itself what it becomes when for u, v, w we substitute $-u$, $-v$, $-w$, which, it may be remarked, and verified by looking at (5), does not change the value of p.

15. We shall now suppose the initial motion to consist of a laminar motion $[f(y), 0, 0]$ superimposed on a homogeneous and isotropic distribution $(\mathfrak{u}_0, v_0, w_0)$; so that we have

$$\text{when } t = 0, \quad u = f(y) + \mathfrak{u}_0, \quad v = v_0, \quad w = w_0 \text{ ......(28)};$$

and we shall endeavour to find such a function, $f(y, t)$, that at any time t the velocity-components shall be

$$f(y, t) + \mathfrak{u}, v, w \text{......................(29)},$$

where $\mathfrak{u}$, v, w are quantities of each of which every large enough average is zero, so that particularly, for example,

$$0 = \text{xzav}\,\mathfrak{u} = \text{xzav}\,v = \text{xzav}\,w \text{..............(30)}.$$

16. Substituting (29) for u, v, w in (2) we find

$$\frac{df(y,t)}{dt} + \frac{d\mathfrak{u}}{dt} = -\left\{f(y,t)\frac{d\mathfrak{u}}{dx} + v\frac{df(y,t)}{dy}\right\}$$
$$-\left(\mathfrak{u}\frac{d\mathfrak{u}}{dx} + v\frac{d\mathfrak{u}}{dy} + w\frac{d\mathfrak{u}}{dz} + \frac{dp}{dx}\right) \text{...(31)}.$$

Take now xzav of both members. The second term of the first member and the second term of the second member disappear, each in virtue of (30). The first and last terms of the second member disappear, each in virtue of (18) alone, and also each in virtue of (30). There remains

$$\frac{df(y,t)}{dt} = -\,\text{xzav}\left(\mathfrak{u}\frac{d\mathfrak{u}}{dx} + v\frac{d\mathfrak{u}}{dy} + w\frac{d\mathfrak{u}}{dz}\right) \text{ ......(32)}.$$

To simplify, add to the second member [by (1)]

$$0 = -\,\text{xzav}\left(\mathfrak{u}\frac{d\mathfrak{u}}{dx} + \mathfrak{u}\frac{dv}{dy} + \mathfrak{u}\frac{dw}{dz}\right) \text{ ....(33)};$$

and, the first and third pair of terms of the thus-modified second member vanishing by (18), find

$$\frac{df(y,t)}{dt} = -\text{xzav}\frac{d(\mathfrak{u}v)}{dy} \quad\ldots\ldots\ldots\ldots\ldots\ldots(34).$$

It is to be remarked that this result involves, besides (1), no other condition respecting ($\mathfrak{u}$, v, w) than (30); no isotropy, no homogeneousness in respect to y; and only homogeneousness of *régime* with respect to y and z, with no mean translational motion.

The x-translational mean component of the motion is wholly represented by $f(y, t)$, and, so far as our establishment of (34) is concerned, may be of any magnitude, great or small relatively to velocity-components of the turbulent motion. It is a fundamental formula in the theory of the turbulent motion of water between two planes; and I had found it in endeavouring to treat mathematically my brother Prof. James Thomson's theory of the "Flow of Water in Uniform *Régime* in Rivers and other Open Channels"* In endeavouring to advance a step towards the law of distribution of the laminar motion at different depths, I was surprised to discover the seeming possibility of a law of propagation as of distortional waves in an elastic solid, which constitutes the conclusion of my present communication, on the supposition of § 15 that the distribution $\mathfrak{u}_0$, v_0, w_0 is isotropic, and that $df(y,t)/dy$, divided by the greatest value of $f(y, t)$, is infinitely small in comparison with the smallest values of m, n, q, in the Fourier-formulæ (6), (7), (8) for the turbulent motion.

17. By (34) we see that, if the turbulent† motion remained, through time, isotropic† as at the beginning, $f(y, t)$ would remain constantly at its initial value $f(y)$. To find whether the turbulent motion does remain isotropic†, and, if it does not, to find what we want to know of its deviation from isotropy, let us find xzav $d(\mathfrak{u}v)/dt$, by (2) and (3), as follows:—First, by multiplying (31) by v, and (3) by $\mathfrak{u}$, and adding, we find

$$v\frac{df(y,t)}{dt} + \frac{d(\mathfrak{u}v)}{dt} = -\left\{f(y,t)\frac{d(\mathfrak{u}v)}{dx} + v^2\frac{df(y,t)}{dy}\right\}$$
$$-\left\{\mathfrak{u}\frac{d(\mathfrak{u}v)}{dx} + v\frac{d(\mathfrak{u}v)}{dy} + w\frac{d(\mathfrak{u}v)}{dz} + v\frac{dp}{dx} + \mathfrak{u}\frac{dp}{dy}\right\}\ldots(35).$$

* *Proc. Roy. Soc.* Aug. 15, 1878.

† [Modify this. W. T.]

Taking xzav of this, and remarking that the first term of the first member disappears by (30), and the first term of the second member by (18), we find, with V^2, as in §§ 8, 9, to denote the average y-component-velocity of the turbulent motion,

$$\frac{d}{dt}\{\text{xzav}(\mathfrak{u}v)\} = -V^2\frac{df(y,t)}{dy} - Q \text{ ....(36)},$$

where

$$Q = \text{xzav}\left\{\mathfrak{u}\frac{d(\mathfrak{u}v)}{dx} + v\frac{d(\mathfrak{u}v)}{dy} + w\frac{d(\mathfrak{u}v)}{dz} + v\frac{dp}{dx} + \mathfrak{u}\frac{dp}{dy}\right\} \text{...(37)}.$$

18. Let

$$p = \mathfrak{p} + \varpi \text{........................(38)},$$

where $\mathfrak{p}$ denotes what p would be if f were zero. We find, by (5),

$$-\nabla^2\varpi = 2\frac{df(y,t)}{dy}\frac{dv}{dx} \text{ ....(39)},$$

and, by (27) and (37),

$$Q = \text{xzav}\left(v\frac{d\varpi}{dx} + \mathfrak{u}\frac{d\varpi}{dy}\right) \text{................(40)}.$$

So far we have not used either the supposition of initial isotropy for the turbulent motion, or of the infinitesimalness of df/dy. We now must introduce and use both suppositions.

19. To facilitate the integration of (39), we now use our supposition that $d/dy\,.f(y,t)$, divided by the greatest value of $f(y,t)$, is infinitely small in comparison with m, n, q, which, as is easily proved, gives

$$\varpi = 2\frac{df(y,t)}{dy}\frac{1}{-\nabla^2}\frac{dv}{dx} \text{ ....(41)},$$

by which (40) becomes

$$Q = -2\frac{df(y,t)}{dy}\text{xzav}\left(v\frac{d}{dx} + \mathfrak{u}\frac{d}{dy}\right)\nabla^{-2}\frac{dv}{dx} \text{ ......(42)}.$$

Now, by (x, z) isotropy, we have

$$2\text{xzav}\left(v_0\frac{d}{dx} + \mathfrak{u}_0\frac{d}{dy}\right)\nabla^{-2}\frac{dv_0}{dx}$$

$$= \text{xzav}\left\{v_0\left(\frac{d^2}{dx^2} + \frac{d^2}{dz^2}\right) + \left(\mathfrak{u}_0\frac{d}{dx} + w_0\frac{d}{dz}\right)\frac{d}{dy}\right\}\nabla^{-2}v_0 \text{...(43)}$$

Performing integrations by parts for the last two terms of the second member, and using (1), we find

$$\text{xzav}\left(\mathfrak{u}_0\frac{d}{dx}+w_0\frac{d}{dz}\right)\frac{d}{dy}\nabla^{-2}v_0=-\,\text{xzav}\left(\frac{d\mathfrak{u}_0}{dx}+\frac{dw_0}{dz}\right)\frac{d}{dy}\nabla^{-2}v_0$$

$$=\text{xzav}\,\frac{dv_0}{dy}\frac{d}{dy}\nabla^{-2}v_0\,;$$

and so we find, by (43) and (42),

$$Q_0=-\frac{df(y,t)}{dy}\,\text{xzav}\left\{v_0\left(\frac{d^2}{dx^2}+\frac{d^2}{dz^2}\right)+\frac{dv_0}{dy}\frac{d}{dy}\right\}\nabla^{-2}v_0\,\text{...(44)}.$$

20. Using now the Fourier expansion (7) for v_0, we find

$$-\nabla^{-2}v_0=\sum^{e}\sum^{f}\sum^{g}\sum_{m}\sum_{n}\sum_{q}\beta^{(e,f,g)}_{(m,n,q)}\frac{\cos(mx+e)\sin(ny+f)\cos(qz+g)}{m^2+n^2+q^2}$$

$$\text{.........(45)}.$$

Hence we find (with suffixes &c. dropped),

$$\text{xzav}\,\frac{dv_0}{dy}\frac{d}{dy}\nabla^{-2}v_0=-\tfrac{1}{8}\Sigma\Sigma\Sigma\Sigma\Sigma\Sigma\,\frac{n^2\beta^2}{m^2+n^2+q^2}\quad\text{...(46)*},$$

and

$$\text{xzav}\,v_0\left(\frac{d^2}{dx^2}+\frac{d^2}{dz^2}\right)\nabla^{-2}v_0=\tfrac{1}{8}\Sigma\Sigma\Sigma\Sigma\Sigma\Sigma\,\frac{(m^2+q^2)\beta^2}{m^2+n^2+q^2}\,\text{.........(47)}.$$

Now, in virtue of the average uniformity of the constituent terms implied in isotropy and homogeneousness (§§ 7, 8, 9), the second member of (46) is equal to $-\tfrac{1}{8}\Sigma\Sigma\Sigma\Sigma\Sigma\Sigma\,\tfrac{1}{3}\beta^2$, and therefore (§ 9) equal to $-\tfrac{1}{9}R^2$; and similarly we see that the second member of (47) is equal to $+\tfrac{2}{9}R^2$. Hence, finally, by (44),

$$Q_0=-\tfrac{1}{9}R^2\frac{df(y,t)}{dy}\,\text{....................(48)};$$

and (36) for $t=0$, with $\tfrac{1}{3}R^2$ for V^2 on account of isotropy, becomes

$$\left\{\frac{d}{dt}\,\text{xzav}\,(\mathfrak{u}v)\right\}_{t=0}=-\tfrac{2}{9}R^2\left\{\frac{df(y,t)}{dy}\right\}_{t=0}\text{.........(49)}.$$

The deviation from isotropy, which this equation shows†, is very small, because of the smallness of df/dy; and (27) does not need isotropy, but holds in virtue of (30). Hence (49) is not confined to the initial values (values for $t=0$) of the two members, because we neglect an infinitesimal deviation from $\tfrac{2}{9}R^2$ in the

* Here and henceforth an averaging through y-spaces so small as to cover no sensible differences of $f(y,t)$, but infinitely large in proportion to n^{-1}, is implied.

† [See note to § 13. W. T., Aug. 10, 1889.]

first factor of the second member, considering the smallness of the second factor. Hence, for all values of t, unless so far as the "random" character referred to at the end of § 13 may be lost by a rearrangement of vortices vitiating (27),

$$\frac{d}{dt}\,\text{xzav}\,(\mathfrak{u}v) = -\tfrac{2}{9}R^2\frac{df(y,t)}{dy} \quad\text{..............(50).}$$

21. Eliminating the first member from this equation, by (34), we find

$$\frac{d^2f}{dt^2} = \tfrac{2}{9}R^2\frac{d^2f}{dy^2} \quad\text{......................(51).}$$

Thus we have the very remarkable result that laminar disturbance is propagated according to the well-known mode of waves of distortion in a homogeneous elastic solid; and that the velocity of propagation is $R\sqrt{2}/3$, or about ·47 of the average velocity of the turbulent motion of the fluid. This might seem to go far towards giving probability to the vortex theory of the luminiferous ether, were it not for the doubtful proviso at the end of § 20.

22. If the undisturbed condition of the medium be a stable symmetrical distribution of vortex-rings the suggested vitiation by "rearrangement" cannot occur. For example, let it be such as is represented in fig. 1, where the small white and black circles

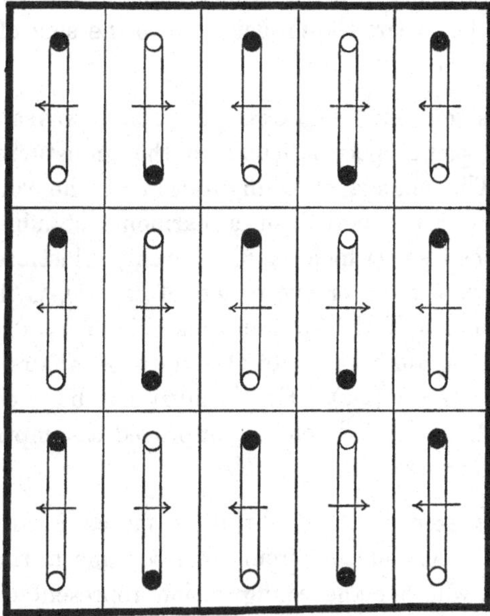

Fig. 1.

represent cross sections of the rings: the white where the rotation is opposite to, and the black where it is in the same direction as, the rotation of the hands of a watch placed on the diagram facing towards the spectator. Imagine first each vortex-ring to be in a portion of the fluid contained within a rigid rectangular box, of which four sides are indicated by the fine lines crossing one another at right angles throughout the diagram; and the other pair are parallel to the paper, at any distance asunder we like to imagine. Supposing the volume of the rotationally moving portion of the fluid constituting the ring to be given, there is clearly one determinate shape, and diametral magnitude, in which it must be given in order that the motion may be steady. Let it be so given, and fill space with such rectangular boxes of vortices arranged facing one another oppositely in the manner shown in the diagram. Annul now the rigidity of the sides of the boxes. The motion continues unchangedly steady. But is it stable, now that the rigid partitions are done away with? No proof has yet been given that it is. If it is, laminar waves, such as waves of light, could be propagated through it; and the velocity of propagation would be $R\sqrt{2/3}$ if the sides of the ideal boxes parallel to the undisturbed planes of the rings are square (which makes ave $\mathfrak{u}^2 =$ ave w^2), and if the distance between the square sides of each box bears the proper ratio to the side of the square to make ave $v^2 =$ ave $\mathfrak{u}^2 =$ ave w^2.

23. Consider now, for example, plane waves, or laminar vibrations, in planes perpendicular to the undisturbed planes of the rings. The change of configuration of the vortices in the course of a quarter period of a harmonic standing vibration, $f(y, t) = \cos \omega t \sin \kappa y$ (which is more easily illustrated diagrammatically than a wave or succession of waves), is illustrated in fig. 2, for a portion of the fluid on each side of $y = 0$. The upper part of the diagram represents the state of affairs when $t = 0$; the lower when $t = \pi/(2\omega)$. But it must not be overlooked, that all this §§ 22, 23 depends on the unproved assumption that the symmetrical arrangement is *stable.*

24. It is exceedingly doubtful, so far as I can judge after much anxious consideration from time to time during these last twenty years, whether the configuration represented in fig. 1, or any other symmetrical arrangement, is stable when the rigidity

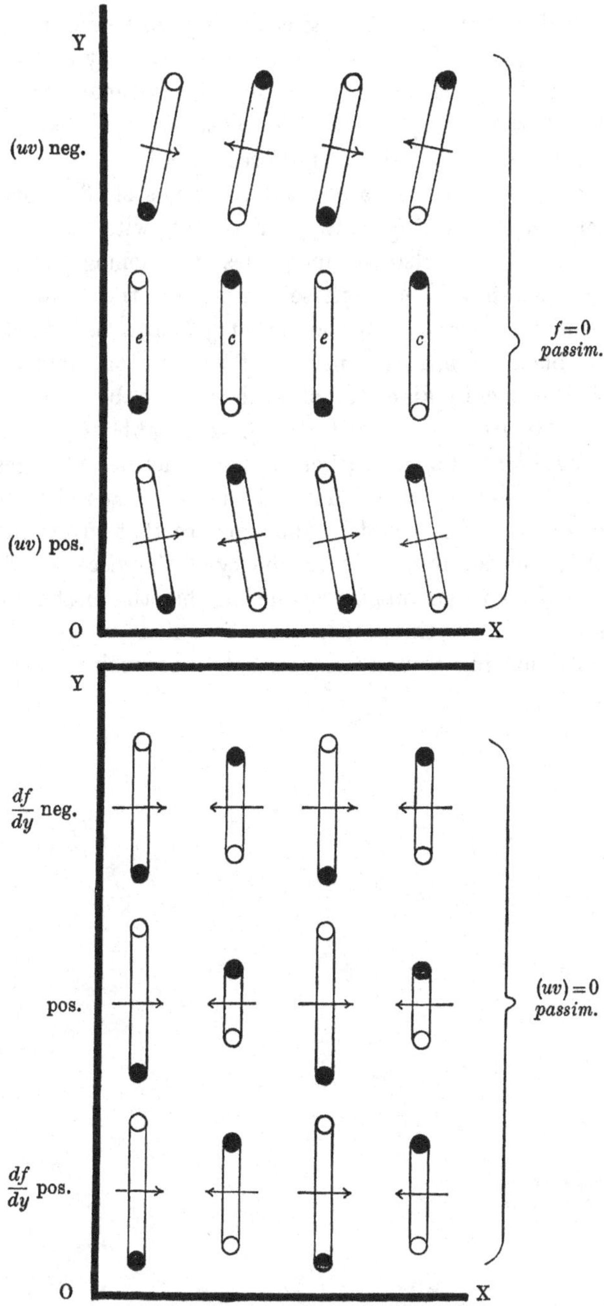

Fig. 2.

Here (uv) means an average of the kind described in the footnote on (46);
e, e are rings which are being expanded;
and c, c are rings which are being contracted.

of the ideal partitions enclosing each ring separately is annulled throughout space. It is possible that the rigidity of two, three, or more of the partitions may be annulled without vitiating the stability of the steady symmetric motion; but that if it be annulled through the whole of space, for all the partitions, the symmetric motion is unstable, and the rings shuffle themselves into perpetually varying relative positions, with *average homogeneousness*, like the ultimate molecules of a homogeneous liquid. I cannot see how, under these conditions, the "vitiating rearrangement" referred to at the end of § 20 can be expected not to take place within the period of a wave or vibration. To suppose the overall diameter of each ring to be very small in proportion to its average distances from neighbours, so that the crowd would be analogous rather to the molecules of a gas than to those of a liquid, would not help us to escape the vitiating rearrangement which would be analogous to that investigated by Maxwell in his admirable kinetic theory of the viscosity of gases. I am thus driven to admit, in conclusion, that the most favourable verdict I can ask for the propagation of laminar waves through a turbulently moving inviscid liquid is the Scottish verdict of *not proven.*

34. Rectilineal Motion of Viscous Fluid between two Parallel Planes.

[From the *Philosophical Magazine*, Vol. xxiv. August, 1887, pp. 188—196; having been read before the Royal Society of Edinburgh, July 15, 1887.]

27. Since the communication of the first of this series of articles to the Royal Society of Edinburgh in April, and its publication in the *Philosophical Magazine* in May and June*, the stability or instability of the steady motion of a viscous fluid has been proposed as subject for the Adams Prize of the University of Cambridge for 1888†. The present communication (§§ 27—40) solves the simpler of the two cases specially referred to by the Examiners in their announcement, and prepares the way for the investigation of the less simple by a preliminary laying down, in §§ 27—29, and equations (7) to (12) below, of the fundamental equations of motion of a viscous fluid kept moving by gravity between two infinite plane boundaries inclined to the horizon at any angle I, and given with any motion deviating infinitely little from the determinate steady motion which would be the unique and essentially stable solution if the viscosity were sufficiently large. It seems probable, almost certain indeed, that analysis similar to that of §§ 38 and 39 will demonstrate that the steady motion is stable for any viscosity, however small; and that the practical unsteadiness pointed out by Stokes forty-four years ago, and so admirably investigated experimentally five or six years ago by Osborne Reynolds, is to be explained by limits of stability becoming narrower and narrower the smaller is the viscosity.

Let OX be chosen in one of the bounding planes, parallel to the direction of the rectilineal motion; and OY perpendicular to

* See *Phil. Mag.* July, 1887, p. 142.

† [Reprinted *supra*, pp. 166 *seq.* The numbering of the sections is continuous with that paper.]

the two planes. Let the x-, y-, z-, component velocities, and the pressure, at (x, y, z, t), be denoted by $U+u$, v, w and p respectively; U denoting a function of (y, t). Then, calling the density of the fluid unity, and the viscosity μ, we have, as the equations of motion*,

$$\frac{du}{dx}+\frac{dv}{dy}+\frac{dw}{dz}=0 \quad\ldots\ldots\ldots\ldots\ldots\ldots(1);$$

$$\left.\begin{aligned}\frac{d}{dt}(U+u)+(U+u)\frac{du}{dx}+v\frac{d}{dy}(U+u)+w\frac{du}{dz}&=\mu\nabla^2(U+u)-\frac{dp}{dx}+g\sin I\\ \frac{dv}{dt}+(U+u)\frac{dv}{dx}+v\frac{dv}{dy}+w\frac{dv}{dz}&=\mu\nabla^2 v-\frac{dp}{dy}-g\cos I\\ \frac{dw}{dt}+(U+u)\frac{dw}{dx}+v\frac{dw}{dy}+w\frac{dw}{dz}&=\mu\nabla^2 w-\frac{dp}{dz}\end{aligned}\right\}$$

$$\ldots\ldots\ldots(2);$$

where ∇^2 denotes the "Laplacian" $\frac{d^2}{dx^2}+\frac{d^2}{dy^2}+\frac{d^2}{dz^2}$.

28. If we have $u=0$, $v=0$, $w=0$; $p=C-g\cos I\,.\,y$; the four equations are satisfied identically; except the first of (2), which becomes

$$\frac{dU}{dt}=\mu\frac{d^2U}{dy^2}+g\sin I\ldots\ldots\ldots\ldots\ldots\ldots(3).$$

This is reduced to

$$\frac{dv}{dt}=\mu\frac{d^2v}{dy^2}\ldots\ldots\ldots\ldots\ldots\ldots\ldots\ldots\ldots\ldots(4),$$

if we put

$$U=v+\tfrac{1}{2}g\sin I/\mu\,.\,(b^2-y^2)\quad\ldots\ldots\ldots\ldots(5).$$

For terminal conditions (the bounding planes supposed to be $y=0$ and $y=b$), we may have

$$\left.\begin{aligned}v&=F(t)\ \text{when}\ y=0\\ v&=\mathfrak{F}(t)\quad\ ,,\quad\ y=b\end{aligned}\right\}\quad\ldots\ldots\ldots\ldots\ldots(6),$$

where F and $\mathfrak{F}$ denote arbitrary functions. These equations (4) and (6) show (what was found forty-two years ago by Stokes) that the diffusion of velocity in parallel layers, *provided it is exactly in parallel layers*, through a viscous fluid, follows Fourier's law of the "linear" diffusion of heat through a homogeneous solid. Now, towards answering the highly important and interesting question which Stokes raised,—Is this laminar motion unstable

* Stokes's *Collected Papers*, Vol. I. p. 93.

in some cases?—go back to (1) and (2), and in them suppose u, v, w to be each infinitely small: (1) is unchanged; (2), with U eliminated by (5), become

$$\frac{du}{dt}+[v+\tfrac{1}{2}c(b^2-y^2)]\frac{du}{dx}+v\left(\frac{dv}{dy}-cy\right)=\mu\nabla^2u-\frac{dp}{dx}\quad\ldots(7),$$

$$\frac{dv}{dt}+[v+\tfrac{1}{2}c(b^2-y^2)]\frac{dv}{dx}\qquad\qquad=\mu\nabla^2v-\frac{dp}{dy}\quad\ldots(8),$$

$$\frac{dw}{dt}+[v+\tfrac{1}{2}c(b^2-y^2)]\frac{dw}{dx}\qquad\qquad=\mu\nabla^2w-\frac{dp}{dz}\quad\ldots(9);$$

where $$c=g\sin I/\mu\quad\ldots\ldots\ldots\ldots\ldots\ldots(10),$$

and, for brevity, p denotes, instead of as before the pressure simply, now the pressure $+g\cos I.y$.

We still suppose v to be a function of y and t determined by (4) and (6). Thus (1) and (7), (8), (9) are four equations which, with proper initial and boundary conditions, determine the four unknown quantities u, v, w, p, in terms of x, y, z, t.

29. It is convenient to eliminate u and w; by taking d/dx, d/dy, d/dz of (7), (8), (9), and adding. Thus we find, in virtue of (1),

$$2\left(\frac{dv}{dy}-cy\right)\frac{dv}{dx}=-\nabla^2p\quad\ldots\ldots\ldots\ldots\ldots\ldots(11).$$

This and (8) are two equations for the determination of v and p. Eliminating p between them, we find

$$\frac{d\nabla^2v}{dt}-\left(\frac{d^2v}{dy^2}-c\right)\frac{dv}{dx}+[v+\tfrac{1}{2}c(b^2-y^2)]\frac{d\nabla^2v}{dx}=\mu\nabla^4v\ldots(12),$$

a single equation which, with proper initial and boundary conditions, determines the one unknown, v. When v is thus found, (8), (7), (9) determine p, u, and w.

30. An interesting and practically important case is presented by supposing one or both of the bounding planes to be kept oscillating in its own plane; that is, F and $\mathfrak{F}$ of (6) to be periodic functions of t. For example, take

$$F=a\cos\omega t,\ \mathfrak{F}=0\quad\ldots\ldots\ldots\ldots\ldots\ldots(13).$$

The corresponding periodic solution of (4) is

$$v=a\,\frac{\epsilon^{(b-y)\sqrt{\omega/2\mu}}-\epsilon^{-(b-y)\sqrt{\omega/2\mu}}}{\epsilon^{b\sqrt{\omega/2\mu}}-\epsilon^{-b\sqrt{\omega/2\mu}}}\cos\left(\omega t-y\sqrt{\frac{\omega}{2\mu}}\right)\ldots(14).$$

In connexion with this case there is no particular interest in supposing a current to be maintained by gravity; and we shall therefore take $c=0$, which reduces (7) (8), (9), (11), (12), to

$$\frac{du}{dt}+v\frac{du}{dx}+\frac{dv}{dy}v=\mu\nabla^2 u-\frac{dp}{dx} \quad \text{..........(15)},$$

$$\frac{dv}{dt}+v\frac{dv}{dx} \quad =\mu\nabla^2 v-\frac{dp}{dy} \quad \text{..........(16)},$$

$$\frac{dw}{dt}+v\frac{dw}{dx} \quad =\mu\nabla^2 w-\frac{dp}{dz} \quad \text{..........(17)},$$

$$2\frac{dv}{dy}\frac{dv}{dx} \quad = -\nabla^2 p \quad \text{..............(18)},$$

$$\frac{d\nabla^2 v}{dt}-\frac{d^2 v}{dy^2}\frac{dv}{dx}+v\frac{d\nabla^2 v}{dx}=\mu\nabla^4 v \text{..................(19)};$$

in all of which v is the function of (y, t) expressed by (14).

These equations (15)—(19) are of course satisfied by $u=0$, $v=0$, $w=0$, $p=0$. The question of stability is, Does every possible solution of them come to this in time? It seems to me probable that it does; but I cannot, at present at all events, enter on the investigation. The case of $b=\infty$ is specially important and interesting.

31. The present communication is confined to the much simpler case in which the two bounding planes are kept moving relatively with constant velocity; including as sub-case, the two planes held at rest, and the fluid caused by gravity to move between them. But we shall first take the much simpler sub-case, in which there is relative motion of the two planes, and no gravity. This is the very simplest of all cases of the general question of the Stability or Instability of the Motion of a Viscous Fluid. It is the second of the two cases prescribed by the Examiners for the Adams Prize of 1888. I have ascertained, and I now give (§§ 32—39 below) the proof, that in this sub-case the steady motion is wholly stable, however small or however great be the viscosity; and this without limitation to two-dimensional motion of the admissible disturbances.

32. In our present sub-case, let βb be the relative velocity of the two planes; so that in (6) we may take $F=0$, $\mathfrak{F}=\beta b$; and the corresponding steady solution of (4) is

$$v=\beta y \quad \text{....................(20)}.$$

Thus equation (19) becomes reduced to

$$\left.\begin{aligned}\frac{d\sigma}{dt}+\beta y\frac{d\sigma}{dx}&=\mu\nabla^2\sigma,\\ \text{where}\qquad \sigma&=\nabla^2 v\end{aligned}\right\}\quad\text{.................(21);}$$

and (18), (15), (16), (17) become

$$2\beta\frac{dv}{dx}=-\nabla^2 p\quad\text{......................(22),}$$

$$\frac{du}{dt}+\beta y\frac{du}{dx}+\beta v=\mu\nabla^2 u-\frac{dp}{dx}\quad\text{.................(23),}$$

$$\frac{dv}{dt}+\beta y\frac{dv}{dx}\qquad=\mu\nabla^2 v-\frac{dp}{dy}\quad\text{.................(24),}$$

$$\frac{dw}{dt}+\beta y\frac{dw}{dx}\qquad=\mu\nabla^2 w-\frac{dp}{dz}\quad\text{.................(25).}$$

It may be remarked that equations (22)—(25) imply (1), and that any four of the five determine the four quantities u, v, w, p. It will still be convenient occasionally to use (1). We proceed to find the complete solution of the problem before us, consisting of expressions for u, v, w, p satisfying (22)—(25) for all values of x, y, z, t; and the following initial and boundary conditions:—

$$\left.\begin{aligned}\text{when } t=0:\ u,\ v,\ w \text{ to be arbitrary functions}\\ \text{of } x,\ y,\ z, \text{ subject only to (1)}\end{aligned}\right\}\text{...(26);}$$

$$\left.\begin{aligned}&u=0,\ v=0,\ w=0, \text{ for } y=0 \text{ and all values of } x,\ z,\ t\\ &u=0,\ v=0,\ w=0, \text{ for } y=b \qquad\quad ,, \qquad\quad ,,\end{aligned}\right\}\text{...(27).}$$

33. First let us find a particular solution **u**, **v**, **w**, **p**, which shall satisfy the initial conditions (26), irrespectively of the boundary conditions (27), except as follows:—

$$\left.\begin{aligned}\mathbf{v}=0, \text{ when } t=0 \text{ and } y=0\\ \mathbf{v}=0, \text{ when } t=0 \text{ and } y=b\end{aligned}\right\}\quad\text{...........(28).}$$

Next, find another particular solution, $\mathfrak{u}$, $\mathfrak{v}$, $\mathfrak{w}$, $\mathfrak{p}$, satisfying the following initial and boundary equations:—

$$\mathfrak{u}=0,\ \mathfrak{v}=0,\ \mathfrak{w}=0, \text{ when } t=0\quad\text{...........(29);}$$

$$\left.\begin{aligned}\mathfrak{u}+\mathbf{u}=0,\ \mathfrak{v}+\mathbf{v}=0,\ \mathfrak{w}+\mathbf{w}=0, \text{ when } y=0\\ \text{and when } y=b\end{aligned}\right\}\quad\text{...(30).}$$

The required complete solution will then be

$$u=\mathfrak{u}+\mathbf{u},\quad v=\mathfrak{v}+\mathbf{v},\quad w=\mathfrak{w}+\mathbf{w}\text{...........(31).}$$

34. To find **u**, **v**, **w**, remark that, if μ were zero, the complete integral of (21) would be

$$\sigma = \text{arb. func.}\,(x - \beta y t)\,;$$

and take therefore as a trial for a type-solution with μ not zero,

$$\sigma = T\epsilon^{\iota[mx+(n-m\beta t)y+qz]} \quad\text{..................(32)};$$

where T is a function of t, and ι denotes $\sqrt{-1}$. Substituting accordingly in (21), we find

$$\frac{dT}{dt} = -\mu\,[m^2 + (n - m\beta t)^2 + q^2]\,T \quad\text{...........(33)};$$

whence, by integration,

$$T = C\epsilon^{-\mu t[m^2+n^2+q^2-nm\beta t+(m^2/3)\beta^2t^2]} \quad\text{...........(34)}.$$

By the second of (21), and (32), we find

$$v = -T\,\frac{\epsilon^{\iota[mx+(n-m\beta t)y+qz]}}{m^2+(n-m\beta t)^2+q^2} \quad\text{..............(35)};$$

whence, by (22),

$$p = -2\beta m\iota T\,\frac{\epsilon^{\iota[mx+(n-m\beta t)y+qz]}}{[m^2+(n-m\beta t)^2+q^2]^2} \quad\text{.........(36)}.$$

Using this in (25), and putting

$$w = W\epsilon^{\iota[mx+(n-m\beta t)y+qz]} \quad\text{..................(37)},$$

we find

$$\frac{dW}{dt} = -\mu\,[m^2+(n-m\beta t)^2+q^2]\,W - \frac{2\beta mqT}{[m^2+(n-m\beta t)^2+q^2]^2}\text{...(38)},$$

which, integrated, gives W.

Having thus found v and w, we find u by (1), as follows:—

$$u = -\frac{(n-m\beta t)\,v + qw}{m} \quad\text{.................(39)}.$$

35. Realizing, by adding type-solutions for $\pm\iota$ and $\pm n$, with proper values of C, we arrive at a complete real type-solution with, for v, the following—in which K denotes an arbitrary constant:—

$$v = \tfrac{1}{2}K\left\{\frac{\epsilon^{-\mu t[m^2+n^2+q^2-nm\beta t+(m^2/3)\beta^2t^2]}}{m^2+(n-m\beta t)^2+q^2}\,{\cos \atop \sin}\,[mx+(n-m\beta t)\,y+qz]\right.$$

$$\left.-\frac{\epsilon^{-\mu t[m^2+n^2+q^2+nm\beta t+(m^2/3)\beta^2t^2]}}{m^2+(n+m\beta t)^2+q^2}\,{\cos \atop \sin}\,[mx-(n+m\beta t)y+qz]\right\}\text{...(40)}.$$

This gives, when $t=0$,

$$v=\frac{\mp K}{m^2+n^2+q^2}\sin ny \frac{\sin}{\cos}(mx+qz) \quad \text{........(41)},$$

which fulfils (28) if we make

$$n=i\pi/b \quad \text{......................(42)};$$

and allows us, by proper summation for all values of i from 1 to ∞, and summation or integration with reference to m and q, with properly determined values of K, after the manner of Fourier, to give any arbitrarily assigned value to $v_{t=0}$ for every value of x, y, z,

$$\left.\begin{array}{llll} \text{from} & x=-\infty & \text{to} & x=+\infty \\ \text{,,} & y=0 & \text{,,} & y=b \\ \text{,,} & z=-\infty & \text{,,} & z=+\infty \end{array}\right\} \quad \text{...........(43)}.$$

The same summation and integration applied to (40) gives $\mathbf{v}$ for all values of t, x, y, z; and then by (38), (37), (39) we find corresponding determinate values of w and u.

36. To give now an arbitrary initial value, $\mathbf{w}_0$, to the z-component of velocity, for every value of x, y, z, add to the solution (u, $\mathbf{v}$, w), which we have now found, a particular solution (u', v', w') fulfilling the following conditions:—

$$\left.\begin{array}{l} v'=0 \text{ for all values of } t, x, y, z \\ w'=\mathbf{w}_0-w_0 \text{ for } t=0, \text{ and all values of } x, y, z \end{array}\right\} \quad \text{...(44)},$$

and to be found from (25) and (1), by remarking that $v'=0$ makes, by (22), $p'=0$, and therefore (23) and (25) become

$$\frac{du'}{dt}+\beta y\frac{du'}{dx}=\mu\nabla^2 u' \quad \text{..............(45)},$$

$$\frac{dw'}{dt}+\beta y\frac{dw'}{dx}=\mu\nabla^2 w' \quad \text{................(46)}.$$

Solving (46); just as we solved (21), by (32), (33), (34); and then realizing and summing to satisfy the arbitrary initial condition, as we did for v in (40), (41), (42), we achieve the determination of w'; and by (1) we determine the corresponding u', *ipso facto* satisfying (45). Lastly, putting together our two solutions, we find

$$\mathbf{u}=u+u', \quad \mathbf{v}=\mathbf{v}, \quad \mathbf{w}=w+w' \quad \text{...........(47)}$$

as a solution of (26) without (27), in answer to the first requisition of § 33. It remains to find $\mathfrak{u}$, $\mathfrak{b}$, $\mathfrak{w}$, in answer to the second requisition of § 33.

37. This we shall do by first finding a real (simple harmonic) periodic solution of (21), (22), (23), (25), fulfilling the condition

$$\left.\begin{aligned} u &= A\cos\omega t + B\sin\omega t\\ v &= C\cos\omega t + D\sin\omega t\\ w &= E\cos\omega t + F\sin\omega t \end{aligned}\right\}\text{ when } y=0$$
$$\left.\begin{aligned} u &= \mathfrak{A}\cos\omega t + \mathfrak{B}\sin\omega t\\ v &= \mathfrak{C}\cos\omega t + \mathfrak{D}\sin\omega t\\ w &= \mathfrak{E}\cos\omega t + \mathfrak{F}\sin\omega t \end{aligned}\right\}\text{ when } y=b \quad\ldots\ldots\ldots\ldots(48),$$

where A, B, C, D, E, F, $\mathfrak{A}$, $\mathfrak{B}$, $\mathfrak{C}$, $\mathfrak{D}$, $\mathfrak{E}$, $\mathfrak{F}$ are twelve arbitrary functions of (x, z). Then, by taking $\int_0^\infty d\omega f(\omega)$ of each of these after the manner of Fourier, we solve the problem of determining the motion produced throughout the fluid, by giving to every point of each of its approximately plane boundaries an infinitesimal displacement of which each of the three components is an arbitrary function of x, z, t. Lastly, by taking these functions each $=0$ from $t=-\infty$ to $t=0$, and each equal to minus the value of **u**, **v**, **w** for every point of each boundary, we find the $\mathfrak{u}$, $\mathfrak{b}$, $\mathfrak{w}$ of § 33. The solution of our problem of § 32 is then completed by equations (31). To do all this is a mere routine after an imaginary type-solution is provided as follows.

38. To satisfy (21) assume

$$\begin{aligned} v &= \epsilon^{\iota(\omega t+mx+qz)}\mathcal{V}\\ &= \epsilon^{\iota(\omega t+mx+qz)}\Bigg\{H\epsilon^{y\sqrt{(m^2+q^2)}} + K\epsilon^{-y\sqrt{(m^2+q^2)}}\\ &\quad+\frac{1}{2\sqrt{(m^2+q^2)}}\Bigg[\epsilon^{y\sqrt{(m^2+q^2)}}\int_0^y dy\,\epsilon^{-y\sqrt{(m^2+q^2)}}\left[Lf(y)+MF(y)\right]\\ &\quad-\epsilon^{-y\sqrt{(m^2+q^2)}}\int_0^y dy\,\epsilon^{y\sqrt{(m^2+q^2)}}\left[Lf(y)+MF(y)\right]\Bigg]\Bigg\} \ldots(49), \end{aligned}$$

where H, K, L, M are arbitrary constants and f, F any two particular solutions of

$$\iota(\omega+m\beta y)\,\sigma = \mu\left[\frac{d^2\sigma}{dy^2}-(m^2+q^2)\,\sigma\right] \ldots\ldots\ldots(50).$$

This equation, if we put

$$m\beta/\mu = \gamma, \text{ and } m^2 + q^2 + \iota\omega/\mu = \lambda \quad \text{.........(51)},$$

becomes

$$\frac{d^2\sigma}{dy^2} = (\lambda + \iota\gamma y)\,\sigma \quad \text{....................(52)};$$

which, integrated in ascending powers of $(\lambda + \iota\gamma y)$, gives two particular solutions, which we may conveniently take for our f and F, as follows:—

$$\left.\begin{aligned} f(y) &= 1 - \frac{\gamma^{-2}(\lambda + \iota\gamma y)^3}{3.2} + \frac{\gamma^{-4}(\lambda + \iota\gamma y)^6}{6.5.3.2} - \frac{\gamma^{-6}(\lambda + \iota\gamma y)^9}{9.8.6.5.3.2} + \&c. \\ F(y) &= \lambda + \iota\gamma y - \frac{\gamma^{-2}(\lambda + \iota\gamma y)^4}{4.3} + \frac{\gamma^{-4}(\lambda + \iota\gamma y)^7}{7.6.4.3} - \frac{\gamma^{-6}(\lambda + \iota\gamma y)^{10}}{10.9.7.6.4.3} + \&c. \end{aligned}\right\} \quad \text{.........(53)}.$$

39. *These series are essentially convergent for all values of* y. Hence in (49) we have a solution continuous from $y=0$ to $y=b$; and by its four arbitrary constants we can give any prescribed values to $\mathcal{V}$, and $d\mathcal{V}/dy$, for $y=0$ and $y=b$. This done, find p determinately by (24); and then integrate (25) for w in *an essentially convergent series* of ascending powers of $\lambda + \iota\gamma y$, which is easily worked out, but need not be written down at present, except in abstract as follows:—

$$w = \mathcal{W}\epsilon^{\iota(\omega t + mx + qz)} \quad \text{.....................(54)};$$

where

$$\left.\begin{aligned} \mathcal{W} = H\mathfrak{F}_1(\lambda + \iota\gamma y) + K\mathfrak{F}_2(\lambda + \iota\gamma y) + L\mathfrak{F}_3(\lambda + \iota\gamma y) \\ + M\mathfrak{F}_4(\lambda + \iota\gamma y) + P\epsilon^{y\sqrt{(m^2+q^2)}} + Q\epsilon^{-y\sqrt{(m^2+q^2)}} \end{aligned}\right\} \text{...(55)}.$$

Here P and Q are the two fresh constants, due to the integration for w. By these we can give to $\mathcal{W}$ any prescribed values for $y=0$ and $y=b$. Lastly, by (1), with (49), we have

$$\left.\begin{aligned} u &= \mathcal{U}\epsilon^{\iota(\omega t + mx + qz)} \\ \text{where} \quad \mathcal{U} &= -\left(\frac{1}{m\iota}\frac{d\mathcal{V}}{dy} + \frac{q}{m}\mathcal{W}\right) \end{aligned}\right\} \quad \text{.................(56)}.$$

Our six arbitrary constants, H, K, L, M, P, Q, clearly allow us to give any prescribed values to each of $\mathcal{U}$, $\mathcal{V}$, $\mathcal{W}$, for $y=0$ and for $y=b$. Thus the completion of the realized problem with real data of arbitrary functions, as described in § 37, becomes a mere affair of routine.

40. Now remark that the (**u**, **v**, **w**) solution of § 34 comes essentially to nothing, asymptotically as time advances, as we see by (33), (34), and (38). Hence the ($\mathfrak{u}$, $\mathfrak{v}$, $\mathfrak{w}$) of § 37, which rise gradually from zero at $t=0$, comes asymptotically to zero again as t increases to ∞. We conclude that the steady motion is stable*.

BROAD RIVER FLOWING DOWN AN INCLINED PLANE BED.

[From the *Philosophical Magazine*, Vol. XXIV. September, 1887, pp. 272—278.]

41. Consider now the second of the two cases referred to in § 27—that is to say, the case of water on an inclined plane bottom, under a fixed parallel plane cover (ice, for example), both planes infinite in all directions and gravity everywhere uniform. We shall include, as a sub-case, the icy cover moving with the water in contact with it, which is particularly interesting, because, as it annuls tangential force at the upper surface, it is, for the steady motion, the same case as that of a broad open river flowing uniformly over a perfectly smooth inclined plane bed. It is not the same, except when the motion is steadily laminar, the difference being that the surface is kept rigorously plane, but not

* [It would seem (cf. Lord Rayleigh, "On the Question of the Stability of the Flow of Fluids," *Phil. Mag.* XXXIV. 1892, pp. 59—70: *Scientific Papers*, IV. p. 582) that, in addition to the forced motion determined in § 40, the fluid is capable of a set of free motions in each of which the velocity at the boundary is null. In the text, annulling u, v, w (or what is the same, $\mathcal{W}$, $\mathcal{V}$ and $d\mathcal{V}/dy$ in § 39), at each boundary leads, as only the ratios of the six constants are involved, to a period equation in ω, introducing for each free period normal types of motion whose scale of magnitude is undetermined: imaginary values of these free periods might involve instability. A similar criticism is applied by Lord Rayleigh himself to the argument of the sections next following. The experimental investigation of Osborne Reynolds, referred to *infra*, appear to show however that within certain limits of the velocity of flow, the steady flow is practically stable. One suggestion, mentioned by Lord Rayleigh, *loc. cit.*, is that, as there is no continuous transition from steady motion with small viscosity to motion of perfect fluid with no viscosity, the actual motion of a fluid of small viscosity may involve instabilities in a very thin layer along the boundary (cf. *infra*). In any case, the investigations in the text would perhaps, in addition to the interesting general remarks, still retain an application as determining how far steady laminar motion, if somehow established, is susceptible to disturbance by the action of outside forces.

In reply to an inquiry, Lord Rayleigh now refers to a paper by Prof. W. McF. Orr, *Proc. Roy. Irish Acad.* XXVII. No. 3, 1907, extending his own previous criticism somewhat as above, with which he is disposed to agree. In that paper, however, arguments are given (pp. 72, 74, 99) which are held to make it probable that the free internal motions fade away exponentially, and that the forced oscillation determined in the text is the actual solution; it is urged that it does in fact satisfy

free from tangential force, by a rigid cover, while the open surface is kept almost but not quite rigorously plane by gravity, and rigorously free from tangential force. But, provided the bottom is smooth, the smallness of the dimples and little round hollows which we see on the surface, produced by turbulence (when the motion is turbulent), seems to prove that the motion must be very nearly the same as it would be if the upper surface were kept rigorously plane, and free from tangential force.

42. The sub-case described in § 31 having been disposed of in §§ 32—40, we now take the including case, described in the first half-sentence of § 31; for which we have, as steady solution, according to (5),

$$U = \beta y - \tfrac{1}{2}cy^2 \quad\text{.....................(57),}$$

if we reckon y from the bottom upwards. Thus (7), (8), (9), (11), (12) become

$$\frac{du}{dt} + (\beta y - \tfrac{1}{2}cy^2)\frac{du}{dx} + (\beta - cy)\,v = \mu\nabla^2 u - \frac{dp}{dx} \quad\text{......(58),}$$

$$\frac{dv}{dt} + (\beta y - \tfrac{1}{2}cy^2)\frac{dv}{dx} = \mu\nabla^2 v - \frac{dp}{dy} \quad\text{......(59),}$$

$$\frac{dw}{dt} + (\beta y - \tfrac{1}{2}cy^2)\frac{dw}{dx} = \mu\nabla^2 w - \frac{dp}{dz} \quad\text{......(60),}$$

$$2(\beta - cy)\frac{dv}{dx} = -\nabla^2 p \quad\text{...(61),}$$

$$\frac{d\nabla^2 v}{dt} + c\frac{dv}{dx} + (\beta y - \tfrac{1}{2}cy^2)\frac{d\nabla^2 v}{dx} = \mu\nabla^4 v \quad\text{...........(62).}$$

43. We have not now any such simple partial solution as that of §§ 34, 35, 36 for the sub-case there dealt with; and we proceed at once to the virtually inclusive* investigation specified in § 37, and, as in § 38, assume

$$v = \epsilon^{\iota(\omega t + mx + qz)}\,\mathfrak{V} \quad\text{....................(63).}$$

(29) as is here tacitly assumed. There also discussions of problems of this type by O. Reynolds' energy method are given in pp. 122—138, with an account of previous work of that kind. In a previous part (*loc. cit.* No. 2, §§ 3A, 5, 8) Prof. Orr gives reasons for modifying Lord Rayleigh's conclusion that in the absence of viscosity the laminar motion would be stable, in the sense that this stability would exist only for very small disturbances; cf. *supra*, § 27.]

* The Fourier-Sturm-Liouville analysis (Fourier, *Théorie de la Chaleur*; Sturm and Liouville, Liouville's *Journal* for the year 1836; and Lord Rayleigh's *Theory of Sound*, § 142, Vol. I.) shows how to express an arbitrary function of x, y, z by summation of the type-solutions of §§ 37, 39 above and § 43 (63), (67), (70) here, and so to complete, whether for our present case or former sub-case, the fulfilment of the conditions (26), (27), without using the method of §§ 34, 35, 36.

This gives $\frac{d}{dt} = \iota\omega$, $\frac{d}{dx} = \iota m$, and $\nabla^2 = \frac{d^2}{dy^2} - m^2 - q^2$......(64);

and (62) becomes therefore

$$\mu \frac{d^4 \mathcal{V}}{dy^4} - \{2\mu(m^2+q^2) + \iota[\omega + m(\beta y - \tfrac{1}{2}cy^2)]\} \frac{d^2\mathcal{V}}{dy^2} + \{\mu(m^2+q^2)^2 + \iota[\omega + m(\beta y - \tfrac{1}{2}cy^2)](m^2+q^2) - \iota cm\}\mathcal{V} = 0 \quad \text{...(65)},$$

or, for brevity,

$$\mu \frac{d^4\mathcal{V}}{dy^4} + (e + fy + gy^2)\frac{d^2\mathcal{V}}{dy^2} + (h + ky + ly^2)\mathcal{V} = 0 \quad \text{...(66)}.$$

To integrate this, assume

$$\mathcal{V} = c_0 + c_1 y + c_2 y^2 + c_3 y^3 + c_4 y^4 + \text{\&c.} \quad \text{........(67)};$$

and, by equating to zero the coefficient of y^i in (66), we find

$$(i+4)(i+3)(i+2)(i+1)\mu c_{i+4} + (i+2)(i+1)ec_{i+2} + (i+1)ifc_{i+1} + [i(i-1)g + h]c_i + kc_{i-1} + lc_{i-2} = 0 \quad \text{...(68)}.$$

Making now successively $i = 0$, $i = 1$, $i = 2$, ..., and remembering that c with any negative suffix is zero, we find

$$\left.\begin{array}{l} 4.3.2.1.\mu c_4 + 2.1.ec_2 + hc_0 = 0, \\ 5.4.3.2.\mu c_5 + 3.2.ec_3 + 2.1.fc_2 + hc_1 + kc_0 = 0, \\ 6.5.4.3.\mu c_6 + 4.3.ec_4 + 3.2.fc_3 + (2.1.g + h)c_2 + kc_1 + lc_0 = 0, \\ 7.6.5.4.\mu c_7 + 5.4.ec_5 + 4.3.fc_4 + (3.2.g + h)c_3 + kc_2 + lc_1 = 0, \\ \quad \text{\&c.} \qquad \text{\&c.} \qquad \text{\&c.} \end{array}\right\} \quad \text{.........(69)}.$$

These equations, taken in order, give successively $c_4, c_5, c_6, \ldots$, each explicitly as a linear function of c_0, c_1, c_2, c_3; and by using in (67) the expressions so obtained, we find

$$\mathcal{V} = c_0\mathfrak{F}_0(y) + c_1\mathfrak{F}_1(y) + c_2\mathfrak{F}_2(y) + c_3\mathfrak{F}_3(y) \quad \text{......(70)},$$

where c_0, c_1, c_2, c_3 are four arbitrary constants, and $\mathfrak{F}_0, \mathfrak{F}_1, \mathfrak{F}_2, \mathfrak{F}_3$ four functions, each wholly determinate, expressed in a series of ascending powers of y which by (68) we see to be *convergent for all values of* y, unless μ be zero. The essential convergency of these series proves (as in § 39 for the case of no gravity) that *the steady motion* ($u = 0$, $v = 0$, $w = 0$) *is stable, however small be* μ, *provided it is not zero.*

44. The less is μ, the less the convergence. When μ is very small there is divergence for many terms, but ultimate convergence.

45. In the case of $\mu=0$, the differential equation (65), or (66), becomes reduced from the 4th to the 2nd order, and may be written as follows:—

$$\frac{d^2\mathfrak{V}}{dy^2}=\left\{m^2+q^2-\frac{cm}{\omega+m(\beta y-\frac{1}{2}cy^2)}\right\}\mathfrak{V}\;\;\ldots\ldots\ldots(71).$$

This, for the case of two-dimensional motion ($q=0$), agrees with Lord Rayleigh's result, expressed in the last equation of his paper on "The Stability or Instability of certain Fluid Motions" (*Proc. Lond. Math. Soc.* Feb. 12, 1880). The integral, but now with only two arbitrary constants (c_0, c_1), is still given in ascending powers of y by (67) and (68), which, with $\mu=0$, and the thus-simplified values of e, f, g put in place of these letters, becomes

$$-\iota\left[(i+2)(i+1)\,\omega c_{i+2}+(i+1)\,im\beta c_{i+1}\right]$$
$$+\left[\frac{\iota}{2}i(i-1)\,mc+h\right]c_i+kc_{i-1}+lc_{i-2}=0\;\;\ldots\ldots(72).$$

For very great values of i this gives

$$\omega c_{i+2}+m\beta c_{i+1}-\tfrac{1}{2}mcc_i=0\;\;\ldots\ldots\ldots\ldots\ldots(73),$$

which shows that ultimately, except in the case of one particular value of the ratio c_1/c_0,

$$c_{i+1}/c_i=\zeta^{-1}\;\;\ldots\ldots\ldots\ldots\ldots\ldots\ldots\ldots(74),$$

where ζ denotes the smaller root of the equation

$$\omega+m\beta y-\tfrac{1}{2}mcy^2=0\ldots\ldots\ldots\ldots\ldots\ldots\ldots(75).$$

Hence there is certainly not convergence for values of y exceeding the smaller root of (75), and thus the proof of stability is lost.

46. But the differential equation, simplified in (71) for the case of no viscosity, may no doubt be treated more appropriately in respect to the question of stability or instability, by writing it as follows [ζ', ζ denoting the two roots of (75)],

$$\frac{d^2\mathfrak{V}}{dy^2}=\left\{m^2+q^2+\frac{2}{\zeta'-\zeta}\left(\frac{1}{\zeta-y}-\frac{1}{\zeta'-y}\right)\right\}\mathfrak{V}\;\;\ldots\ldots(76),$$

and integrating with special consideration of the infinities at $y=\zeta$ and $y=\zeta'$. One way of doing this, which I merely suggest at present, and do not follow out for want of time, is to assume

$$\mathfrak{V}=C\;\{\zeta-y+c_2(\zeta-y)^2+c_3(\zeta-y)^3+\&c.\},$$
$$+\,C'\{\zeta'-y+c_2'(\zeta'-y)^2+c_3'(\zeta'-y)^3+\&c.\}\;\;\ldots(77),$$

where C and C' are two arbitrary constants, and $c_2, c_3, \ldots, c_2', c_3', \ldots$ coefficients to be determined so as to satisfy the differential equation. This is very easily done, and when done shows that each series converges for all values of y less than ζ' and exceeding ζ. The working out of this in detail would be very interesting, and would constitute the full mathematical treatment of the problem of finding sinuous stream-lines (curves of sines) throughout the space between the two "cat's-eye" borders (corresponding to $y=\zeta$ and $y=\zeta'$) which I proposed in a short communication to Section A of the British Association at Swansea, in 1880*, "On a Disturbing Infinity in Lord Rayleigh's solution for Waves in a plane Vortex stratum." It is to be remarked that this disturbing infinity vitiates the seeming proof of stability contained in Lord Rayleigh's equations (56), (57), (58)†.

47. Realizing (63), and interpreting the result in connexion with (57), we see that

(*a*) The solution which we have found consists of a wave-disturbance travelling in any (x, z) direction, of which the propagational velocity in the x-direction is $-\omega/m$.

(*b*) The roots (ζ, ζ') of (75) are values of y at places where the velocity of the undisturbed laminar flow is equal to the x-velocity of the wave-disturbance.

Hence, supposing the bounding-planes to be plastic, and force to be applied to either or both of them so as to produce an infinitesimal undulatory corrugation, according to the formula $\cos(\omega t + mx + qz)$, this surface-action will cause throughout the interior a corresponding infinitesimal wave-motion if *ω/m is not equal to the value of U for any plane of the fluid between its boundaries.* But the infinity corresponding to $y=\zeta$ or $y=\zeta'$ will vitiate this solution *if ω/m is equal to the value of U for some one plane of the fluid or for two planes of the fluid*; and the true solution will involve the "cat's-eye pattern" of stream-lines, and the enclosed elliptic whirls, at this plane or these planes.

* Of which an abstract is published in *Nature* for November 11, 1880, and in the British Association volume Report for the year. In this abstract cancel the statement "is stable," with reference to a certain steady motion described in it [*supra*, p. 186; but see next footnote].

† [Lord Rayleigh remarks in reply (*loc. cit.*, *supra*) that when ω is complex there is no disturbing infinity; so that the argument does not fail, regarded as one for excluding complex values of ω, though it may not completely ensure stability. He reverts to the subject in *Proc. Lond. Math. Soc.* XXVII. 1895; *Scientific Papers*, IV. pp. 203—9. The inference in the text, and *supra*, p. 186, is that there cannot be steady motion in these cases unless this band of vortices has been established.]

48. Now let the fluid be given moving with the steady laminar flow between two parallel boundary planes, expressed by (57), which would be a condition of kinetic equilibrium (proved stable in § 43) under the influence of gravity and viscosity; and let both gravity and viscosity be suddenly annulled. The fluid is still in kinetic equilibrium; but is the equilibrium stable? To answer this question, let one or both bounding-surfaces be infinitesimally dimpled in any place and left free to become plane again. The Fourier synthesis of this surface-operation is

$$\int_0^\infty \int_0^\infty dm\,dq\, F(m, q) \cos \omega t \cos mx \cos qz \quad \ldots\ldots(78),$$

or

$$\tfrac{1}{2}\int_0^\infty \int_0^\infty dm\,dq\, F(m, q) \{\cos(\omega t - mx) + \cos(\omega t + mx)\} \cos qz \quad \ldots\ldots(79),$$

where ω is determined as a function of m and q, which implies harmonic surface-undulations travelling in opposite x-directions, with all values from 0 to ∞ of (ω/m), the wave-velocity. Hence (§ 47) the interior disturbance essentially involves elliptic whirls. Thus we see that the given steady laminar motion is *thoroughly* unstable, being ready to break up into eddies in every place, on the occasion of the slightest shock or bump on either plastic plane boundary. The slightest degree of viscosity, as we have seen, makes the laminar motion stable; but the smaller the viscosity with a given value of $g \sin I$, or the greater the value of $g \sin I$ with the same viscosity, the narrower are the limits of this stability. Thus we have been led by purely mathematical investigation to a state of motion agreeing perfectly with the following remarkable descriptions of observed results by Osborne Reynolds (*Phil. Trans.* March 15, 1883, pp. 955, 956):—

"The fact that the steady motion breaks down suddenly, shows that the fluid is in a state of instability for disturbances of the magnitude which cause it to break down. But the fact that in some conditions it will break down for a large disturbance, while it is stable for a smaller disturbance, shows that there is a certain residual stability, so long as the disturbances do not exceed a given amount...."

"And it was a matter of surprise to me to see the sudden force with which the eddies sprang into existence, showing a highly unstable condition to have existed at the time the steady motion broke down."

"This at once suggested the idea that the condition might be one of instability for disturbance of a certain magnitude, and stable for smaller disturbances."

49. The motion investigated experimentally by Reynolds, and referred to in the preceding statements, was that of water in a long straight uniform tube of circular section. It is to be hoped that candidates for the Adams Prize of 1888 may investigate this case mathematically, and give a complete solution for infinitesimal deviations from rectilineal motion. It is probable that for it, and generally for a uniform straight tube of any cross section, including the extreme, and extremely simplified, case of rectilinear motion of a viscous fluid between two parallel *fixed* planes, which I have worked out above, the same general conclusion as that stated at the end of § 40 and in §§ 43—48 will be found true.

50. In the case of no gravity ($g \sin I = 0$), and the viscous fluid kept in "shearing" or "laminar" motion by relative motion of the two parallel planes, there is, when viscosity is annulled, no disturbing instability in the steady uniform shearing motion, with its uniform molecular rotation throughout, which viscosity would produce; and therefore our reason for suspecting any limitation of the excursions within which there is stability, and for expecting possible *permanence* of any kind of turbulent or tumultuous motion between two *perfectly* smooth planes (or between two polished planes with any practical velocities) does not exist in this case. But a great variety of general observation (and particularly Rankine and Froude's doctrine of the "skin-resistance" of ships, and Froude's experimental determination of the resistance experienced by a very smooth, thin, vertical board, 19 inches broad and 50 feet long, moved at different uniform speeds* through water in a broad

* 'Report to the Lords Commissioners of the Admiralty on Experiments for the Determination of the Frictional Resistance of Water on a Surface under various conditions, performed at Chelston Cross (Torquay), under the Authority of their Lordships.' By W. Froude. (London: Taylor and Francis. 1874.)

Froude found that, at a constant velocity of 600 feet per minute, the resistance of the water against one of his smoothest surfaces, at positions two feet abaft of the cutwater and 50 feet abaft of the cutwater, respectively, was ·295 of a pound per square foot, and ·244 of a pound per square foot. Remark that this astonish-

deep tank 278 feet long) makes it certain that if water be given at rest between two infinite planes both at rest, and if one of the planes be suddenly, *or not too gradually,* set in motion, and kept moving uniformly, the motion of the water will be at first turbulent, and the ultimate condition of uniform shearing will be approached by gradual reduction and ultimate annulment of the turbulence. I hope to make a communication on this subject to Section A of the British Association in Manchester, and to have it published in the October number of the *Philosophical Magazine.* Corresponding questions must be examined with reference to the corresponding tubular problem, of an infinitely long, straight, solid bar kept moving in water within an infinitely long fixed tube. It is to be hoped that the 1888 Adams Prize will bring out important investigations on this subject.

ingly great force of a quarter of a pound per square foot (!!) is the resistance due to uniform laminar flow of water between two parallel planes $\frac{1}{30}$ of a centimetre ($\frac{1}{900}$ of a foot!) asunder, when one of the planes is moving relatively to the other at 10 feet (300 centimetres) per second, if the water be at the temperature 0° Cent., for which the viscosity, calculated from Poiseuille's observations on the flow of water in capillary tubes, is $1{\cdot}34 \times 10^{-5}$ of a gramme weight per square centimetre.

35. ON DEEP-WATER TWO-DIMENSIONAL WAVES PRODUCED BY ANY GIVEN INITIATING DISTURBANCE*.

[From *Proc. Roy. Soc. Edin.* Vol. XXV. read Feb. 1, 1904, pp. 185—196; *Phil. Mag.* Vol. VII. June, 1904, pp. 609—620.]

1. CONSIDER frictionless water in a straight canal, infinitely long and infinitely deep, with vertical sides. Let it be disturbed from rest by any change of pressure on the surface, uniform in every line perpendicular to the plane sides, and left to itself under constant air pressure. It is required to find the displacement and velocity of every particle of the water at any future time. Our initial condition will be fully specified by a given normal component velocity, and a given normal component displacement, at every point of the surface.

2. Taking O, any point at a distance h above the undisturbed water level, draw OX parallel to the length of the canal, and OZ vertically downwards. Let ξ, ζ be the displacement-components and $\dot{\xi}$, $\dot{\zeta}$ the velocity-components of any particle of the water whose undisturbed position is (x, z). We suppose the disturbance infinitesimal; by which we mean that the change of distance between any two particles of water is infinitely small in comparison with their undisturbed distance; and the line joining them experiences changes of direction which are infinitely small in comparison with the radian. Water being assumed incompressible and frictionless, its motion, started primarily from rest by pressure applied to the free surface, is essentially irrotational. Hence we have

$$\xi = \frac{d}{dx}\phi(x, z, t); \quad \zeta = \frac{d}{dz}\phi(x, z, t); \quad \dot{\xi} = \frac{d}{dx}\dot{\phi}; \quad \dot{\zeta} = \frac{d}{dz}\dot{\phi} \quad \ldots(1);$$

where $\phi(x, z, t)$, or ϕ, as we may write it for brevity when convenient, is a function of the variables which may be called the displacement-potential; and $\dot{\phi}(x, z, t)$ is what is commonly called

* [On the problems treated in the following group of papers and their history, cf. the Presidential Address 'On Deep-Water Waves' by Prof. H. Lamb, *Proc. Lond. Math. Soc.* Vol. II. (1904), pp. 371—400.]

the velocity-potential. Thus a knowledge of the function ϕ, for all values of x, z, t, completely defines the displacement and the velocity of the fluid. And, by the fundamentals of hydrokinetics, a knowledge of ϕ for every point of the free surface suffices to determine its value throughout the water, in virtue of the equation

$$\frac{d^2\phi}{dx^2}+\frac{d^2\phi}{dz^2}=0 \quad \ldots\ldots\ldots\ldots\ldots\ldots\ldots(2).$$

The motion being infinitesimal, and the density being taken as unity, another application of the fundamental hydrokinetics gives

$$p-\Pi=g(z-h+\zeta)-\frac{d^2\phi}{dt^2}=g(z-h)+g\frac{d\phi}{dz}-\frac{d^2\phi}{dt^2}\ldots(3);$$

where g denotes gravity; Π the uniform atmospheric pressure on the free surface; and p the pressure at the point $(x, z+\zeta)$ within the fluid.

3. To apply (3) to the water-surface, put in it, $z=h$; it gives

$$g\left(\frac{d\phi}{dz}\right)_{z=h}=\left(\frac{d^2\phi}{dt^2}\right)_{z=h} \ldots\ldots\ldots\ldots\ldots\ldots\ldots(4);$$

and therefore if we could find a solution of this equation for all values of z, with (2) satisfied, we should have a solution of our present problem. Now we *can* find such a solution; by a curiously altered application of Fourier's celebrated solution,

$$\left[(t+c)^{-\frac{1}{2}}\,\epsilon^{\frac{-x^2}{4k(t+c)}}, \quad \text{for} \quad \frac{dv}{dt}=k\frac{d^2v}{dx^2}\right],$$

his equation for the linear conduction of heat. Change $t+c$, x, k, into $z+\iota x$, t, g^{-1} respectively:—we have (4), and we see that a solution of it is

$$\frac{1}{\sqrt{(z+\iota x)}}\,\epsilon^{\frac{-gt^2}{4(z+\iota x)}}\ldots\ldots\ldots\ldots\ldots\ldots\ldots(5);$$

which also satisfies (2) because any function of $z+\iota x$ satisfies (2) if ι denotes $\sqrt{-1}$. Hence if $\{RS\}$ denotes a realisation* by taking half the sum of what is written after it with $\pm\,\iota$, we have, as a real solution of (4) for our problem

$${}_1\phi(x,z,t)=\{RS\}\,\frac{1}{\sqrt{(z+\iota x)}}\,\epsilon^{\frac{-gt^2}{4(z+\iota x)}} \quad \ldots\ldots\ldots(6),$$

* A very easy way of effecting the realisations in (6) and (9) is by aid of De Moivre's theorem with, for one angle concerned in it, $\chi=\tan^{-1}x/z$; and another angle $=gt^2x/4(z^2+x^2)$.

$$=\frac{1}{\sqrt{2}\,.\,\rho}\left[\sqrt{(\rho+z)}\cos\frac{gt^2x}{4\rho^2}+\sqrt{(\rho-z)}\sin\frac{gt^2x}{4\rho^2}\right]\epsilon^{\frac{-gt^2z}{4\rho^2}}\quad\ldots(7)^*,$$

$$=\sqrt{\frac{1}{\rho}}\sin\left(\frac{gt^2x}{4\rho^2}+\theta\right)\epsilon^{\frac{-gt^2z}{4\rho^2}}\quad\ldots\ldots\ldots\ldots\ldots\ldots(8),$$

where $\rho^2=z^2+x^2$ and $\theta=\tan^{-1}\sqrt{\frac{\rho+z}{\rho-z}}$.

The sign of $\sqrt{(\rho-z)}$ changes when x passes through zero.

Going back now to (5), and denoting by $\{RD\}$ the difference of its values for $\pm\iota$ divided by 2ι, we have another solution of our problem essentially different from (6), as follows

$$_2\phi(x,z,t)=\{RD\}\frac{1}{\sqrt{(z+\iota x)}}\epsilon^{\frac{-gt^2}{4(z+\iota x)}}\quad\ldots\ldots\ldots\ldots(9),$$

$$=\frac{1}{\sqrt{2}\,.\,\rho}\left[\sqrt{(\rho+z)}\sin\frac{gt^2x}{4\rho^2}-\sqrt{(\rho-z)}\cos\frac{gt^2x}{4\rho^2}\right]\epsilon^{\frac{-gt^2z}{4\rho^2}}\quad\ldots(10),$$

$$=\sqrt{\frac{1}{\rho}}\sin\left(\frac{gt^2x}{4\rho^2}+\theta-\frac{\pi}{2}\right)\epsilon^{\frac{-gt^2z}{4\rho^2}}\quad\ldots\ldots\ldots\ldots\ldots\ldots(11).$$

4. The annexed diagram, fig. 1, represents for $t=0$ the solutions $_2\phi$ and $_1\phi$ as functions of x, with $z=1$ for convenience in the drawing. The formulas which we find by taking $t=0$ in (7) $\times\sqrt{2}$ and (10) $\times\sqrt{2}$ are†, the minus sign in (10) being omitted for convenience,

$$_1\phi=\frac{\sqrt{[\sqrt{(x^2+z^2)}+z]}}{\sqrt{(x^2+z^2)}};\quad {}_2\phi=\frac{\sqrt{[\sqrt{(x^2+z^2)}-z]}}{\sqrt{(x^2+z^2)}}\quad\ldots(12).$$

Before passing to the practical interpretation of our solutions, remark first that (12) contain full specifications of two distinct initiating disturbances; in each of which ϕ may be taken as a displacement-potential, or as a velocity-potential, or as a horizontal displacement-component or velocity, or as a vertical displacement-component or velocity. Thus we have really preparation for *six* different cases of motion, of which we shall choose one, $-\zeta=\sqrt{2}\times(7)$, for detailed examination.

5. Taking $z=h=1$, for the water-surface, let the two curves of figure 1 represent initial *displacements*, (12), of the water-surface,

* Solution (7) was given first in *Proc. R. S. E.* Jan. 1887, and *Phil. Mag.* Feb. 1887 [*supra*, p. 307: it has not been reprinted here].

† [Prof. Lamb remarks (*loc. cit.*) that the initial disturbance is not entirely localised at the origin, as $\int\phi\,dx$ does not converge. See also *infra*, § 101.]

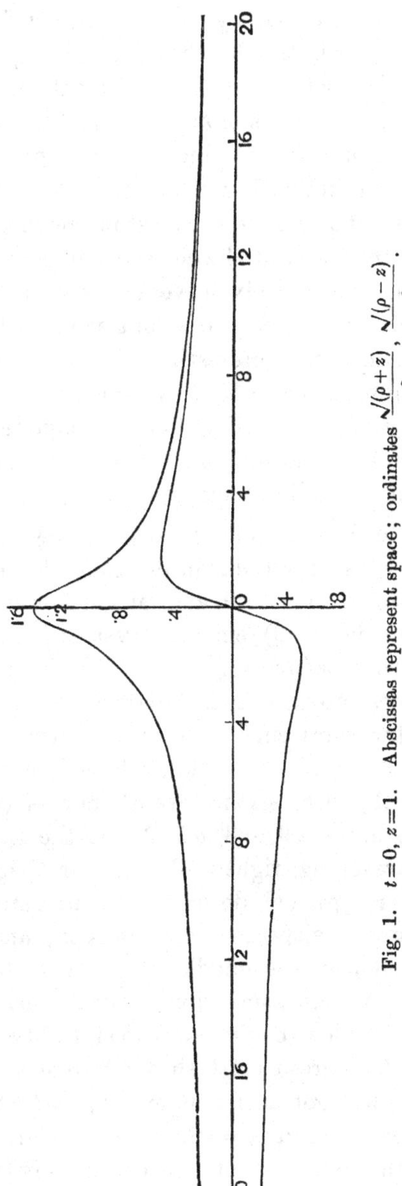

Fig. 1. $t=0$, $z=1$. Abscissas represent space; ordinates $\frac{\sqrt{(\rho+z)}}{\rho}$, $\frac{\sqrt{(\rho-z)}}{\rho}$.

left to itself with the water everywhere at rest. The displacements at any subsequent time t are expressed in real symbols by (7), (10) without the divisor $\sqrt{2}$, and by (8), (11) with a factor $\sqrt{2}$ introduced; either of which may be chosen according to convenience in calculation. One set has thus been calculated from (8), with $g=4$, and $z=1$, for six values of t; $\cdot 5$, 1, 1·5, 2, 2·5, and 5; and for a sufficiently large number of values of x to represent the results by the curves shown in figs. 2 and 3. Except for the time $t=5$, each curve shows sufficiently all the most interesting characteristics of the figure of the water at the corresponding time. The curve for $t=5$ does not perceptibly leave the zero line at distances $x<1{\cdot}8$; but if we could see it, it would show us two and a half wavelets possessing very interesting characteristics; shown in the table of numbers, § 7 below, by which we see that several different curves with scales of ordinates magnified from one to one thousand, and to one million, and to ten thousand million, would be needed to exhibit them graphically.

6. Looking to the curves for $t=0$ and $t=\frac{1}{2}$; we see that at first the water rises at all distances from the middle of the disturbance greater than $x=1{\cdot}9$, and falls at less distances. And we see that the middle ($x=0$) remains a crest (or positive maximum) till a very short time before $t=\frac{1}{2}$, when it begins to be a hollow. A crest then comes into existence beside it and begins to travel outwards. On the third curve, $t=1$, we see this crest, travelled to a distance $x=1{\cdot}7$, from the middle where it came into being; and on the fourth, fifth, sixth, seventh curves (figs. 1, 2) we see it got to distances 2·9, 4·8, 6·5, 22, at the times $1\frac{1}{2}$, 2, $2\frac{1}{2}$, 5. This crest travelling rightwards on our diagrams has its anterior slope very gradual down to the undisturbed level at $x=\infty$. Its posterior slope is much steeper; and ends at the bottom of the hollow in the middle of the disturbance, at times from $t=\frac{1}{2}$ to $t=1\frac{1}{2}$. At some time, which must be very soon after $t=1\frac{1}{2}$, this hollow begins to travel rightwards from the middle, followed by a fresh crest shed off from the middle. At $t=2$, the hollow has got as far as $x=\cdot 9$; at $t=2\frac{1}{2}$, and 5, respectively, it has reached $x=1{\cdot}75$, and $x=6{\cdot}7$. Looking in imagination to the extension of our curves leftwards from the middle of the diagram, we find an exact counterpart of what we have been examining on the right. Thus we see an initial elevation, symmetrical on the two sides of a convex crest, of

height 1·41 above the undisturbed level, sinking in the middle and rising on the two flanks. The crest becomes less and less convex till it gets down to height 1·1, when it becomes concave; and two equal and similar wave-crests are shed off on the two sides, travelling away from it rightwards and leftwards with accelerated velocities, each remaining for ever convex. Thus we see the beginnings of two endless processions of waves travelling outwards in the two directions; originating as infinitesimal wavelets shed off on the two sides of the middle line. Each crest and each hollow travels with increasing velocity. Each wave-length, from crest to crest, or from hollow to hollow, becomes longer and longer as it advances outwards; all this according to law fully expressed in (8) of § 3 above.

7. Here is now the table of numbers promised in § 5 above; it practically defines the forms and magnitudes of the two and a half wavelets, between $x=0$ and $x=2$, which the space-curve for $t=5$ (figs. 2 and 3) fails to show.

$$\rho^2 = x^2 + h^2;\quad h=1;\quad g=4;\quad t=5;\quad -\zeta = \sqrt{\frac{2}{\rho}}\sin\left(\frac{xt^2}{\rho^2}+\theta\right)\epsilon^{\frac{-t^2}{\rho^2}}.$$

Col. 1	Col. 2	Col. 3	Col. 4	Col. 5	Col. 6	Col. 7
x	$\sqrt{\frac{2}{\rho}}$	$\sqrt{\frac{\rho+1}{2\rho}}=\sin\theta$	Initial Elevation of Water-surface at Distance x. $\sin\theta\,.\sqrt{\frac{2}{\rho}}=-\zeta_0$	$\sin\left(\frac{xt^2}{\rho^2}+\theta\right)$	$\epsilon^{\frac{-t^2}{\rho^2}}$	Elevation of Water-surface at Time t, and Distance x. $-\zeta$
0	1·4142	1·0000	1·4142	1·0000	10^{-10} ·1357	$+10^{-10}$ ·1963
·05	1·4140	·9997	1·4140	·3434	" ·1478	" " ·0717
·064	...	...	...	0	...	0
·10	1·410	·9987	1·409	− ·7541	" ·1778	-10^{-10} ·1891
·15	1·407	·9972	1·403	− ·8997	" ·3066	" " ·3882
·20	1·401	·9952	1·393	− ·0032	" ·3632	" " ·0016
·202	...	...	...	0	...	0
·30	1·384	·9894	1·370	·8997	" 1·094	$+10^{-10}$ 1·362
·363	...	...	...	0	...	0
·40	1·362	·9820	1·338	− ·5451	" 4·366	-10^{-10} 3·243
·60	1·309	·9638	1·262	− ·2341	" 103·9	" " 31·84
·632	...	...	...	0	...	0
·80	1·249	·9437	1·179	·7593	10^{-5} ·02396	$+10^{-5}$ ·0227
1·00	1·190	·9239	1·099	·8962	" ·2958	" " ·3152
1·25	1·118	·9015	1·007	·6831	" 5·793	" " 4·424
1·50	1·053	·8817	·9287	·4923	" 45·63	" " 23·67
1·517	...	...	...	0	...	0
1·75	·9961	·8651	·8616	− ·6832	" 212·5	-10^{-5} 144·6

TABLE (*continued*).

$$h=1\,;\quad g=4\,;\quad t=5\,;\quad -\zeta=\sqrt{\frac{2}{\rho}}\,\sin\left(\frac{xt^2}{\rho^2}+\theta\right)\epsilon^{\frac{-t^2}{\rho^2}}$$

Col. 1	Col. 2	Col. 3	Col. 4	Col. 5	Col. 6	Col. 7
x	$\sqrt{\frac{2}{\rho}}$	$\sqrt{\frac{\rho+1}{2\rho}}=\sin\theta$	Initial Elevation of Water-surface at Distance x. $\sin\theta\cdot\sqrt{\frac{2}{\rho}}=-\zeta_0$	$\sin\left(\frac{xt^2}{\rho^2}+\theta\right)$	$\epsilon^{\frac{-t^2}{\rho^2}}$	Elevation of Water-surface at Time t, and Distance x. $-\zeta$
2·00	·9456	·8506	·8045	− ·9997	10^{-5} 848·2	-10^{-5} 801·9
2·50	·8612	·8243	·7142	− ·1633	·03180	„ „ 447·3
2·54	...	...	...	0	...	0
3·00	·7952	·8113	·6452	·8296	·08210	·0542
3·50	·7411	·7980	·5917	·9473	·1516	·1064
4·0	·6965	·7882	·5490	·4856	·2298	·07771
4·41	...	...	...	0	...	0
4·5	·6588	·7798	·5139	− ·0944	·3083	− ·01917
5·0	·6262	·7733	·4843	− ·5584	·3823	− ·1336
5·5	·5981	·7678	·4592	− ·8457	·4493	− ·2273
6·0	·5733	·7629	·4375	− ·9781	·5122	− ·2872
6·5	·5513	·7587	·4185	− ·9956	·5641	− ·3096
7·0	·5318	·7555	·4018	− ·9374	·6066	− ·3024
7·5	·5150	·7522	·3868	− ·8333	·6462	− ·2773
8·0	·4980	·7494	·3734	− ·7053	·6808	− ·2392
9·0	·4699	·7451	·3501	− ·4289	·7372	− ·1486
10	·4461	·7416	·3308	− ·1679	·6346	− ·04753
10·62	...	...	...	0	...	0
11	·4255	·7385	·3142	·05698	·8147	·01975
12	·4076	·7359	·2999	·2428	·8416	·08375
13	·3916	·7339	·2874	·3940	·8644	·1334
14	·3775	·7318	·2762	·5175	·8808	·1721
15	·3648	·7302	·2663	·6163	·8954	·2013
16	·3533	·7286	·2574	·6953	·9082	·2231
18	·3331	·7266	·2420	·8098	·9260	·2498
20	·3160	·7256	·2290	·8831	·9396	·2622
22	·3014	·7230	·2179	·9313	·9497	·2666
24	·2885	·7216	·2082	·9627	·9579	·2661
26	·2772	·7206	·1998	·9815	·9638	·2622
28	·2672	·7193	·1923	·9915	·9685	·2565
30	·2581	·7187	·1855	·9979	·9727	·2505
32	·2500	·7181	·1795	·9999	·9759	·2439
34	·2425	·7173	·1740	·9993	·9786	·2371
38	·2294	·7163	·1643	·9933	·9828	·2240
42	·2182	·7155	·1561	·9840	·9847	·2188
46	·2084	·7147	·1490	·9734	·9883	·2005
50	·2000	·7141	·1428	·9623	·9902	·1905
55	·1906	·7135	·1360	·9486	·9917	·1794
60	·1826	·7129	·1302	·9361	·9931	·1697
70	·1690	·7120	·1203	·9125	·9949	·1535
80	·1581	·7114	·1125	·8931	·9961	·1407
100	·1415	·7108	·1005	·8626	·9977	·1217
∞	0	·7071	0	·7071	1·0000	0

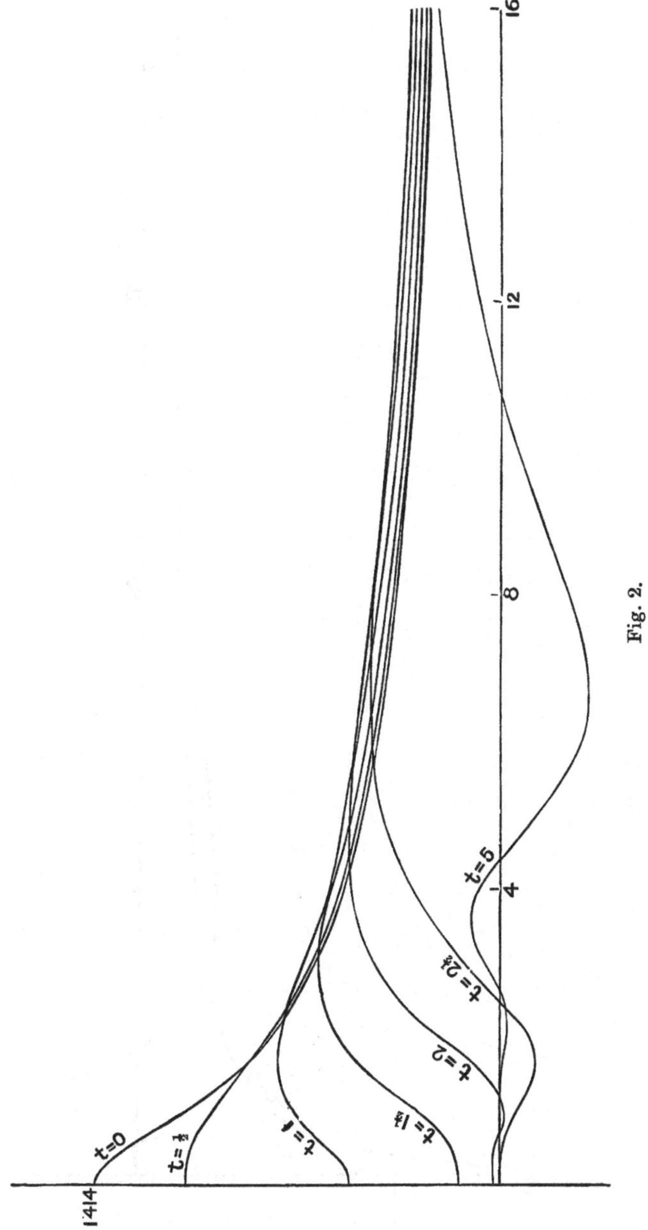

Fig. 2.

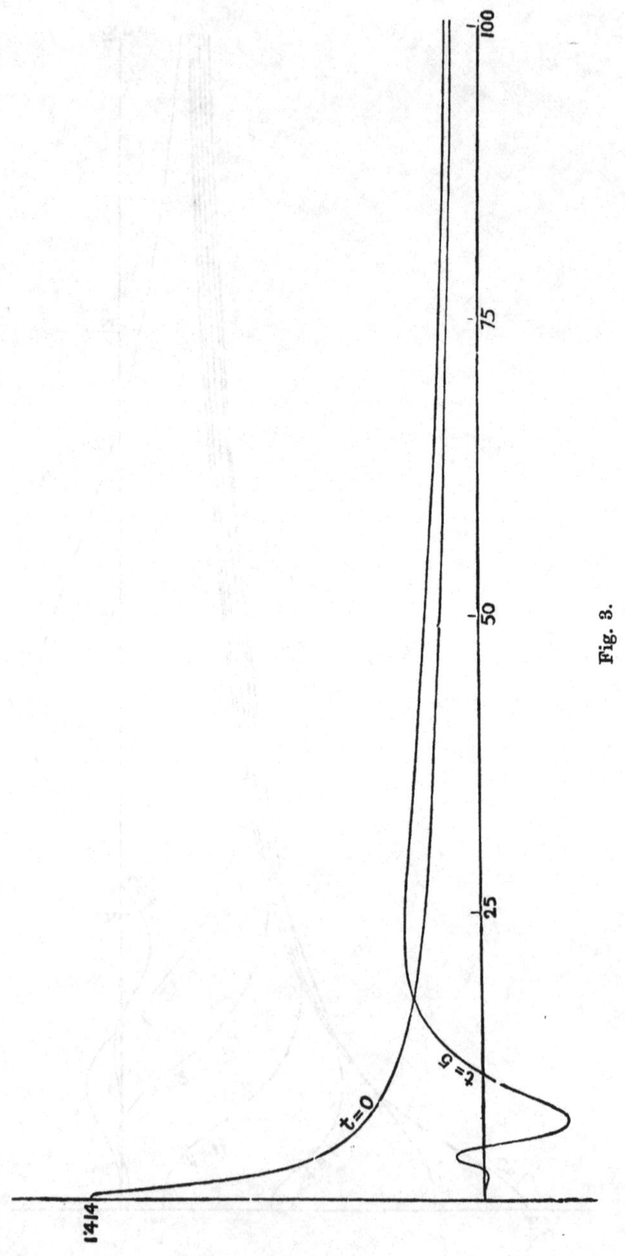

Fig. 3.

8. Look at the values shown in the previous table for the three factors which constitute $-\zeta$;—we see that the first factor (col. 2) decreases slowly from $x=0$ to $x=\infty$; the second factor (col. 5) alternates between $+1$ and -1 with increasing distances (semi-wave-lengths) from zero to zero as x increases. The third factor (col. 6) increases gradually from ϵ^{-t^2/h^2} at $x=0$, to 1 at $x=\infty$. At $x=50h$, the third factor is ·99, which is so nearly unity that the diminution of amplitude is, for all greater values of x, practically given by the first factor alone, which diminishes from ·2 at $x=50h$, to 0 at $x=\infty$.

9. The diagrams hitherto given, figs. 1, 2, 3, may be called space-curves, as on each of them abscissas represent distance from the centre of the disturbance. Fig. 4 is a time-curve (abscissas representing time) for $x=2h$. It represents a very gradual rise, from $t=0$ to $t=·6$, followed by a fall to a minimum at $t=2·8$, and a succession of alternations, with smaller and smaller maximum elevations and depressions, and shorter and shorter times from zero to zero, on to $t=\infty$. The same words with altered figures describe the changes of water level at any fixed position farther from the centre of disturbance than $x=2$. The following table shows, for the case $x=100h$, all the times of zero less than $71h$, and the elevations and depressions at the intermediate times when the second factor (col. 5 of § 7) has its maximum and minimum values (± 1). These elevations and depressions are very approximately the greatest in the intervals between the zeros, because the third factor (col. 6, § 7) varies but slowly, as shown in the first column of the present table.

10. Our assumption $h=1$ for the free surface involves no restriction of our solution to a particular case of the general formula (7). Our assumption $g=4$ merely means that our unit of abscissas is half the space fallen through in our unit of time. The fundamental formulas of § 3 may be geometrically explained by, as in § 2, taking O, our origin of co-ordinates, at a height h above the water level, and defining ρ as the distance of any particle of the fluid from it. When, as in §§ 5—9, we are only concerned with particles in the free surface (that is to say when $z=h$), we see that if x is a large multiple of z, $\rho \doteqdot x$. See for example the heading of the table of § 9. And if we are concerned with particles below the surface, we still have $\rho \doteqdot x$, if x is a

$$h=1;\quad x=100;\quad \rho=100{\cdot}005\,.\,h;\quad \theta=\tan^{-1}\sqrt{\frac{101}{99}}=45^\circ\,18'$$

$\epsilon^{\frac{-t^2}{\rho^2}}$	Times of Zero and of Approximate Maximum Elevation and Depression	Approximate Maximum Elevations and Depressions	$\epsilon^{\frac{-t^2}{\rho^2}}$	Times of Zero and of Approximate Maximum Elevation and Depression	Approximate Maximum Elevations and Depressions
·9922	8·333	+·1403	·7718	50·90	+·1091
...	15·33	0	...	52·42	0
·9616	19·80	−·1360	·7478	53·90	−·1058
...	23·43	9	...	55·34	0
·9317	26·57	+·1317	·7247	56·74	+·1025
...	29·38	0	...	58·10	0
·9031	31·94	−·1277	·7023	59·45	−·0993
...	34·31	0	...	60·75	0
·8750	36·54	+·1237	·6806	62·03	+·0962
...	38·62	0	...	63·29	0
·8480	40·61	−·1199	·6595	64·51	−·0933
...	42·50	0	...	65·72	0
·8219	44·31	−·1162	·6392	66·90	+·0904
...	46·04	0	...	68·07	0
·7964	47·72	−·1157	·6195	69·21	−·0876
...	49·34	0	...	70·34	0

large multiple of z. Thus we have the following approximation for (7) of § 3:—

$$_1\phi(x,z,t) \doteqdot \frac{1}{\sqrt{2}\,.\,x}\left[\sqrt{(x+z)}\cos\frac{gt^2}{4x}+\sqrt{(x-z)}\sin\frac{gt^2}{4x}\right]\epsilon^{\frac{-gt^2z}{4x^2}} \quad\text{...(13).}$$

Suppose now $d_1\phi/dt$ to represent $-\zeta$ (instead of $-\dot{\zeta}$, as in §§ 5—9); we have

$$-\zeta=\frac{d}{dt}\,{_1\phi}(x,z,t) \quad\text{.....................(14),}$$

which is easily found from (13) without farther restrictive suppositions. But if we suppose that z is negligibly small in comparison with x; and farther that

$$\frac{gt^2z}{4x^2}\doteqdot 0 \quad\text{..........................(15),}$$

we find by (14)

$$-\zeta\doteqdot\frac{gt}{2\sqrt{2}\,.\,x^{3/2}}\left(\cos\frac{gt^2}{4x}-\sin\frac{gt^2}{4x}\right) \quad\text{...........(16).}$$

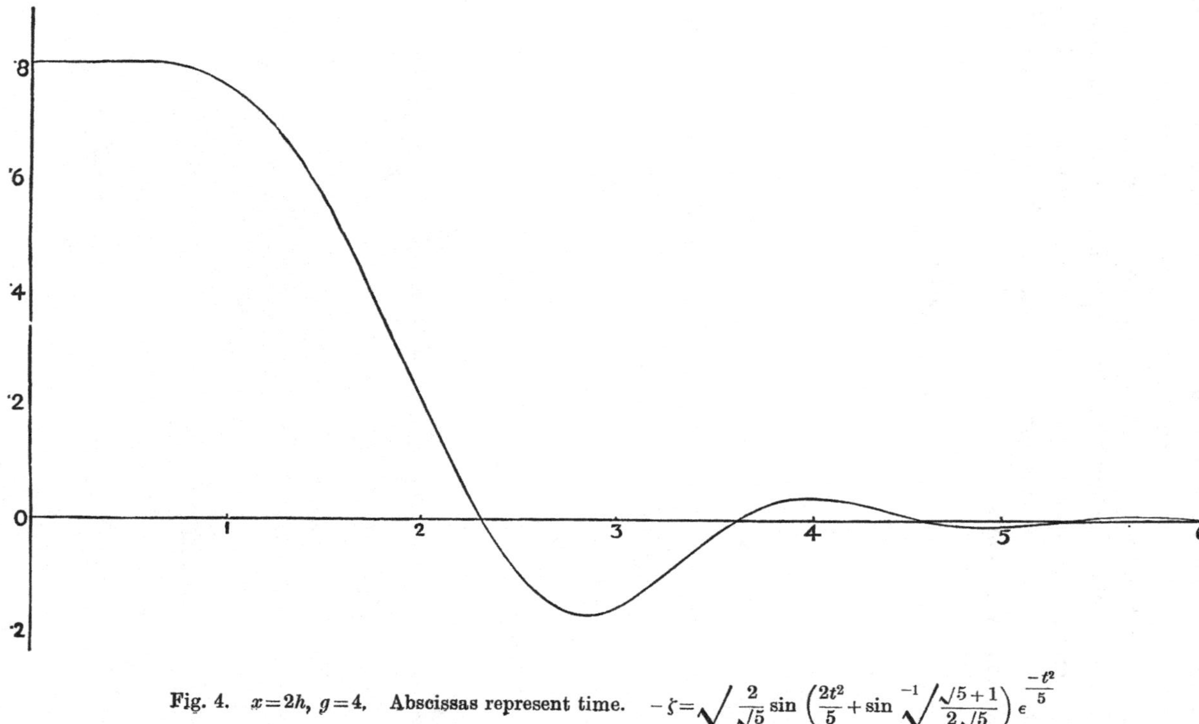

Fig. 4. $x=2h$, $g=4$. Abscissas represent time. $-\zeta=\sqrt{\frac{2}{\sqrt{5}}}\sin\left(\frac{2t^2}{5}+\sin^{-1}\sqrt{\frac{\sqrt{5}+1}{2\sqrt{5}}}\right)\epsilon^{\frac{-t^2}{5}}$

This, except the sign $-$ instead of $+$, is Cauchy's solution*; of which he says that when the time has advanced so much as to violate a condition equivalent to (15), "le mouvement change avec la méthode d'approximation." The remainder of his Note XVI. (about 100 pages) is chiefly devoted to very elaborate efforts to obtain definite results for the larger values of t. This object is thoroughly attained by the exponential factor in (8) of § 3 above, without the crippling restriction $z/x \doteqdot 0$ which vitiates (16) for small values of x.

* *Œuvres*, Vol. I. note XVI. p. 193.

36. On the Front and Rear of a Free Procession of Waves in Deep Water.

[From *Proc. Roy. Soc. Edin.* June 20, 1904 ; *Phil. Mag.* Vol. VIII. Oct. 1904, pp. 454—470.]

11. The present communication is substituted for another bearing the same title, which was read before the Royal Society of Edinburgh on January 7th, 1887, because the result of that paper was rendered imperfect and unsatisfactory by omission of the exponential factor referred to in § 10 of my paper of February 1st, 1904. I shall refer henceforth to the last-mentioned paper as §§ 1—10 above. [See *supra*, p 307.]

12. I begin by considering processions produced by superposition of static initiating disturbances, of the type expressed in (12) of § 4 above; graphically represented by fig. 1; and leading to motion investigated in §§ 1—3, 5—10. The particular type of that solution which I now choose, is that chosen at the end of § 4, which we, with a slight but useful modification*, may now write as follows:—

$$\left.\begin{array}{l} -\zeta = \phi(x, t) = \sqrt{\dfrac{2}{\rho}} \cos\left(\dfrac{gt^2 x}{4\rho^2} - \tfrac{1}{2}\chi\right) \epsilon^{\frac{-gt^2 z}{4\rho^2}} \\ \text{where} \qquad \rho = \sqrt{(z^2 + x^2)}, \text{ and } \chi = \tan^{-1}(x/z) \end{array}\right\} \text{......(17).}$$

Here $-\zeta$ denotes the upward vertical component of the displacement of the fluid at time t from its undisturbed position at point (x, z), which may be either in the free surface or anywhere below it. Taking $t = 0$ in (17), we have, for the initial height of the free surface above the undisturbed level,

$$-\zeta_0 = \phi(x, 0) = \sqrt{\frac{2}{\rho}} \cdot \sqrt{\frac{\rho + z}{2\rho}} = \frac{\sqrt{(\rho + z)}}{\rho} \text{......(18).}$$

* The substitution of $\frac{1}{2}\chi$, for $\frac{1}{2}\pi - \tan^{-1}\sqrt{\dfrac{\rho+z}{\rho-z}}$, saves considerable labour and use of logarithms; especially when, as in our calculations, $z = 1$.

13. We shall first take, as initiating disturbance, a row extending from $-\infty$ to $+\infty$ of superpositions of (18); alternately positive and negative; and placed at equal successive distances $\frac{1}{2}\lambda$: so that we now have

$$-\zeta_0 = P(x, 0) = \sum_{i=-\infty}^{i=+\infty} (-1)^i \phi\left(x + i\frac{\lambda}{2}, 0\right) \quad \text{......(19)},$$

or, as we may write it,

$$-\zeta_0 = P(x, 0) = \sum_{i=-\infty}^{i=+\infty} D(x + i\lambda, 0) \quad \text{........(19')},$$

where

$$D(x, 0) = \phi(x, 0) - \phi\left(x + \frac{\lambda}{2}, 0\right) \quad \text{...........(20)}.$$

In (19), P denotes a space-periodic function, with λ for its period. This formula, with t substituted for 0, represents $-\zeta_t$, being the elevation of the surface above undisturbed level at time t, in virtue of initial disturbance represented by (19).

14. Remark now that whatever function be represented by ϕ, the formula for P in (19) implies that

$$P(x + \lambda, 0) = P(x, 0) \quad \text{.................(21)},$$

which means that P is a space-periodic function with λ for period. And (19) also implies that

$$P(x + \tfrac{1}{2}\lambda, 0) = -P(x, 0) \quad \text{.............(22)};$$

which includes (21). And with the actual function, (18), which we have chosen for $\phi(x, 0)$, the fact that $\phi(x, 0) = \phi(-x, 0)$ makes

$$P(x, 0) = P(-x, 0) \text{.....................(23)}.$$

Thus (19) has a graph of the character fig. 5, symmetrical on each

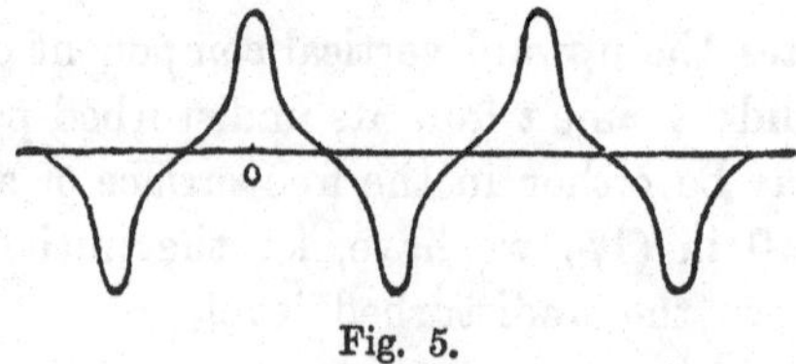

Fig. 5.

side of each maximum and minimum ordinate. The Fourier harmonic analysis of $P(x, 0)$, when subject to (22) and (23), gives

$$P(x, 0) = A_1 \cos\frac{2\pi x}{\lambda} + A_3 \cos 3\frac{2\pi x}{\lambda} + A_5 \cos 5\frac{2\pi x}{\lambda} + \dots \text{(24)}.$$

15. *Digression on periodic functions generated by addition of values of any function for equidifferent arguments.* Let $f(x)$ be any function whatever, periodic or non-periodic; and let

$$P(x) = \sum_{i=-\infty}^{i=+\infty} f(x+i\lambda) \quad \text{..................(25)};$$

which makes
$$P(x) = P(x+\lambda) \quad \text{.....................(26)}.$$

Let the Fourier harmonic expansion of $P(x)$ be expressed as follows:—

$$\left.\begin{aligned} P(x) = A_0 + A_1 \cos\alpha + A_2 \cos 2\alpha + A_3 \cos 3\alpha + \ldots \\ + B_1 \sin\alpha + B_2 \sin 2\alpha + B_3 \sin 3\alpha + \ldots \end{aligned}\right\} \text{ where } \alpha = \frac{2\pi x}{\lambda} \quad \text{.........(27)}.$$

Denoting by j any integer, we have by Fourier's analysis

$$\tfrac{1}{2}\lambda \begin{matrix} A_j \\ B_j \end{matrix} = \int_0^\lambda dx P(x) \begin{matrix} \cos \\ \sin \end{matrix} j \frac{2\pi x}{\lambda} \quad \text{...........(28)};$$

which gives

$$\left.\begin{aligned} \tfrac{1}{2}\lambda A_j &= \sum_{i=-\infty}^{i=+\infty} \int_0^\lambda dx f(x+i\lambda) \cos j \frac{2\pi x}{\lambda} = \int_{-\infty}^{+\infty} dx f(x) \cos j \frac{2\pi x}{\lambda} \\ \tfrac{1}{2}\lambda B_j &= \sum_{i=-\infty}^{i=+\infty} \int_0^\lambda dx f(x+i\lambda) \sin j \frac{2\pi x}{\lambda} = \int_{-\infty}^{+\infty} dx f(x) \sin j \frac{2\pi x}{\lambda} \end{aligned}\right\} \quad \text{......(29)}.$$

16. Take now in (29), as by (19′), (20),

$$f(x) = \phi(x, 0) - \phi\left(x + \frac{\lambda}{2}, 0\right) \quad \text{...........(30)}.$$

This reduces all the B's to zero; reduces the A's to zero for even values of j; and for odd values of j gives, in virtue of (22),

$$\tfrac{1}{2}\lambda A_j = 2\int_{-\infty}^{+\infty} dx \phi(x, 0) \cos j \frac{2\pi x}{\lambda} \quad \text{...........(31)}.$$

Go back now to §§ 3, 4, (6), (12), above; and, according to the last lines of § 4, take

$$\phi(x, 0) = \left\{RS\right\} \frac{\sqrt{2}}{\sqrt{(z+\iota x)}} = \frac{\sqrt{(\rho+z)}}{\rho} \quad \text{.........(32)}.$$

Hence, for the harmonic expansion (24) of $P(x, 0)$, we have

$$A_j = \frac{4}{\lambda}\int_{-\infty}^{+\infty} dx \frac{\sqrt{(\rho+z)}}{\rho} \cos j \frac{2\pi x}{\lambda} = \frac{4}{\lambda}\left\{RS\right\}\int_{-\infty}^{+\infty} dx \frac{\sqrt{2}}{\sqrt{(z+\iota x)}} \cos j \frac{2\pi x}{\lambda} \quad \text{......(33)}.$$

The imaginary form of the last member of this equation facilitates the evaluation of the integral. Instead of $\cos j\dfrac{2\pi x}{\lambda}$ in the last factor, substitute

$$\cos j\frac{2\pi x}{\lambda}+\iota\sin\frac{2\pi x}{\lambda},\quad \text{or } \epsilon^{\iota j\frac{2\pi x}{\lambda}} \text{ ....(34).}$$

The alternative makes no difference in the summation $\int_{-\infty}^{+\infty}dx$, because the sine term disappears for the same reason that the sine terms in (29) disappear because of (30). Thus (33) becomes

$$A_j=\frac{4}{\lambda}\left\{RS\right\}\int_{-\infty}^{+\infty}dx\frac{\sqrt{2}}{\sqrt{(z+\iota x)}}\epsilon^{j\frac{2\pi\iota x}{\lambda}} \text{ ....(35);}$$

put now $\sqrt{(z+\iota x)}=\iota\sigma$;

whence $\dfrac{dx}{\sqrt{(z+\iota x)}}=2d\sigma$, and $\iota x=-\sigma^2-z$..........(36).

Using these in (35) we may omit the instruction $\{RS\}$ because nothing imaginary remains in the formula: thus we find

$$\left.\begin{aligned}A_j&=\frac{8\sqrt{2}}{\lambda}\int_{-\infty}^{+\infty}d\sigma\,\epsilon^{-\frac{2\pi j}{\lambda}\sigma^2}.\,\epsilon^{-\frac{2\pi jz}{\lambda}}=\epsilon^{-\frac{2\pi jz}{\lambda}}.\frac{8\sqrt{2}}{\lambda}.\sqrt{\frac{\lambda}{2\pi j}}.\sqrt{\pi}\\ &=\epsilon^{-\frac{2\pi jz}{\lambda}}.\frac{8}{\sqrt{j\lambda}}\end{aligned}\right\} \text{ ......(37).}$$

The transition in (37) is made in virtue of Laplace's celebrated discovery $\int_{-\infty}^{+\infty}d\sigma\,\epsilon^{-m\sigma^2}=\sqrt{\dfrac{\pi}{m}}$.

17. Equation (37) allows us readily to see how near to a curve of sines is the graph of $P(x, 0)$, for any particular value of λ/z. It shows that

$$A_1=\frac{8}{\sqrt{\lambda}}\epsilon^{-\frac{2\pi z}{\lambda}};\ A_3/A_1=\sqrt{\tfrac{1}{3}}.\epsilon^{-\frac{4\pi z}{\lambda}};\ A_5/A_3=\sqrt{\tfrac{3}{5}}.\epsilon^{-\frac{4\pi z}{\lambda}} \text{ ...(38).}$$

Suppose for example $\lambda=4z$; we have

$$\epsilon^{-\frac{4\pi z}{\lambda}}=\epsilon^{-\pi}=\cdot043214;\ A_3/A_1=\cdot02495;\ A_5/A_3=\cdot03347 \text{ ...(39).}$$

Thus we see that A_3 is about $\frac{1}{40}$ of A_1; and A_5, about $\frac{1}{30}$ of A_3. This is a fair approach to sinusoidality; but not quite near enough for our present purpose. Try next $\lambda=2z$; we have

$$A_1=\frac{8}{\sqrt{\lambda}}\ 043214;\ \epsilon^{-2\pi}=\cdot001867;\ A_3/A_1=\cdot001078 \text{...(40).}$$

Thus A_3 is about a thousandth of A_1; and A_5 about $1\frac{1}{3}\times 10^{-6}$ of A_1. This is a quite good enough approximation for our present purpose: A_5 is imperceptible in any of our calculations: A_3 is negligible, though perceptible if included in our calculations (which are carried out to four significant figures): but it would be utterly imperceptible in our diagrams. Henceforth we shall occupy ourselves chiefly with the free surface, and take $z=h$, the height of O, the origin of coordinates above the undisturbed level of the water.

18. To find the water-surface at any time t after being left free and at rest, displaced according to any periodic function $P(x)$ expressed Fourier-wise as in (27); take first, for the initial motionless surface-displacement, a simple sinusoidal form,

$$-\zeta_0 = A\cos(mx-c) \quad \text{.................}(41).$$

Going back to (2), (3), and (4) above, let $w(z, x, t)$ be the downwards vertical component of displacement. We thus have, as the differential equations of the motion,

$$g\frac{dw}{dz}=\frac{d^2w}{dt^2} \quad \text{.......................}(42),$$

$$\frac{d^2w}{dx^2}+\frac{d^2w}{dz^2}=0 \quad \text{...................}(43).$$

These are satisfied by

$$-w = C\epsilon^{-mz}\cos(mx-c)\cos t\sqrt{gm} \quad \text{...........}(44),$$

which expresses the well-known law of two-dimensional periodic waves in infinitely deep water. And formula (44) with $C\epsilon^{-mh}=A$ and $t=0$, agrees with (41). Hence the addition of solutions (44), with jm for m; with A successively put equal to A_1, $A_2 \ldots$, B_1, $B_2 \ldots$; and, with $c=0$ for the A's, and $=\frac{1}{2}\pi$ for the B's, gives us, for time t, the vertical component-displacement at depth $z-h$ below the surface, if at time $t=0$ the water was at rest with its surface displaced according to (27). Thus, with (38), and (44), we have $P(x, t)$.

19. Looking to (44) and (27), and putting $m=2\pi/\lambda$, we see that the component motion due to any one of the A's or B's in the initial displacement is an endless infinite row of standing waves, having wave-lengths equal to λ/j and time-periods expressed by

$$\tau_j=\frac{2\pi}{\sqrt{gm}}=\sqrt{\frac{2\pi\lambda}{jg}} \quad \text{..................}(45).$$

The whole motion is not periodic because the periods of the constituent motions, being inversely as $\sqrt{j}$, are not commensurable. But by taking $\lambda = 2h$ as proposed in § 17, which, according to (40), makes A_3, for the free surface, only a little more than 1/1000 of A_1, we have so near an approach to sinusoidality that in our illustrations we may regard the motion as being periodic, with period (45) for $j = 1$. This makes $\tau = \sqrt{\pi}$ when, as in § 5, we, without loss of generality (§ 10), simplify our numerical statements by taking $g = 4$; and $h = 1$, which makes the wave-length $= 2$.

20. Toward our problem of "front and rear," remark now that the infinite number of parallel straight standing sinusoidal waves which we have started everywhere over an infinite plane of originally undisturbed water, may be ideally resolved into two processions of sinusoidal waves of half their height travelling in contrary directions with equal velocities $2/\sqrt{\pi}$.

Instead now of covering the whole surface with standing waves, cover it only on the negative side of the line (not shown in our diagrams) YOY', that is the left side of O the origin of coordinates; and leave the water plane and motionless on the right side to begin. At all *great* distances on the left side of O, there will be in the beginning, standing waves equivalent to two trains of progressive waves, of wave-length 2, travelling rightwards and leftwards with velocity $2/\sqrt{\pi}$. The smooth water on the right of O is obviously invaded by the rightward procession.

21. Our investigation proves that the extreme perceptible rear of the leftward procession (marked R in fig. 10 below) does not, through the space OR on the left side of O, broadening with time, nor anywhere on the right of O, perceptibly disturb the rightward procession.

22. Our investigation also proves that the surface at O has simple harmonic motion through all time. It farther shows that the rightward procession is very approximately sinusoidal, with simple harmonic motion, through a space OF (fig. 9) to the right of O, broadening with time; and that, at any particular distance rightwards from O, this approximation becomes more and more nearly perfect as time advances. What I call the front of the rightward procession, is the wave disturbance beyond the point F, at a not strictly defined distance rightwards from O, where the approximation to sinusoidality of shape, and simple harmonic

quality of motion, is only just perceptibly at fault. We shall find that beyond F the waves are, as shown in fig. 9, less and less high, and longer and longer, at greater and greater distances from O, at one and the same time; but that the wave-height does not at any time or place come abruptly to nothing. The propagational velocity of the beginning of the disturbance is in reality infinite, because we regard the water as infinitely incompressible.

23. Thus we see that the front of the rightward procession, with sinusoidal waves following it from O, is simply given by the calculation, for positive values of x, of the motion due to an initial motionless configuration of sinusoidal furrows and ridges on the left side of O. Fig. 8 represents a static initial configuration, which we denote by $Q(x, 0)$, approximately realising the condition stated in § 20. Fig. 9 represents on the same scale of ordinates the surface displacement at the time 25τ in the subsequent motion due to that initial configuration; which, for any time t, we denote by $Q(x, t)$ defined as follows:—

$$Q(x, t) = \tfrac{1}{2}\phi(x, t) - \phi(x+1, t) + \phi(x+2, t) - \dots \textit{ ad inf.} \dots (46),$$

where ϕ is the function defined by (17), with $z = 1$ and $g = 4$.

24. The wave-height, at all distances so far leftward from O that the influence of the rear of the leftward procession has not yet reached them at any particular time, t, after the beginning, is simply the $P(x, t)$ of § 13 calculated according to §§ 18, 17; and the motion there is still merely standing waves, ideally resolvable into rightward and leftward processions. Let I, beyond the leftward range of fig. 10, be the point of the ideally extended diagram, not precisely defined, where the leftward procession at any particular time, t, becomes sensibly influenced by its own rear. Between I and R the whole motion is transitional in character, from the regular sinusoidal motion $P(x, t)$ of the water on the left side of I, to regular sinusoidal motion of wave-height $\tfrac{1}{2}P(x, t)$, from R to O; and on to F of fig. 9, the beginning of the front of the disturbance in the rightward procession. Hence to separate ideally the leftward procession from the whole disturbance due to the initial configuration, we have only to subtract $\tfrac{1}{2}P(x, t)$ from $Q(x, t)$ calculated for negative values of x. Thus the expression for the whole of the leftward procession is

$$Q(x, t) - \tfrac{1}{2}P(x, t) \text{ for negative values of } x \dots\dots (47).$$

Fig. 10 represents the free surface thus found for the leftward procession alone at time $t=25\tau$.

25. The function $D(x, t)$, which appears in § 13 as an item in one of the modes of summing shown for $P(x, 0)$ in (19′), and indicated for $P(x, t)$ at the end of § 13, and which has been used in some of our summations for $Q(x, t)$; is represented in figs. 6 and 7, for $t=0$, and $t=25\tau$ respectively.

26. Except for a few of the points of fig. 6, representing $D(x, 0)$, the calculation has been performed solely for integral values of x. It seemed at first scarcely to be expected that a fair graphic representation could be drawn from so few calculated points; but the curves have actually been drawn by Mr Witherington with no other knowledge than these points, except information as to all zeros (curve cutting the line of abscissas), through the whole range of each curve. The calculated points are marked on each curve: and it seems certain that, with the knowledge of the zeros, the true curve must lie very close in each case to that drawn by Mr Witherington.

27. The calculation of $Q(x, t)$, for positive integral values of x, is greatly eased by the following arrangements for avoiding the necessity for direct summation of a sluggishly convergent infinite series shown in (46), by use of our knowledge of $P(x, t)$. We have, by (46) and (19),

$$Q(0, t) = \tfrac{1}{2}\phi(0, t) - \phi(1, t) + \phi(2, t) - \dots \textit{ ad. inf.} \quad \dots(48),$$

$$P(0, t) = \sum_{i=-\infty}^{i=+\infty} (-1)^i \phi(i, t) \quad \dots\dots\dots\dots(49).$$

Hence, in virtue of $\phi(-i, t) = \phi(i, t)$,

$$P(0, t) = 2Q(0, t) \quad \dots\dots\dots\dots\dots\dots(50).$$

Again going back to (46), we have

$$Q(x, t) = \tfrac{1}{2}\phi(x, t) - \phi(x+1, t) + \phi(x+2, t) - \phi(x+3, t) + \dots$$

$$Q(x+1, t) = \qquad \tfrac{1}{2}\phi(x+1, t) - \phi(x+2, t) + \phi(x+3, t) - \dots$$

By adding these we find

$$Q(x+1, t) + Q(x, t) = \tfrac{1}{2}[\phi(x, t) - \phi(x+1, t)] = \tfrac{1}{2}D(x, t) \dots(51).$$

By successive applications of this equation, we find

$$2Q(x+i, t) = (-1)^i\, 2Q(x, t) - (-1)^i D(x, t) \pm \dots + D(x+i-1, t) \quad \dots\dots(52).$$

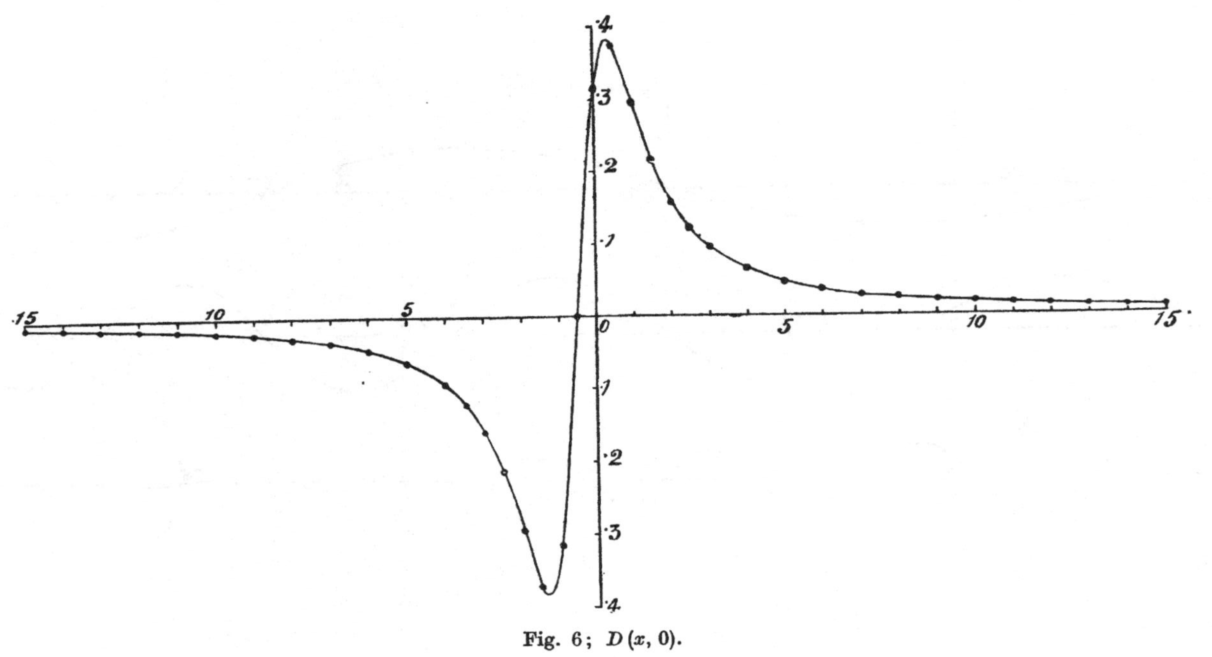

Fig. 6; $D(x, 0)$.

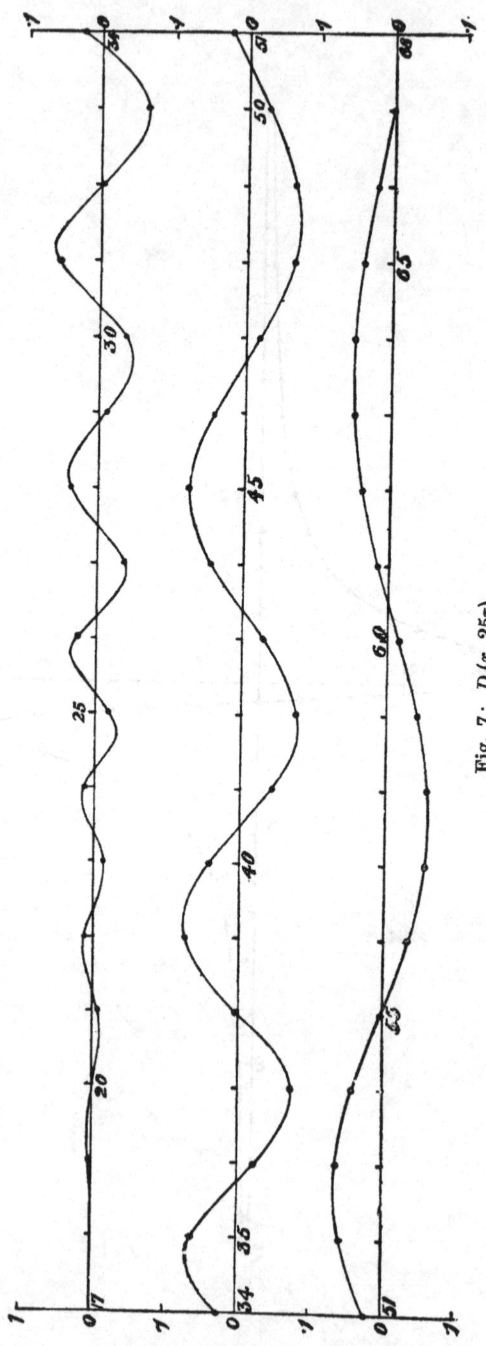

Fig. 7; $D(x, 25\tau)$.

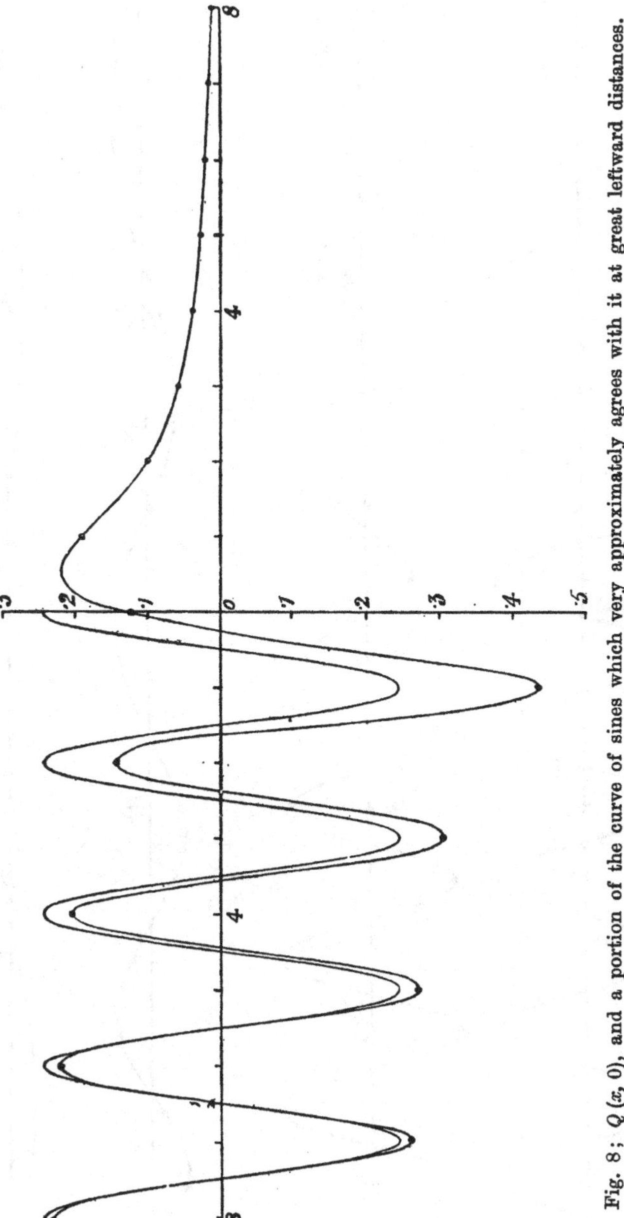

Fig. 8; $Q(x, 0)$, and a portion of the curve of sines which very approximately agrees with it at great leftward distances.

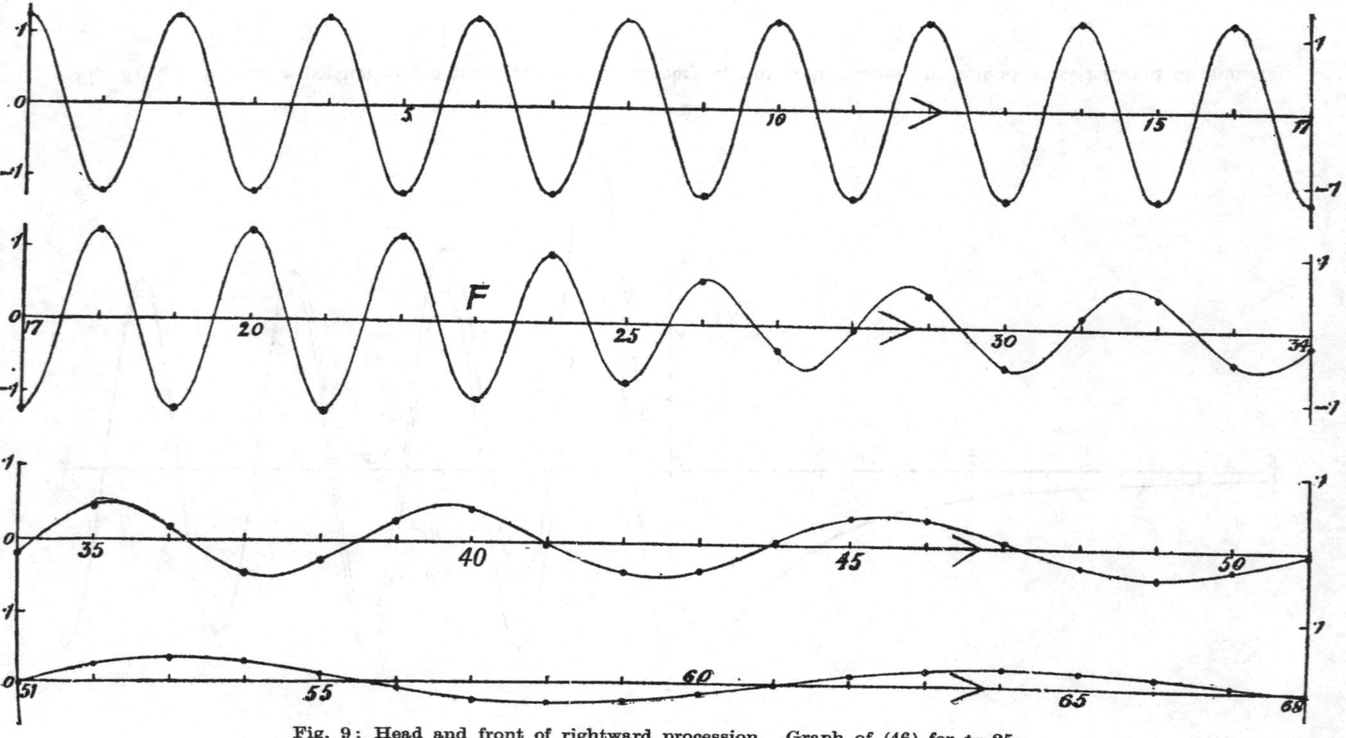

Fig. 9; Head and front of rightward procession. Graph of (46) for $t=25\tau$.

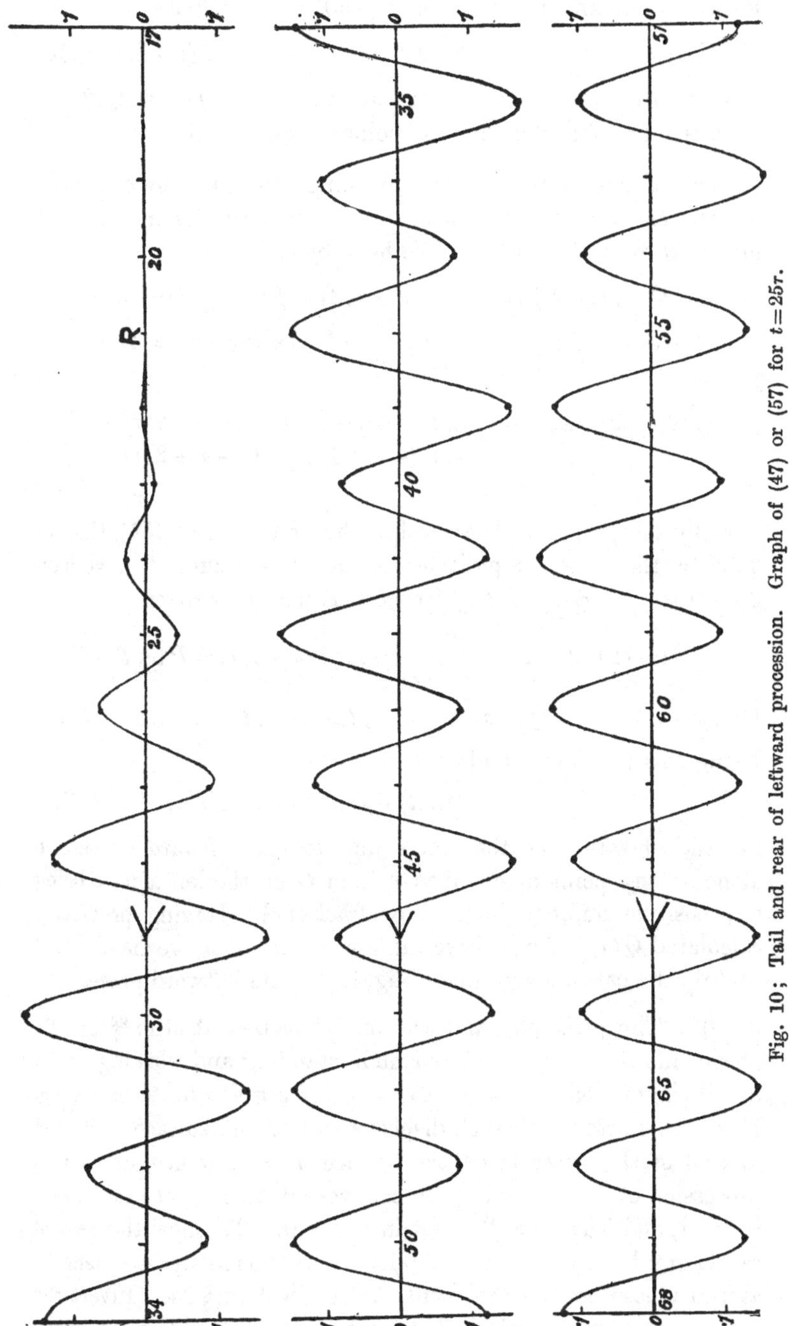

Fig. 10; Tail and rear of leftward procession. Graph of (47) or (57) for $t=25\tau$.

Hence by putting $x=0$, and using (50), we find finally

$$2Q(i,t)=(-1)^i P(0,t)-(-1)^i D(0,t)\pm\ldots+D(i-1,t) \quad (53).$$

This is thoroughly convenient to calculate $Q(1,t)$, $Q(2,t)$... successively; for plotting the points shown in fig. 9.

28. For fig. 10, instead of assuming as in (47) the calculation of $Q(x,t)$ for negative values of x, a very troublesome affair, we may now evaluate it thus. We have by (46)

$$Q(x,t)=\tfrac{1}{2}\phi(x,t)-\phi(x+1,t)+\phi(x+2,t)-\ldots,$$

$$Q(-x,t)=\tfrac{1}{2}\phi(-x,t)-\phi(-x+1,t)+\phi(-x+2,t)-\ldots.$$

Hence

$$\left.\begin{aligned}Q(x,t)+Q(-x,t)=\phi(x,t)-\phi(x+1,t)+\phi(x+2,t)-\ldots\\ -\phi(-x+1,t)+\phi(-x+2,t)-\ldots\end{aligned}\right\} \quad \ldots\ldots(54).$$

Now by the property of ϕ, used in the first term of (54), that its value is the same for positive and negative values of x, we have $\phi(-x+i,t)=\phi(x-i,t)$. Hence (54) may be written

$$Q(x,t)+Q(-x,t)=\sum_{i=-\infty}^{i=+\infty}(-1)^i\phi(x+i,t)=P(x,t) \quad (55).$$

Hence $$Q(-x,t)=P(x,t)-Q(x,t) \quad \ldots\ldots\ldots\ldots\ldots(56).$$

Using this in (47) we find

$$\tfrac{1}{2}P(x,t)-Q(x,t) \quad \ldots\ldots\ldots\ldots\ldots\ldots\ldots(57),$$

for the elevation of the water due to the leftward procession alone at any point at distance x from O on the left side, x being any positive number, integral or fractional. Having previously calculated $Q(x,t)$ for positive integral values of x, we have found by (57) the calculated points of fig. 10 for the leftward procession.

29. The principles and working plans described in §§ 11—28 above, afford a ready means for understanding and working out in detail the motion, from $t=0$ to $t=\infty$, of a given finite procession of waves, started with such displacement of the surface, and such motion of the water below the surface, as to produce, at $t=0$, a procession of a thousand or more waves advancing into still water in front, and leaving still water in the rear. To show the desired result graphically, extend fig. 10 leftwards to as many wave-lengths as you please beyond the point, I, described in § 24. Invert the diagram thus drawn relatively to right and left, and fit it on to the

diagram, fig. 9, extended rightwards so far as to show no perceptible motion; say to $x=200$, or 300, of our scale. The diagram thus compounded represents the water surface at time 25τ after a commencement correspondingly compounded from fig. 8, and another similar figure drawn to represent the rear of the finite (two-ended) procession which we are now considering.

30. Direct attack on the problem thus indirectly solved, gives, for the case of 1000 wave-crests in the beginning, the following explicit solution,

$$-\zeta = \sum_{i=0}^{i=2000} (-1)^i \psi(x-i, t) \quad \text{..............(58)},$$

where ψ is a function found according to the principles indicated in § 4 above, to express the same surface-displacement as our function ϕ of § 12, and the proper velocities below the surface to give, in the sum, a rightward procession of waves. Our present solution shows how rapidly the initial sinusoidality of the head and front of a one-ended infinite procession, travelling rightwards, is disturbed in virtue of the hydrokinetic circumstances of a procession invading still water. Our solution, and the item towards it represented in figs. 6 and 7, and in fig. 2 of § 6 above, show how rapidly fresh crests are formed. The whole investigation shows how very far from finding any definite "group-velocity" we are, in any initially given group of two, three, four, or any number, however great, of waves. I hope in some future communication to the Royal Society of Edinburgh to return to this subject in connection with the energy principle set forth by Osborne Reynolds*, and the interferential theory of Stokes† and Rayleigh‡ giving an absolutely definite group-velocity in their case of an infinite number of mutually supporting groups. But my first hydrokinetic duty, the performance of which I hope may not be long deferred, is to fulfil my promises regarding ship-waves, and circular waves travelling in all directions from a place of disturbance in water.

31. The following tables show some of the most important numbers which have been calculated, and which may be useful in farther prosecution of the subject of the present paper.

* *Nature*, Vol. XVI. 1877, pp. 343—4.

† Smith's Prize Paper, *Camb. Univ. Calendar*, 1876.

‡ *Sound*, ed. 1, Vol. I. 1877, pp. 246—7.

TABLE I.

$$t=0\,;\ \phi(x,0)=\frac{\sqrt{(\rho+z)}}{\rho};\ D(x,0)=\phi(x,0)-\phi(x+1,0).$$

x	$\phi(x, 0)$	$D(x, 0)$	x	$\phi(x, 0)$	$D(x, 0)$
−1	1·0987	−·3155	34	·1740	·0026
0	1·4142	·3155	35	·1714	·0025
1	1·0987	·2942	36	·1689	·0023
2	·8045	·1593	37	·1666	·0023
3	·6452	·0962	38	·1643	·0022
4	·5490	·0647	39	·1621	·0021
5	·4843	·0468	40	·1600	·0020
6	·4375	·0357	41	·1580	·0019
7	·4018	·0284	42	·1561	·0019
8	·3734	·0232	43	·1542	·0018
9	·3502	·0194	44	·1524	·0017
10	·3308	·0166	45	·1507	·0017
11	·3142	·0143	46	·1490	·0016
12	·2999	·0125	47	·1474	·0016
13	·2874	·0111	48	·1458	·0015
14	·2763	·0100	49	·1443	·0015
15	·2663	·0089	50	·1428	·0014
16	·2574	·0081	51	·1414	·0014
17	·2493	·0073	52	·1400	·0014
18	·2420	·0068	53	·1386	·0013
19	·2352	·0062	54	·1373	·0013
20	·2290	·0058	55	·1360	·0012
21	·2232	·0053	56	·1348	·0012
22	·2179	·0050	57	·1336	·0012
23	·2129	·0047	58	·1324	·0011
24	·2082	·0043	59	·1313	·0011
25	·2039	·0041	60	·1302	·0011
26	·1998	·0039	61	·1291	·0011
27	·1959	·0036	62	·1280	·0010
28	·1923	·0035	63	·1270	·0010
29	·1888	·0033	64	·1260	·0010
30	·1855	·0031	65	·1250	·0010
31	·1824	·0029	66	·1240	·0009
32	·1795	·0028	67	·1231	·0009
33	·1767	·0027			

TABLE II.

$$t=25\tau;\ \tau=\sqrt{\pi};\ \chi=\tan^{-1}\frac{x}{z};\ \phi=\sqrt{\frac{2}{\rho}}\cos\left[x\left(\frac{25}{\rho}\right)^2\pi-\frac{1}{2}\chi\right]\epsilon^{-\left(\frac{25}{\rho}\right)^2\pi}$$

x	$\left(\frac{xt^2}{\rho^2}+\theta\right)$	$\epsilon^{\frac{-t^2}{\rho^2}}$	$\phi(x, 25\tau)$	$D(x, 25\tau)$
15	$41\pi+133°\ 43'$	·0002	·0000	+ ·0001
16	$39\pi+\ 30°\ 41'$	·0005	− ·0001	− ·0002
17	$36\pi+161°\ 31'$	·0011	+ ·0001	− ·0002
18	$34\pi+157°\ 22'$	·0024	+ ·0003	+ ·0006
19	$33\pi+\ 11°\ \ 3'$	·0044	− ·0003	+ ·0020
20	$31\pi+\ 77°\ 25'$	·0075	− ·0023	− ·0018
21	$29\pi+171°\ 23'$	·0118	− ·0005	− ·0055
22	$28\pi+109°\ 24'$	·0174	+ ·0050	+ ·0117
23	$27\pi+\ 68°\ 20'$	·0246	− ·0067	− ·0136
24	$26\pi+\ 45°\ 33'$	·0333	+ ·0069	+ ·0146
25	$25\pi+\ 39°\ \ 6'$	·0434	− ·0077	− ·0188
26	$24\pi+\ 46°\ 38'$	·0550	+ ·0111	+ ·0281
27	$23\pi+\ 67°\ \ 0'$	·0679	− ·0170	− ·0386
28	$22\pi+\ 98°\ 45'$	·0820	+ ·0216	+ ·0377
29	$21\pi+140°\ 41'$	·0917	− ·0161	− ·0101
30	$21\pi+\ 11°\ 47'$	·1131	− ·0060	− ·0372
31	$20\pi+\ 71°\ 11'$	·1299	+ ·0312	+ ·0558
32	$19\pi+138°\ \ 6'$	·1472	− ·0246	− ·0032
33	$19\pi+\ 31°\ 50'$	·1651	− ·0214	− ·0626
34	$18\pi+111°\ 49'$	·1832	+ ·0412	+ ·0267
35	$18\pi+\ 17°\ 29'$	·2016	+ ·0145	+ ·0637
36	$17\pi+108°\ 23'$	·2201	− ·0492	− ·0266
37	$17\pi+\ 24°\ \ 6'$	·2385	− ·0226	− ·0713
38	$16\pi+124°\ 14'$	·2569	+ ·0487	+ ·0021
39	$16\pi+\ 48°\ 27'$	·2752	+ ·0466	+ ·0728
40	$15\pi+156°\ 27'$	·2934	− ·0262	+ ·0425
41	$15\pi+\ 87°\ 58'$	·3112	− ·0687	− ·0410
42	$15\pi+\ 22°\ 44'$	·3287	− ·0277	− ·0751
43	$14\pi+140°\ 32'$	·3459	+ ·0474	− ·0290
44	$14\pi+\ 81°\ \ 8'$	·3629	+ ·0764	+ ·0434
45	$14\pi+\ 24°\ 24'$	·3794	+ ·0330	+ ·0741
46	$13\pi+150°\ \ 6'$	·3956	− ·0411	+ ·0429
47	$13\pi+\ 98°\ \ 9'$	·4112	− ·0840	− ·0190
48	$13\pi+\ 48°\ 20'$	·4267	− ·0650	− ·0642
49	$13\pi+\ \ 0°\ 32'$	·4416	− ·0008	− ·0657
50	$12\pi+134°\ 40'$	·4560	+ ·0649	− ·0282
51	$12\pi+\ 90°\ 36'$	·4702	+ ·0931	+ ·0224
52	$12\pi+\ 48°\ 10'$	·4840	+ ·0707	+ ·0582
53	$12\pi+\ \ 7°\ 25'$	·4973	+ ·0125	+ ·0643
54	$11\pi+148°\ \ 9'$	·5101	− ·0518	+ ·0417
55	$11\pi+110°\ 18'$	·5226	− ·0935	+ ·0035
56	$11\pi+\ 73°\ 48'$	·5348	− ·0970	− ·0332
57	$11\pi+\ 38°\ 35'$	·5464	− ·0638	− ·0556
58	$11\pi+\ \ 4°\ 34'$	·5580	− ·0082	− ·0578
59	$10\pi+151°\ 43'$	·5690	+ ·0496	− ·0421
60	$10\pi+119°\ 57'$	·5797	+ ·0917	− ·0152
61	$10\pi+\ 89°\ 14'$	·5900	+ ·1069	+ ·0141
62	$10\pi+\ 59°\ 30'$	·6001	+ ·0928	+ ·0373
63	$10\pi+\ 30°\ 43'$	·6098	+ ·0555	+ ·0501
64	$10\pi+\ \ 2°\ 50'$	·6193	+ ·0054	+ ·0506
65	$9\pi+155°\ 48'$	·6284	− ·0452	+ ·0403
66	$9\pi+129°\ 35'$	·6372	− ·0855	+ ·0226
67	$9\pi+104°\ \ 9'$	·6459	− ·1081	+ ·0022
68	$9\pi+\ 79°\ 28'$	·6540	− ·1103	

37. Deep Water Ship-Waves.

[From the *Proceedings of the Royal Society of Edinburgh*, June 20, 1904; *Phil. Mag.* Vol. IX. June, 1905, pp. 733—757.]

§§ 32—64. Canal Ship-Waves.

32. To avoid the somewhat cumbrous title "Two-dimensional," I now use the designation "Canal* Waves" to denote waves in a canal with horizontal bottom and vertical sides, which, if not two-dimensional in their source, become more and more approximately two-dimensional at greater and greater distances from the source. In the present communication the source is such as to render the motion two-dimensional throughout; the two dimensions being respectively perpendicular to the bottom, and parallel to the length of the canal: the canal being straight.

33. The word "deep" in the present communication and its two predecessors (§§ 1—31) is used for brevity to mean infinitely deep; or so deep that the motion does not differ sensibly from what it would be if the water, being incompressible, were infinitely deep. This condition is practically fulfilled in water of finite depth if the distance between every crest (point of maximum elevation), and neighbouring crest on either side, is less than one-half or one-third of its distance from the bottom.

34. By "ship-waves" I mean any waves produced in open sea or in a canal by a moving generator; and for simplicity I suppose the motion of the generator to be rectilineal and uniform. The generator may be a ship floating on the water, or a submarine ship or a fish moving at uniform speed below the surface; or,

* This designation does not include an interesting class of canal waves of which the dynamical theory was first given by Kelland in the *Trans. Roy. Soc. Edin.* for 1839; the case in which the wave length is very long in comparison with the depth and breadth of the canal, and the transverse section is of any shape other than rectangular with horizontal bottom and vertical sides.

as suggested by Rayleigh, an electrified body moving above the surface. For canal ship-waves, if the motion of the water close to the source is to be two-dimensional, the ship or submarine must be a pontoon having its sides (or a submerged bar having its ends) plane and fitting to the sides of the canal, with freedom to move horizontally. The submerged surface must be cylindric with generating lines perpendicular to the sides.

35. The case of a circular cylindric bar of diameter small compared with its depth below the surface, moving horizontally at a constant speed, is a mathematical problem which presents interesting difficulties, worthy of serious work for anyone who may care to undertake it. The case of a floating pontoon is much more difficult, because of the discontinuity between free surface of water and water-surface pressed by a rigid body of given shape, displacing the water.

36. Choosing a much easier problem than either of those, I take as wave generator a forcive* consisting of a given continuous distribution of pressure at the surface, travelling over the surface at a given speed. To understand the relation of this to the pontoon problem, imagine the rigid surface of the pontoon to become flexible; and imagine applied to it, a given distribution Π of pressure, everywhere perpendicular to it. Take O, any point at a distance h above the undisturbed water-level, draw OX parallel to the length of the canal and OZ vertically downwards. Let ξ, ζ be the displacement-components of any particle of the water whose undisturbed position is (x, z). We suppose the disturbance infinitesimal; by which we mean that the change of distance between any two particles of water is infinitely small in comparison with their undisturbed distance; and that the line joining them experiences changes of direction which are infinitely small in comparison with the radian. For liberal interpretation of this condition see § 61 below. Water being assumed frictionless, its motion, started primarily from rest by pressure applied to the free surface, is essentially irrotational. But we need not assume this at present: we see immediately that it is proved by our equations of motion, when in them we suppose the motion to be

* "Forcive" is a very useful word introduced, after careful consultation with literary authorities, by my brother the late Prof. James Thomson, to denote *any system of force.*

infinitesimal. The equations of motion, when the density of the liquid is taken as unity, are

$$\left.\begin{aligned}\frac{d^2\xi}{dt^2}+\dot\xi\frac{d\xi}{dx}+\dot\zeta\frac{d\xi}{dz}&=\quad-\frac{dp}{dx}\\ \frac{d^2\zeta}{dt^2}+\dot\xi\frac{d\zeta}{dx}+\dot\zeta\frac{d\zeta}{dz}&=g-\frac{dp}{dz}\end{aligned}\right\}\quad\text{.............(59),}$$

where g denotes the force of gravity and p the pressure at (x, z, t). Assuming now the liquid to be incompressible, we have

$$\frac{d\dot\xi}{dx}+\frac{d\dot\zeta}{dz}=0\quad\text{........................(60).}$$

37. The motion being assumed to be infinitesimal, the second and third terms of the first members of (59) are negligible, and the equations of motion become

$$\left.\begin{aligned}\frac{d^2\xi}{dt^2}&=\quad-\frac{dp}{dx}\\ \frac{d^2\zeta}{dt^2}&=g-\frac{dp}{dz}\end{aligned}\right\}\quad\text{.......................(61).}$$

This, by taking the difference of two differentiations, gives

$$\frac{d}{dt}\left(\frac{d\dot\xi}{dz}-\frac{d\dot\zeta}{dx}\right)=0\quad\text{....................(62),}$$

which shows that if at any time the motion is zero or irrotational it remains irrotational for ever.

38. If at any time there is rotational motion in any part of the liquid, it is interesting to know what becomes of it. Leaving for a moment our present restriction to canal waves, imagine ourselves on a very smooth sea in a ship, kept moving uniformly at a good speed by a tow-rope above the water. Looking over the ship's side we see a layer of disturbed motion, showing by dimples in the surface innumerable little whirlpools. The thickness of this layer increases from nothing perceptible near the bow to perhaps 10 or 20 cms. near the stern; more or less according to the length and speed and smoothness of the ship. If now the water suddenly loses viscosity and becomes a perfect fluid, the dynamics of vortex motion tells us that the rotationally moving water gets left behind by the ship, and spreads out in the more and more distant wake and becomes lost*; without, however,

* It now seems to me certain that if any motion be given within a finite portion of an infinite incompressible liquid originally at rest, its fate is necessarily dissi-

losing its kinetic energy, which becomes reduced to infinitely small velocities in an infinitely large portion of liquid. The ship now goes on through the calm sea without producing any more eddies along its sides and stern, but leaving within an acute angle on each side of its wake, smooth ship-waves with no eddies or turbulence of any kind. The ideal annulment of the water's viscosity diminishes considerably the tension of the tow-rope, but by no means annuls it; it has still work to do on an ever increasing assemblage of regular waves extending farther and farther right astern, and over an area of 19° 28′ ($\tan^{-1}\sqrt{\tfrac{1}{8}}$) on each side of mid-wake, as we shall see in about § 80 below. Returning now to two-dimensional motion and canal waves: we, in virtue of (62), put

$$\dot{\xi} = \frac{d\phi}{dx}, \quad \dot{\zeta} = \frac{d\phi}{dz} \quad \text{.....................(63)},$$

where ϕ denotes what is commonly called the "velocity-potential"; which, when convenient, we shall write in full $\phi(x, z, t)$. With this notation (61) gives by integration with respect to x and z,

$$\frac{d\phi}{dt} = -p + g(z + C) \quad \text{.....................(64)}.$$

And (60) gives

$$\frac{d^2\phi}{dx^2} + \frac{d^2\phi}{dz^2} = 0 \quad \text{.........................(65)}.$$

Following Fourier's method, take now

$$\phi(x, z, t) = -k\epsilon^{-mz}\sin m(x - vt) \quad \text{...........(66)},$$

which satisfies (65) and expresses a sinusoidal wave-disturbance, of wave-length $2\pi/m$, travelling x-wards with velocity v.

39. To find the boundary-pressure Π, which must act on the water-surface to get the motion represented by (66), when m, v, k are given, we must apply (64) to the boundary. Let $z = 0$ be the undisturbed surface; and let $\mathbf{d}$ denote its depression, at (x, o, t), below undisturbed level; that is to say,

$$\dot{\mathbf{d}} = \dot{\zeta}(x, o, t) = \frac{d}{dz}\phi(x, z, t)_{z=0} = mk\sin m(x - vt) \quad \text{...(67)},$$

pation to infinite distances with infinitely small velocities everywhere; while the total kinetic energy remains constant. After many years of failure to prove that the motion in the ordinary Helmholtz circular ring is stable, I came to the conclusion that it is essentially unstable, and that its fate must be to become dissipated as now described. I came to this conclusion by extensions not hitherto published of the considerations described in a short paper entitled: "On the stability of steady and of periodic fluid motion," in the *Phil. Mag.* for May 1887. [Reprinted *supra*, pp. 166 *seq.*]

whence by integration with respect to t,

$$\mathrm{d} = \frac{k}{v} \cos m (x - vt) \text{ ....(68).}$$

To apply (64) to the surface, we must, in gz, put $z = \mathrm{d}$; and in $d\phi/dt$ we may put $z = 0$, because d, k, are infinitely small quantities of the first order, and their product is neglected in our problem of infinitesimal displacements. Hence with (66) and (68), and with Π taken to denote surface-pressure, (64) becomes

$$kmv \cos m (x - vt) = \frac{g}{v} k \cos m (x - vt) - \Pi + gC \text{ ...(69);}$$

whence, with the arbitrary constant C taken $= 0$,

$$\Pi = kv \left(\frac{g}{v^2} - m\right) \cos m (x - vt) \text{ ....(70);}$$

and, eliminating k by (68), we have finally,

$$\Pi = (g - mv^2)\, \mathrm{d} \text{ ....(71).}$$

Thus we see that if $v = \sqrt{g/m}$, we have $\Pi = 0$, and therefore we have a train of free sinusoidal waves having wave-length equal to $2\pi/m$. This is the well-known law of relation between velocity and length of free deep-sea waves. But if v is not equal to $\sqrt{g/m}$, we have forced waves with a surface-pressure $(g - mv^2)\,\mathrm{d}$ which is directed with or against the displacement according as $v <$ or $> \sqrt{g/m}$.

40. Let now our problem be:—given Π, a sum of sinusoidal functions, instead of a single one, as in (70);—required d the resulting displacement of the water-surface. We have by (71) and (70), with properly altered notation,

$$\Pi = \Sigma B \cos m (x - vt + \beta) \text{ ....(72),}$$

$$\mathrm{d} = \Sigma \frac{B}{g - mv^2} \cos m (x - vt + \beta) + A \cos \frac{g}{v^2} (x - vt + \gamma) \text{ (73),}$$

where B, m, β are given constants having different values in the different terms of the sums; and v is a given constant velocity. The last term of (73) expresses, with two arbitrary constants (A, γ), a train of free waves which we may superimpose on any solution of our problem.

41. It is very interesting and instructive in respect to the dynamics of water-waves, to apply (72) to a particular case of

Fourier's expansion of periodic arbitrary functions such as a distribution of alternate constant pressures, and zeros, on equal successive spaces, travelling with velocity v. But this must be left undone for the present, to let us get on with ship-waves; and for this purpose we may take as a case of (72), (73),

$$\Pi = gc\,\{\tfrac{1}{2} + e\cos\theta + e^2\cos 2\theta + \text{etc.}\} = gc\,\frac{\tfrac{1}{2}(1-e^2)}{1-2e\cos\theta+e^2} \quad \text{.........(74)},$$

$$\mathbf{d} = Jc\left\{\frac{1}{2J} + \frac{e}{J-1}\cos\theta + \frac{e^2}{J-2}\cos 2\theta + \text{etc.}\right\} \quad \text{.........(75)};$$

where

$$\theta = \frac{2\pi}{a}(x - vt + \beta) \quad \text{.....................(76)};$$

$$v^2 = \frac{g\lambda}{2\pi}; \quad J = \frac{a}{\lambda} = \frac{ga}{2\pi v^2} \quad \text{..............(77)};$$

and e may be any numeric <1. Remark that when $v=0$, $J=\infty$, and we have by (75) and (74), $\mathbf{d} = \Pi/g$, which explains our unit of pressure.

42. To understand the dynamical conditions thus prescribed, and the resulting motion:—remark first that (74), with (76), represents a space-periodic distribution of pressure on the surface, travelling with velocity v; and (75) represents the displacement of the water-surface in the resulting motion, when space-periodic of the same space-period as the surface-pressure. Any motion whatever, consequent on any initial disturbance and no subsequent application of surface-pressure, may be superimposed on the solution represented by (75), to constitute the complete solution of the problem of finding the motion in which the surface-pressure is that given in (74).

43. To understand thoroughly the constitution of the forcive-datum (74) for Π, it is helpful to know that, n denoting any positive or negative integer, we have

$$2\pi(\tfrac{1}{2} + e\cos\theta + e^2\cos 2\theta + \text{etc.}) = \sum_{n=-\infty}^{n=\infty} \frac{ba}{b^2 + (x-na)^2} \quad \text{...(78)},$$

if

$$\left.\begin{aligned} b &= \frac{a}{2\pi}\log(1/e) \\ x &= \frac{a}{2\pi}\theta \end{aligned}\right\} \quad \text{.....................(79)}.$$

This we find by applying § 15 above to the periodic function represented by the second member of (78).

The equality of the two members of (78) is illustrated by fig. 11;

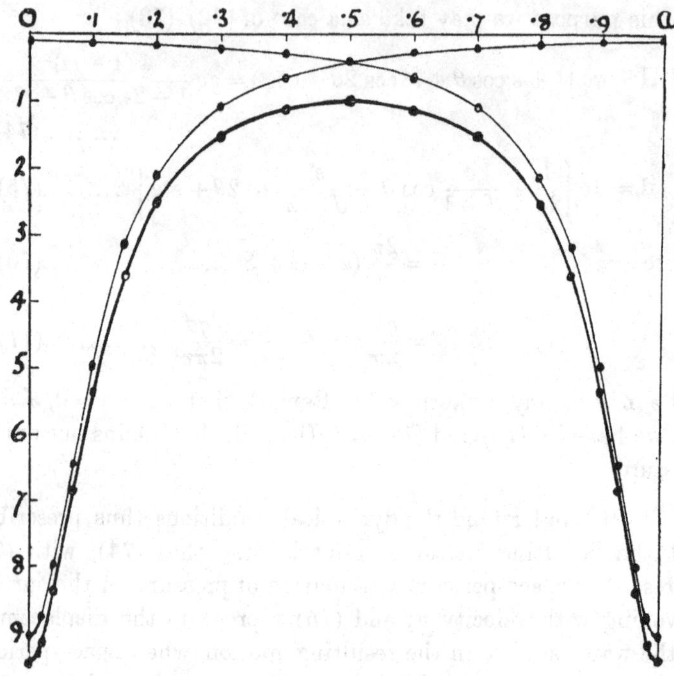

Fig. 11; $e = \cdot 5$.

in which; for the case $e = \cdot 5$ and consequently, by (79), $b/a = \cdot 1103$; the heavy curve represents the first member, and the two light curves represent two terms of the second member; which are as many as the scale of the diagram allows to be seen on it. There is a somewhat close agreement between each of the light curves, and the part of the heavy curve between a maximum and the minimum on each side of it. Thus we see that even with e so small as ·5, we have a *not very rough* approximation to equality between successive half periods of the first member of (78) and a single term of its second member. If e is <1 by an infinitely small difference this approximation is infinitely nearly perfect. It is so nearly perfect for $e = \cdot 9$ that fig. 12 cannot show any deviation from it, on a scale of ordinates one-tenth of that of fig. 11. The tendency to agreement between the first member of (78) and a single term of its second member with values of e approaching to 1, is well shown by the following modification of the last member of (74):—

$$\Pi = gc\frac{\frac{1}{2}(1-e^2)}{1-2e\cos\theta+e^2} = gc\frac{\frac{1}{2}(1-e^2)}{(1-e)^2+4e\sin^2\frac{1}{2}\theta} \quad \text{...(80).}$$

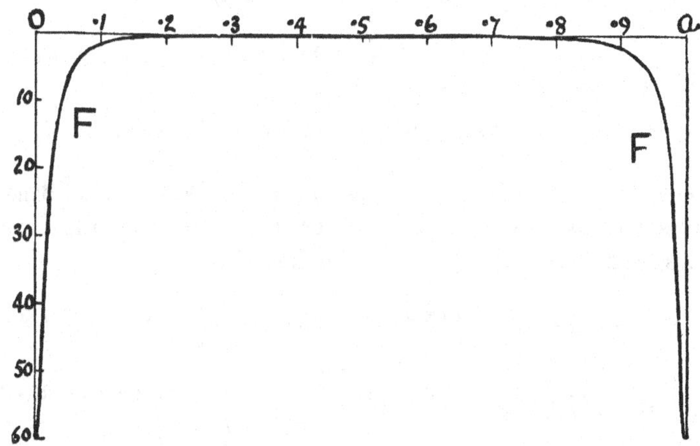

Fig. 12; $e = \cdot 9$.

Thus we see that if $e \risingdotseq 1$, Π is very great when θ is very small; and Π is very small *unless* θ is very small (or very nearly $= 2i\pi$). Thus when $e \risingdotseq 1$, we have

$$\frac{1}{gc}\Pi \risingdotseq \frac{\frac{1}{2}(1-e^2)}{(1-e)^2+e\theta^2} \quad \text{..................(81);}$$

which means expressing Π approximately by a single term of the second member of (78).

44. Return to our dynamical solution (75); and remark that if J is an integer, one term of (75) is infinite, of which the dynamical meaning is clear in (70). Hence to have every term of (75) finite we must have $J = j + \delta$, where j is an integer and δ is < 1; and we may conveniently write (75) as follows:

$$\begin{aligned}\mathbf{d} = c(\delta+j)\Big\{\tfrac{1}{2}\frac{1}{\delta+j} + \frac{e\cos\theta}{\delta+j-1} + \frac{e^2\cos 2\theta}{\delta+j-2} + \dots + \frac{e^j\cos j\theta}{\delta} \\ - \frac{e^{j+1}\cos(j+1)\theta}{1-\delta} - \frac{e^{j+2}\cos(j+2)\theta}{2-\delta} - \text{ad inf.}\Big\} \quad \text{...(82);}\end{aligned}$$

or

$$\mathbf{d} = \mathscr{F} + \mathscr{I} \quad \text{.......................(83),}$$

where $\mathscr{F}$ and $\mathscr{I}$ denote finite and infinite series shown in (82).

45. We are going to make $\delta = \frac{1}{2}$; and in this case $\mathscr{I}$ can be summed, in finite terms, as follows. First multiply each term by $e^{j+\delta} e^{-j-\delta}$; and we find

$$\mathscr{I} = -c(\delta + j) e^{j+\delta} \left[\frac{e^{1-\delta}}{1-\delta} \cos(j+1)\theta + \frac{e^{2-\delta}}{2-\delta} \cos(j+2)\theta + \text{etc.} \right]$$

$$= -c(\delta + j) e^{j+\delta} \int de \left[e^{-\delta} \cos(j+1)\theta + e^{1-\delta} \cos(j+2)\theta + \text{etc.} \right]$$

$$= -c(\delta + j) e^{j+\delta} \int de \, e^{-\delta} \{RS\} q^{j+1} (1 + eq + e^2 q^2 + \text{etc.});$$

where q denotes $\epsilon^{\iota\theta}$; and, as in § 3 above, $\{RS\}$ denotes realisation by taking half sum for $\pm \iota$. Summing the infinite series, and performing $\int de$, for the case $\delta = \frac{1}{2}$, we find

$$\mathscr{I} = -c(j + \tfrac{1}{2}) e^{j+\frac{1}{2}} \{RS\} q^{j+\frac{1}{2}} \log \frac{1 + \sqrt{qe}}{1 - \sqrt{qe}} \quad \text{...............(84)},$$

$$= -c(j + \tfrac{1}{2}) e^{j+\frac{1}{2}} \{RS\} q^{j+\frac{1}{2}} \log \frac{1 + \sqrt{e} \cos \frac{1}{2}\theta + \iota \sqrt{e} \sin \frac{1}{2}\theta}{1 - \sqrt{e} \cos \frac{1}{2}\theta - \iota \sqrt{e} \sin \frac{1}{2}\theta}$$

$$= -c(j + \tfrac{1}{2}) e^{j+\frac{1}{2}} \{RS\} q^{j+\frac{1}{2}} \left[\log \sqrt{\frac{1 + 2\sqrt{e} \cos \frac{1}{2}\theta + e}{1 - 2\sqrt{e} \cos \frac{1}{2}\theta + e}} + \iota (\psi - \psi') \right]$$

where

$$\psi = \tan^{-1} \frac{\sqrt{e} \sin \frac{1}{2}\theta}{1 + \sqrt{e} \cos \frac{1}{2}\theta}, \quad \psi' = \tan^{-1} \frac{-\sqrt{e} \sin \frac{1}{2}\theta}{1 - \sqrt{e} \cos \frac{1}{2}\theta} \quad \text{...(85)},$$

and therefore $$\psi - \psi' = \tan^{-1} \frac{2\sqrt{e} \sin \frac{1}{2}\theta}{1 - e}.$$

Hence finally

$$\mathscr{I} = c(j + \tfrac{1}{2}) e^{j+\frac{1}{2}} \left\{ -\cos(j + \tfrac{1}{2})\theta \log \sqrt{\frac{1 + 2\sqrt{e} \cos \frac{1}{2}\theta + e}{1 - 2\sqrt{e} \cos \frac{1}{2}\theta + e}} + \sin(j + \tfrac{1}{2})\theta \tan^{-1} \frac{2\sqrt{e} \sin \frac{1}{2}\theta}{1 - e} \right\} \quad \text{...(86)}.$$

For our present case, of $\delta = \frac{1}{2}$, (82) gives

$$\mathscr{K} = c(j + \tfrac{1}{2}) \left\{ \frac{1}{2} \frac{1}{j + \frac{1}{2}} + \frac{e \cos \theta}{j - \frac{1}{2}} + \frac{e^2 \cos 2\theta}{j - \frac{3}{2}} + \ldots + \frac{e^j \cos j\theta}{\frac{1}{2}} \right\} \quad \text{...(87)}.$$

With $\mathscr{I}$ and $\mathscr{K}$ thus expressed, (83) gives the solution of our problem.

46. In all the calculations of §§ 46—61 I have taken $e = \cdot 9$, as suggested for hydrokinetic illustrations in Lecture X. of my Baltimore Lectures, pp. 113, 114, from which fig. 12, and part of

fig. 11 above, are taken. Results calculated from (83), (86), (87), are represented in figs. 13—16, all for the same forcive, (74) with $e = \cdot 9$, and for the four different velocities of its travel, which correspond to the values 20, 9, 4, 0, of j. The wave-lengths of free waves having these velocities are [(77) above] $2a/41$, $2a/19$, $2a/9$, and $2a$. The velocities are inversely proportional to $\sqrt{41}$, $\sqrt{19}$, $\sqrt{9}$, $\sqrt{1}$. Each diagram shows the forcive by one curve, a repetition of fig. 12; and shows by another curve the depression, **d**, of the water-surface produced by it, when travelling at one or other of the four speeds.

47. Taking first the last, being the highest, of those speeds, we see by fig. 16 that the forcive travelling at that speed produces maximum displacement *upwards* where the *downward* pressure is greatest; and maximum *downward* displacement where the pressure (everywhere downward) is least. Judging dynamically it is easy to see that greater and greater speeds of the forcive would still give displacements above the mean level where the downward pressure of the forcive is greatest, and below the mean level where it is least; but with diminishing magnitudes down to zero for infinite speed.

And in (75) we have, for all positive values of $J < 1$, a series always convergent (though sluggishly when $e \fallingdotseq 1$), by which the displacement can be exactly calculated for every value of θ.

48. Take next fig. 15, for which $J = 4\frac{1}{2}$, and therefore, by (77), $v = \sqrt{ga/9\pi}$, and $\lambda = a/4\cdot 5$. Remark that the scale of ordinates is, in fig. 15, only $1/2\cdot 5$ of the scale in fig. 16; and see how enormously great is the water-disturbance now in comparison with what we had with the same forcive, but three times greater speed and nine times greater wave-length ($v = \sqrt{ga/\pi}$, $\lambda = 2a$). Within the space-period of fig. 15 we see four complete waves, very approximately sinusoidal, between M, M, two maximums of depression which are *almost exactly* (but very slightly less than) quarter wave-lengths between C and C. Imagine the curve to be exactly sinusoidal throughout, and continued sinusoidally to cut the zero line at CC.

We should thus have in CC a train of $4\frac{1}{2}$ sinusoidal waves; and if the same is continued throughout the infinite procession ... CCC ... we have a discontinuous periodic curve made up of continuous portions each $4\frac{1}{2}$ periods of sinusoidal curve beginning

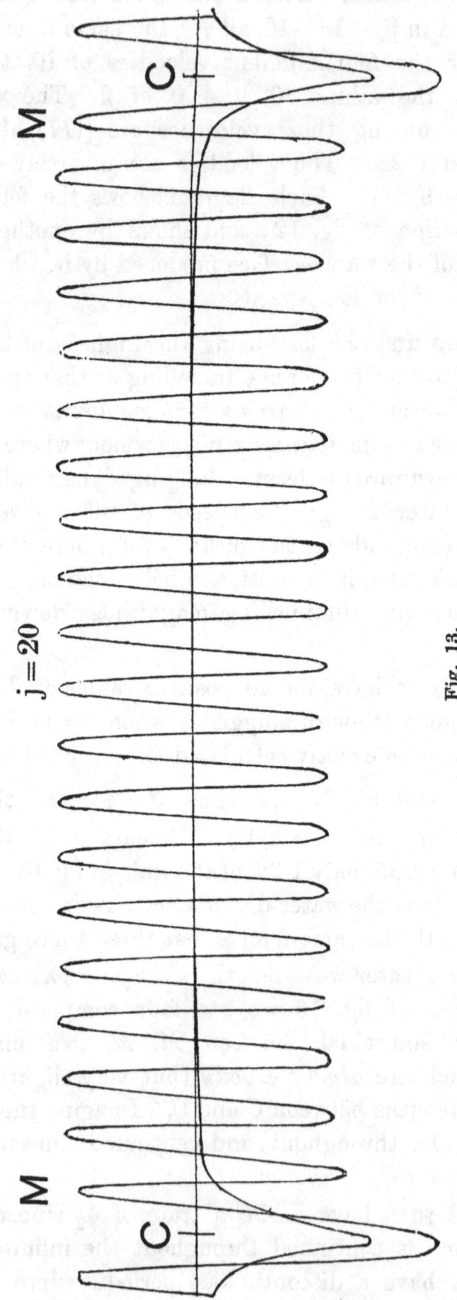

Fig. 13.

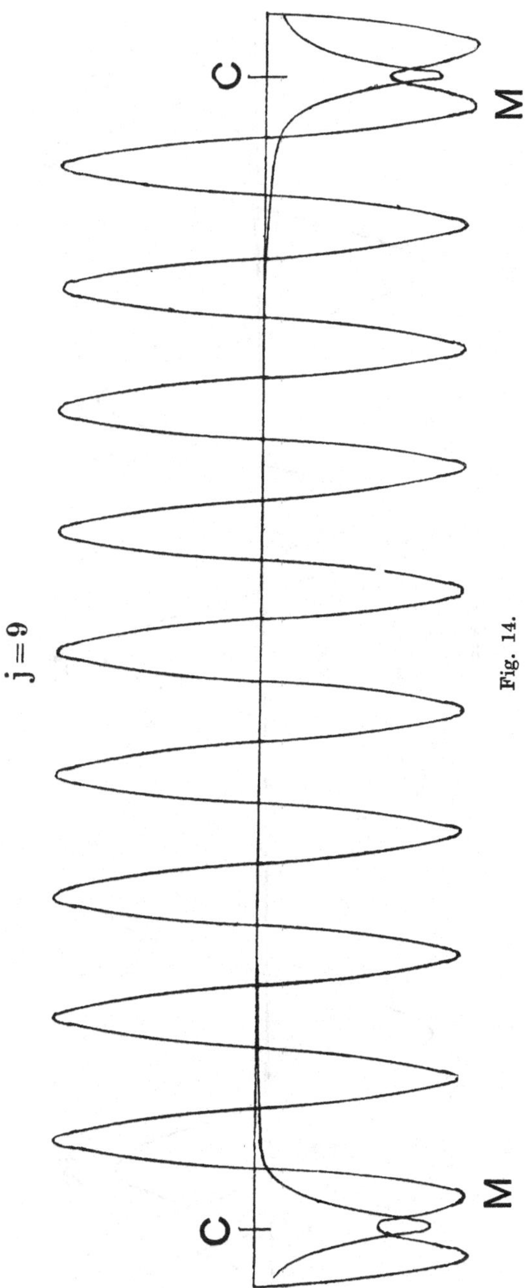

Fig. 14.

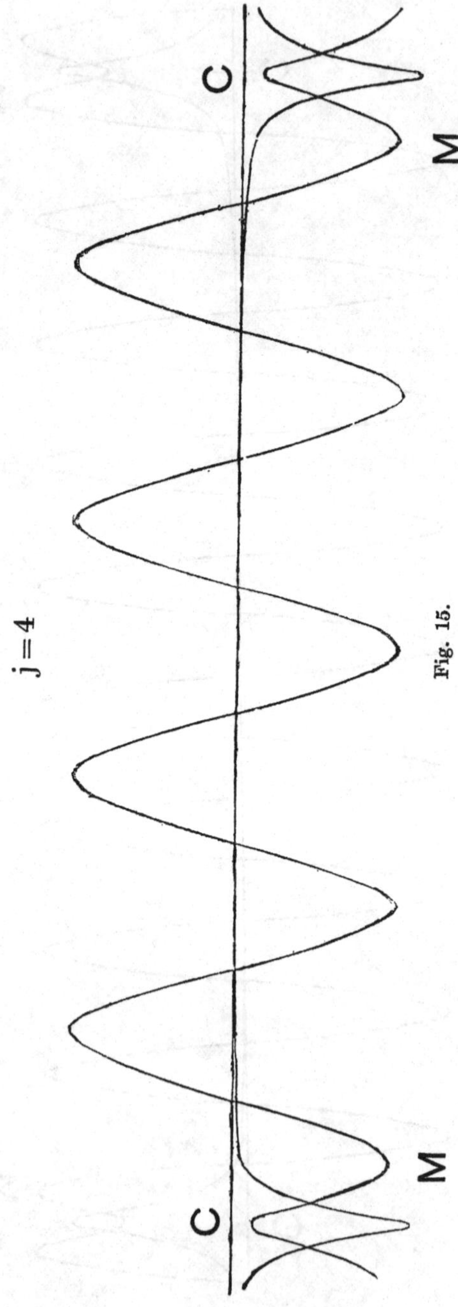

Fig. 15.

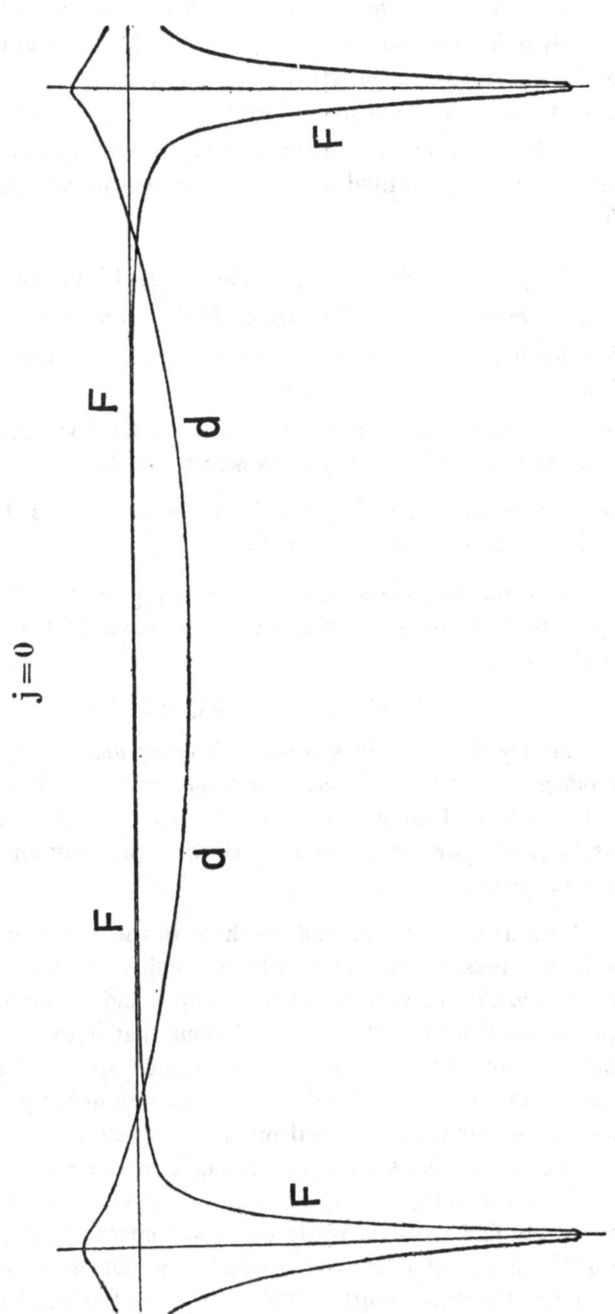
j = 0
F
d
F
d
F
F

and ending with zero. The change at each point of discontinuity C is merely a half-period change of phase. A slight alteration of this discontinuous curve within 60° on each side of each C, converts it into the continuous wavy curve of fig. 15, which represents the water-surface due to motion of speed $\sqrt{ga/9\pi}$ of the pressural forcive represented by the other continuous curve of fig. 15.

49. Every word of § 48 is applicable to figs. 14 and 13 except references to *speed* of the forcive, which is $\sqrt{ga/19\pi}$ for fig. 14 and $\sqrt{ga/41\pi}$ for fig. 13; and other statements requiring modification as follows:—

For $4\frac{1}{2}$ "periods" or "waves," in respect to fig. 15; substitute $9\frac{1}{2}$ in respect to fig. 14, and $20\frac{1}{2}$ in respect to fig. 13.

For "depression" in defining MM in respect to figs. 15, 14; substitute *elevation* in the case of fig. 13.

50. How do we know that, as said in § 48, the formula {(83), (86), (87)} gives for a wide range of about 120° on each side of $\theta = 180°$,

$$\mathrm{d}(\theta) \doteqdot (-1)^j\, \mathrm{d}(180°)\,.\,\sin(j+\tfrac{1}{2})\,\theta \quad \text{.........(88)},$$

which is merely §§ 48, 49 in symbols? *it being understood that* j *is any integer not* < 4; *and that* e *is* ·9, *or any numeric between* ·9 *and* 1? I wish I could give a *short* answer to this question without help of hydrokinetic ideas! Here is the only answer I can give at present.

51. Look at figs. 12—16, and see how, in the forcive defined by $e = {·}9$, the pressure is almost wholly confined to the spaces $\theta < 60°$ on each side of each of its maximums, and is very nearly null from $\theta = 60°$ to $\theta = 300°$. It is obvious that if the pressure were perfectly annulled in these last-mentioned spaces, while in the spaces within 60° on each side of each maximum the pressure is that expressed by (74), the resulting motion would be sensibly the same as if the pressure were throughout the whole space CC ($\theta = 0°$ to $\theta = 360°$), exactly that given by (74). Hence we must expect to find through nearly the whole space of 240°, from 60° to 300°, an almost exactly sinusoidal displacement of water-surface, having the wave-length $360°/(j+\frac{1}{2})$ due to the translational speed of the forcive.

52. I confess that I did *not* expect so small a difference from sinusoidality through the *whole* 240°, as calculation by {(83), (86), (87)} has proved; and as is shown in figs. 18, 19, 20, by the D-curve on the right-hand side of C, which represents in each case the value of

$$D(\theta) = \mathbf{d}(\theta) - (-1)^j \mathbf{d}(180°) \cdot \sin(j + \tfrac{1}{2})\,\theta \ldots\ldots(89),$$

being the difference of $\mathbf{d}(\theta)$ from one continuous sinusoidal curve. The exceeding smallness of this difference for distances from C exceeding 20° or 30°, and therefore through a range between CC of 320°, or 300°, is very remarkable in each case.

53. The dynamical interpretation of (88), and figs. 18, 19, 20, is this:—Superimpose on the solution {(83), (86), (87)} a "free wave" solution according to (73), taken as

$$-(-1)^j \mathbf{d}(180°) \cdot \sin(j + \tfrac{1}{2})\,\theta \ldots\ldots\ldots\ldots\ldots(90).$$

This approximately annuls the approximately sinusoidal portion between C and C shown in figs. 13, 14, 15; and approximately doubles the approximately sinusoidal displacement in the corresponding portions of the spaces CC, and CC on the two sides of CC. This is a very interesting solution of our problem § 36; and, though it is curiously artificial, it leads direct and short to *the* determinate solution of the following general problem of canal ship-waves:—

54. Given, as forcive, the isolated distribution of pressure defined in fig. 12, travelling at a given constant speed; required *the* steady distribution of displacement of the water in the place of the forcive, and before it and behind it; which becomes established after the motion of the forcive has been kept steady for a sufficiently long time. Pure synthesis of the special solution given in §§ 1—10 above, solves not only the problem now proposed, but gives the whole motion from the instant of the application of the moving forcive. This synthesis, though easily put into formula, is not easily worked out to any practical conclusion. On the other hand, here is my present short but complete solution of the problem of stable steady motion for which we have been preparing, and working out illustrations in §§ 32—53.

Continue leftward, indefinitely, as a curve of sines, the D-curve of each of figs. 18, 19, 20; leaving the forcive curve, F, isolated, as shown already in these diagrams. Or, analytically stated:—

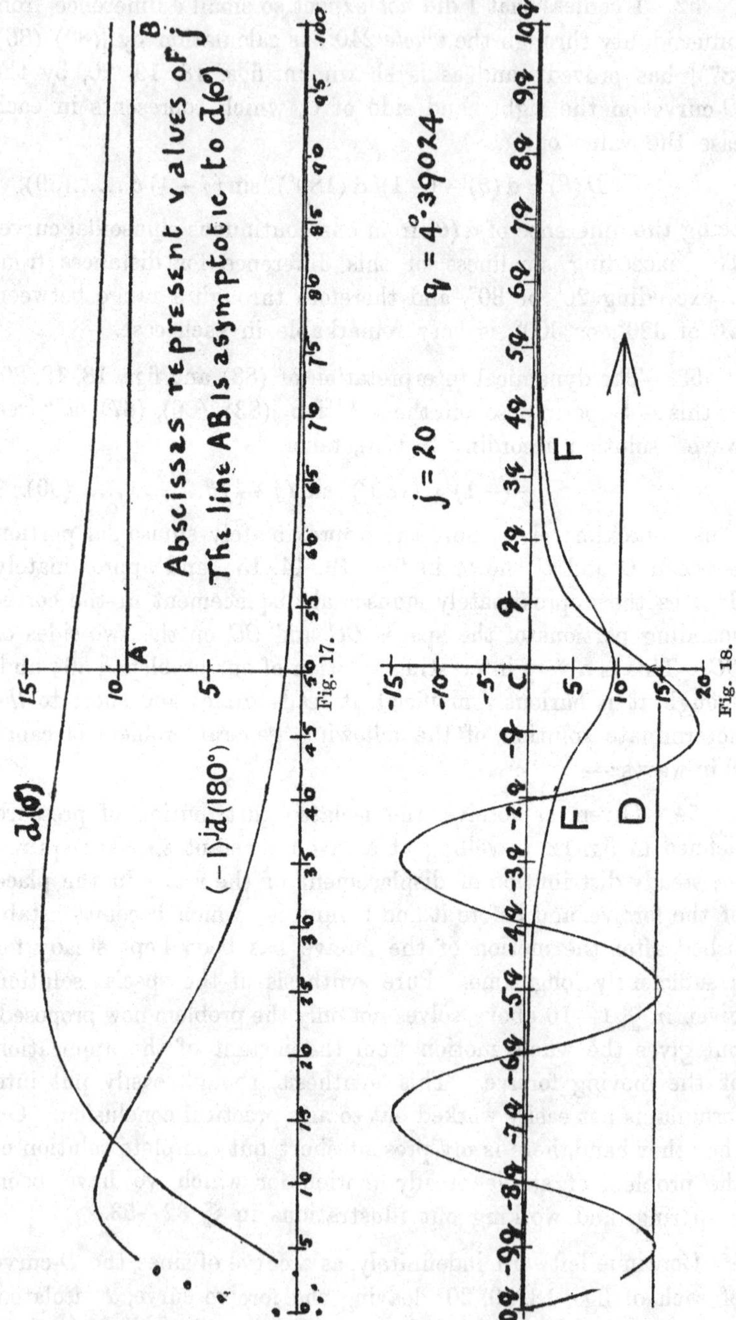
d(0°)
(−1)^j d(180°)
Abscissæ represent values of j
The line AB is asymptotic to d(0°)
A
B
j = 20
q = 4°·39024
F
D

Fig. 17.

Fig. 18.

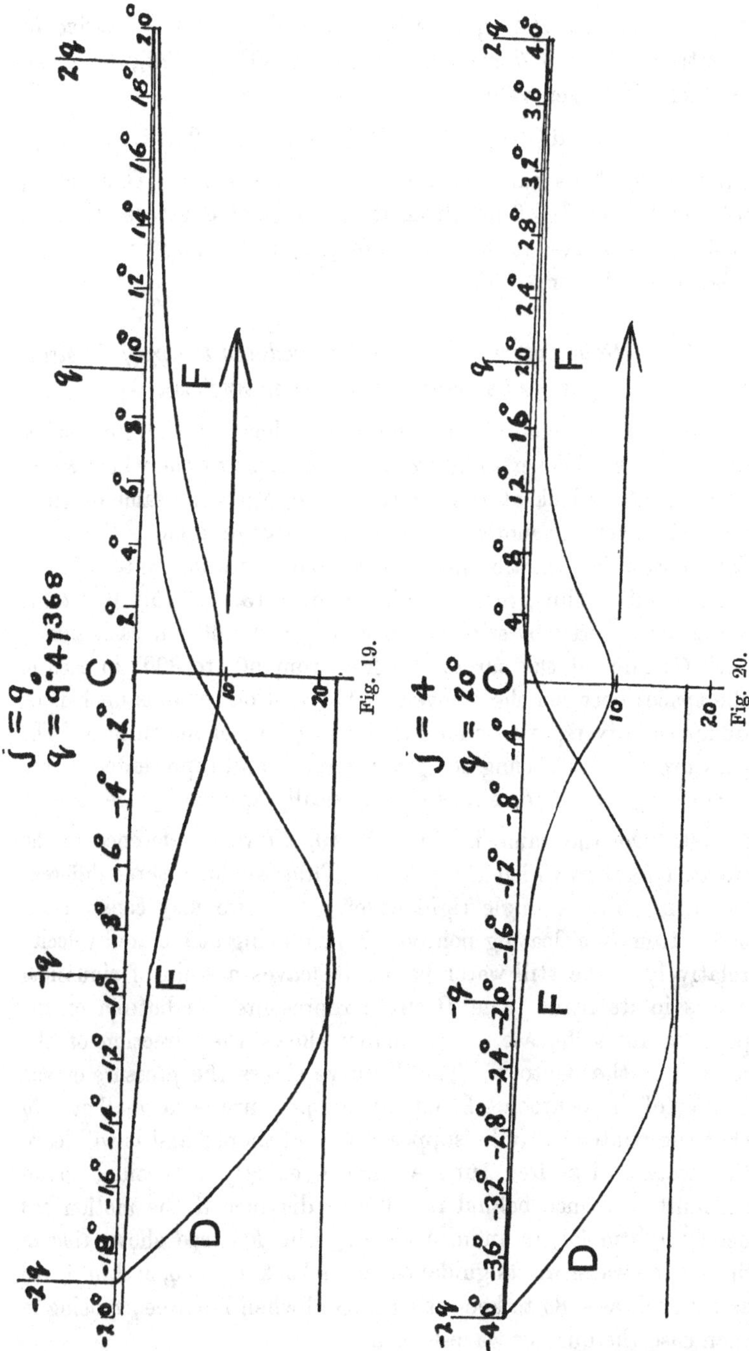

Fig. 19.

Fig. 20.

in (89) calculate the equal values of $\mathrm{d}(\theta)$ for equal positive and negative values of θ from 0° to 40° or 50° by {(83), (86), (87)}; and for all larger values of θ take

$$\mathrm{d}(\theta) \doteqdot (-1)^j \, \mathrm{d}(180°) \sin (j+\tfrac{1}{2}) \theta \text{ ....(91)},$$

where $\mathrm{d}(180°)$ is calculated by {(83), (86), (87)}. This used in (89), makes $D(\theta) \doteqdot 0$ for all positive values of θ greater than 40° or 50°; and makes it the double of (91) for all negative values of θ beyond − 40° or − 50°.

55, 56. *Rigid Covers or Pontoons, introduced to apply the given forcive (pressure on the water-surface).*

55. In any one of our diagrams showing a water-surface imagine a rigid cover to be fixed, fitting close to the whole water-surface. Now look at the forcive curve, F, on the same diagram, and wherever it shows no sensible pressure remove the cover. The motion (non-motion in some parts) of the whole water remains unchanged. Thus, for example, in figs. 13, 14, 15, 16, let the water be covered by stiff covers fitting it to 60° on each side of each C; and let the surface be free from 60° to 300° in each of the spaces between these covers. The motion remains unchanged under the covers, and under the free portions of the surface. The pressure Π constituting the given forcive, and represented by the F curve in each case, is now automatically applied by the covers.

56. Do the same in figs. 18, 19, 20 with reference to the isolated forcives which they show. Thus we have three different cases in which a single rigid cover, which we may construct as the bottom of a floating pontoon, kept moving at a stated velocity relatively to the still water before it, leaves a train of sinusoidal waves in its rear. The D curve represents the bottom of the pontoon in each case. The arrow shows the direction of the motion of the pontoon. The F curve shows the pressure on the bottom of the pontoon. In fig. 20 this pressure is so small at $-2q$ that the pontoon may be supposed to end there; and it will leave the water with free surface almost exactly sinusoidal to an indefinite distance behind it (infinite distance if the motion has been uniform for an infinite time). The F curve shows that in fig. 19 the water wants guidance as far back as $-3q$, and in fig. 18 as far back as $-8q$ to keep it sinusoidal when left free; q being in each case the quarter wave-length.

57—60. *Shapes for Waveless Pontoons, and their Forcives.*

57. Taking any case such as those represented in figs. 18, 19, 20; we see obviously that if any two equal and similar forcives are applied, with a distance $\frac{1}{2}\lambda$ between corresponding points, and if the forcive thus constituted is caused to travel at speed equal to $\sqrt{g\lambda/2\pi}$, being, according to (77) above, the velocity of free waves of length λ, the water will be left waveless (at rest) behind the travelling forcive.

58. Taking for example the forcives and speeds of figs. 18, 19, 20, and duplicating each forcive in the manner defined in § 57, we find (by proper additions of two numbers, taken from our tables of numbers calculated for figs. 18, 19, 20) the numbers which give the depressions of the water in the three corresponding waveless motions. These results are shown graphically in fig. 21, on scales arranged for a common velocity. The free wave-length for this velocity is shown as $4q$ in the diagram.

59. The three forcives, and the three waveless water-shapes produced by them, are shown in figs. 22, 23, 24 on different scales, of wave-length, and pressure, chosen for the convenience of each case.

60. As most interesting of the three cases take that derived from $j = 9$ of our original investigation. By looking at fig. 23 we see that a pontoon having its bottom shaped according to the D curve from $-3q$ to $+3q$, $1\frac{1}{2}$ free wave-lengths, will leave the water sensibly flat and at rest if it moves along the canal at the velocity for which the free wave-length is $4q$. And the pressure of the water on the bottom of the pontoon is that represented hydrostatically by the F curve.

61. Imagine the scale of abscissas in each of the four diagrams, figs. 21—24, to be enlarged tenfold. The greatest steepnesses of the D curve in each case are rendered sufficiently moderate to allow it to fairly represent a real water-surface under the given forcive. The same may be said of figs. 15, 16, 18, 19, 20; and of figs. 13, 14 with abscissas enlarged twentyfold. In respect to mathematical hydrokinetics generally; it is interesting to remark that a very liberal interpretation of the condition of infinitesimality (§ 36 above) is practically allowable. Inclinations to the horizon of as much as $\frac{1}{10}$ of a radian ($5°\cdot7$; or, say, $6°$), in any real case of

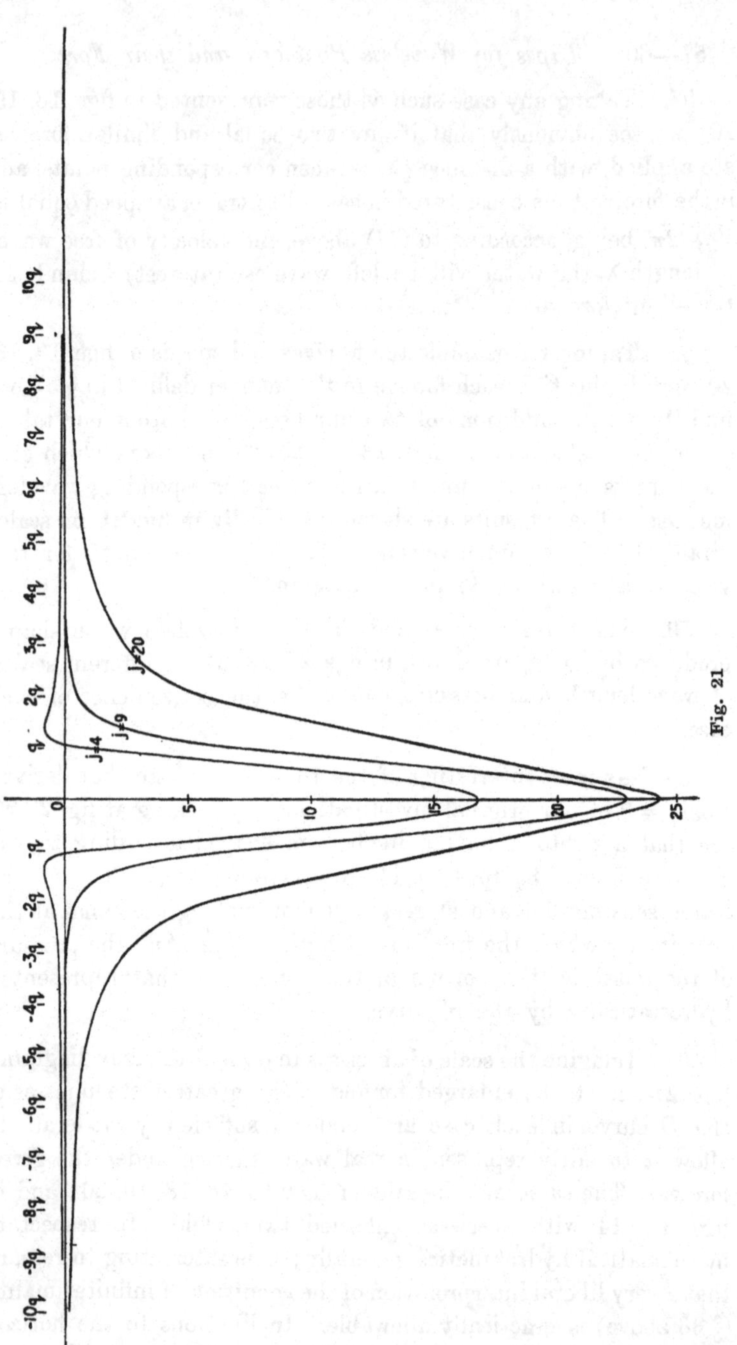

Fig. 21

water-waves or disturbances, will not seriously vitiate the mathematical result.

62. Fig. 17 represents the calculations of $\mathbf{d}(0^\circ)$ and $(-1)^j\mathbf{d}(180^\circ)$ for twenty-nine integral values of j; 0, 1, 2, 3, ... 19, 20, 30, 40, ... 90, 100; from the following formulas, found by putting $\theta=0^\circ$ and $\theta=180^\circ$; and with $e=\cdot 9$ in each case, and $c=1$

$$\mathbf{d}_j(0^\circ)=(2j+1)\,e^j\left[-\tfrac{1}{2}e^{\frac{1}{2}}\log\frac{1+\sqrt{e}}{1-\sqrt{e}}+1+\frac{e^{-1}}{3}+\frac{e^{-2}}{5}+\dots\right.$$
$$\left.+\frac{e^{-j+1}}{2j-1}+\tfrac{1}{2}\frac{e^{-j}}{2j+1}\right]\quad\dots(92),$$

$$\mathbf{d}_j(180^\circ)=(-1)^j(2j+1)\,e^j\left[\tfrac{1}{2}e^{\frac{1}{2}}\tan^{-1}\frac{2\sqrt{e}}{1-e}+1-\frac{e^{-1}}{3}+\frac{e^{-2}}{5}+\dots\right.$$
$$\left.+(-1)^{j-1}\frac{e^{-j+1}}{2j-1}+(-1)^j\tfrac{1}{2}\frac{e^{-j}}{2j+1}\right]\quad\dots(93).$$

The asymptote of $\mathbf{d}(0^\circ)$ shown in the diagram is explained by remarking that when j is infinitely great, the travelling velocity of the forcive is infinitely small; and therefore, by end of § 41, the depression is that hydrostatically due to the forcive pressure. This, at $\theta=0^\circ$, is equal to

$$\tfrac{1}{2}\frac{1+e}{1-e}c=\frac{1\cdot 9}{\cdot 2}c=9\cdot 5\,.\,c.$$

63. The interpretation of the curves of fig. 17 for points between those corresponding to integral values of j is exceedingly interesting. We shall be led by it into an investigation of the disturbance produced by the motion of a single forcive, expressed by

$$\Pi=\frac{gcb}{b^2+x^2}\quad\dots\dots\dots\dots(94);$$

but this must be left for a future communication, when it will be taken up as a preliminary to sea ship-waves.

64. The plan of solving by aid of periodic functions the two-dimensional ship-wave problem for infinitely deep water, adopted in the present communication, was given in Part III. of a series of papers on Stationary Waves in Flowing Water, published in the *Philosophical Magazine*, October 1886 to January 1887, with

analytical methods suited for water of finite depths. The annulment of sinusoidal waves in front of the source of disturbance (a bar across the bottom of the canal), by the superposition of a train of free sinusoidal waves which double the sinusoidal waves in the rear, was illustrated (December 1886) by a diagram [p. 295 *supra*] on a scale too small to show the residual disturbance of the water in front, described in § 53 above, and represented in figs. 18, 19, 20.

In conclusion, I desire to thank Mr J. de Graaff Hunter for his interested and zealous co-operation with me in all the work of the present communication, and for the great labour he has given in the calculation of results, and their representation by diagrams.

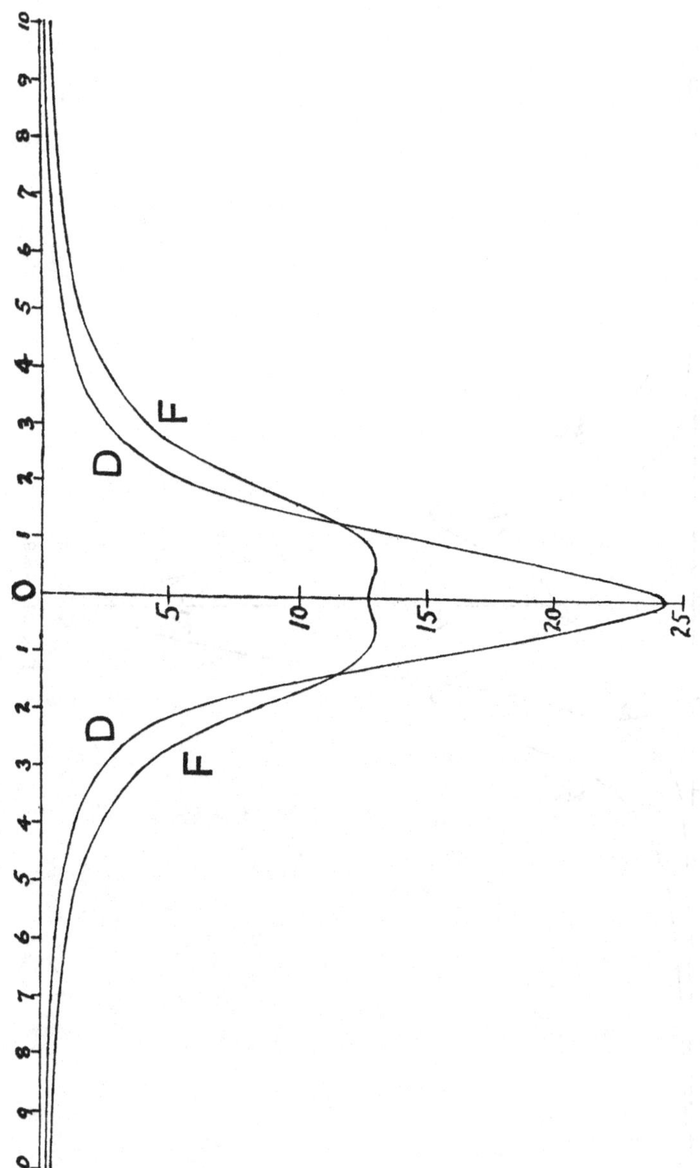

Fig. 22; $j=20$. Scale of abscissas is quarter-wave-lengths.

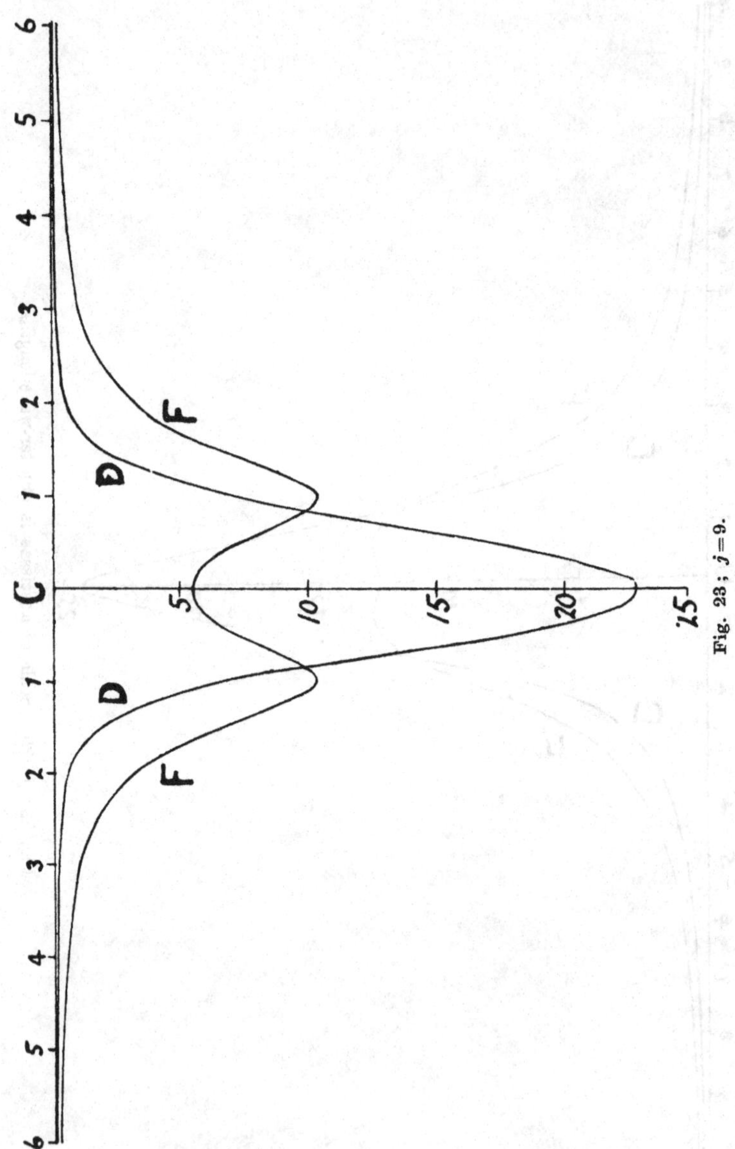

Fig. 23; $j=9$.

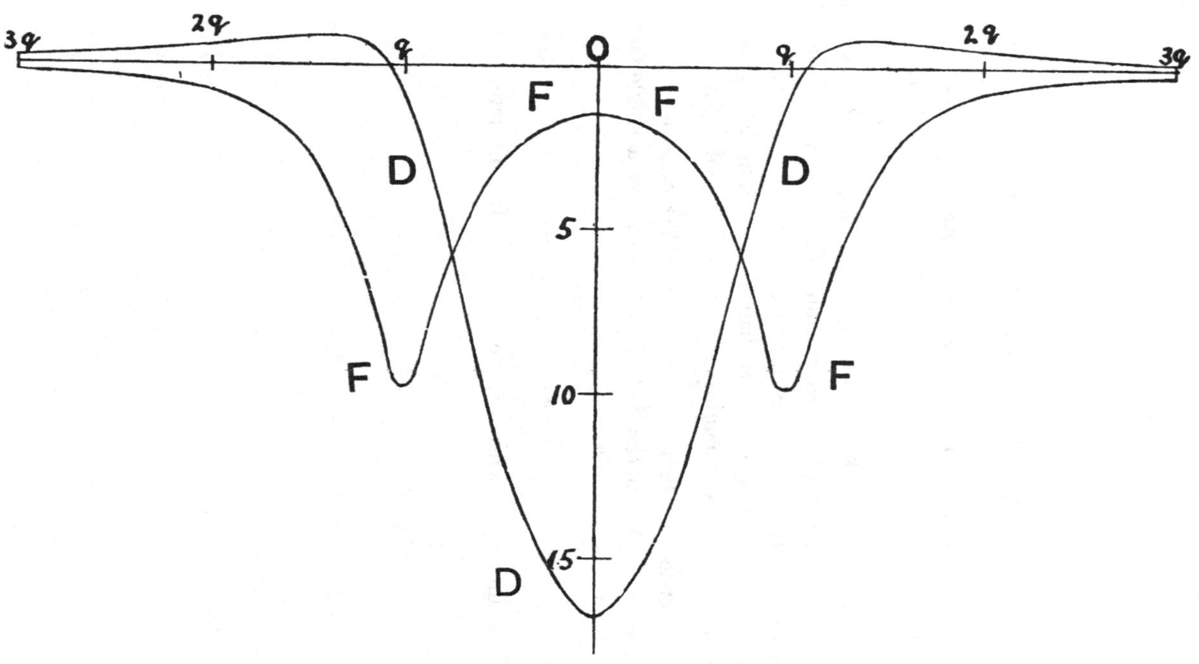

Fig. 24; $j=4$. Scale of abscissas is quarter-wave-lengths.

38. Deep Sea Ship-Waves.

[From the *Proceedings of the Royal Society of Edinburgh*, July 17, 1905; *Philosophical Magazine*, Vol. XI. January, 1906, pp. 1--25.]

65. Referring to § 63, we must, for the present, as time presses, leave detailed interpretation of the curves of fig. 17: merely remarking that, according to § 44, if $\delta = 0$ (which means that J is an integer), the disturbance, **d**, is infinitely great; of which the dynamical meaning is clear in (70) of § 39.

66. Let us now find the depression of the water at distance x from the origin, when the disturbance is due to a single forcive, expressed by the formula*

$$\Pi(x) = g\frac{kb^2}{x^2 + b^2} \quad \text{.....................(95)},$$

travelling uniformly with any velocity v. If this forcive were

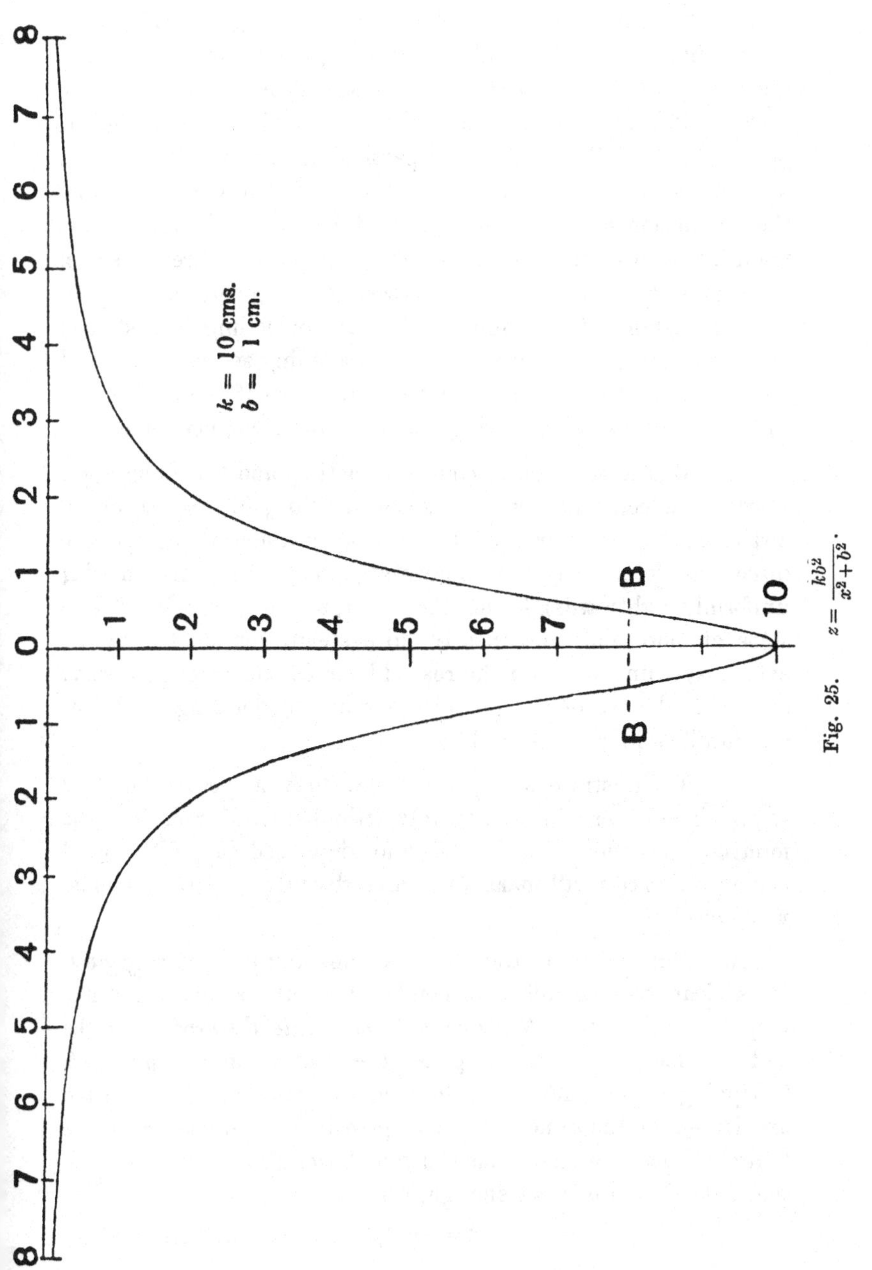

Fig. 25. $z = \frac{kb^2}{x^2 + b^2}$.

67. Now let the forcive be suddenly set in motion, and kept moving uniformly with any velocity v in the rightward direction of our diagrams. This will produce a great commotion, settling ultimately into more and more nearly steady motion through greater and greater distances from O. The investigation of §§ 1—10 above (Feb. 1904), and particularly the results described in §§ 5, 6, and illustrated in figs. 2, 3, show that in our present case the commotion, however violent, even if including *splashes* *, divides itself into two parts which travel away in the two directions from O, ultimately at wave-speed increasing in proportion to square root of distance (according to the law of falling bodies); and leaving in their rears, through ever broadening spaces, what would be more and more nearly absolute quiescence if the forcive were suddenly to cease after having acted for any time, long or short.

68. But if the forcive continues acting, and travelling rightwards with constant speed, v, according to § 67, the travelling away of the two parts of the initial commotion in the two directions from O (itself merely a point of reference, moving uniformly rightwards), leaves the water, as shown by fig. 26, in a state of more and more nearly quite steady motion through an ever broadening space on the rear side of O, and through a small space in advance of O; provided certain moderating conditions are fulfilled in respect to k, b, v.

69. To illustrate and prove § 68; first suppose v infinitely small. The water will be infinitely little disturbed from the static forcive-curve shown in fig. 25, and described in § 66. Small enough velocities will make very small disturbance with any finite value of kb.

70. But now go to the other extreme and let v be very great. It is clear, on dynamical principles without calculation, that v may be *great enough to make but very little disturbance* of the water-surface, however steep be the static forcive curve. A "skipping stone" and a ricochetting cannon shot, illustrate the application of the same dynamical principle in three-dimensional hydrokinetics. By mathematical calculation (§ 79 below) we shall see that, when v is great enough, we have

$$h \doteq 4\pi A/\lambda \quad \text{.......................(97)},$$

* However sudden and great the commotion is, the motion of the liquid is, and continues to be, irrotational throughout.

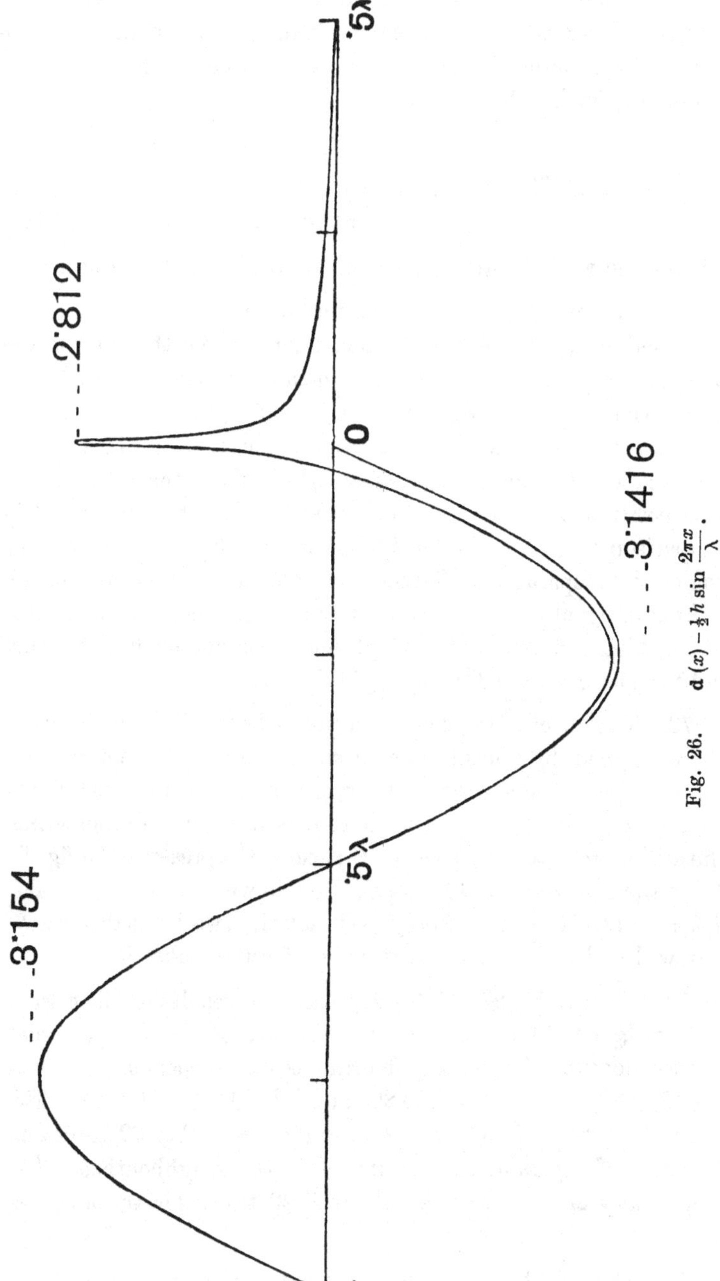

Fig. 26. $\mathbf{d}(x) - \frac{1}{2}h \sin \frac{2\pi x}{\lambda}$.

where h denotes the height of crests above mean water-level in the train of sinusoidal free waves left in the rear of the travelling forcive; A denotes the area of the forcive curve (fig. 25); being given in § 66 by the equation

$$A = \pi bk \quad \text{.........................(98):}$$

and λ, given [§ 39, (71)] by

$$\lambda = 2\pi v^2/g \quad \text{.......................(99),}$$

denotes the wave-length of free waves travelling with velocity v.

71. A very important theorem in respect to ship-waves is expressed by (97). Without calculation we see that, *if* λ *is very great* in comparison with πb (the "mean breadth" of the forcive curve according to § 66), h must be simply proportional to A, for different forcives travelling at the same speed. This we see because, for the same value of b, h/k is the same, and because superposition of different forcives within any breadth small in comparison with λ, gives for h the sum of the values which they would give separately. Farther without calculation, we can see, by imagining altered the scale of our diagrams, that $h\lambda/A$ must be constant. But without calculation I do not see how we could find the factor 4π of (97), as in § 79 below.

72. The effect of the condition prescribed in § 71 is illustrated and explained by considering cases in which it is not fulfilled. For example, let two forcives be superposed with their middles at distance $\frac{1}{2}\lambda$; they will give $h = 0$, that is to say no train of waves. The displaced water surface for this case is represented in fig. 27. Or let their distance be $\frac{1}{3}\lambda$ or $\frac{2}{3}\lambda$; the two will give the same value of h as that given by one only. Or let the two be at distance λ; they will make h twice as great as one forcive makes it.

73. In figs. 26, 27, 29, 30, representing results of the calculations of §§ 78, 79 below, the abscissas are all marked according to wave-length. The scale of ordinates corresponds, in each of figs. 26, 27, 29, to $k = 243{\cdot}89$, and $\pi b = 1{\cdot}0251 \,.\, 10^{-3} \,.\, \lambda$. This makes by (98) and (97) $A = \frac{1}{4}\lambda$, and $h = \pi$. Fig. 30 represents the curve of fig. 29 at the maximum, in the neighbourhood, of O, on a greatly magnified scale: about 1720 times for the abscissas, and 39 times for the ordinates.

74. Fig. 26 shows, on the right-hand side, the water slightly heaped up in front of the travelling forcive, which is a distribution

of *downward* pressure whose middle is at O. On the left side of O, we see the water surface not differing perceptibly from a curve of sines beyond half a wave-length rearwards from O. A small portion of a wave-length of a true curve of sines in the diagram shows how little the water's surface differs from the curve of sines at even so small a distance from O as a quarter-wave-length.

It must be remembered that in reality the water surface is everywhere very nearly level; and in considering, as we shall have to do later, the work done by the forcive, we must interpret properly the enormous exaggeration of slopes shown in the diagrams. It is interesting to remark that the static *depression*, k, which the forcive if at rest would produce, is about 87 times the *elevation* actually produced above O by the forcive, travelling at the speed at which free waves, of the wave-length shown in the diagrams, travel. It is interesting also to remark that the limitation to very small slopes is not binding on the static forcive curve. Thus for example, a distribution of static pressure, everywhere perpendicular to the free surface, producing static depression exactly agreeing with fig. 25, would, if caused to travel at a speed for which the free wave-length is very large in comparison with b, produce a disturbance, represented by fig. 26 with waves of moderate slopes: and, as said in § 69 above, would produce no disturbance at all if the speed of travelling were infinitely great.

75. Fig. 27 is interesting as showing the waveless disturbance produced by two equal and similar forcives with their middles at distance equal to half the wave-length. This disturbance is essentially symmetrical in front and rear of the middle between the two forcives. By dynamical considerations of the equilibrium of downward pressures, we see that the area of fig. 27 (portion above line of abscissas being reckoned as negative) must be exactly equal to $2A$, the sum of the areas of the two forcives, representing their integral amount of downward pressure. This area, being $2\pi bk$, with the numerical data of § 73, is numerically $\frac{1}{2}\lambda$; that is to say a rectangle whose length is $\frac{1}{2}\lambda$, and breadth the unit of our vertical scale. Approximate mensuration, with a very rough estimate of the area beyond the range of the diagram, continued to infinity on the two sides, verifies this conclusion.

76. Fig. 28 is designed on the same plan as fig. 27 but with eleven half-wave-lengths as the distance between the two forcives

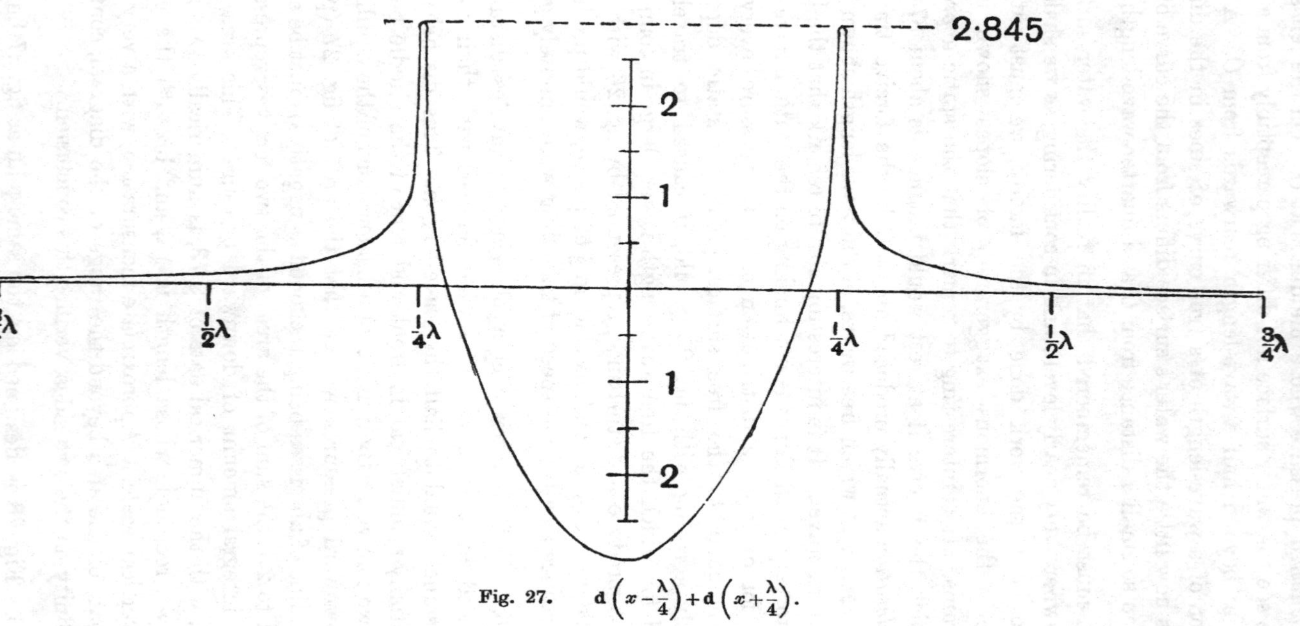

Fig. 27. $d\left(x-\frac{\lambda}{4}\right)+d\left(x+\frac{\lambda}{4}\right)$.

instead of one half-wave-length. Like fig. 27, it is symmetrical on the two sides of the middle of the diagram; but, instead of being waveless, as is fig. 27, it shows four and a half waves, all very approximately sinusoidal, with two depressional halves of waves at their two ends, and elevations coming asymptotically to zero beyond the two ends of the diagram. The curve represented by fig. 26 is very accurately the right-hand extreme of fig. 28: and the same figure, turned right to left, is the left-hand extreme of fig. 28. If we commence with the water wholly at rest, and start the forcives at the proper speed, with force gradually (or somewhat suddenly) increasing up to the prescribed amount, the motion produced will be that represented by fig. 28, with, superimposed upon it, a disturbance quickly disappearing in ever lengthening waves of diminishing amplitude, travelling away in both directions from our field. If now, with the regular *régime* represented by fig. 28, we suddenly cease to apply the forcives, we have left a free procession of four and a half very approximately sinusoidal waves, between a front and a rear deviating from sinusoidality as shown in the diagram. From the instant of being left free, the front of this procession and its rear will rapidly become modified: while for three periods the central part of the procession will have travelled three wave-lengths, with very little deviation from sinusoidality. But, after four or five periods from the instant of being left free, the whole procession will have got into confusion. After twenty or thirty or forty periods, the water will be sensibly quiescent, not only through the space where the procession was, but through a considerable part of the space over which it would have travelled if its front and rear had been kept guarded by the continued action of the two travelling forcives. At no time after the cessation of the forcives can we reasonably or conveniently assign a "group velocity" to the group or procession of waves with which we are concerned*. A prevalent idea is, I believe, that such a group of deep sea waves could be regarded as travelling with half the "wave-velocity" of waves of the length given in the original group. In § 30 above, reasons are given for accepting the theory of "group velocity" only in the case of mutually supporting groups, given by Stokes in his Smith's Prize examination paper, published in the *Cambridge University Calendar* for 1876: and for rejecting it for the case of any single group of waves. In reality the front

[* On this subject see references on p. 304 *supra*.]

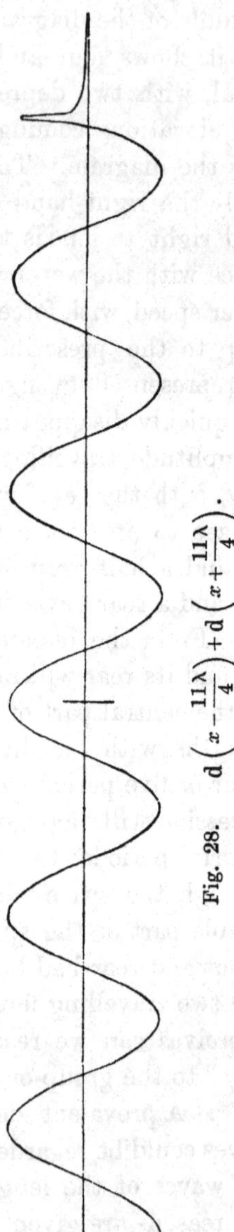

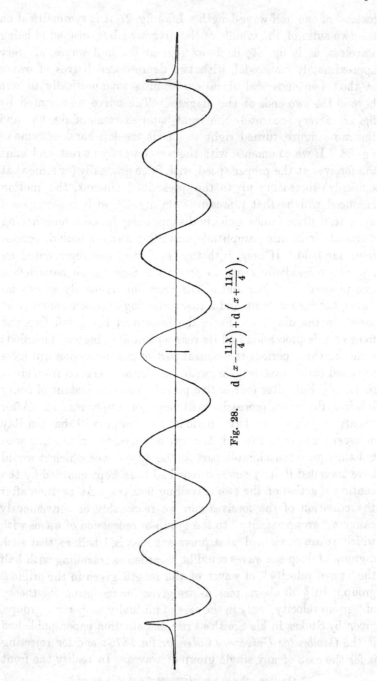

Fig. 28. $d\left(x-\frac{11\lambda}{4}\right)+d\left(x+\frac{11\lambda}{4}\right)$.

of a group, left to itself, travels with accelerated velocity exceeding the velocity of periodic waves of the given wave-length, instead of with half that velocity.

77. Fig. 29 shows *the steady motion*, symmetrical in front and rear of a single travelling forcive, which is a solution of our problem; but it is an unstable solution (as probably are the solutions of the problem of § 45 above, shown in figs. 13, 14, 15). If any large finite portion of the water is given in motion according to fig. 29, say, for example, 50 wave-lengths preceding O (the forcive) and 50 wave-lengths following O, the front of the whole procession, to the right of O, will become dissipated into non-periodic waves travelling rightwards and leftwards with increasing wave-lengths and increasing velocities; and the approximately steady periodic portion of it will shrink backwards relatively to the forcive. Thus before the forcive has travelled fifty wave-lengths, the periodic waves in front of it are all gone: but there is still irregular disturbance both before and behind it. After the forcive has travelled a hundred wave-lengths, the whole motion in advance of it, and the motion for perhaps 30 wave-lengths or more in its rear, will have settled to nearly the condition represented by fig. 26, in which there is a small regular elevation in advance of the forcive, and a regular train of approximately sinusoidal waves in its rear; these waves being of double the wave-height given originally. This motion, as said above in § 68, will go on, leaving behind the forcive a train of steady periodic waves, increasing in number; and behind these an irregular train of waves, shorter and shorter, and less and less high the farther rearward we look for them (see R in fig. 10 of §§ 26, 27 above). It is an interesting, but not at all an easy problem, to investigate the extreme rear (with practically motionless water behind it) of the train of waves in the wake of a forcive travelling uniformly for ever. I hope to return to this subject when we come to consider the work done by the travelling forcive.

78. Pass now to the investigation of the formulas by the calculation of which figs. 26, 27, 28, 29, 30 have been drawn, and the theorem of (97) is proved. Go back to the problem of § 41 above: but instead of taking $e = \cdot 9$, as in §§ 46—61, take $e = 1 - 10^{-4}$; and $c = 1/(2j+1)$. By (86) and (87) of § 45 we have the following solution

$$\mathrm{d} = \mathcal{I} + \mathcal{J} \qquad (100),$$

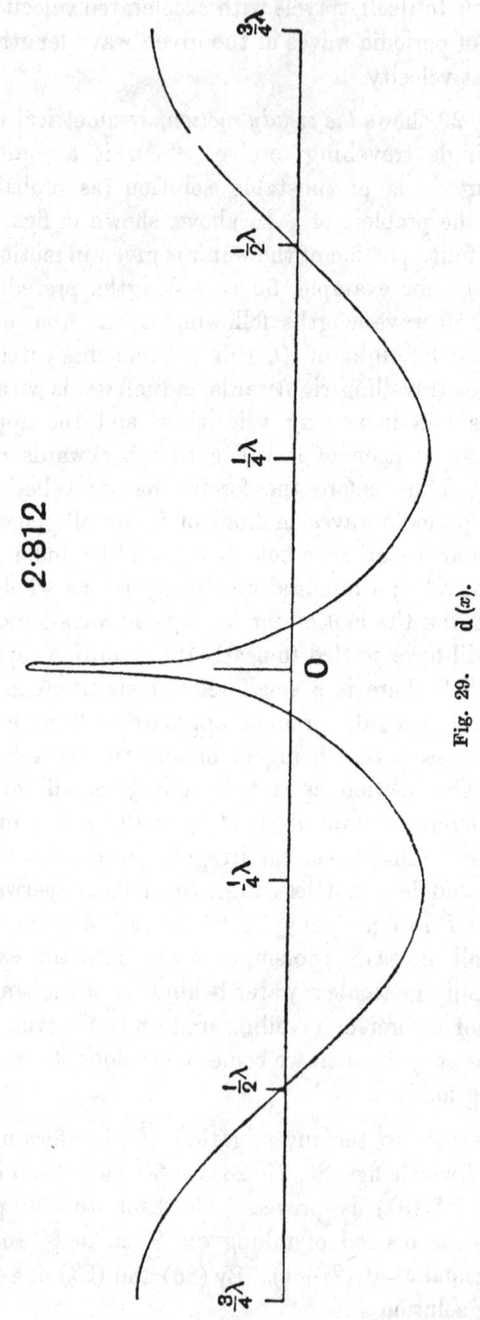

Fig. 29. d (x).

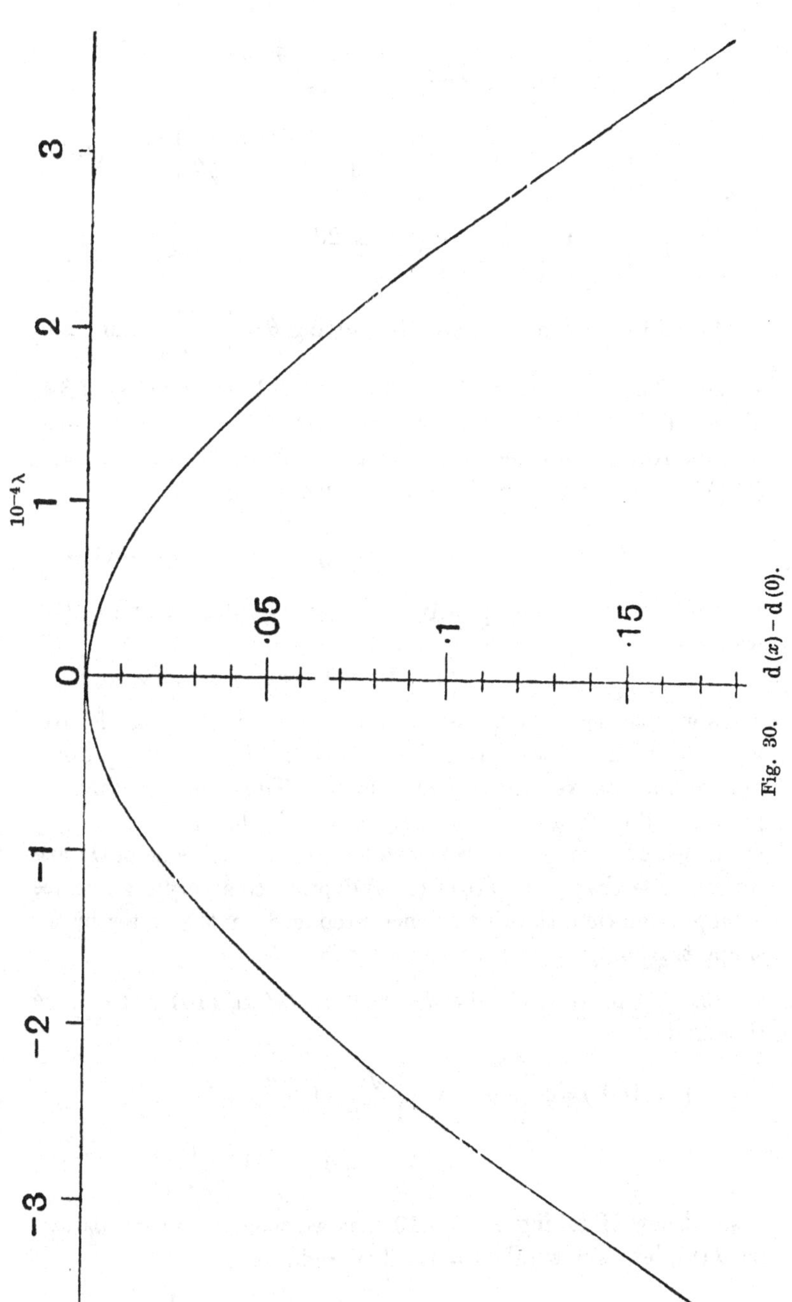

Fig. 30. d (x) − d (0).

where

$$\mathscr{I} = e^{j+\frac{1}{2}} \left\{ \tfrac{1}{2} \sin (j+\tfrac{1}{2})\theta \tan^{-1} \frac{2\sqrt{e}\sin\frac{1}{2}\theta}{1-e} \right.$$

$$\left. - \tfrac{1}{4} \cos (j+\tfrac{1}{2})\theta \log \frac{1+2\sqrt{e}\cos\frac{1}{2}\theta + e}{1-2\sqrt{e}\cos\frac{1}{2}\theta + e} \right\} \quad \text{...(101)}$$

and

$$\mathscr{K} = \tfrac{1}{2} \cdot \frac{1}{2j+1} + \frac{e\cos\theta}{2j-1} + \frac{e^2\cos 2\theta}{2j-3} + \ldots + e^j \cos j\theta \quad \text{...(102).}$$

Fig. 29 has been calculated by putting $\theta = \frac{x}{\lambda} \cdot \frac{360°}{j+\frac{1}{2}}$, and taking $j = 20$. The explanation is that, as we shall see by (78) of § 43 above, (100), (101), (102), express the water disturbance due to an infinite row of forcives at consecutive distances each equal to $(20\frac{1}{2})\lambda$; the expression for each forcive being

$$\frac{gcba/2\pi}{b^2+(x-na)^2} \quad \text{.....................(103),}$$

where n is zero or any positive or negative integer; and by (79) we have

$$b = 20{\cdot}5 \,.\, 10^{-4} \,.\, \lambda/2\pi \quad \text{..................(104).}$$

Thus we see that the pressure at O due to each of the forcives next to O, on the two sides, is $1/\{1+(2\pi\,.\,10^4)^2\}$ of the pressure due to the forcive whose centre is O. Thus we see that the pressures due to all the forcives, except the last mentioned, may be neglected through several wave-lengths on each side of O: and we conclude that (100), (101), (102) express, to a very high degree of approximation, the disturbance produced in the water by the single travelling forcive whose centre is at O.

79. To prove (97) take $\theta = 180°$ in (100), (101), (102); we thus find

$$(-1)^j \mathbf{d}(180°) = e^j \left\{ \tfrac{1}{2}\sqrt{e} \tan^{-1} \frac{2\sqrt{e}}{1-e} + 1 - \frac{e^{-1}}{3} + \frac{e^{-2}}{5} \ldots \right.$$

$$\left. + \tfrac{1}{2}(-e)^{-j} \cdot \frac{1}{2j+1} \right\} \quad \text{...(105).}$$

Instead now of taking $e = 1 - 10^{-4}$, as we took in our calculations for $\mathbf{d}(0)$, let us now take $e = 1$. This reduces (105) to

$$(-1)^j \mathbf{d}(180°) = \frac{\pi}{4} + 1 - \tfrac{1}{3} + \tfrac{1}{5} \ldots + \frac{(-1)^{j-1}}{2j-1} + \tfrac{1}{2}\frac{(-1)^j}{2j+1} \quad \text{...(106).}$$

Lastly take j an infinitely great odd or even integer, and we find

$$\mathrm{d}\,(180°) = (-1)^j \tfrac{1}{2}\pi \quad \text{..................(107).}$$

Now fig. 26 is, as we have seen, found by superimposing on the motion represented by fig. 29 an infinite train of periodic waves represented by $-\frac{1}{2}h \,.\, \sin(2\pi x/\lambda)$, and therefore $h = \pi$, which proves (97).

80. To pass now from the two-dimensional problem of canal-ship-waves to the three-dimensional problem of sea-ship-waves, we shall use a synthetic method given by Rayleigh at the end of his paper on "The form of standing waves on the surface of running water," communicated to the London Mathematical Society in December 1883*. In an infinite plane expanse of water, consider two or more forcives, such as that represented by (95) of § 66, with their horizontal medial generating lines in different directions through one point O, travelling with uniform velocity, v, in any direction. The superposition of these forcives, and of the disturbances of the water which they produce, each calculated by an application of (100), (101), (102), gives us the solution of a three-dimensional wave problem; which becomes the ship-wave-problem if we make the constituents infinitely small and infinitely numerous. Rayleigh took each constituent forcive as confined to an infinitely narrow space, and combated the consequent troublesome infinity by introducing a resistance to be annulled in interpretation of results for points not infinitely near to O. I escape from the trouble in the two-dimensional system of waves, by taking (95) to express the distribution of pressure in the forcive, and making b as small as we please. Thus, as indicated in §§ 79, 73, 76, by taking $b = 10\frac{1}{4}\lambda/(10^4 \,.\, \pi)$ we calculated a finite value for $\mathrm{d}\,(0)$. But for values of x, considerably greater than half a wave-length, we were able to simplify the calculations by taking $b = 0$.

81. For the three-dimensional system let, in fig. 31, ψ be the inclination to OX of the rearward wave-normal of one of the constituent systems of waves. This is also the inclination to OY of the medial line of the travelling forcive to which that set of waves is due. Take now for the forcive obtained by the

* *Proc. Lond. Math. Soc.*, 1883: republished in Rayleigh's *Scientific Papers*, Vol. II., Art. 109.

superposition of an infinite number of constituents, as described in § 80,

$$\frac{1}{g}\Pi(x, y) = \int_0^\pi d\psi \frac{b^2 k}{[(x\cos\psi + y\sin\psi)^2 + b^2]} \quad \text{...(108)},$$

where k may be a function of ψ, and b is the same for all values of ψ.

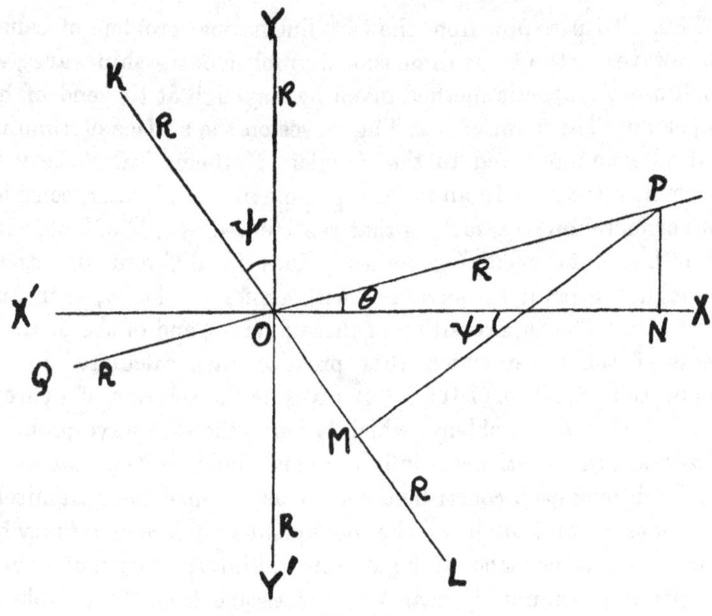

Fig. 31.

For the case of a circular forcive system we must take k constant; and we find

$$\frac{1}{g}\Pi(r) = \frac{\pi bk}{\sqrt{(r^2 + b^2)}} \text{ where } r^2 = x^2 + y^2 \quad \text{......(109)}.$$

82. Let now the forcive, whether circular or not, be kept travelling in the direction of x negative* with velocity v: and let λ denote the corresponding free wave-length given by the formula $2\pi v^2/g$. This is the wave-length of the constituent train of waves corresponding to $\psi = 0$. For the ψ-constituent, the component velocity perpendicular to the front is $v\cos\psi$, and the wave-length is $\lambda\cos^2\psi$. Looking now to fig. 26, with $\lambda\cos^2\psi$

* This is opposite to the direction of the motion of the forcive in fig. 26.

instead of λ; and to fig. 31; and to equations (97), (98); we see that the portion of the depression at (x, y) due to the constituent of forcive shown under the integral in (108) is

$$\frac{4\pi^2 bk \,.\, d\psi}{\lambda \cos^2\psi} \sin \frac{2\pi(x\cos\psi + y\sin\psi)}{\lambda\cos^2\psi} \quad \text{.........(110)},$$

provided $x\cos\psi + y\sin\psi$ is considerably greater than $\frac{1}{2}\lambda\cos^2\psi$. Hence for the depression at (x, y) due to the whole travelling forcive, we have

$$\mathbf{d}(x, y) = 4\pi^2 b \int_{-(\frac{1}{2}\pi-\theta)}^{\frac{1}{2}\pi} \frac{kd\psi}{\lambda\cos^2\psi} \sin \frac{2\pi(x\cos\psi + y\sin\psi)}{\lambda\cos^2\psi} \quad \text{.........(111)}.$$

83. The reason for choosing the limits $-(\frac{1}{2}\pi - \theta)$ to $\frac{1}{2}\pi$ is that each constituent forcive gives a train of sinusoidal waves in its rear, and no perceptible disturbance in its front at distances from it exceeding half a wave-length. Look now to fig. 31, and consider the infinite number of medial lines of the forcives included in the integrals (108), (111); all as lines passing through O. Four examples, QP, $Y'Y$, LK, XX' of these lines are shown in the diagram: corresponding respectively to $\psi = -(\frac{1}{2}\pi - \theta)$, $\psi = 0$, $\psi =$ any positive acute angle, $\psi = \frac{1}{2}\pi$. On each of the first three of these lines RR indicates the rear. The fourth, XX', is in the direction of the motion, and has neither front nor rear. The integral (111) must include all, and only all, the medial lines which have rears towards P. Hence QP is one limit of ψ in (111) because it passes through P; XX' is the other limit because it has neither front nor rear. Thus all the lines included in the integral, lie in the obtuse angle POX'. Thus the integral (111) expresses the depression at $P(x, y)$ due to the joint action of all the constituent forcives, because none except those whose medial lines lie in the angle POX' contribute anything to the disturbance of the water at P.

84. For interpreting and approximately evaluating the definite integral, we may conveniently put

$$r = \sqrt{x^2 + y^2}, \quad \text{and} \quad u = \frac{\cos(\psi - \theta)}{\cos^2\psi} \quad \text{........(112)},$$

and write (111) as follows:

$$\mathbf{d}(x, y) = 4\pi^2 b \int_{-(\frac{1}{2}\pi-\theta)}^{\frac{1}{2}\pi} \frac{kd\psi}{\lambda\cos^2\psi} \sin \frac{2\pi r u}{\lambda} \quad \text{...(113)}.$$

Now if we suppose r/λ very great, there will be exceedingly rapid transitions between equal positive and negative values of $\sin(2\pi ru/\lambda)$, *which will cause cancelling of all portions of the integral except those, if any there are, for which $du/d\psi$ vanishes.* We shall see presently that *there are* two such values, ψ_1, ψ_2, both real if $\tan\theta < \sqrt{\frac{1}{8}}$; u being a maximum (u_1) for one of them, and a minimum (u_2) for the other; and that, when θ has any value between $\tan^{-1}\sqrt{\frac{1}{8}}$ and $2\pi - \tan^{-1}\sqrt{\frac{1}{8}}$, the values of ψ_1, ψ_2 are both imaginary. Consideration of this last-mentioned case shows that, in the whole area of sea in advance of two lines through the centre of the travelling forcive inclined at equal angles of $\tan^{-1}\sqrt{\frac{1}{8}}$ (or 19° 28′), on each side of the mid-wake, there is no perceptible disturbance at distances of much more than a half wave-length from the centre of the forcive. The main disturbance by ship-waves, therefore, lies in the rearward angular space between these two lines. It is illustrated by fig. 32, as we now proceed to prove by the proper interpretation of (113). Expanding the argument of the sin in (113) by Taylor's theorem for values of ψ differing from ψ_1 by small fractions of a radian, we find

$$\frac{2\pi ru}{\lambda} \doteq \frac{2\pi r}{\lambda}\left[u_1 + \tfrac{1}{2}\left(\frac{d^2u}{d\psi^2}\right)_1(\psi - \psi_1)^2\right] = \alpha_1 - q_1^2 \ldots (114),$$

where

$$\alpha_1 = \frac{2\pi ru_1}{\lambda}, \quad \text{and} \quad q_1 = (\psi - \psi_1)\sqrt{\frac{\pi r}{\lambda}\left(-\frac{d^2u}{d\psi^2}\right)_1} \quad \ldots (115).$$

From the second of (115) we find $d\psi = dq_1/(\beta_1\sqrt{\pi})$, where

$$\beta_1 = \sqrt{\frac{r}{\lambda}\left(-\frac{d^2u}{d\psi^2}\right)_1} \quad \ldots\ldots\ldots\ldots\ldots\ldots (116).$$

Dealing similarly in respect to ψ_2 and values of ψ differing but little from it, we take $+q_2^2$ instead of the $-q_1^2$ of (114), and $(d^2u/d\psi^2)_2$ instead of the $-(d^2u/d\psi^2)_1$ of (115); because u_1 is the maximum and u_2 the minimum. Calling k_1, k_2 the values of k corresponding to ψ_1, ψ_2, and using these expressions properly in (113), we find, for the depression of the water at (x, y),

$$\mathsf{d}(x, y) = \frac{4b\pi^{3/2}}{\lambda}\left[\frac{k_1}{\beta_1\cos^2\psi_1}\int_{-\infty}^{\infty} dq_1 \sin(\alpha_1 - q_1^2) + \frac{k_2}{\beta_2\cos^2\psi_2}\int_{-\infty}^{\infty} dq_2 \sin(\alpha_2 + q_2^2)\right] \ldots (117).$$

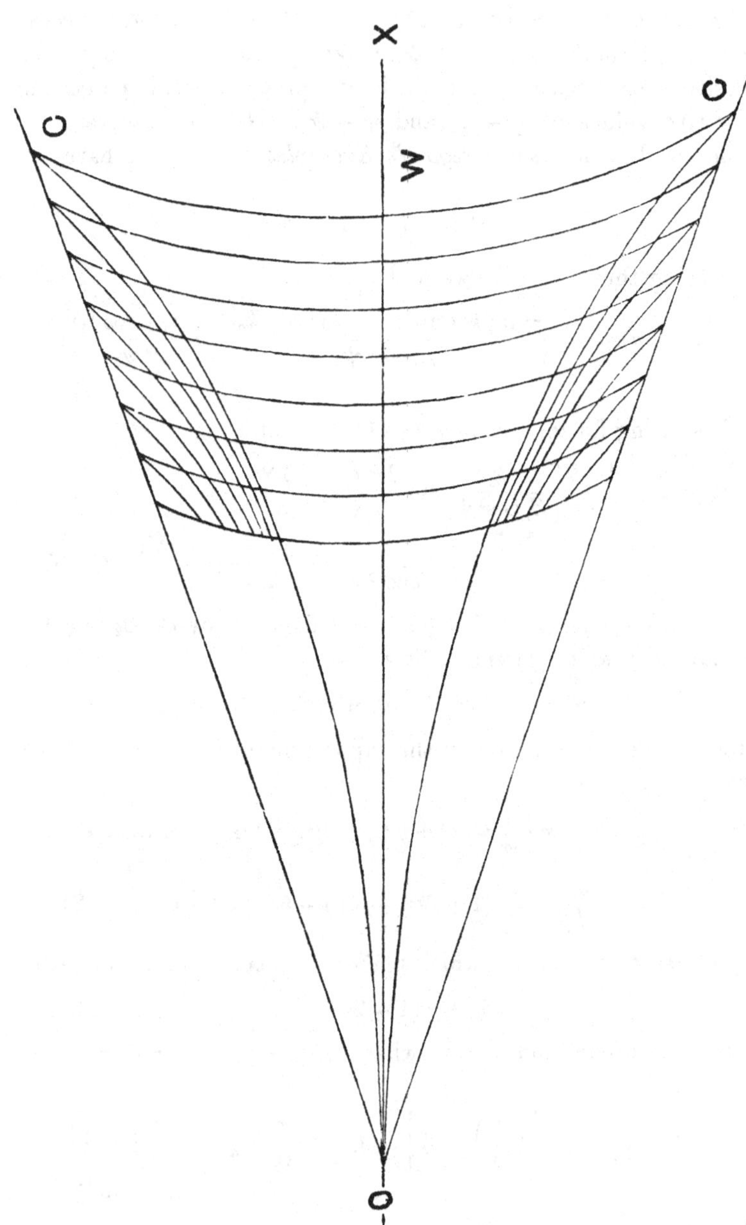

Fig. 32. Isophasal Curves.

The limits ∞, $-\infty$ are assigned to the integrations relatively to q_1 and q_2 because the greatness of r/λ in (115) and the corresponding formula relative to ψ_2, makes q_1 and q_2 each very great, (positive or negative), for moderate properly small positive or negative values of $\psi - \psi_1$ and $\psi - \psi_2$. Now as discovered by Euler or Laplace (see *Gregory's Examples*, p. 479), we have

$$\int_{-\infty}^{\infty} dq \sin q^2 = \int_{-\infty}^{\infty} dq \cos q^2 = \sqrt{\pi/2},$$

and using these in (117) we find

$$\mathbf{d}(x, y) = \frac{2\sqrt{2\pi^2 b}}{\lambda}\left[\frac{k_1(\sin\alpha_1 - \cos\alpha_1)}{\beta_1\cos^2\psi_1} + \frac{k_2(\sin\alpha_2 + \cos\alpha_2)}{\beta_2\cos^2\psi_2}\right] \quad \text{.........(118).}$$

Substituting for α_1, α_2 values by (115) we find

$$\mathbf{d}(x, y) = \frac{4\pi^2 b}{\lambda}\left[\frac{k_1}{\beta_1\cos^2\psi_1}\sin\frac{2\pi}{\lambda}\left(ru_1 - \frac{\lambda}{8}\right) + \frac{k_2}{\beta_2\cos^2\psi_2}\sin\frac{2\pi}{\lambda}\left(ru_2 + \frac{\lambda}{8}\right)\right] \text{...(118)}'.$$

85. To determine the quantities denoted by β_1, β_2 in (116)—(118)′, we write (112) as follows:—

$$ru = (x + yt)\sqrt{1 + t^2}, \text{ where } t = \tan\psi \text{.........(119).}$$

Hence, by differentiation on the supposition of x, y, r constant, we find

$$r\frac{du}{d\psi} = [xt + y(1 + 2t^2)]\sqrt{1 + t^2} \quad \text{.........(120).}$$

$$r\frac{d^2u}{d\psi^2} = [x(1 + 2t^2) + yt(5 + 6t^2)]\sqrt{1 + t^2} \text{......(121).}$$

By (120) we find for t_m, which makes u a maximum or minimum,

$$xt_m + y(1 + 2t_m^2) = 0 \text{.................(122)};$$

a quadratic equation which, when $(y/x)^2 < \frac{1}{8}$, has real roots as follows:—

$$t_1 = -\frac{x}{4y} - \sqrt{\left[\left(\frac{x}{4y}\right)^2 - \frac{1}{2}\right]}, \quad t_2 = -\frac{x}{4y} + \sqrt{\left[\left(\frac{x}{4y}\right)^2 - \frac{1}{2}\right]} \quad \text{.........(123).}$$

And substituting t_m (either of these) for t in (121), we find

$$r\left(\frac{d^2u}{d\psi^2}\right)_m = [x(1 - t_m^2) + 2yt_m]\sqrt{1 + t_m^2} \text{......(124),}$$

or with simplification by (119),

$$r\left(\frac{d^2u}{d\psi^2}\right)_m = 2ru_m - x(1+t_m^2)^{3/2} \qquad\text{.........(124)}'.$$

Eliminating t_m^2 from the first factor of (124) by (122) we find

$$r\left(\frac{d^2u}{d\psi^2}\right)_m = \left[\frac{3x}{2} + t_m\left(2y + \frac{x^2}{2y}\right)\right]\cdot\sqrt{1+t_m^2} \quad\text{...(124)}'',$$

which, with $m=1$, and $m=2$, gives β_1 and β_2 by (116).

86. Using (123) we see that $(d^2u/d\psi^2)_m$ vanishes when $x = y\sqrt{8}$, and that it is negative for t_1, and positive for t_2, when $x > y\sqrt{8}$. Hence t_1 makes $d^2u/d\psi^2$ negative. Therefore u_1 is the maximum; and t_2 makes it positive. Therefore u_2 is the minimum; and (119) gives for these maximum and minimum values

$$ru_1 = (x+yt_1)\sqrt{1+t_1^2}, \quad ru_2 = (x+yt_2)\sqrt{1+t_2^2} \quad\text{...(125)}.$$

By (122), (123) we see that when $y/x = 0$, we have $-t_1 = +\infty$, and $-t_2 = 0$. If we increase y from 0 to $+x/\sqrt{8}$, $-t_1$ falls continuously from ∞ to $\sqrt{\tfrac{1}{2}}$, and $-t_2$ rises continuously from 0 to $\sqrt{\tfrac{1}{2}}$. Thus $-t_1$ and $-t_2$ become, each of them, $\sqrt{\tfrac{1}{2}}$; which is the tangent of 35° 16′.

Geometrical digression on a system of autotomic, mono-parametric co-ordinates.* §§ 87—90.

87. In (119) put $\qquad ru = a \qquad$........................(126),

where a denotes the parameter OW of the curve OCC, fig. 32, which we are about to describe; being the curve given intrinsically by (119) and (122) with suffix 'm' omitted from t. In the present paper these curves may be called isophasals, because the argument of the sine in (130) below is the same for all points on any one of them.

Solving (119) and (122) for x and y, we find†

$$x = a\frac{1+2t^2}{(1+t^2)^{3/2}}, \quad y = a\frac{-t}{(1+t^2)^{3/2}} \qquad\text{...........(127)}.$$

* Of this kind of co-ordinates in a plane, we have a well-known case in the elliptic co-ordinates consisting of confocal ellipses and hyperbolas.

† [In the lecture on Ship-Waves (*supra*, No. 32, p. 307; see Pop. Lect. III., p. 482) a diagram of "echelon curves" is given, like fig. 32, with the line of cusps (naturally) at the same inclination 19° 28′; but the equations are different from (127), as representing the effect of a different travelling distribution of pressure. See J. H. Michell, *Phil. Mag.*, Vol. XLV., 1898, pp. 106—123; Lamb, *Hydrodynamics*, 3rd ed., 1906, § 253, also new matter in the German translation; and T. H. Havelock, *Proc. Roy. Soc.*, Aug., 1908, as *supra*, p. 304, who states that his

The largest of the eight curves shown in fig. 32 has been described according to values of x, y calculated from these two equations, by giving to $-t$ values tan 0°, tan 10°, tan 20°, ... tan 90°. The seven other isophasals partially shown in fig. 32, all similar to the largest, have been drawn to correspond to seven equidifferent smaller values, 19λ, 18λ ... 13λ, of the parameter a, if we make the largest equal to 20λ.

88. It is seen in the diagram that every two of these isophasals cut one another in two points, at equal distances on the two sides of OW. If we continue the system down to parameter 0, every point within the angle COC is the intersection of two and only two of the curves given by (127), with two different values of the parameter a. If we are to complete each curve algebraically, we must duplicate our diagram by an equal and similar pattern on the left of O: and the doubled pattern, thus obtained, would show a system of waves, equal and similar in the front and rear, which (§ 77 above) is possible but instable. We are, however, at present only concerned with the stable ship-waves contained in the angle $\pm 19° 28'$ on the two sides of the mid-wake; and we leave the algebraic extension with only the remark that all points in the angle COC of the diagram, and the opposite angle leftward of O, can be specified by real values of the parameter a: while imaginary values of it would specify real points in the two obtuse angles.

89. By differentiation of (127), we find

$$dx/dy = -t = -\tan\psi \qquad (128);$$

which proves that $\tan^{-1} t$ is the angle measured anti-clockwise from OY to the tangent to the curve at any point (x, y), in the lower half of the diagram. Elimination of t between the two equations of (127) gives, as the cartesian equation of our curve,

$$(x^2 + y^2)^3 + a^2(8y^4 - 20x^2y^2 - x^4) + 16a^4y^2 = 0 \qquad (129).$$

But the implicit equations (127) are much more convenient for all our uses. It is interesting to verify (129) for the case $-t = \pm\sqrt{\tfrac{1}{2}}$ in (127), corresponding to either of the two cusps shown in the diagram.

90. Going back now to § 86 and the continuous variations considered in it, we see that $-t_1$ and $-t_2$ are respectively the

equations of form are equivalent to Lord Kelvin's. The original report of the lecture included models expressing the law of amplitude of the waves, which are not reproduced in the reprint.]

tangents of the inclinations, reckoned from OY clockwise, of portions of the long arc OC and of the short arc WC, in the upper half of the diagram. Thus, if we carry a point from O to C in the long arc, and from C to W in the short arc, we have the change of inclinations to OY represented continuously by the decrease of $\tan^{-1}(-t_1)$ from 90° to 35° 16′, while y increases from 0 to $x\sqrt{8}$; and the farther decrease of $\tan^{-1}(-t_2)$ from 35° 16′ to 0°, while y diminishes from $x\sqrt{8}$ to 0 again. The inclination to OY of the two branches meeting in the cusp, C, is 35° 16′ (or $\tan^{-1}\sqrt{\frac{1}{2}}$). For any point in the short arc CWC of the curve u or $\cos(\psi-\theta)/\cos^2\psi$, is a minimum. In each of the long arcs u is a maximum. At every point of the curve the value of u, whether minimum or maximum, is a/r. Hence for different points of the curve, u is inversely proportional to the radius vector from O.

91. Going back to (118)′ we now see that for all points on any one of our curves, ru_1 and ru_2 have both the same value, being the parameter OW of the curve. The first part of (118)′ is one constituent of the depression at any point on either of the long arcs; and the second part of (118)′ is one constituent of the depression at any point on the short arc. Taking for example the largest of the curves shown in fig. 32, we now see that for any point of either of its long arcs, the second constituent of the depression of the water is to be calculated from the second part of (118); while for any point of its short arc, the second constituent of the depression is to be calculated from the first part of (118)′.

92. Explaining quite similarly the determination of $\mathbf{d}(x, y)$ for every point of each of the smaller curves which we see in the diagram cutting the longer arcs of the largest curve, we arrive at the following conclusions as *the complete solution of our problem.*

The whole system of standing waves in the wake of the travelling forcive is given by the superposition of constituents calculated according to (127), with greater and smaller values of the parameter a with infinitely small successive differences. Hence, what we see in looking at the waves from above is exactly a system of crossing hills and valleys, with ridges and beds of hollows, all shaped according to the isophasal curves shown in fig. 32. Looking at any one of the short arc-ridges and following it through the cusps, we find it becoming the middle line of a valley in each of the long arcs of the curve. And following a short arc mid-valley through the cusps, we find, in the continuation of the curve, two long ridges. Every ridge, long or short, is furrowed by valleys. All the curved ridges and valleys are parts of one continuous system of curves, illustrated by fig. 32 and expressed by the algebraic equation (129).

With these explanations we may write (118)′ as follows:

$$\mathrm{d}(x, y) = \frac{4\pi^2 bk \sec^2\psi}{\lambda\beta} \sin\frac{2\pi}{\lambda}\left(ru \pm \frac{\lambda}{8}\right) \quad\text{......(130)},$$

where

$$\beta = \sqrt{\frac{r}{\lambda}\left(\pm\frac{d^2u}{d\psi^2}\right)} \quad\text{..........................(131)}.$$

93. An important, perhaps the most important, feature of the wave-system which we actually see on the two sides of the mid-wake of a steamer travelling through smooth water at sea, or of a duckling* swimming as fast as it can in a pond, is the steepness of the waves in two lines which we know to be inclined at 19° 28′ to the mid-wake. The theory of this feature is expressed by the coefficient of the sine in (130), and is well illustrated by the calculation of $\sqrt{\frac{a}{\lambda}}\cdot\frac{\sec^2\psi}{\beta}$ for eleven points of any one of the curves of fig. 32, the results of which are shown in column 6 of the following table. They express the depression below, and elevation above mid-level, due to one constituent of the system of

Col. 1	Col. 2	Col. 3	Col. 4	Col. 5	Col. 6
$-\psi$	$\frac{x}{a}$	$\frac{y}{a}$	u	$\frac{r}{a}\frac{d^2u}{d\psi^2}$	$\sqrt{\frac{a}{\lambda}}\cdot\frac{\sec^2\psi}{\beta}$
0°	1·0000	0·0000	1·00000	1·00000	1·0000
10°	1·0145	·1685	·97239	·93782	1·0647
20°	1·0497	·3201	·91587	·73497	1·3210
30°	1·0825	·3750	·87290	·33333	2·3094
35° 16′	1·0887	·3849	·86602	0·00000	∞
40°	1·0826	·3773	·87225	− ·40830	2·6660
50°	1·0201	·3166	·93624	− 1·84070	1·7839
60°	·8750	·2165	1·10941	− 5·00003	1·7888
70°	·6441	·1100	1·53041	− 14·0987	2·2793
80°	·3421	·0297	2·91222	− 63·3341	4·1672
90°	0·0000	0·0000	∞	− ∞	∞

* In the case of even the highest speed attained by a duckling, this angle is perhaps perceptibly greater than 19° 28′, because of the dynamic effect of the capillary surface tension of water. See *Baltimore Lectures*, p. 593 (letter to Professor Tait, of date 23rd Aug. 1871) and pp. 600, 601 (letter to William Froude, reprinted from *Nature* of 26th Oct. 1871).

crossing hills and valleys described in § 92. Column 1 is $-\psi$. Columns 2, 3 are x/a and y/a, calculated from (127). Column 4 is u, calculated by (126) from columns 2, 3. Column 5 is $r/a \,.\, d^2u/d\psi^2$, calculated from (124)′ and columns 2, 4. Column 6 is $\sqrt{\frac{a}{\lambda}} \cdot \frac{\sec^2\psi}{\beta}$, calculated from (131) and column 1. u, being, as we have seen, a minimum for values of $-\psi$ from 0 to 35° 16′, and a maximum for values from this to 90°, we see that the proper suffix in columns 4, 6, for the first four lines of each column is 2, and for the last six lines is 1.

94. In (130), k is generally a function of ψ; but if the forcive is circular (§ 81 above), k is a constant, and for points on one of the isophasal curves (a = constant) the only variable coefficients of the sine are $\sec^2\psi$, and β^{-1}. But for different isophasal curves the coefficient in (130) expressing the magnitude of the range above and below mean level, varies inversely as $\sqrt{a}$. For mid-wake ($\psi = 0$) a is simply the distance from the forcive: and we conclude, not merely for our point-forcive, but for a great ship, that the waves at a very large number of wave-lengths right astern, are smaller in height inversely as the square root of the distance from the forcive or from the middle of the ship.

95. The infinity for $\psi = \pm 35° 16'$ represents a feature analogous to a caustic in optics. There is in nature no infinity for either case, if the source is finite and distributed, not infinitely intense and confined to an infinitely small space. According to the methods followed in §§ 1—80 above, we have in every case a finite intensity of source, or of forcive, except in § 80 where we have supposed b infinitely small in comparison with λ, and we avoid the infinity shown in column 6: and can, by great labour, calculate a table of mitigated numbers, rising to a very large maximum at $\psi = \pm 35° 16'$; but not to infinity; and so arrive mathematically at an expression for the very high waves seen on the two bounding lines of the wave-disturbance, inclined at 19° 28′ to the mid-wake. But it is interesting to remember that we see in reality a considerable number of white-capped waves (would-be infinities) before the well-known large glassy waves which form so interesting a feature of the wave-disturbances.

§§ 80—95 of the present paper are merely a working out of the simple problem of purely gravitational waves with no surface-tension on the principle given by Rayleigh* in 1883 for the much more complex problem of capillary waves in front, in which surface-tension is the chief constituent of the forcive, and waves in the rear, in which the chief constituent of the forcive is gravitational.

In all the work arithmetical, algebraic, graphic of §§ 32—95 above, I have had much valuable assistance from Mr J. de Graaff Hunter; who has just now been appointed to a post in the National Physical Laboratory.

* *Proc. Lond. Math. Soc.*, Vol. xv., pp. 69—78, 1883; reprinted in Lord Rayleigh's *Scientific Papers*, Vol. II., pp. 258—267.

39. Initiation of Deep-Sea Waves of Three Classes: (1) from a Single Displacement; (2) from a Group of Equal and Similar Displacements; (3) by a Periodically Varying Surface-Pressure.

[From *Proc. Roy. Soc. Edin.*, Jan. 22, 1906; *Phil. Mag.*, Vol. XIII., Jan. 1907, pp. 1—36.]

(1) *Disturbance due to an Initiational Form more convenient than that of* §§ 3—31 *of previous Papers on Waves.* §§ 96—113.

96. The investigations of §§ 5—31, including the "front and rear" of infinitely long free processions of waves in deep water, are all founded on initiational disturbances, according to the first of two typical forms described in §§ 3, 4. In this form the initial disturbance is everywhere elevation or everywhere depression, and its amount, at great distances from the origin varies inversely as the square root of the distance ρ, from a point at a small height h above the water-surface in the middle of the disturbance. In the present paper a new form of type-disturbance is derived indifferently from either the first or the second, of the forms of §§ 3, 4: from the first, by double differentiation with reference to time, t; from the second, by single differentiation with reference to space, x.

97 (being a repetition of §§ 1, 2, slightly modified with respect to notation). Consider a frictionless incompressible liquid (called water for brevity) in a straight canal, infinitely long and infinitely deep, with vertical sides. Let it be disturbed from its level by any change of pressure on the surface, uniform in every line perpendicular to the plane sides, and let it be left to itself under constant air pressure. It is required to find the displacement

and velocity of every particle of water at any future time. Our initial condition will be fully specified by a given normal component of velocity, and a given normal component of displacement, at every point of the surface.

Taking O, any point at a distance h above the undisturbed water level, draw OX parallel to the length of the canal, and OZ vertically downwards. Let ξ, ζ be the displacement components, and $\dot{\xi}$, $\dot{\zeta}$ the velocity components, of any particle of the water whose undisturbed position is (x, z). We suppose the disturbance infinitesimal; by which we mean that the change of distance between any two particles of water is infinitely small in comparison with their undisturbed distance; and the line joining them experiences changes of direction which are infinitely small in comparison with the radian. Water being assumed incompressible and frictionless, its motion, started primarily from rest by pressure applied to the free surface, is essentially irrotational. Hence we have

$$\xi = \frac{d}{dx} F(x, z, t); \quad \zeta = \frac{d}{dz} F(x, z, t); \quad \dot{\xi} = \frac{d}{dx} \dot{F}; \quad \dot{\zeta} = \frac{d}{dz} \dot{F} \quad \text{.........(132)},$$

where $F(x, z, t)$, or F as we may write it for brevity when convenient, is a function which may be called the displacement-potential; and $\dot{F}(x, z, t)$ is what is commonly called the velocity-potential. Thus a knowledge of the function F, for all values of x, z, t, completely defines the displacement and the velocity of the fluid. And towards the determination of F we have, in virtue of the incompressibility of the fluid,

$$\frac{d^2F}{dx^2} + \frac{d^2F}{dz^2} = 0 \quad \text{....................(133)}.$$

In virtue of this equation, the well-known primary theory of Gauss and Green shows that, if F is given for every point of the free surface of the water, and is zero at every point infinitely distant from it the value of F is determinate throughout the fluid. The motion being infinitesimal, and the density being taken as unity, an application of fundamental hydrokinetics gives

$$p - \Pi = g(z - h + \zeta) - \frac{d^2F}{dt^2} = g(z - h) + g\frac{dF}{dz} - \frac{d^2F}{dt^2} \quad \text{...(134)},$$

where g denotes gravity; Π the uniform atmospheric pressure on the free surface; and p the pressure at the point $(x, z+\zeta)$ within the fluid.

98. Suppose now that $F(x, z, t)$ is a function which, besides satisfying (133), satisfies also the equation

$$g\frac{dF}{dz}=\frac{d^2F}{dt^2}\text{.....................(135)};$$

we see by (134) that the corresponding fluid motion of which F is the displacement-potential (132), has constant pressure over every surface $(z+\zeta)$; that is to say, every surface which was level when the water was undisturbed. Thus our problem of finding any possible infinitesimal irrotational motion of the fluid, in which the free surface is under any constant pressure, is solved by finding solutions of (133) and (135).

99. Now by differentiation we verify that, as found in § 3 above,

$$F=(z+\iota x)^{-\frac{1}{2}}\epsilon^{\frac{-gt^2}{4(z+\iota x)}}\text{ ....(136)}$$

satisfies (133) and (135). By changing ι into $-\iota$, and by integrations or differentiations performed on (136), according to the symbol $\frac{d^{i+j+k}}{dt^i dx^j dz^k}$, where i, j, k are any integers positive or negative, we can derive from (136) any number of imaginary solutions. And by addition of these, with constant coefficients, we can find any number of realised solutions. If, as in § 97, we regard any one of the formulas thus obtained as a displacement-potential, then by taking d/dz of it we find ζ, the vertical component displacement, which we shall take as the most convenient expression in each case for the solutions with which we are concerned. Or we may, if we please, take any solution of (135) as representing, not a displacement-potential, but a velocity-potential, or a horizontal component of displacement or velocity, or a vertical component of displacement or velocity.

100. Thus it was that in § 12 we took

$$\left.\begin{aligned} -\zeta = \{RS\}\sqrt{2}\,(z+\iota x)^{-\frac{1}{2}}\,\epsilon^{\frac{-gt^2}{4(z+\iota x)}} & \\ =\phi\,(x, z, t) \qquad = \sqrt{\frac{2}{\rho}}\cos\left(\frac{gt^2 x}{4\rho^2}-\tfrac{1}{2}\chi\right)\epsilon^{\frac{-gt^2 z}{4\rho^2}} & \\ \text{where } \rho=\sqrt{(z^2+x^2)}, \text{ and } \chi=\tan^{-1}(x/z) & \end{aligned}\right\}\text{(137)},$$

and in all of §§ 1—31, this notation ϕ and $-\zeta$ was consistently used, with $-\zeta$ to denote, when positive, upward displacement of the water (represented by upward ordinates in the drawings). In the two curves of § 4, fig. 1, that which has its maximum over O represents (137), for $t=0$. The other curve of fig. 1, with positive and negative ordinates on the two sides of O, represents (137), with $-\{RD\}$ instead of $\{RS\}$. The symbols $\{RS\}$ and $\{RD\}$ were introduced in § 3 above; $\{RS\}$ to denote a realisation by taking half the sum of what is written after it with $\pm\iota$, and $\{RD\}$ to denote a realization by taking $1/2\iota$ of the formula written after it minus $1/2\iota$ of the same formula with $+\iota$ changed into $-\iota$. A new curve in which the ordinates are numerically equal to $\frac{1}{\sqrt{2}}\frac{d}{dx}$ of the ordinates of the second of the old curves of fig. 1, is now given in the accompanying diagram, fig. 33; and close above it the first of the old curves of fig. 1 is reproduced, with ordinates reduced in the ratio $2\sqrt{2}$ to 1, for the sake of comparison with the new curve. This new curve represents the more convenient initiational form referred to in the title of the present paper.

Its equation, found by taking $t=0$ in (139) or in (144) [most easily from the imaginary form of (139)], is as follows:

$$\psi\,(x, z, 0) = \frac{1}{2\sqrt{2}}\frac{\sqrt{(\rho+z)}}{\rho^3}(2z-\rho)\,\ldots\ldots\ldots(138).$$

101. The original derivation of the new particular solution (which we shall call ψ) from the primary (136), as indicated in § 100, is shown by the following formula:

$$\left.\begin{aligned}\psi(x, z, t) &= \{RD\} \frac{d}{dx} \frac{-1}{\sqrt{(z+\iota x)}} \epsilon^{\frac{-gt^2}{4(z+\iota x)}} \\ &= \frac{1}{2\rho^{3/2}} \left\{\cos\left(\frac{gt^2 x}{4\rho^2} - \tfrac{3}{2}\chi\right) - \tfrac{1}{2}\frac{gt^2}{\rho}\cos\left(\frac{gt^2 x}{4\rho^2} - \tfrac{5}{2}\chi\right)\right\} \epsilon^{\frac{-gt^2 z}{4\rho^2}} \\ \text{where}\quad \rho &= \sqrt{(z^2 + x^2)}, \text{ and } \chi = \tan^{-1}(x/z)\end{aligned}\right\} \quad \text{.........(139).}$$

An equivalent formula for the same derivation, which will be found more convenient in §§ 135—157 below, is as follows:

$$\psi(x, z, t) = \{RS\} \frac{1}{g} \frac{d^2}{dt^2} \frac{-1}{\sqrt{(z+\iota x)}} \epsilon^{\frac{-gt^2}{4(z+\iota x)}} = \frac{-1}{g\sqrt{2}} \frac{d^2}{dt^2} \phi(x, z, t) \quad \text{.........(140).}$$

The equivalence of (139) and (140) is easily proved by remarking that by (133) and (135),

$$\frac{dF}{dx} = \iota \frac{dF}{dz} = \frac{\iota}{g} \frac{d^2F}{dt^2} \quad \text{..................(141),}$$

and therefore

$$\{RD\} \frac{d}{dx} \frac{-1}{\sqrt{(z+\iota x)}} \epsilon^{\frac{-gt^2}{4(z+\iota x)}} = \{RS\} \frac{1}{g} \frac{d^2}{dt^2} \frac{-1}{\sqrt{(z+\iota x)}} \epsilon^{\frac{-gt^2}{4(z+\iota x)}} \quad \text{...(142).}$$

102. Look now to fig. 33, and see within how narrow a space, say from $x = -2$ to $x = +2$, in the new curve, the main initial disturbance is confined, while in the old curve it spreads so far and wide that at $x = \pm 20$ it amounts to about ·16 of the maximum disturbance in the middle, and according to the law of inverse proportion to square root of distance, which holds for large values of x for the old curve, at $x = 80$ it would still be as much as ·1 of the maximum. The comparative narrowness of the initial disturbance represented by the new curve, and the ultimate law of decrease according to $x^{-\frac{3}{2}}$ (instead of $x^{-\frac{1}{2}}$ for the old curve) are great advantages of the new curve in the applications and illustrations of the theory to be given in §§ 135—157 below.

103. Remark also that the total area of the old curve from $-\infty$ to $+\infty$ is infinitely great, while it is zero for the new curve.

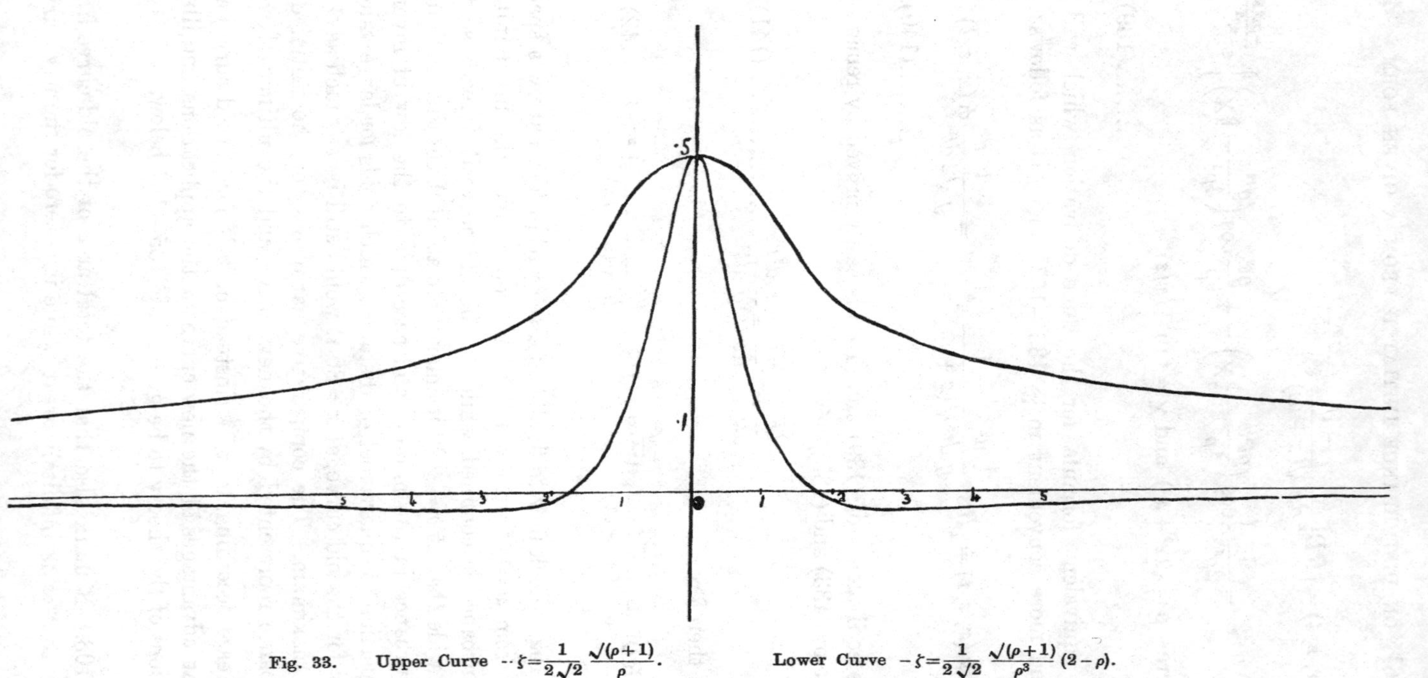

Fig. 33. Upper Curve $-\zeta = \frac{1}{2\sqrt{2}} \frac{\sqrt{(\rho+1)}}{\rho}$. Lower Curve $-\zeta = \frac{1}{2\sqrt{2}} \frac{\sqrt{(\rho+1)}}{\rho^3}(2-\rho)$.

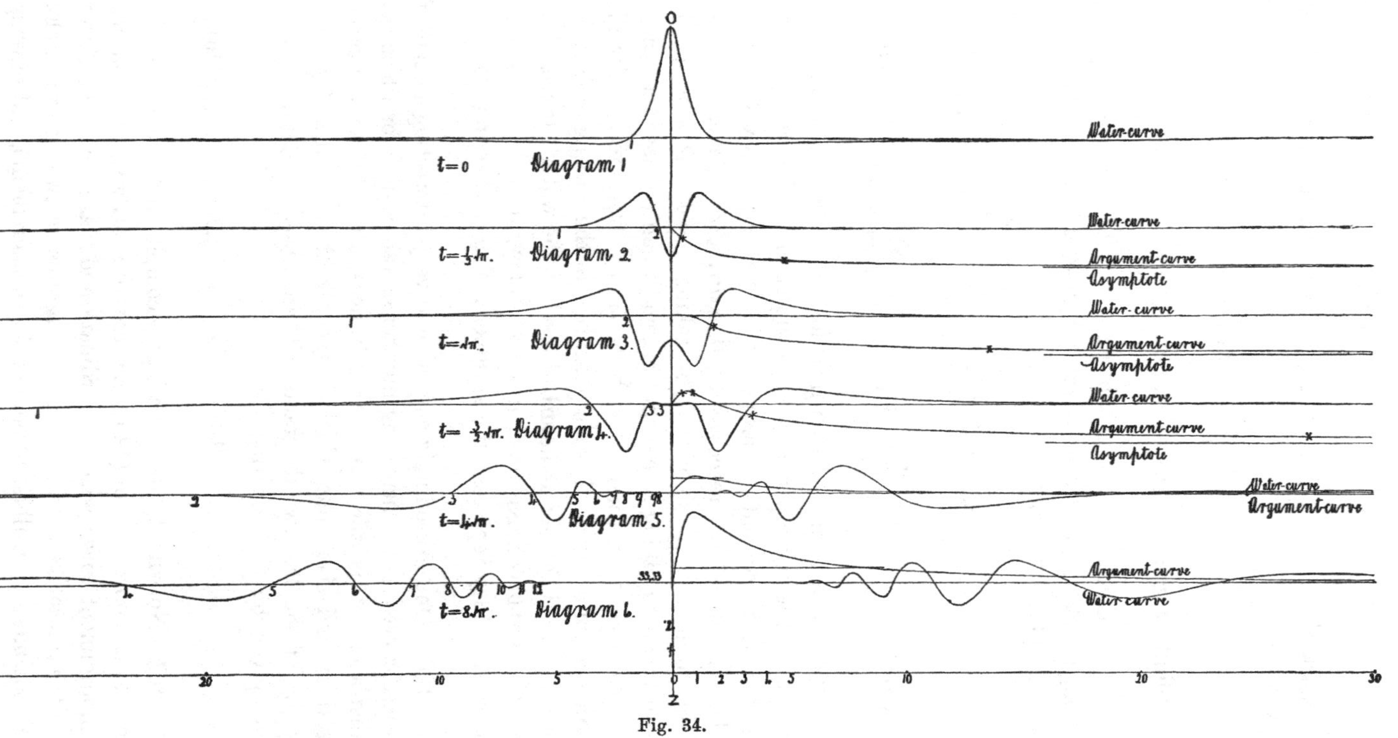

Water-curve
t=0 Diagram 1
Water-curve
Argument-curve
Asymptote
Diagram 2
Water-curve
Argument-curve
Asymptote
Diagram 3
Water-curve
Argument-curve
Asymptote
Diagram 4
Water-curve
Argument-curve
Diagram 5
Argument-curve
Water-curve
Diagram 6

Fig. 34.

Remark also that the potential energy of the initial disturbance, being

$$\tfrac{1}{2}g\int_{-\infty}^{+\infty} dx\,[\zeta(x, 1, 0)]^2 \quad\text{.................. (143)},$$

is infinitely great for the old curve, while for the new it is finite.

104. Equation (139) may be written in the following modified form, which is more convenient for some of our interpretations and graphic constructions:

$$\psi(x, z, t) = -\frac{\sqrt{\{x^2 + (\tfrac{1}{2}gt^2 - z)^2\}}\,\epsilon^{\frac{-gt^2 z}{4\rho^2}}}{2\rho^{5/2}}\cos A \quad\text{...... (144)},$$

where

$$A = \frac{gt^2 x}{4\rho^2} - \tfrac{3}{2}\chi - \tan^{-1}\frac{gt^2\sin\chi}{gt^2\cos\chi - 2\rho} \quad\text{...... (145)}.$$

105. The main curves, which for brevity we shall call water-curves in the accompanying six diagrams of fig. 34, represent the surface displacements according to our new solution $\psi(x, z, t)$ for the six values of t respectively, 0, $\tfrac{1}{2}\sqrt{\pi}$, $\sqrt{\pi}$, $\tfrac{3}{2}\sqrt{\pi}$, $4\sqrt{\pi}$, $8\sqrt{\pi}$. The formulas are simplified by taking $g = 4$. This is merely equivalent to taking as our unit of length half the space descended in one second of time, by a body falling from rest under the influence of gravity. For simplification in the writing of formulas we take $z = 1$ for the undisturbed level of the water-surface. The subsidiary curves, explained in § 107 below, are called argument-curves, as they represent the argument of the cosine in (144).

106. One exceedingly curious and very interesting feature of these curves is the increasing number of values of x for which the displacement is zero as time advances, and the large figures, sixteen and sixty-four, which it reaches at the times, $4\sqrt{\pi}$ and $8\sqrt{\pi}$, of the last two diagrams. These zeros, for any value of t, are given by the equation

$$A = \tfrac{1}{2}(2i + 1)\pi \quad\text{.................... (146)}.$$

107. Notwithstanding the highly complicated character of the function represented in (145), the zeros are easily found by tracing an argument-curve, with A as ordinate, and x as abscissa (as shown on the x-positive halves of the six diagrams on two different scales chosen merely for illustration, not for measurement), and drawing

parallels to the abscissa line at distances from it representing $-\frac{3}{2}\pi$, $-\frac{1}{2}\pi$, $\frac{1}{2}\pi$, $\frac{3}{2}\pi$, $\frac{5}{2}\pi$, etc. A parallel at distance $-\frac{7}{4}\pi$ is an asymptote to each of the argument-curves, and is shown in diagrams 2, 3, 4, on one scale of ordinates. The parallel corresponding to distance $\frac{13}{2}\pi$ is shown in the fifth and sixth diagrams, on the smaller scale of ordinates used in their argument-curves.

108. The first diagram shows zeros at $x = \pm\sqrt{3}$, of which that at $x = -\sqrt{3}$ is marked 1. In the second diagram the argument-curve indicates zeros for the $-\frac{3}{2}\pi$ and $-\frac{1}{2}\pi$ parallels, which are seen distinctly on the water-curve. The zero corresponding to the $-\frac{1}{2}\pi$ parallel was formed at the origin at the time when $\frac{1}{2}gt^2$ was equal to z, that is, when t was $1/\sqrt{2}$, or ·707. It is a coincidence of two zeros for x-positive and x-negative.

Diagram No. 3 shows that, shortly before its time, a maximum has come into existence in the argument-curve, which still indicates only two zeros. These are marked by crosses.

Diagram No. 4 shows that, in the interval between its time and the time of No. 3, two zeros of the water-curve for x-positive have come into existence. These and the corresponding zeros for x-negative are seen distinctly on the water-curve; and their indications for x-positive are marked by four crosses on the argument-curve.

Diagram No. 5 shows that, between its time and that of No. 4, twelve fresh zeros have come into existence on each side of OZ, one pair of which is indicated for example on the argument-curve by the parallel $\frac{13}{2}\pi$. Nine only out of all the sixteen zeros on either side are perceptible on the water-curve. The seven imperceptible zeros, on each side, all lie between $x = 0$ and $x = \pm 1$.

Diagram No. 6 shows that, between its time and that of No. 5, forty-eight fresh zeros for x-positive have come into existence, one pair of which is indicated by the parallel $\frac{13}{2}\pi$. Fourteen only out of all the sixty-four zeros on each side are perceptible on the water-curve. Thirty-one of the fifty imperceptible zeros on each side lie between $x = 0$ and $x = \pm 1$.

109. After the time $1/\sqrt{2}$, the zeros originate in pairs on the two sides of the origin* (x-positive and x-negative): those on the positive side by the two intersections of one of the parallels corresponding to $(2i+1)\pi/2$ with the argument-curve. The maximum of the argument-curve travels slowly in the outward direction towards $x=1$ as time advances to infinity. At times $4\sqrt{\pi}$ and $8\sqrt{\pi}$, of diagrams 5 and 6, it has reached so close to $x=1$ that this point has been regarded as the actual position of the maximum, both for the purpose of drawing the curve, and for the determination of the total number of zeros.

110. Each zero which originates according to an intersection on the outward side of the argument-curve travels outwards with increasing velocity to infinity, as time advances. Each of the others of the pairs of zeros, that is to say, each zero originating according to an intersection on the inward side of the argument-curve, travels very slowly inwards with velocity diminishing to nothing as time advances to infinity. Thus the motion of the water in the space between $x=-1$ and $x=+1$ becomes more and more nearly an increasing number of inward travelling waves, with lengths slowly diminishing to zero; and, as we see by the exponential factor in (144), with amplitudes and with slopes also slowly diminishing to zero: as time advances to infinity.

111. The semi-period of one of these quasi standing waves is, as we find from (139), approximately equal to $\frac{2\pi\rho^2}{gtx}$ when the time is so far advanced that $\frac{1}{2}gt^2$ is very great in comparison with ρ. Thus we see that the period is infinite at the origin. This agrees with the history of the whole motion at the origin, which, as we see by putting $x=0$ in (139), with $z=1$ and $g=4$, is expressed by the formula

$$-\zeta = \tfrac{1}{2}(1-2t^2)\,\epsilon^{-t^2} \quad \text{.................}(147).$$

The motion of the water in the space between $x=-1$ and $x=+1$ is of a very peculiar and interesting character. Towards a full

* If we continue the argument-curve to the side of the origin for x-negative, we must include large negative values of i in (146): but for simplicity we have confined the argument-curve to positive values of x.

understanding of it, it may be convenient to study the simplified approximate solution

$$-\zeta \doteqdot -\frac{t^2}{\rho^{5/2}}\cos\left(\frac{t^2x}{\rho^2}-\tfrac{5}{2}\tan^{-1}x\right)\epsilon^{\frac{-t^2}{\rho^2}} \quad\text{.........(148),}$$

which the realised part of (139) gives when $\frac{1}{2}gt^2$ is very large in comparison with ρ.

112. The outward travelling zeros on the two sides, beyond the distances ± 1 from the origin, divide the water into consecutive parts, in each of which it is wholly elevated or depressed. These parts we may call half-waves. They travel outwards with ever-increasing length and propagational velocity. Each of the half-waves developed after $t=\sqrt{\pi}$, as it travels outward, increases at first to a maximum elevation or maximum depression, and after that diminishes to zero as time advances to infinity.

113. It is interesting to trace the progress of each of the zeros in the intervals between the times of our six diagrams. This is facilitated by the numbers marked on several of the zeros in the different diagrams. Thus, confining our attention to the left-hand side of fig. 34, we see in diagram 1 a single zero numbered 1. The future zeros are to be numbered in the order of their coming into existence, 2; 3, 3; 4, 4; ...; 10, 10; ...; 33, 33; ... all in pairs after zero 2. Thus diagram 2 shows zero 1 considerably advanced leftwards (that is, outwards); and zero 2 beginning its outward progress. Diagram 3 shows zeros 1 and 2 each advanced farther outwards, 1 farther than 2. Diagram 4 shows all the zeros which have come into existence at time $\frac{3}{2}\sqrt{\pi}$. These are zeros 1 and 2, both farther outwards than at time $\sqrt{\pi}$, and a pair, 3, 3, which have come into existence shortly before the time $\frac{3}{2}\sqrt{\pi}$. The outer of these two travels outwards and the inner inwards. Some time later 4, 4 come into existence between 3 and 3: later still 5, 5 come into existence between 4 and 4.

In diagram 5, zero 1 has passed out of range leftwards: but we see distinctly the outward zeros 2, 3, 4, 5, 6, 7, 8, 9, and indications of the inward zeros 9, 8. The whole train of zeros for time $4\sqrt{\pi}$, shown and ideally continued to the middle by numbers, is 1, 2, 3, 4, 5, 6, 7, 8, 9, 9, 8, 7, 6, 5, 4, 3; sixteen in all.

Zero 3 has passed out of the range of diagram 6, but we see in it distinctly the outward zeros 4, 5, ... 12, and an indication of the pair 33, 33, which has come into existence before the time $8\sqrt{\pi}$. The whole train of zeros for time $8\sqrt{\pi}$, indicated by numbers, is 1, 2, ... 32, 33, 33, 32, ... 4, 3; sixty-four in all.

(2) *Illustrations of the Indefinite Extension and Multiplication of a Group of Two-Dimensional, Deep-Sea Waves Initially Finite in Number.* §§ 114—117.

114. The water is left at rest and free, after being initially displaced to a configuration of a finite number of sinusoidal mountains and valleys—five mountains and four valleys; in the diagrams placed before the Society. The initial group of waves, shown in diagram 1, of fig. 35, is formed by placing side by side, at distances equal to z (taken as unity), nine of the curves of diagram 1, fig. 34, alternately positive and negative. Diagrams 2 and 3, of fig. 35, are made by corresponding superpositions of the curves of diagrams 5 and 6, of fig. 34. Thus what, according to the known law of deep-sea periodic waves (§ 19 above), would be definitely and precisely the wave-length, if the numbers of crests and hollows were infinitely great, would be 2; and as we are taking $g=4$, the period would be $\sqrt{\pi}$, and the propagational velocity would be $2/\sqrt{\pi}$.

115. Immediately after the water is left free, the disturbance begins analysing itself into two groups of waves, seen travelling in contrary directions from the middle line of the diagram. The perceptible fronts of these two groups extend rightwards and leftwards from the end of the initial single static group, far beyond the "hypothetical fronts," supposed to travel at half the wave-velocity, which (according to the dynamics of Osborne Reynolds and Rayleigh, in their important and interesting consideration of the work required to feed a uniform procession of water-waves) would be the actual fronts *if* the free groups remained uniform. How far this *if* is from being realised is illustrated by the diagrams of fig. 35, which show a great extension outwards in each direction far beyond distances travelled at half the "wave-velocity." While there is this great extension of the fronts

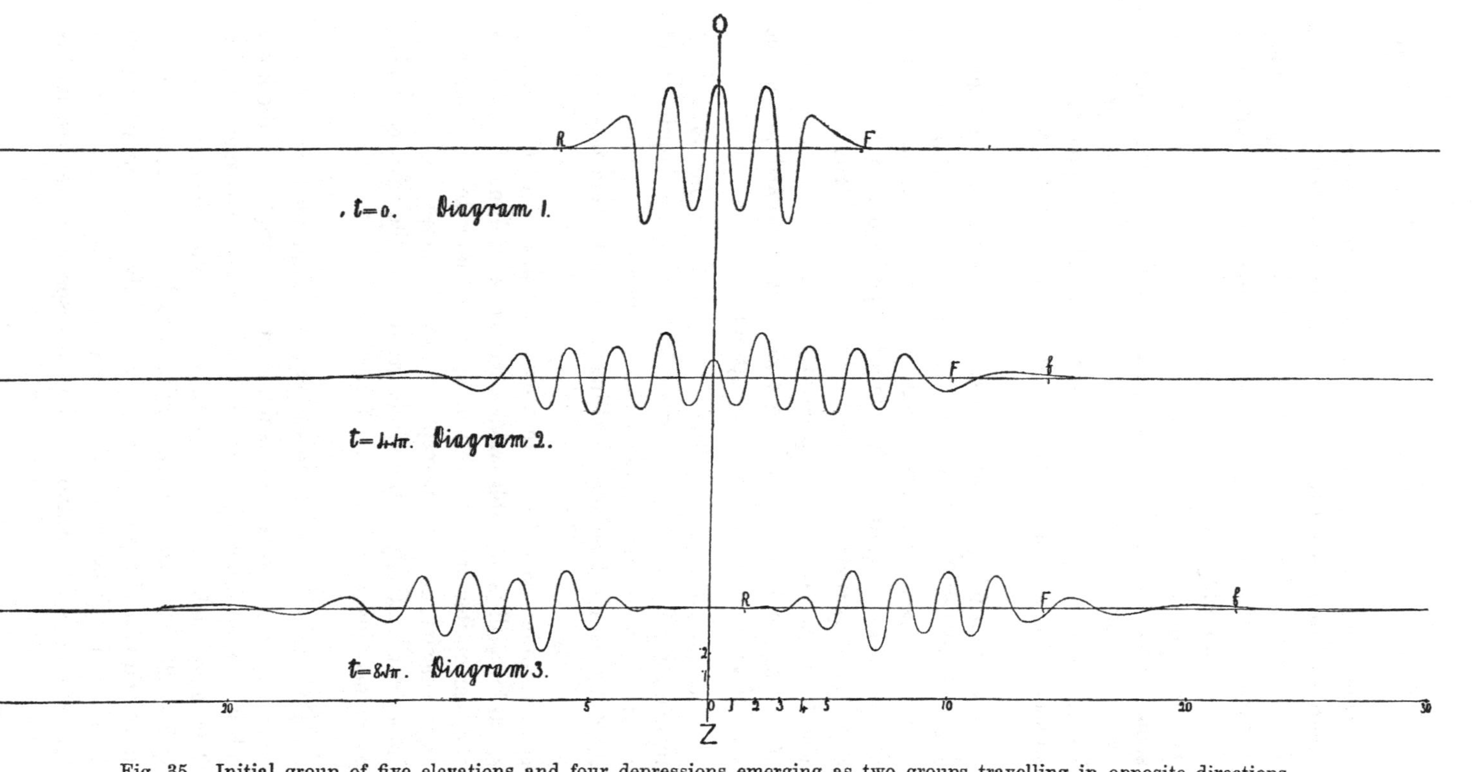

Fig. 35. Initial group of five elevations and four depressions emerging as two groups travelling in opposite directions.

outward from the middle, we see that the two groups, after emergence from co-existence in the middle, travel with their rears leaving a widening space between them of water not perceptibly disturbed, but with very minute wavelets in ever-augmenting number following slower and slower in the rear of each group. The extreme perceptible rear travels at a speed closely corresponding to the "half wave-velocity," found by Stokes as exactly the group-velocity of his uniform succession of groups, produced by the interference of two co-existent infinite processions of sinusoidal waves, having slightly different wave-lengths.

116. Our fairly uniform rear velocity is illustrated in diagrams 1 and 3, of fig. 35. In diagram 1, R indicates the perceptible rear of the component group commencing its rightward progress at $t=0$. In diagram 3, R shows the position reached at time $8\sqrt{\pi}$ (eight periods) by an ideal point travelling rightwards from the R of diagram 1 at a speed of half the wave-velocity. This R of diagram 3 corresponds to a fairly well-marked perceptible rear of the rightward travelling group.

Look now to F, F, F, in the three diagrams of fig. 35, and f, f, in diagrams 2 and 3. In diagram 1, F marks a perceptible front for the rightward travelling component group. In diagrams 2 and 3, F, f show ideal points travelling rightwards from it at speeds respectively, the half wave-velocity, and the wave-velocity. We see a manifest wave-disturbance far in advance of F, F; and very small but still perceptible wave-disturbance in front of f, f. Thus the perceptible front travels at speed actually higher than the wave-velocity, and this perceptible front becomes more and more important relatively to the whole group with the advance of time, as we may judge from fig. 9 of § 20 above.

117. It is interesting to see by these diagrams how nearly the hypothetical group-velocity is found in the rears: while the fronts advance with much greater and with ever-increasing velocity. The more elaborate calculations and graphical constructions of §§ 20—29 above led to corresponding conclusions in respect to the front and rear of a procession, given initially as an infinitely great number of regular sinusoidal waves travelling in one direction. The diagrams, figs. 9 and 10, showed respectively, at twenty-five

periods after a sinusoidal commencement, a front extending forward indefinitely, and a perceptible rear lagging scarcely two wave-lengths behind a point, travelling from the initial position of the rear at a speed of half the wave-velocity.

(3) *The Initiation and Continued Growth of a Train of Two-Dimensional Waves due to the Sudden Commencement of a Stationary, Sinusoidally Varying, Surface-Pressure.* §§ 118—158.

118. A forcive consisting of a finite sinusoidally varying pressure is applied, and kept through all time applied, to the surface of the water within a finite practically limited space on each side of the middle line of the disturbance. In the beginning the water was everywhere at rest and its surface horizontal. The problem solved is, to find the elevation or depression of the water at any distance from the mid-line of the working forcive, and at any time after the forcive began to act.

119. As a preliminary (§§ 119—126) let us consider the energy in a uniform procession of sinusoidal waves, in a straight canal, infinitely long and infinitely deep, with vertical sides. If the water is disturbed from rest by any pressure on its upper surface, and afterwards left to itself under constant air pressure, we know by elementary hydrokinetics that its motion will be irrotational throughout the whole volume of the water: and if, at any subsequent time, the surface is brought to rest, suddenly or gradually, all the water at every depth will come to rest at the instant when the whole surface is brought to rest. This, as we know from Green, is true even if the initial disturbance is so violent as to cause part of the water to break away in drops: and it would be true separately for each portion of the water detached from the main volume in the canal, as well as for the water remaining in the canal, if stoppage of surface motion is made for every detached portion before it falls back into the canal.

120. Because the motion of the water is irrotational, we have

$$\dot{\xi} = \frac{d\dot{F}}{dx}; \quad \dot{\zeta} = \frac{d\dot{F}}{dz} \quad \text{.................(149)},$$

where $\dot{F}$ denotes the velocity-potential, F having been taken as the displacement-potential (§ 97 above). And by dynamics for infinitesimal motion, as in (64) of § 38 above,

$$p = -\frac{d}{dt}\dot{F}(x, z, t) + g(z - 1 + C) \quad \text{.........(150)}.$$

To express the surface condition, let $z = 1$ be the undisturbed level; and let ζ_1 denote the vertical component displacement of a surface particle of the water, taken positive when downwards; and let Π denote constant surface-pressure, and take $\frac{\Pi}{g}$ as the value of the arbitrary constant, C. Thus (150) gives, at the disturbed surface,

$$0 = -\frac{d}{dt}\dot{F}(x, 1 + \zeta_1, t) + g\zeta_1 = -\frac{d}{dt}\dot{F}(x, 1, t) + g\zeta_1 \text{...(151)}.$$

The equality between the second and third members of this formula is due to the disturbance being infinitely small, which makes $\frac{d}{dt}\dot{F}(x, 1 + \zeta_1, t) - \frac{d}{dt}\dot{F}(x, 1, t)$ an infinitely small quantity of the second order, negligible in comparison with $g\zeta_1$, which is an infinitely small quantity of the first order.

121. For a sinusoidal wave-disturbance of wave-length $2\pi/m$, travelling x-wards with velocity v, we have as in (66) above,

$$F(x, z, t) = -k\epsilon^{-m(z-1)} \sin m(x - vt) \text{.........(152)}.$$

For surface-equation (151) becomes

$$0 = kmv \cos m(x - vt) - g\zeta_1 \text{..............(153)}.$$

This gives as the equation of the free surface

$$\zeta_1 = h \cos m(x - vt) \text{..................(154)},$$

where

$$h = kmv/g \text{........................(155)}.$$

Now by (149) and (152) with $z = 1$, we find

$$\zeta_1 = kv^{-1} \cos m(x - vt) \text{..................(156)}.$$

Comparison of this with (154) gives

$$v^2 = g/m = \lambda g/2\pi \text{.....................(157)}.$$

122. Let us now find A (activity), the rate of doing work by the pressure of the water on one side upon the water on the other side of a vertical plane (x). We have

$$A = \int_1^\infty dz p\dot{\xi} = \int_1^\infty dz\dot{\xi}\left[-\frac{d\dot{F}}{dt} + g(z - 1 + C)\right] \text{...(158)}.$$

Eliminating from this $\dot{\xi}$ and $\dot{F}$ by (149) and (152), we find

$$A = -km \cos m(x - vt) \int_1^\infty dz\, \epsilon^{-m(z-1)} [-kmv\epsilon^{-m(z-1)} \cos m(x - vt) + g(z - 1 + C)] \quad \ldots(159).$$

Hence, performing the operations $\int_1^\infty dz$, we find

$$A = -km \cos m(x - vt) \left[-\frac{kv}{2} \cos m(x - vt) + g\left(\frac{1}{m^2} + \frac{C}{m}\right)\right] \quad (160).$$

123. Remarking now that $2\pi/mv$ is the periodic time of the wave, and denoting by W the total work per period, done by the water on the negative side of the plane (x) upon the water on the positive side, we have

$$W = \int_0^{2\pi/mv} dt \,.\, A = \frac{\pi}{mv} \,.\, \tfrac{1}{2} \,.\, k^2 mv = \tfrac{1}{2}\pi k^2 \quad \ldots\ldots(161).$$

124. We are going to compare this with the total energy, kinetic and potential, $K + P$, per wave-length. In the first place we shall find separately the kinetic energy, K, and the potential energy, P. We have (the density of the water being taken as unity)

$$K = \tfrac{1}{2}\int_0^\lambda dx \int_1^\infty dz\, (\dot{\xi}^2 + \dot{\zeta}^2) \quad \ldots\ldots\ldots\ldots\ldots(162);$$

$$P = \tfrac{1}{2} g \int_0^\lambda dx\, \zeta_1^2 \quad \ldots\ldots\ldots\ldots\ldots\ldots\ldots(163),$$

where ζ_1 denotes the surface displacement.

By (149) and (152) we find

$$\dot{\xi} = -mk\epsilon^{-m(z-1)} \cos m(x - vt) \quad \ldots\ldots\ldots\ldots(164);$$

$$\dot{\zeta} = mk\epsilon^{-m(z-1)} \sin m(x - vt) \quad \ldots\ldots\ldots\ldots\ldots(165);$$

$$\zeta_1 = \frac{k}{v} \cos m(x - vt) \quad \ldots\ldots\ldots\ldots\ldots(156) \text{ repeated.}$$

Hence,
$$K = \tfrac{1}{2}\int_0^\lambda dx \frac{m^2 k^2}{2m} = \tfrac{1}{4} mk^2 \lambda = \tfrac{1}{2}\pi k^2 \quad \ldots\ldots\ldots(166);$$

$$P = \tfrac{1}{2} g \frac{k^2}{v^2} \tfrac{1}{2}\lambda = \tfrac{1}{2}\pi k^2 \quad \ldots\ldots\ldots\ldots\ldots\ldots(167),$$

where v^2 is eliminated by (157).

125. Thus we see that the kinetic energy per wave-length, and the potential energy per wave-length, are each equal to the

work done per period by the water on the negative side, upon the water on the positive side, of any vertical plane perpendicular to the length and sides of the canal. Thus we arrive at the remarkable and well-known conclusion that in a regular procession of deep-sea waves, the work done on any vertical plane is only half the total energy per wave-length. This is only half enough to feed a regular procession, advancing to infinity with abruptly ending front, travelling *with the wave-velocity* v. It is exactly enough to feed an ideal procession of regular periodic waves, coming abruptly to nothing at a front travelling *with half the "wave-velocity"* v; which is Osborne Reynolds'* important contribution to the ideal doctrine of "group-velocity."

126. The dynamical conclusion of § 125 is very important and interesting in the theory of two-dimensional ship-waves. It shows that the approximately regular periodic train of waves in the rear of a travelling forcive, investigated in §§ 48—54 and 65—79 above, cannot be as much as half the space travelled by the forcive, from the commencement of its motion; but that it would be exactly that half-space if some modifying pressure were so applied to the water-surface in the rear as to cause the waves to remain uniformly periodic to the end of the train; without, on the whole, either doing work on them, or taking work from them.

A corresponding statement is applicable to our present subject, as we shall see in §§ 156, 157 below.

127. Go back to § 118; and first, instead of a sinusoidally varying pressure, imagine applied a series of impulsive pressures, each of which superimposes a certain velocity-potential upon that due to all the previous impulses; and let it be required to find the resulting velocity-potential at any time t, after some, or after all, of the impulses. Consider first a single impulse at time $t-q$; that is to say, at a time preceding the time t by an interval q. Let the velocity-potential at time t, due to that single impulse applied at the earlier time $t-q$, be denoted by

$$CV(x, z, q) \quad\ldots\ldots\ldots\ldots\ldots\ldots\ldots(168).$$

According to this notation the instantaneously generated velocity-potential is $CV(x, z, 0)$, and the value of this at the bounding

* *Nature*, Aug. 1877, and *Brit. Ass. Report*, 1877.

surface of the water is $CV(x, 1, 0)$. Hence, by elementary hydrokinetics, if I denotes the impulsive surface-pressure, we have

$$I = -CV(x, 1, 0) \quad \text{.....................(169).}$$

128. Considering now successive impulses at time preceding the time t, by amounts $q_1, q_2, \ldots q_i$; and denoting by $S(x, z, t)$ the sum of the resulting velocity-potentials at time t, we find

$$S(x, z, t) = C_1 V(x, z, q_1) + C_2 V(x, z, q_2) + \ldots C_i V(x, z, q_i) \ldots \text{(170).}$$

Supposing now the impulses to be at infinitely short intervals of time, we translate the formula (170) into the language of the integral calculus as follows:—

$$S(x, z, t) = \int_0^t dq f(t-q) V(x, z, q) \quad \text{.........(171),}$$

where $f(t-q)$ denotes an arbitrary function of $(t-q)$, according to which the surface-pressure, arbitrarily applied at time $(t-q)$, is as follows:—

$$\Pi(t-q) = -f(t-q) V(x, 1, 0) \ldots\ldots\ldots\ldots \text{(172).}$$

Hence the pressure applied to the surface at time t, denoted by $\Pi(x, 1, t)$, is as follows:

$$\Pi(x, 1, t) = -f(t) V(x, 1, 0) \quad \text{...........(173).}$$

129. The solution (170) or (171) gives the velocity-potential throughout the liquid which follows determinately from the dynamical data described in §§ 127, 128. From it, by differentiations with reference to x and z, and integrations with respect to t, we can find the displacement components ξ, ζ of any particle of the liquid whose co-ordinates were x, z when the fluid was given at rest. But we can find them more directly, and with considerably less complication of integral signs, by direct application of the same plan of summing as that used in (170), (171). Thus if, instead of $V(x, z, q)$ in (171), we substitute $\frac{d}{dx} V(x, z, q)$, and again $\frac{d}{dx} V(x, z, q)$, we find $\dot{\xi}$ and $\dot{\zeta}$. And if we take

$$\int_0^q dq \frac{d}{dx} V(x, z, q) \text{ and } \int_0^q dq \frac{d}{dz} V(x, z, q) \quad \text{......(174)}$$

in place of $V(x, z, q)$ in (171), we find the two components ξ, ζ of the displacement of any particle of the fluid. Confining our attention to vertical displacements, and using (179) below, we thus find

$$\zeta(x, z, t) = \frac{1}{g}\int_0^t dq f(t-q)\frac{d}{dq}V(x, z, q) \text{ ......(175).}$$

130. To illustrate the meaning of the notation and analytical expressions in (171), (173), (175), take the simplest possible example, $f(t-q) = 1$. This makes Π the same for all values of t; and (173) becomes

$$\Pi = -V(x, 1, 0) \text{ ....(176);}$$

and by integration (175) becomes

$$\zeta(x, z, t) = g^{-1}[V(x, z, t) - V(x, z, 0)] \text{ ......(177).}$$

Putting now in this $z = 1$, and using (176), we find

$$\zeta(x, 1, t) = g^{-1}[V(x, 1, t) - V(x, 1, 0)] = g^{-1}[V(x, 1, t) + \Pi] \text{ ....(178).}$$

The interpretation of this, as t increases from 0 to ∞, is that the sudden application and continued maintenance of a pressure $-V(x, 1, 0)$ over the whole fluid surface, initially plane and level, produces a depression, ζ, which gradually increases from 0, at $t = 0$, to its hydrostatic value Π/g, at $t = \infty$. The gradual subsidence of the difference from the static condition, as time advances from 0 to ∞, is illustrated by the diagrams of fig. 34, for the case in which we choose for $V(x, 1, 0)$ the $\psi(x, 1, 0)$ of §§ 100—104 above.

131. To understand thoroughly the meaning of $V(x, z, q)$ as defined in § 127; remark first that it is the velocity-potential of a possible motion of water, under the influence of gravity, with no surface-pressure, or with merely a pressure uniform over its infinite free surface. This is equivalent to saying that $V(x, z, q)$ fulfils the equations

$$\frac{d^2V}{dx^2} + \frac{d^2V}{dz^2} = 0, \text{ and } g\frac{dV}{dz} = \frac{d^2V}{dq^2} \text{ ....(179).}$$

Secondly, remark that at the instant $q = 0$, there is no surface displacement; hence $V(x, z, q)$ is the velocity-potential at time q, due to an instantaneous impulsive pressure, $-V(x, 1, 0)$, applied to the surface of the fluid at rest and in equilibrium, at time $q = 0$. Now, allowing negative values of q, think of a state of motion from which our actual condition of no displacement, and of velocity-potential equal to $V(x, z, 0)$, would be reached and passed through when q passes from negative to positive. It is clear that the values of $V(x, z, q)$ are equal for equal positive and negative values of q. Hence, when $q=0$, we have

$$\frac{d}{dq} V(x, z, 0) = 0 \quad\text{.....................(180).}$$

132. Consideration of the $V(x, z, q)$, defined in § 127, which allows $V(x, 1, 0)$ to be any arbitrary function of x, but requires dV/dq to be zero when $q=0$, suggests an allied hydrokinetic problem:—to find W fulfilling (179) with W in place of V; and, at time $q=0$, having $W=0$ and dW/dq any arbitrary function of x. We assume, as is convenient for our present purpose, that for large values of x

$$V(x, z, 0) = 0, \text{ and } W(x, z, 0) \doteqdot 0 \quad\text{.........(181).}$$

This implies that for all values of x and z, large or small, but for large values of q,

$$V(x, z, q) = 0, \text{ and } W(x, z, q) \doteqdot 0 \quad\text{.........(182).}$$

133. In the V-problem the initiational condition is:—displacement zero and initiational velocity *virtually given* throughout the fluid as the determinate result of an arbitrarily distributed impulsive pressure on the surface.

In the W-problem the initiational condition is:—the fluid held at rest with its surface kept to any arbitrarily prescribed shape by fluid pressure, and then left free by sudden and permanent annulment of this pressure.

Without going into the question of a complete solution of this (V, W) problem for any arbitrary initiational data, we find a class of thoroughly convenient solutions in a formula originally given in the *Proceedings of the Royal Society of Edinburgh*, January, 1887; republished in the *Phil. Mag.*, February, 1887; and used in § 3 and § 99 above. We may now write that formula in the following comprehensive realised expression for V or W:—

$$\left.\begin{aligned} \{RS\} \text{ or } \{RD\} \frac{d^{i+j+k}}{dt^i dx^j dz^k} \frac{1}{\sqrt{(z+\iota x)}} \epsilon^{\frac{-gt^2}{4(z+\iota x)}} & \\ = V(x, z, t), \text{ when } i \text{ is even} & \\ = W(x, z, t), \text{ when } i \text{ is odd} & \end{aligned}\right\} \text{...(183).}$$

By using (179) we may, instead of (183), take the following as equally comprehensive:—

$$\left.\begin{aligned}\{RS\} \text{ or } \{RD\}\left(A + B\frac{d}{dx}\right)\frac{d^i}{dt^i}\frac{1}{\sqrt{(z+\iota x)}}\,\epsilon^{\frac{gt^2}{4(z+\iota x)}}\\ = V(x, z, t), \text{ when } i \text{ is even}\\ = W(x, z, t), \text{ when } i \text{ is odd}\end{aligned}\right\}\ldots(183').$$

134. Going back to (171) and (175), remark that integration by parts gives

$$\int_0^t dq\, f(t-q)\frac{d}{dq}V(x, z, q) = f(0)\, V(x, z, t) - f(t)\, V(x, z, 0) + \int_0^t dq\, f(t-q)\, V(x, z, q)\ldots(184).$$

This shows that if by quadrature or otherwise we have calculated the velocity-potentials $S(x, z, t)$, as given by (171), for both forms of $f(t-q)$ in (186) below, we can find the vertical component displacements of any particle of the liquid by (175), without farther integration. The formula (184) also shows how by successive integration by parts we can reduce

$$\int_0^t dq\, f(t-q)\frac{d^i}{dq^i}V(x, z, q) \;\ldots\ldots\ldots\ldots\ldots(185)$$

to the primary integral $S(x, z, t)$, as expressed in (171).

135. Going back now to §§ 128, 127, 118: to make the applied forcive a sinusoidally varying pressure put

$$f(t-q) = {\cos \atop \sin}\,\omega(t-q) \;\ldots\ldots\ldots\ldots\ldots(186);$$

which, by (173), makes

$$\Pi(x, 1, t) = -{\cos \atop \sin}\,\omega t\, V(x, 1, 0)\ldots\ldots\ldots\ldots(187).$$

And now let us arrange to fully work out our problem for two cases of surface distribution of pressure, corresponding to the two initiational forms ϕ, ψ, described in §§ 96—113 above. For this purpose take, with the notation of § 101,

$$V(x, z, t) = \phi(x, z, t);$$

or $$V(x, z, t) = \psi(x, z, t) = -\frac{1}{g\sqrt{2}}\frac{d^2}{dt^2}\phi(x, z, t)\ldots(188).$$

For brevity we shall call these two cases case ϕ and case ψ. Thus, in these cases (171) and (175), expressing respectively the velocity-potential at, and the vertical component displacement of, any point of the fluid at any time, become

$$S_\phi(x, z, t) = \int_0^t dq \begin{matrix}\cos\\ \sin\end{matrix} \omega(t-q)\,\phi(x, z, q);$$

$$S_\psi(x, z, t) = \int_0^t dq \begin{matrix}\cos\\ \sin\end{matrix} \omega(t-q)\,\psi(x, z, q) \quad \ldots(189);$$

$$\zeta_\phi(x, z, t) = \frac{1}{g}\int_0^t dq \begin{matrix}\cos\\ \sin\end{matrix} \omega(t-q)\,\frac{d}{dq}\phi(x, z, q);$$

$$\zeta_\psi(x, z, t) = \frac{1}{g}\int_0^t dq \begin{matrix}\cos\\ \sin\end{matrix} \omega(t-q)\,\frac{d}{dq}\psi(x, z, q) \quad \ldots(190).$$

136. The illustrations in figs. 36, 37, 38 are time-curves in which the ordinates have been calculated by continuous quadrature from one or other of the four formulas (189), (190).

137. The curves in fig. 39, being space curves in which the ordinates are vertical component displacements of the water-surface, are therefore pictures of the water-surface (greatly exaggerated in respect to slopes of course), and may be shortly named water-surface curves. Their ordinates have been calculated by an analytical method described in § 151 below. They cannot be calculated continuously for successive values of x by the method of continuous quadratures; if that were the method employed, the value of the ordinate for each value of x would need to be calculated by an independent quadrature $\left(\int_0^t dq\right)$ from 0 to the particular value of t for which the water-surface is represented by the curve. The values of t chosen for fig. 39 are respectively $i\tau$, $(i+1/8)\tau$, $(i+2/8)\tau$, $(i+3/8)\tau$, $(i+4/8)\tau$, where i is any very large integer, and τ denotes $2\pi/\omega$, the period of the varying surface-pressure to which the fluid motion considered is due.

In all our illustrations we have taken $\omega = \sqrt{\pi}$, which makes $\tau = 2\sqrt{\pi}$, and, with $g = 4$ as in § 105, makes the wave-length $\lambda = 8$.

138. In figs. 36 and 37, all the curves correspond to $\cos\omega(t-q)$ in the formulas. In fig. 38, all the curves correspond to $\sin\omega(t-q)$ in the formulas.

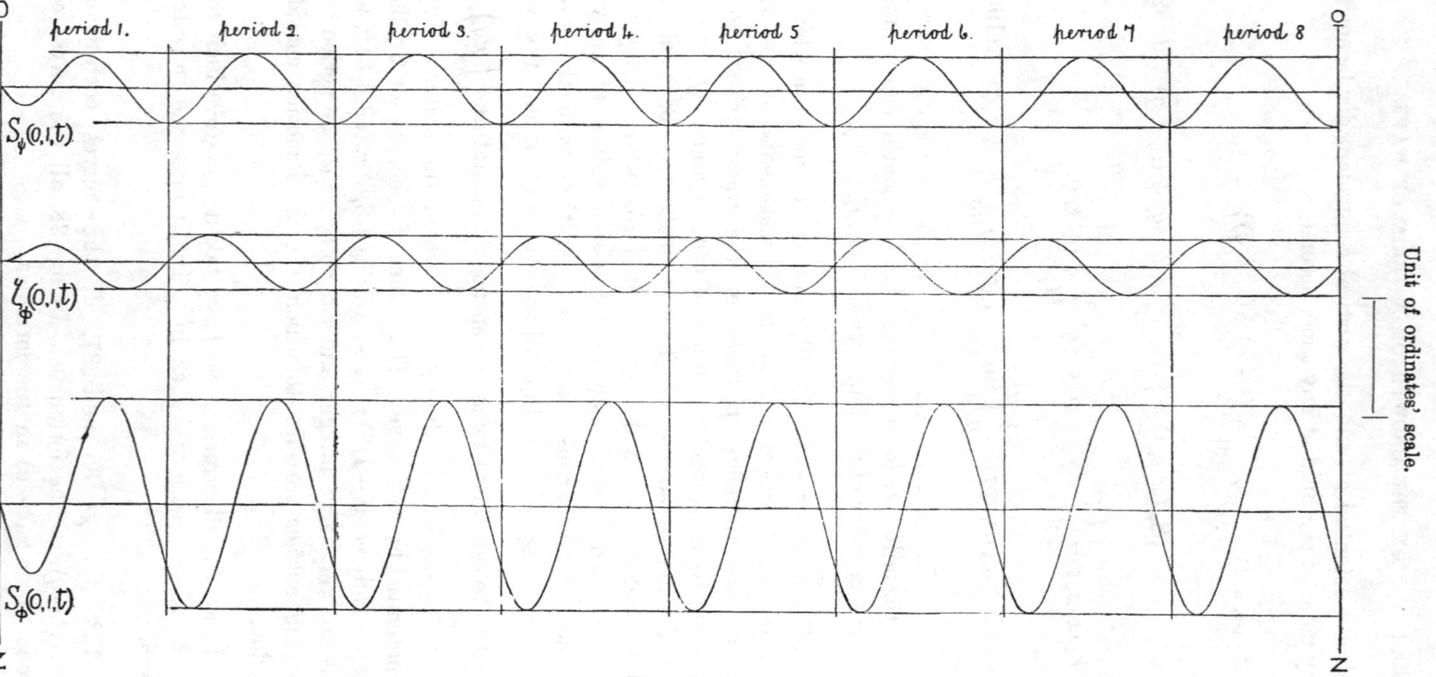

O
O
period 1.
period 2.
period 3.
period 4.
period 5
period 6.
period 7
period 8
$S_\psi(0,1,t)$
$\zeta_\phi(0,1,t)$
$S_\phi(0,1,t)$
Z
Z
Unit of ordinates' scale.

Fig. 36.

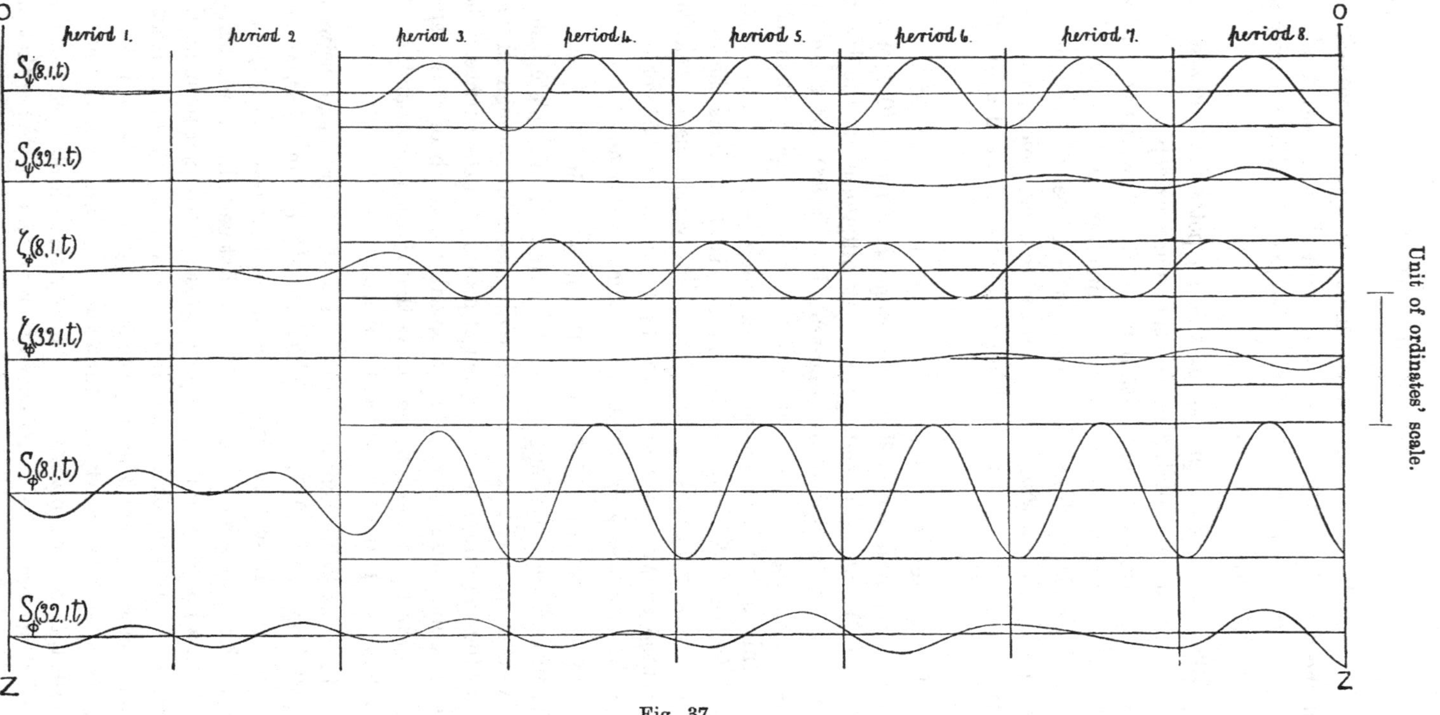
O
O
period 1.
period 2
period 3.
period 4.
period 5.
period 6.
period 7.
period 8.
$S_\psi(8,1,t)$
$S_\psi(32,1,t)$
$\zeta_\phi(8,1,t)$
$\zeta_\phi(32,1,t)$
$S_\phi(8,1,t)$
$S_\phi(32,1,t)$
Unit of ordinates' scale.
Z
Z

Fig. 37.

In fig. 39, the inscriptions of times correspond to $\cos \omega (t - q)$ in the formulas. The same curves, with the inscriptions altered to $(i + 2/8)\tau$, $(i + 3/8)\tau$, $(i+4/8)\tau$, $(i+5/8)\tau$, $(i+6/8)\tau$, correspond to $\sin \omega (t - q)$ in the formulas.

139. In fig. 36, representing velocity-potentials and a surface displacement, none of the curves shows any perceptible deviation from sinusoidality except within period 1. Towards the end of period 1 the numbers found by the quadratures show deviations from sinusoidality diminishing to about 1/10 per cent., and imperceptible in the drawings. This proves that sinusoidality is exact within 1/10 per cent. through all time after the end of the first period.

It is interesting to see, in period 1, how nearly the rise from the initial zero follows the same law for $S_\phi (0, 1, t)$ and $S_\psi (0, 1\ t)$: notwithstanding the vast difference in the law of initiating surface-pressure, represented by (188), for these two cases. In fig. 36, the initiating surface-pressure commences suddenly at its negative maximum value, $-\sqrt{2}$ for case ϕ, and -5 for case ψ, of which the former is 2·83 times the latter. The semi-amplitudes of the subsequent variations of velocity-potential shown in the first and third curves are ·954 for case ϕ and ·318 for case ψ, of which the former is 3·00 times the latter.

140. The first, third, and fifth curves of fig. 37 show, at a distance of one wave-length from the origin, the complete history of velocity-potential and of surface displacement through all time from the beginning of application of pressure to the surface. The very approximately accurate sinusoidality of each of these three curves through periods 6, 7, 8, shows that the continuation through endless time is in each case sinusoidal.

In remarkable contrast with the initial agreement between $S_\psi (0, 1, t)$ and $S_\phi (0, 1, t)$, to which we alluded in § 139, we find very instructively a remarkable contrast between $S_\psi (8, 1, t)$ and $S_\phi (8, 1, t)$ throughout the whole of the first period. Remembering that in a liquid of unit density the pressure is equal to minus the rate of augmentation of the velocity-potential per unit of time, and remarking that the displacement $\zeta_\phi (0, 1, t)$ is, as is shown in its curve, very nearly zero throughout the first period, and that $\zeta_\psi (0, 1, t)$ is certainly still more nearly zero throughout the first period, though we have no curve to represent it, we see that the

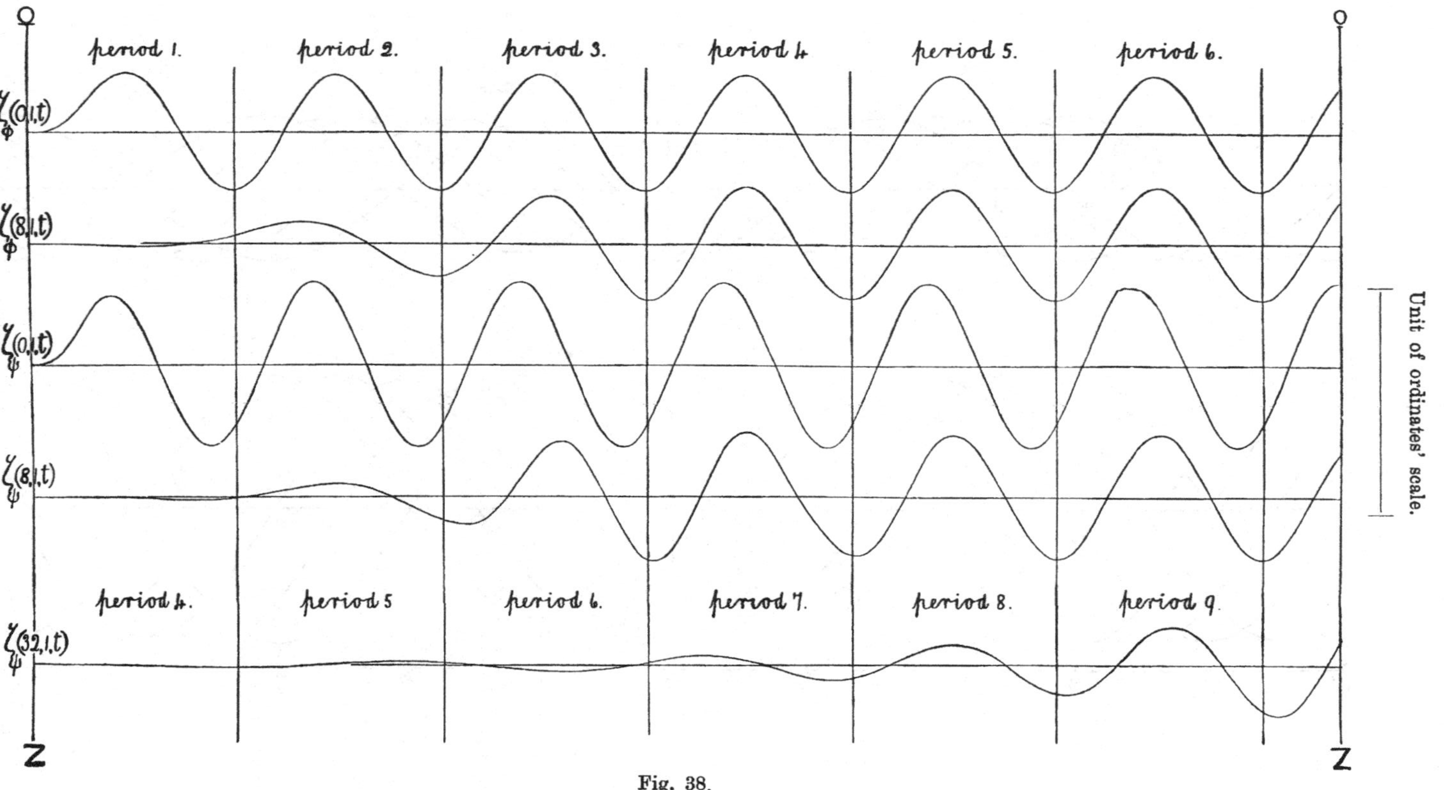

O
O
period 1.
period 2.
period 3.
period 4
period 5.
period 6.
$\zeta_\phi(0,1,t)$
$\zeta_\phi(8,1,t)$
$\zeta_\psi(0,1,t)$
$\zeta_\psi(8,1,t)$
Unit of ordinates' scale.
period 4.
period 5
period 6.
period 7.
period 8.
period 9.
$\zeta_\psi(32,1,t)$
Z
Z

Fig. 38.

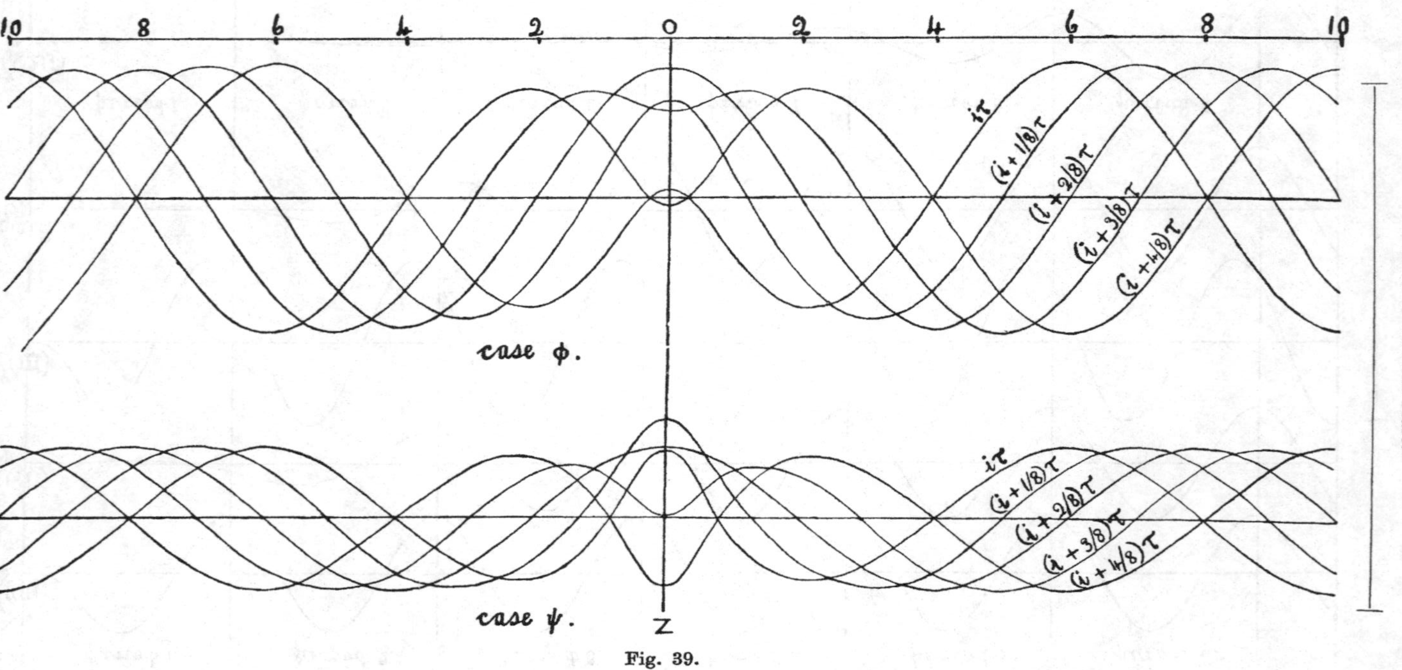
10
8
6
4
2
0
2
4
6
8
10
iτ
(i + 1/8)τ
(i + 2/8)τ
(i + 3/8)τ
(i + 4/8)τ
case φ.
iτ
(i + 1/8)τ
(i + 2/8)τ
(i + 3/8)τ
(i + 4/8)τ
case ψ.
Z
Unit of vertical scale.

Fig. 39.

negatives of the tangents of the slopes in the curves for $S_\psi(8, 1, t)$ and $S_\phi(8, 1, t)$ represent very nearly the values of the applied surface-pressures during the whole of the first period*. Look now to fig. 33; see how near to zero is $\psi(8, 1, 0)$, and how far from zero is $\phi(8, 1, 0)$; and we see dynamically how it is that $S_\psi(8, 1, t)$ is very nearly zero throughout the first period, and $S_\phi(8, 1, t)$ is very far from zero, and is somewhat near to being sinusoidal.

141. We have also a very instructive comparison between $\zeta_\phi(8, 1, t)$ and $S_\phi(8, 1, t)$. In the ϕ case, for values of x as large as 8, or larger, we approach somewhat nearly to the case of a sinusoidally varying uniform surface-pressure over an infinite plane area of water, in which there would be no surface displacement, and the pressure at and below the surface would be at every instant equal to the applied surface-pressure plus the gravitational augmentation of pressure below the surface. Thus we see why it is that, with a great periodic variation of applied surface-pressure, at $x = 8$, there is scarcely any rise and fall of the surface level there, until after a period and a half from the beginning of the motion, as shown in the curves for $\zeta_\phi(8, 1, t)$.

142. The second, fourth, and sixth curves of fig. 37 represent the arrival of three classes of disturbance, S_ψ, ζ_ϕ, S_ϕ, at $x = 32$, four wave-lengths from the origin. If the front of the disturbance travelled at exactly the wave-velocity, the disturbances of the different kinds would all commence suddenly at the end of period 4. In the cases of $S_\psi(32, 1, t)$ and $\zeta_\phi(32, 1, t)$ the diagram shows that they are quite imperceptible at the end of period 4, and begin to be considerable at the end of period 8, which would be the exact time of arrival if there was a definite "group-velocity" equal to half the wave-velocity. The largeness of $S_\phi(32, 1, t)$, approximately uniform throughout the first four periods, is explained in § 140. Its gradual augmentation through periods 5, 6, 7, 8, depends on the wave propagation of disturbances from the origin, as shown for $S_\psi(32, 1, t)$ and $\zeta_\phi(32, 1, t)$ in the second and fourth curves.

143. The $\zeta_\phi(0, 1, t)$ curve of fig. 38 may be compared with the curve of the same designation in fig. 36. They differ because of a quarter period difference in the phase of commencement of the disturbing pressure, which commences suddenly at its

* Remember that downward ordinates in all the curves of figs. 36, 37, 38, 39, correspond to positive values of the quantities represented.

maximum for all the curves of fig. 36, and commences at zero for all the curves of fig. 38. If the S_ψ, S_ϕ curves for initiating pressure commencing at zero were drawn, they would differ from the first and third curves of fig. 36 in being at the commencement tangential to the line of abscissas, instead of being inclined to it in the positive direction, as shown in fig. 36. The ζ curves are all initially tangential to the line of abscissas, but the tangency is only of the first order in fig. 36, while it is of the second order in fig. 38.

144. The third and fourth curves* of fig. 38 show the whole history for the points, $x = 0$, and $x = \lambda$, of the surface displacement expressed by the formulas

$$\zeta_\psi (x, 1, t) = g^{-1} \int_0^t dq \sin \omega (t-q) \frac{d}{dq} \psi (x, 1, q) \ldots (191),$$

which expresses the surface displacement due to surface-pressure expressed by

$$\Pi (x, 1, t) = -\sin \omega t \psi (x, 1, 0) \ldots\ldots\ldots\ldots (192).$$

The fifth curve of fig. 38 shows the history, after period 3, to almost half a period after period 9, of the disturbance at the place $x = 32$. The disturbance has not yet become sinusoidal, but would certainly become almost exactly sinusoidal after a few more periods.

145. In fig. 39, two sets of five curves show, for case ϕ and case ψ, the periodically varying water-surface on each side of the middle, at any long enough time after the beginning of the motion, to give a regular regime of sinusoidal vibration as far as two or three wave-lengths on each side of the middle. The third curve in each case is a curve of sines. The first curve represents the surface at the beginning of a period from $i\tau$ to $(i+1)\tau$. The fifth curve, being the first curve inverted, represents the water-surface at the middle of the period. The other two curves may be described as components of the first and third, according to the following formula:

$$\zeta (x, 1, t) = P \sin \omega t - Q \cos \omega t \ldots\ldots\ldots\ldots (193),$$

where $$P = -A \cos 2\pi x/\lambda \ldots\ldots\ldots\ldots\ldots\ldots (194),$$

* The scale of ordinates of the third, fourth, and fifth curves of fig. 38 is double that of the first and second, indicated on the figure.

and Q is a continuous transcendental function of x, having equal values for $\pm x$, expressed by (195) for positive or negative values of x, exceeding a wave-length.

$$\left.\begin{array}{ll}\text{For } x \text{ positive,} & Q=-A\sin 2\pi x/\lambda \\ \text{for } x \text{ negative,} & Q=+A\sin 2\pi x/\lambda\end{array}\right\}\quad\ldots\ldots\ldots\ldots\ldots\ldots(195),$$

where A denotes the semi-amplitude of the vibration, at any time long enough after the beginning, and place far enough from the middle of the disturbance, to have very approximately sinusoidal motion. The determination of the transcendental function Q, and the calculation of A, for both P and Q, will be virtually worked out in § 151 below.

146. We have now an exceedingly interesting and suggestive analysis of the circumstances represented in fig. 39. Consider separately the two motions corresponding to $P\sin\omega t$ alone, and to $-Q\cos\omega t$ alone. The motion $P\sin\omega t$, if at any instant given from $x=-\infty$ to $x=+\infty$, would continue for ever, as an infinite series of standing waves, without any surface-pressure. Hence our application of surface-pressure is only required for the Q-motion: and if this motion be at any instant given from $x=-\infty$ to $x=+\infty$, it will go on for ever, provided the pressure $-\cos\omega t \, \begin{smallmatrix}\phi\\ \psi\end{smallmatrix}\,(x, 1, 0)$ is applied and kept applied to the surface.

147. The plan of § 146 may be generalised as follows:—Displace the water according to the formula (193) with P omitted, and with Q any arbitrary function of x for moderately great positive or negative values of x, gradually changing into the formula (195) for positive and negative values outside any arbitrarily chosen length MON (MO not necessarily equal to ON). Find mathematically the sinusoidally varying surface-pressure, $F(x)\cos\omega t$, required to cause the motion to continue according to this law. Superimpose, upon the motion thus guided by surface-pressure, the motion $-A\cos 2\pi x/\lambda \, . \, \sin\omega t$, which needs no surface-pressure. In the motion thus compounded, we have equal sinusoidal waves travelling outwards in the two directions beyond MN (semi-amplitude A): and, in the space MN, we have a varying water-surface found by superimposing on the motion $P\sin\omega t$ an arbitrary shape of surface, varying sinusoidally according to the formula $-Q\cos\omega t$.

148. A curiously interesting dynamical consideration is now forced upon us. The P-component of motion needs, as we have seen, no surface-pressure. The Q-component of motion is kept correct by the surface-pressure $F(x)\cos \omega t$, which, in a period, does no total of work on the Q-motion; but work must be done to supply energy for the two trains of waves travelling outwards in the two directions. Hence this work is done by the activity of the surface-pressure upon the P-component of the motion.

149. Another curious question is forced upon us. Our solution of §§ 135—145 has given us determinately and unambiguously, in every variety of the cases considered, the motion of every particle of the water throughout the space occupied. The synthetic method of quadratures which we have used could lead to no other motion at any instant due to the applied surface-pressure; but now, in § 147, we have considered a Q-motion alone, kept correct by the applied surface-pressure. Would this motion be unstable? and, if unstable, would it in a sufficiently long time subside into the motion expressed in the determinate solution of §§ 135—145? The answer is Yes and No. At any instant, say at $t = 0$, let the whole motion be the Q-component alone of § 148. Let now the surface-pressure, $F(x)\cos \omega t$, be suddenly commenced and continued for ever after. It will, according to §§ 135—145, produce determinately a certain compound motion (P, Q) which will be superimposed upon the motion existing at time $t = 0$; and this last-mentioned motion, given with its infinite amount of energy distributed from $x = -\infty$ to $x = +\infty$, and left with no surface-pressure, would clearly never come approximately to quiescence, through any range of distance from O on the two sides. Thus we see that, though the Q-motion alone of § 148 is essentially unstable, the condition of the fluid does not subside into the determinate solution of §§ 135—145. It would so subside, if it were given initially only through any finite space however great, on each side of O. In fact, any given distribution of disturbance through any finite space however great on each side of O, left to itself without any application of surface-pressure, becomes dissipated away to infinity on the two sides; and leaves, as illustrated in §§ 96—113, an ever-broadening space on each side of O, through which the motion becomes smaller and smaller as time advances.

150. It remains only to look into some of the analytical details concerned in the practical working out of our solutions

(189), (190). Taking $\cos\omega(t-q)$ in the formulas, and taking case ϕ, we find by (190)

$$S_\phi(x, z, t) = P\cos\omega t + Q\sin\omega t \quad\text{...........(196)};$$

where

$$P = \int_0^t dq\cos\omega q\phi(x, z, q);\ \text{and}\ Q = \int_0^t dq\sin\omega q\dot{\phi}(x, z, q)\text{...(197)}.$$

When P and Q have been thus found by quadratures, for all values of t, and any particular value of x, by integration by parts on the plan of § 134, we readily find, without farther quadratures, or integrations, expressions for the seven other formulas included in (189), (190).

151. Let us first find P and Q for $t=\infty$. Using the exponential form for ϕ, given by (137), we find

$$P = \{RS\}\sqrt{\frac{8m}{g}}\int_0^\infty dq\cos\omega q\epsilon^{-mq^2};$$

and

$$Q = \{RS\}\sqrt{\frac{8m}{g}}\int_0^\infty dq\sin\omega q\epsilon^{-mq^2}\text{...........(198)},$$

where $m = \frac{1}{4}g/(z+\iota x)$.

Hence, according to an evaluation given by Laplace in 1810 *, we find, taking $g=4$,

$$P = \{RS\}\sqrt{\frac{\pi}{2}}\,\epsilon^{\frac{-\omega^2}{4m}} \quad\text{.................(199)}.$$

The definite integral for Q is a transcendent function of ω and m, not expressible finitely in terms of trigonometrical functions or exponentials. By using the series for $\sin\omega q$ in terms of $(\omega q)^{2i+1}$, and evaluating $\int_0^\infty dq\,q^{2i+1}\epsilon^{-mq^2}$ by integrations by parts, we find the following convergent series for the evaluation of Q, for $t=\infty$; and $g=4$:—

$$\left.\begin{aligned} Q &= \{RS\}\frac{1}{\sqrt{2}}\left[\frac{\omega}{\sqrt{m}} - \frac{1}{2.1.3}\left(\frac{\omega}{\sqrt{m}}\right)^3 + \frac{1}{2^2.1.3.5}\left(\frac{\omega}{\sqrt{m}}\right)^5 \right.\\ &\qquad\left. - \frac{1}{2^3.1.3.5.7}\left(\frac{\omega}{\sqrt{m}}\right)^7 + \text{etc.}\right]\\ &= \frac{1}{\sqrt{2}}\left[\omega\sqrt{\rho}\cos\frac{\chi}{2} - \frac{(\omega\sqrt{\rho})^3}{2.1.3}\cos\tfrac{3}{2}\chi + \frac{(\omega\sqrt{\rho})^5}{2^2.1.3.5}\cos\tfrac{5}{2}\chi \right.\\ &\qquad\left. - \frac{(\omega\sqrt{\rho})^7}{2^3.1.3.5.7}\cos\tfrac{7}{2}\chi + \text{etc.}\right] \end{aligned}\right\}\text{...(200)},$$

where, as in §§ 100—113 above, $\rho = \sqrt{(z^2+x^2)}$, and $\chi = \tan^{-1}(x/z)$.

* *Mémoires de l'Institut*, 1810. See Gregory's *Examples*, p. 480.

This series converges for every value of $\omega\sqrt{\rho}$ however great. But for values of $\omega\sqrt{\rho}$ greater than 4, it diverges to large alternately positive and negative terms before it begins to converge. The largest value of $\omega\sqrt{\rho}$ for which we have used it is $\omega\sqrt{\rho} = 5{\cdot}03$, corresponding to $x = 8$, and requiring, for the accuracy we desire, twenty-one terms of the series. But for this value of $\omega\sqrt{\rho}$ and for all larger values, we have used the ultimately divergent series (208), found in expressing analytically, not merely for $t = \infty$ as in (198), (199), (200), but for all positive values of t great and small, the growth to its final condition when $t = \infty$, of the disturbance produced by our periodically varying application of pressure to the surface of the water initially ($t = 0$) at rest. The curve for $i\tau$ in fig. 39 has been actually calculated by (200) for values of x up to 8, and by the ultimately divergent series for values of x from 5 to 10. The agreement between those of the values which were calculated both by (200) and by the ultimately divergent series (208), was quite satisfactory: so also was the agreement between values of Q found by quadratures for $x = 1$ and $x = 8$, with values found by (200) for $x = 1$ and by (208) for $x = 8$. It is also satisfactory that the values of P found by quadratures, for $x = 1$, and $x = 8$, agreed well with their exact values given by (199), for $t = \infty$.

152. Going back now to the expressions (197) for P and Q, we see that, by an obvious analytical method of treatment, we can reduce them, and therefore (§ 150) all our other formulas, to expressions in terms of a function defined as follows:—

$$E(\sigma) = \int_0^\sigma d\sigma \epsilon^{-\sigma^2} \quad\text{...................(201)},$$

a function well known to mathematicians* through the last hundred and fifty or two hundred years, in the mathematical theory of Astronomical Refraction, and in the theory of Probabilities. I have taken E as an abbreviation of Glaisher's† notation "Erfc," signifying what he calls "Error Function Complement,"

* The beautiful mathematical discovery, $\int_0^\infty d\sigma\epsilon^{-\sigma^2} = \frac{1}{2}\sqrt{\pi}$, seems to have been made by Euler about 1730.

† *Phil. Mag.*, October 1871.

which he uses in connection with his name "Error Function," defined by

$$\mathrm{Erf}(\sigma) = \int_\sigma^\infty d\sigma\,\epsilon^{-\sigma^2} = \tfrac{1}{2}\sqrt{\pi} - \mathrm{Erfc}(\sigma) \quad \ldots\ldots.(202).$$

Using the imaginary expression for ϕ in (137), we find

$$P = \{RS\}\sqrt{\frac{2m}{g}}\,\epsilon^{\frac{-\omega^2}{4m}}\int_0^t dq\left[\epsilon^{-\left(\sqrt{m}q - \iota\frac{\omega}{2\sqrt{m}}\right)^2} + \epsilon^{-\left(\sqrt{m}q + \iota\frac{\omega}{2\sqrt{m}}\right)^2}\right] \quad \ldots\ldots\ldots(203);$$

$$Q = \{RS\}\frac{1}{\iota}\sqrt{\frac{2m}{g}}\,\epsilon^{\frac{-\omega^2}{4m}}\int_0^t dq\left[\epsilon^{-\left(\sqrt{m}q - \iota\frac{\omega}{2\sqrt{m}}\right)^2} - \epsilon^{-\left(\sqrt{m}q + \iota\frac{\omega}{2\sqrt{m}}\right)^2}\right] \quad \ldots\ldots\ldots(204),$$

where $m = \frac{1}{4}g/(z + \iota x)$, as in § 151.

Taking advantage now of the notation (201), we reduce these two expressions to the following:—

$$P = \{RS\}\sqrt{\frac{2}{g}}\,\epsilon^{\frac{-\omega^2}{4m}}\left[E\left(\sqrt{m}t - \iota\frac{\omega}{2\sqrt{m}}\right) + E\left(\sqrt{m}t + \iota\frac{\omega}{2\sqrt{m}}\right)\right] \quad \ldots\ldots\ldots(205);$$

$$Q = \{RS\}\frac{1}{\iota}\sqrt{\frac{2}{g}}\,\epsilon^{\frac{-\omega^2}{4m}}\left[E\left(\sqrt{m}t - \iota\frac{\omega}{2\sqrt{m}}\right) - E\left(\sqrt{m}t + \iota\frac{\omega}{2\sqrt{m}}\right) + 2E\left(\iota\frac{\omega}{2\sqrt{m}}\right)\right] \quad \ldots(206).$$

153. Remark first in passing that, when $\sqrt{m}t$ is infinitely great in comparison with $\omega/(2\sqrt{m})$, these two expressions agree with the expressions, (198), for P and Q with $t = \infty$, which we used in connection with the explanation of fig. 39.

154. And now, with a view to finding P and Q for any chosen values of x, z, t, we have the following known series*:—

$$E(\sigma) = \sigma - \frac{\sigma^3}{3} + \frac{1}{1.2}\frac{\sigma^5}{5} - \frac{1}{1.2.3}\frac{\sigma^7}{7} + \ldots\ldots\ldots\ldots\ldots\ldots\ldots(207),$$

$$E(\sigma) = \frac{1}{2}\sqrt{\pi} - \frac{\epsilon^{-\sigma^2}}{2\sigma}\left[1 - \frac{1}{2\sigma^2} + \frac{1.3}{(2\sigma^2)^2} - \frac{1.3.5}{(2\sigma^2)^3} + \ldots\right] \ldots(208).$$

* See Glaisher, "On a Class of Definite Integrals," *Phil. Mag.*, October 1871; and Burgess, "On the Definite Integral $\frac{2}{\sqrt{\pi}}\int_0^t \epsilon^{-t^2} dt$," *Trans. Roy. Soc. Edin.*, 1898.

The series (207) converges for all values of σ, great or small, real or imaginary: (208) converges in its first i terms, if $2\sigma^2 > 2i - 3$ (modulus understood if σ^2 is imaginary), and after that it diverges, the true value being intermediate between the sum of the convergent terms and this sum with the first term of the divergent series added. The proper rule of procedure to find the result with any desired degree of accuracy, is to first calculate by the ultimately divergent series, and see whether or not it gives the result accurately enough. If it does not, use the convergent series (207), which, by sufficient expenditure of arithmetical labour, will certainly give the result with any degree of accuracy resolved upon.

155. As a guide, not only for numerical calculation, but for judging the character of the desired result without calculation, it is convenient to find the moduluses of the three complex arguments of the function E, in (205), and (206). They are as follows:—

$$\text{mod}\left(\sqrt{m}t - \iota\frac{\omega}{2\sqrt{m}}\right) = \sqrt{\left(\frac{t^2 g}{4\rho} + \frac{t\omega x}{\rho} + \frac{\omega^2\rho}{g}\right)}$$

$$\fallingdotseq t\sqrt{\frac{g}{4x}} + \omega\sqrt{\frac{x}{g}}, \text{ when } x \fallingdotseq \infty \text{ ......(209)};$$

$$\text{mod}\left(\sqrt{m}t + \iota\frac{\omega}{2\sqrt{m}}\right) = \sqrt{\left(\frac{t^2 g}{4\rho} - \frac{t\omega x}{\rho} + \frac{\omega^2\rho}{g}\right)}$$

$$\fallingdotseq t\sqrt{\frac{g}{4x}} - \omega\sqrt{\frac{x}{g}}, \text{ when } x \fallingdotseq \infty \text{ ......(210)};$$

$$\text{mod}\left(\iota\frac{\omega}{2\sqrt{m}}\right) = \omega\sqrt{\frac{\rho}{g}} \quad \text{..................................(211)}.$$

156. The very interesting questions regarding the front of the procession of waves in either direction, of which we have found illustrations in figs. 36, 37, 38, and which we had under consideration in §§ 11—31, 114—117 above, are now answerable in a thoroughly satisfactory mathematical manner, by aid of the formulas (205), (206), (209), (210), (211). When, in the arguments of E, in (205), and (206), $\sqrt{m}t$ is very great in comparison with $\omega/(2\sqrt{m})$, the two added terms in (205) are approximately equal, and (206) is reduced approximately to its last term; and all the solutions (189), (190), become approximately sinusoidal, in respect to t. This is the case when $t\sqrt{\frac{g}{4x}}$ is very great in

comparison with unity, and in comparison with $\omega\sqrt{\frac{x}{g}}$, as we see by looking at the moduluses shown in (209), (210), (211). This allows us to neglect ω in the arguments of E in (205), (206), and makes P and Q constant relatively to t.

157. When t is small or large, and x not so small as to give preponderance to the first terms of the moduluses (209), (210), we have in (205), (206), (189), (190) a full representation of the whole circumstances of the wave-front, extending from $x=\infty$ back to the largest value of x that allows preponderance of $t\sqrt{\frac{g}{4x}}$ over $\omega\sqrt{\frac{x}{g}}$, in the moduluses, (209), (210). Let, for example,

$$t\sqrt{\frac{g}{4x}}=\omega\sqrt{\frac{x}{g}} \quad \text{.................(212)}.$$

This gives

$$x=\frac{gt}{2\omega}=\text{half the wave-velocity}\times\text{time} \quad \text{......(213)}.$$

The moving point thus defined is what in my first paper to the Royal Society of Edinburgh (January 1887), "On the Front and Rear of a Free Procession of Waves in Deep Water*," I called the "Mid-Front," defined in (45) of that paper, which agrees with our present (213). The following passage was the conclusion of that old paper:—"The rear of a wholly free procession of waves may be quite readily studied after the constitution of the front has been fully investigated, by superimposing an annulling surface-pressure upon the originating pressure represented by (12) above [this is a case of (173) of our present paper], after the originating pressure has been continued so long as to produce a procession of any desired number of waves." The instruction thus given with reference to the relation between front and rear has been virtually followed, though with some differences of detail, in §§ 20—24 of my second Royal Society paper, on the same subject, and under the same title (June 20, 1904)†. That second paper contained a first instalment of the "calculations and graphic representations" promised in the old first; the present paper contains, in figs. 35, 36, 37, 38, a further instalment of such illustrations.

158. Throughout my work, §§ 96—157, I have had most

* [No. 31, *supra*—title only.] † [No. 36, *supra*.]

valuable assistance from Mr George Green, not only in the very long and laborious calculations and drawings, which have been wholly made by him, but also in many interesting and difficult questions which occurred in the fundamental mathematics of the subject*.

We hope to apply before long the method of § 128 to calculate, by aid of the formula $\int_0^t dq\dot{\psi}(x - vq, z, t - q)$, the initiation and continued growth of Canal Ship-waves, due to the sudden commencement and continued application of a moving, steady surface-pressure, $\psi(x, 1, 0)$. We hope also to apply (139) of the present paper to the fulfilment of my old promise (§ 30, June 20, 1904, *R.S.E.*) to deal with the beautifully varying procession seen circling outwards from the place of a stone thrown into deep water.

* [The features of the various graphs, calculated in these papers as representing the spread of different types of disturbance, have been analysed and verified in detail by Mr G. Green, from the point of view of group velocity or Lord Kelvin's principle of stationary phase (*supra*, p. 304), in two papers "On Group Velocity and on the Propagation of Waves in a Dispersive Medium," *Proc. R. S. Edin.* Vol. XXIX. 1909, pp. 445—470, and "On Waves in a Dispersive Medium resulting from a Limited Initial Disturbance," *Proc. R. S. Edin.* Vol. XXX. 1909, pp. 1—12.]

40. Physical Explanation of the Mackerel Sky.

[From *British Association Report*, 1876, Pt. II. p. 54; reprinted from Symons's *Monthly Meteorological Magazine*, Vol. XI. 1876, p. 131.]

Sir Wm. Thomson explained the relation of the clouds and their movements, and that it was not essential to the formation of a mackerel sky that there should be two different temperatures. All that was essential was that portions of air should be moving up and down; and further, that the up and down motion should seem as though it resulted from the slipping of one stratum of air upon another and the production thereby of waves; and the second essential was that one or other of the two portions of air should be very near the point of saturation—that it would be clear when down at its lowest point and cloudy when up at its highest*.

* [Cf. Helmholtz, "Ueber atmosphärische Bewegungen" (1888) and following papers, in his *Collected Papers*, Vol. III. p. 289 *seq.*]

GENERAL DYNAMICS.

41. On some Kinematical and Dynamical Theorems.

[From the *Proceedings of the Royal Society of Edinburgh*, Vol. v. April 6, 1863, pp. 113—115.]

In the course of investigations which the author had been led to make in connection with a Treatise on Natural Philosophy which he and Professor Tait are about to publish, he met with some remarkable theorems, which appear to be new and of considerable importance. As the details of the investigations will soon be published, a very brief sketch only is given here.

I. *Twist* of a wire. If a straight wire, of uniform section, have a side line of reference traced on its surface parallel to its axis; and if a perpendicular to this line from any point of the axis be called a *transverse*, the amount of torsion or twist of the wire, when bent into any form, may be determined by the following construction:—

Parallel to the tangent to the axis of the wire, at a point moving along it, let a radius of an unit sphere be drawn, cutting the spherical surface in a curve. From points of this curve draw parallels to the transverses at the corresponding points of the bar. The excess of the change of direction from one point to the other in the curve, above the increase of its inclination to the transverse, is equal to the twist in the corresponding part of the wire*.

From this some very curious consequences follow, of which one is as follows:—If a wire be bent along any curve on a spherical surface, so that a side line of reference lies all along in contact

* [See Thomson and Tait's *Natural Philosophy*, § 123.]

with the sphere, it acquires *no twist*; so that when an apple supposed spherical) is peeled, there is no twist in the peel.

Again, if an infinitely narrow ribband be laid on a surface along a geodetic line, its twist is at every point equal to the tortuosity of its axis.

II.* Given any material system at rest, and subjected to an impulse of any given magnitude and in any specified direction, it will move off so as to take the greatest amount of kinetic energy which the specified impulse can give it.

Cor. If a set of material points be struck independently by impulses, each given in amount, more kinetic energy is generated if the points are perfectly free to move each independently of all the others than if they are connected in any way.

III.† Given any material system at rest. Let any parts of it be set in motion suddenly with given velocities, the other parts being influenced only by their connections with those which are set in motion, the whole system will move so as to have less kinetic energy than belongs to any other motion fulfilling the given velocity conditions.

* [See Thomson and Tait's *Nat. Phil.* § 311.]

† [See Thomson and Tait's *Nat. Phil.* § 312.]

42. On a New Form of Centrifugal Governor.

[From the *Inst. of Engineers in Scotland, Transactions*, Vol. xii. Nov. 25, 1868, pp. 67—69.]

The most obvious idea for a centrifugal governor is to use the increase of centrifugal force produced by increase of speed, without change of radius, as the force to produce the requisite regulating action. And the simplest way of using this force for the purpose is to make it the normal pressure for a frictional arrangement directly and simply resisting the rotatory motion.

The governor now shown to the Institution is of this perfectly rudimentary type, and presents no novelty except in some details as to arrangement and proportion of its parts. It consists of two heavy lead masses, *MM* (see Plate III.*), each suspended from a stiff horizontal frame, *H*, attached to the shaft, *S*, and turning with it round its axis, which is vertical. These masses are prevented from flying out by centrifugal force by a stout ring of gun metal, *R*, of 12 inches internal diameter, fixed horizontally, at about the level of their centres of inertia. But the greater part of the centrifugal force is balanced by powerful springs, *P*, drawing the masses inwards towards the axis. Firm stops, *F*, are placed, level with their centres of inertia, to prevent them being drawn inwards by more than about $\frac{1}{10}$ of an inch from the position which they occupy when rubbing on the gun metal ring. When the machine is set in motion with increasing velocity, the governing masses do not fly out from their stops until the centrifugal force upon them begins to overbalance the force of the springs. A very small increase of velocity above that which first detaches them from their stops causes them to press against the gun metal ring, and gives rise to frictional resistance impeding further augmentation of speed. The bearing springs of each mass are at a very considerable distance apart (5 inches in the instrument exhibited to

* [Not reproduced here.]

the Institution), in a plane perpendicular to the horizontal line from its centre of inertia to the axis. This gives greater firmness to the equilibrium of the suspended mass, very nearly as if it were hung by a rigid horizontal shaft, but without the friction which such a mode of mounting would entail. It allows the horizontal force of friction to act upon the lead mass without sensibly twisting it out of position, and to be transmitted to the rigid rotating frame, so as to resist its motion. The spring which draws each mass towards the axis, is made up of two pieces of stout sheet steel, each curved and tempered properly, placed with their convex sides towards one another, and pressed against one another by forcing their ends together, and uniting them by stout clamps. It acts like a coach spring adapted for pulling instead of pushing. In the instrument shown to the Institution each mass amounts to 26 pounds. The springs are set by an adjusting screw, so that either mass alone (the other being tied in by a cord) begins to press on the ring when one and the same speed is reached. This speed was 120 turns in a minute in the instrument as adjusted when shown to the Institution; and, therefore, as the centre of inertia of each mass is about $4\frac{1}{2}$ inches from the axis, its centrifugal force is about 1·84 times its weight, or 48 pounds weight, which was, therefore, the force with which the spring was adjusted to pull. If, now, the speed is increased by a small percentage above that required to cause the governing mass to begin to press upon the ring, the force with which each will press will exceed the force of the spring by double the same percentage. Thus, if the speed exceed by $\frac{1}{5}$ per cent. that at which the governor begins to act, each mass will press on the ring with a force of 19 of a pound, which will give rise to frictional resistance of $\frac{1}{50}$ of a pound force, if the coefficient of friction be ·105. Thus the whole frictional resistance due to the two masses will be $\frac{1}{25}$ of a pound acting at the distance of half a foot from the axis, and consuming, therefore, $\frac{1}{4}$ of a foot pound per second. To increase the speed further by one per cent. requires so much increase of driving power as to consume $1\frac{1}{4}$ foot pounds per second more. These figures give a sufficient idea of the power of this governor when used simply to consume in friction the additional work done by additions to the driving power without more than a small increase of speed. The conditions to be fulfilled are, that the greatest admissible percentage of increase of speed shall give frictional resistance amounting to more than the greatest permitted change of driving power, and that the part of the driving power spent in friction on the pivots of the governor is small in proportion to the latter. It was designed and constructed for miscellaneous laboratory purposes and lecture illustrations, in which approximately uniform speed is required. The same plan may be useful for chronoscopes in general and for telegraphic apparatus, whether for giving uniform motion to the receiving ribbon of paper, as in the Morse and other recorders, or for mechanical "sending" instruments.

A simple modification allows a plan invented by Professor Fleeming Jenkin, and introduced by him in connection with another form of centrifugal governor, to be applied to the present, by which it will be converted into a powerful steam governor. This consists in unfixing the gun metal ring and supporting it so as to give it freedom to rotate round the same vertical axis as the main shaft of the apparatus. By any convenient mechanism, a rotation of this ring in the same direction as that of the governor, may shut off steam, and rotation in the contrary direction augment the quantity of steam used; and a spring or weight applied to give it rotation in the latter direction when it is not carried the other way by friction of the governing masses. Thus, the instrument exhibited at the Institution gives the means of bringing $1\frac{1}{4}$ foot pounds per second of work to act in cutting off steam, if at any time the speed augments by one per cent.

In the after discussion,

Professor Macquorn Rankine said that this was a governor of a very simple kind; indeed, as Sir Wm. Thomson had said, it was constructed on the simplest of all principles that could be applied to a governor—to make the revolving masses press against the inside of a ring which checked the speed when it became too great. Notwithstanding that the principle on which the governor acted was so simple, it had not previously been applied in practice* There was no doubt of its efficacy in preserving an almost uniform velocity. He had no doubt that it would be practicable to adapt it to steam engines, though not precisely in its present form; but by carrying out certain modifications it could be adapted to that purpose and might be made the means of regulating the speed with that remarkable precision which they had heard stated, and

* Except in Siemens' differential governor.

which he hoped they should see before the meeting broke up. He had had opportunities at the University, in Sir William Thomson's Physical Laboratory, of seeing contrivances of this kind, and had seen that they gave results as to uniformity of speed such as those which they had heard described in the paper. In its present form, which they saw on the table, the governor used up the surplus power in friction. It was easy to understand that by suitable modifications it might be made to act upon a regulating valve or the cut-off of a steam engine.

Sir William Thomson said that with the appliance to which he referred, the steam could be regulated with very great precision; for instance, easily so as to keep the speed within a half per cent. of perfect uniformity. In reference to what Professor Rankine had said as to the possibility of this governor being suitable for marine engines, a governor going at double the velocity would, he thought, not be sensibly disturbed by the rolling of the ship.

Professor Rankine supposed that in using this as a steam governor, an inner ring would be required for the revolving masses to press against when the speed fell short, in order to act in the opposite direction upon the regulator.

Sir William Thomson said there were springs and stops which virtually acted in that way. He might do either what Dr Rankine suggested, or there might be a frictional action of the governor to cut off steam, and a slowly descending weight always throwing on steam, except when the frictional action balances or overcomes it. Or the admission might be effected by a wheel carried round frictionally by a shaft. He thought a weight would be the most convenient way, which would be running down until the full steam was let on; but before full steam was admitted the speed of the governor would cause the masses to press against the ring, which would prevent the weight from running down any further, and so prevent any more steam from getting in.

43. ON A NEW ASTRONOMICAL CLOCK, AND A PENDULUM GOVERNOR FOR UNIFORM MOTION.

[*Roy. Soc. Proc.* Vol. XVII. June 10, 1869, pp. 468—470. Reprinted in *Popular Lectures and Addresses*, II. pp. 387—394.]

44. ON THE PERTURBATIONS OF THE COMPASS PRODUCED BY THE ROLLING OF THE SHIP.

[Read in Section A of the British Association at Belfast, 1874; from the *Philosophical Magazine*, Vol. XLVIII. Nov. 1874, pp. 363—369.]

THE "heeling-error," which has been investigated by Airy and Archibald Smith*, is the deviation of the compass produced by a "steady heel" (as a constant inclination of the ship round a longitudinal axis, approximately horizontal, is called). It depends on a horizontal component of the ship's magnetic force, introduced by the inclination; which, compounded with the horizontal component existing when the ship is upright, gives the altered horizontal component when the ship is inclined. Regarding only the error of direction, and disregarding the change of the intensity of the directing force, we may define the heeling-error as the angle between the directions, for the ship upright and for the ship inclined, of the resultant of the horizontal magnetic forces of earth and ship at the position of the compass. These suppositions would be rigorously realized with the compass supported on a point in the ordinary manner, if the bearing-point were carried by the ship uniformly in a straight line. They are nearly enough realized in a large ship to render inconsiderable the errors due to want of perfect uniformity of the motion of the bearing-point, if this point is placed anywhere in the "axis of rolling"†; for in a large ship the compass, however placed, is not considerably disturbed by pitching, or by the inequalities of the longitudinal translatory motion caused by waves. Hence, supposing the compass placed in the axis of rolling, the perturbation produced in it by the rolling will be solely that due to the

* [See obituary notice of Archibald Smith, *Proc. Roy. Soc.* 1874, to be reprinted in a later volume; also an article in *Popular Lectures and Addresses*, Vol. III. p. 228 (1874), which formed the beginnings of compass investigations.]

† One way, probably the best in practice, of finding by observation the position of the axis of rolling is to hang pendulums from points at different levels in the plane through the keel perpendicular to the deck, till one is found which indicates the same degrees of rolling as those found geometrically by observing a graduated scale (or "batten") seen against the horizon.

variation of the horizontal component of the ship's magnetic force. Such a position of the compass would have one great advantage—that the application of proper magnetic correctors adjusted by trial to do away with the rolling-error, would perfectly correct the heeling-error. To set off against this advantage there are two practical disadvantages:—one, that the axis of rolling (being always below deck) would not be a convenient position for the ordinary modes of using the compass; the other (far more serious), that, at all events in ships with iron decks, the magnetic disturbance produced by the iron of the ship would probably be so much greater at any point of the axis of rolling, than at suitably chosen positions above deck, as to more than counterbalance the grand kinetic advantage of the axial position. But careful trials in ships of various classes ought to be made; and it *may* be found that in some cases the compass may, with preponderating advantage, be placed at the axis of rolling. Hitherto, however, this position for the compass has not been used in ships of any class, and, as we have seen, it is not probable that it can ever be generally adopted for ships of all classes. It is therefore an interesting and important practical problem to determine the perturbations of the compass produced by oscillations or other non-uniform motions of the bearing-point.

The general kinetic problem of the compass is to determine the position at any instant of a rigid body consisting of the needles, framework, and fly-card, which for brevity will be called simply *the compass*, movable on a bearing-point, when this point moves with any given motion. Let the bearing-point experience at any instant a given acceleration α, in any given direction. Let W be the mass (or weight) of the compass, and gW the force of gravity upon it, reckoned in kinetic units. The position of kinetic equilibrium of the compass at that instant is the position in which it would rest under the magnetic forces and a force of *apparent gravity* equal to the resultant of gW and a force αW in the direction opposite to that of α. Now the weight of the compass is so great and its centre of gravity so low that the level of the card is scarcely affected sensibly by the greatest magnetic couple experienced by the needles*. Hence in kinetic

* Generally no adjusting counterpoise for the compass is required when a ship goes from extreme north to extreme south magnetic latitudes.

equilibrium the plane of the compass-card is sensibly perpendicular to the direction of the "apparent gravity" defined above; and the magnetic axis of the needles is in the direction of the resultant of the components, in this plane, of the magnetic forces of earth and ship. Hence it is simply through the *apparent level*, at the place in the ship occupied by the compass, differing from the true gravitation-level, that the problem of the kinetic-equilibrium position of the compass in a rolling ship differs from the problem of the heeling-error referred to above. That we may see the essential peculiarities of our present problem, let there be no magnetic force of the ship herself or cargo. The kinetic-equilibrium position of the magnetic axis of the compass will be simply the line of the component of terrestrial magnetic force in the plane of the apparent level. Let κ be the inclination of this plane to that of the true gravitation-level, and ϕ the azimuth (not greater than 90°) from magnetic north of the line LL' of the intersection of the two planes (a diagram is unnecessary); also let H and Z be the horizontal and vertical components of the terrestrial magnetic force. The component of this force in the plane of apparent level will be the resultant of $H \cos\phi$ along LL' and $H \sin\phi \cos\kappa + Z \sin\kappa$ perpendicular to LL'; and therefore, if $\phi_{,}$ denote the angle at which it is inclined to LL', we have

$$\tan\phi_{,} = \frac{H \sin\phi \cos\kappa + Z \sin\kappa}{H \cos\phi} = \tan\phi \cos\kappa + \frac{Z \sin\kappa}{H \cos\phi}.$$

If, as usual in compass questions, we reckon the *directions* as of forces on *south* magnetic poles (or the northern ends of the compass-needles), the direction of $H \cos\phi$ is along LL' *northwards*, and the direction of $Z \sin\kappa$, when the ship is anywhere north of the magnetic equator, is downwards in the plane of the apparent level.

Now, as we are only considering the effect of rolling, the direction of the given "acceleration" of the bearing-point will always be in a plane perpendicular to the ship's length; and therefore LL' will be parallel to the length. (It will in fact be the line through the "lubber-points" of the compass-bowl.) Hence, and as compass angles are ordinarily read in the plane of the fly-card, the kinetic equilibrium-error of the compass is exactly equal to $\phi_{,} - \phi$. When κ is a small fraction of 57°·3 (the "radian," as the angle whose arc is equal to radius has been called by Professor James Thomson), which is the case except in extreme degrees of rolling when the compass is properly placed*, we have approximately

$$\phi_{,} - \phi = \kappa \frac{Z}{H} \cos \phi.$$

The direction of this error is, for the northern ends of the needles in the northern magnetic hemisphere or for the southern ends in the southern hemisphere, *towards the side on which the apparent level is depressed*—that is (as practically the compass is always *above* the axis of rolling), *towards the elevated side of the ship.* It has its maximum value

$$\kappa Z/H$$

when $\phi = 0$; that is to say, when the ship heads north or south magnetic. To estimate its amount, consider perfectly regular rolling; which in general fulfils approximately the simple harmonic law, so that we may put

$$i = I \sin nt,$$

where i denotes the inclination of the ship at time t, and n and I constants. Let h denote the height of the bearing-point of the compass, vertically above the axis of rolling when the ship is vertical. For the amount of its acceleration we have

$$\alpha = d^2 (hi)/dt^2 = -n^2 hi.$$

Now, if l denote the length of a simple pendulum isochronous with the rolling of the ship, we have

$$n^2 = g/l,$$

and therefore $$\alpha = -gh/l \, . \, i.$$

The direction of α, being tangential to the circle described by the bearing-point, is approximately horizontal; and therefore the direction of apparent gravity will be approximately that of the resultant of

$$g \text{ vertical,}$$

and $$gh/l \, . \, i \text{ horizontal.}$$

Hence $$\kappa = h/l \, . \, i, \text{ approximately.}$$

Hence, when the ship heads north or south, the amount of the kinetic-equilibrium error is approximately

$$Zh/Hl \, . \, i.$$

* The "mast-head compass," perniciously used in too many merchant steamers, may, in moderate enough rolling, experience deviations of apparent level amounting to 20° or 30° on each side of the true gravitation-level.

Suppose, for example, the period of the rolling to be 6 seconds* (or three times the period of the "seconds' pendulum"); l will be 29 feet (or nine times the length of the seconds' pendulum). And suppose the compass to be 14½ feet above the axis of rolling. We have $\kappa = \frac{1}{2}i$ (so that the range of apparent rolling indicated by a pendulum hung from a point in the position of the bearing-point of the compass is greater by half than the true range of the roll). On these suppositions the kinetic-equilibrium error amounts to

$$\tfrac{1}{2}Z/H \,.\, i.$$

About the middle of the British Islands the magnetic dip is 70°, and therefore Z/H (being the natural tangent of the dip) is equal to 2·75 nearly. Hence the kinetic-equilibrium error for the supposed case amounts in this locality to about a degree and three eighths for every degree of roll.

In an iron ship the equilibrium value of the rolling-error will be approximately the sum of the kinetic error investigated above, and a heeling-error found by an investigation readily worked out from that of Archibald Smith in the *Admiralty Compass Manual* (edit. 1869, Section IV. pages 82—89, and Appendix, pages 139—150), with modification to take into account the deviation of the apparent level, at the place of the compass, from the true gravitation-level.

I have used the expression "kinetic-equilibrium error" to distinguish the error investigated above from that actually exhibited by the compass. It is exactly the error which would be shown by an ideal compass with infinitely short period of vibration. A light quick needle (either with silk-fibre suspension, or supported on a point in the ordinary way) having a period of not more than about two seconds, shows the rolling-error very beautifully, taking at every instant almost exactly the position of kinetic equilibrium. I have thus found the rolling and pitching errors so great in a small wooden sailing-vessel that it became very difficult to make exact observations with the quick compass, either in the Frith of Clyde or out at sea on the Atlantic unless when the sea was exceptionally smooth. The well-known

* This would be the case for a ship of any size exposed to regular waves of length 184 feet from crest to crest, and, if moving through the water, moving in a line parallel to the lines of crests.

kinetic theory of "forced oscillations" is readily applied to calculate, whether for a wooden or an iron ship, the actual "rolling-error" of the compass, from the "kinetic-equilibrium error" investigated above. Thus let

u be the deviation of the compass at any instant, from the position it would have if the ship were at rest and upright;

T the period of its natural oscillation if unresisted by any "viscous" influence (the *damping* effect of copper, introduced by Snow Harris and used with good effect in the Admiralty standard compass, being included in this category);

$2f$ a coefficient measuring the amount of viscous resistance;

E the extreme equilibrium value of the rolling-error;

T' the period of the rolling.

For brevity put $n = 2\pi/T$, and $n' = 2\pi/T'$. The differential equation of the motion is

$$\frac{d^2u}{dt^2} + 2f\frac{du}{dt} + n^2u = n^2E\cos n't.$$

The integral of this proper to express the effect of regular rolling is

$$u = -E\frac{n^2\cos(n't+\epsilon)}{\sqrt{\{(n'^2-n^2)^2+4n'^2f^2\}}},$$

where
$$\epsilon = \tan^{-1}\frac{2n'f}{n'^2-n^2}.$$

It would extend the present communication too far to enter on details of this solution. For the present it is enough to say that no admissible degree of viscous resistance can make the rolling-error small enough for practical convenience, unless also the period of the compass is longer than that of any considerable rolling to which the ship may be subjected. Probably a period of from 15 to 30 seconds (such as an ordinary compass has) may be found necessary for general use at sea; and it becomes an important practical question how is this best to be obtained consistently with the smallness of the compass-needles necessary for a thoroughly satisfactory application of the system of magnetic correctors by which Airy proposed to cause the compass in an iron ship to point correct magnetic courses on all points?

45. On a new Form of Astronomical Clock with Free Pendulum and Independently Governed Uniform Motion for Escapement-wheel.

[The main part of paper No. 42 (1869) *supra*, *Popular Lectures and Addresses*, Vol. II. pp. 387—394 is reproduced, with the following addition. From *Brit. Assoc. Report*, 1876, Pt. II. pp. 49—52.]

I AM sorry to say that the hope here expressed has not hitherto been realised. Year after year passed producing only more or less of radical reform in various mechanical details of the governor and of the fine movement, until about six months ago, when, for the first time, I had all except the pendulums in approximately satisfactory condition. By that time I had discovered that my choice of zinc and platinum for the temperature compensation and lead for the weight of the pendulums was a mistake. I had fallen into it about ten years ago through being informed that in Russia the gridiron pendulum had been reverted to because of the difficulty of getting equality of temperature throughout the length of the pendulum; and without stopping to perceive that the right way to deal with this difficulty was to face it and take means of securing practical equality of temperature throughout the length of the pendulum (which it is obvious may be done by simple enough appliances), I devised a pendulum in which the compensation is produced by a stiff tube of zinc and a platinum wire placed nearly parallel each to the other throughout the length of the pendulum; and the two pendulums of the clock shown to the British Association were constructed on this plan. Now it is clear that the materials chosen for compensation should, of all those not otherwise objectionable, be those of greatest and of least expansibility. Therefore, certainly, glass or platinum ought to be one of the materials; and the steel of the ordinary astronomical mercury pendulum is a mistake. Mercury ought to be the other (its cubic expansion being six times the linear expansion of zinc), unless the capillary uncertainty of the mercury surface lead to irregular changes in the rate of the pendulum. The weight of the pendulum ought to be of material of the greatest specific gravity attainable, at all events unless the whole is to be mounted in an air-tight case, because one of the chief errors of the best existing pendulums is that depending on the variations of barometric pressure. The expense of platinum puts it out of the question for the weight of the pendulum, even although the use of mercury for the temperature compensation did not also give mercury for the weight. Thus even though as good compensation could be got by zinc and platinum as by any other means, mercury ought, on account of its superior specific gravity, to be preferred to lead for the weight of the pendulum.

I have accordingly now made several pendulums (for tide-gauges) with no other material in the moving part than glass and mercury, with rounded knife-edges of agate for the fixed support; and I am on the point of making four more for two new clocks which I am having made on the plan which forms the subject of this communication. I have had no opportunity hitherto of testing the performance of any of these pendulums; but their action seems very promising of good results, and the only untoward circumstance which has hitherto appeared in connexion with them has been breakages of the glass in two attempts to have one carried safely to Genoa for a tide-gauge made by Mr White to an order for the Italian Government.

As to the accuracy of my new clock, it is enough to look at the pendulum vibrating with perfect steadiness, from month to month, through a range of half a centimetre on each side of its middle position, with its pallets only touched during $\frac{1}{300}$ of the time by the escapement-tooth, to feel certain that, if the best ordinary astronomical clock owes any of its irregularities to variations of range of its pendulum, or to impulses and friction of its escapement-wheel, the new clock must, when tried with an equally good pendulum, prove more regular. I hope soon to have it tried with a better pendulum than that of any astronomical clock hitherto made; and if it then shows irregularities amounting to $\frac{1}{10}$ of those of the best astronomical clocks, the next step must be to inclose it in an air-tight case kept at constant temperature, day and night, summer and winter.

46. ELASTICITY VIEWED AS POSSIBLY A MODE OF MOTION.

[From *Roy. Institution Proc.* Vol. IX. March 4, 1881, pp. 520, 521; *Popular Lectures and Addresses*, Vol. I. pp. 142—146, or pp. 149—153 in later reprint.]

WITH reference to the title of his discourse the speaker said: "The mere title of Dr Tyndall's beautiful book, *Heat, a Mode of Motion*, is a lesson of truth which has manifested far and wide through the world one of the greatest discoveries of modern philosophy. I have always admired it; I have long coveted it for Elasticity; and now, by kind permission of its inventor, I have borrowed it for this evening's discourse.

"A century and a half ago Daniel Bernoulli shadowed forth the kinetic theory of the elasticity of gases, which has been accepted as truth by Joule, splendidly developed by Clausius and Maxwell, raised from statistics of the swayings of a crowd to observation and measurement of the free path of an individual atom in Tait and Dewar's explanation of Crookes' grand discovery of the radiometer, and in the vivid realisation of the old Lucretian torrents with which Crookes himself has followed up their explanation of his own earlier experiments; by which, less than two hundred years after its first discovery by Robert Boyle, 'the Spring of Air' is ascertained to be a mere statistical resultant of myriads of molecular collisions.

"But the molecules or atoms must have elasticity, and *this* elasticity must be explained by motion before the uncertain sound given forth in the title of the discourse, 'Elasticity viewed as possibly a Mode of Motion,' can be raised to the glorious certainty of 'Heat, a Mode of Motion'."

The speaker referred to spinning-tops, the child's rolling hoop, and the bicycle in rapid motion as cases of stiff, elastic-like firmness produced by motion; and showed experiments with gyrostats in which upright positions, utterly unstable without rotation, were maintained with a firmness and strength and elasticity such as might be by bands of steel. A flexible endless chain seemed rigid when caused to run rapidly round a pulley, and when caused to jump off the pulley, and let fall to the floor, stood stiffly upright for a time till its motion was lost by impact and friction of its links on the floor. A limp disc of indiarubber caused to rotate rapidly seemed to acquire the stiffness of a gigantic Rubens hat-brim. A little wooden ball, which when thrust down under still water jumped up again in a moment, remained down as if embedded in jelly when the water was caused to rotate rapidly, and sprang back, as if the water had elasticity like that of jelly, when it was struck by a stiff wire pushed down through the centre of the cork by which the glass vessel containing the water was filled*. Lastly, large smoke rings discharged from a circular or elliptic aperture in a box were rendered visible, by aid of the electric light, in their progress through the air of the theatre. Each ring was circular, and its motion was steady when the aperture from which it proceeded was circular, and when it was not disturbed by another ring. When one ring was sent obliquely after another the collision or approach to collision sent the two away in greatly changed directions, and each vibrating seemingly like an indiarubber band. When the aperture was elliptic each undisturbed ring was seen to be in a state of regular vibration from the beginning, and to continue so throughout its course across the lecture-room. Here, then, in water and air was elasticity as of an elastic solid, developed by mere motion. May not the elasticity of every ultimate atom of matter be thus explained? But this kinetic theory of matter is a dream, and can be nothing else, until it can explain chemical affinity, electricity, magnetism, gravitation, and the inertia of masses (that is, crowds) of vortices.

Le Sage's theory might give an explanation of gravity and of its relation *to inertia of masses*, on the vortex theory, were it not for the essential æolotropy of crystals, and the seemingly perfect

* [See *supra*, p. 170.]

isotropy of gravity. No finger-post pointing towards a way that can possibly lead to a surmounting of this difficulty, or a turning of its flank, has been discovered, or imagined as discoverable. Belief that no other theory of matter is possible is the only ground for anticipating that there is in store for the world another beautiful book to be called *Elasticity, a Mode of Motion.*

47. Steps towards a Kinetic Theory of Matter.

[From *British Association Report*, Montreal, 1884, pp. 613—622, Presidential Address to Section A. Reprinted in *Popular Lectures and Addresses*, Vol. I. pp. 218—252, or pp. 225—229 in later reprint.]

48. On a Gyrostatic Working Model of the Magnetic Compass.

[From *British Association Report*, 1884, pp. 625—628.]

In my communication to the British Association at Southport*, I explained several methods for overcoming the difficulties which had rendered nugatory, I believe, all previous attempts to realise Foucault's beautiful idea of discovering with perfect definiteness the earth's rotational motion by means of the gyroscope. One of these, which I had actually myself put in practice with partially satisfactory results, was a

Gyrostatic Balance for Measuring the Vertical Component of the Earth's Rotation.

It consisted of one of my gyrostats supported on knife edges attached to its containing case, with their line perpendicular to the axis of the interior flywheel and above the centre of gravity of the flywheel and framework by an exceedingly small height, when the framework is held with the axis of the flywheel and the line of knife edges both horizontal, and the knife edges downwards in proper position for performing their function. The apparatus, when supported on its knife edges with the flywheel not spinning, may be dealt with as the beam of an ordinary balance. Let now the framework bear two small knife edges, or knife-edged holes, like those of the beam of an ordinary balance, giving bearing points for weights in a line cutting the line of the knife edges as nearly as possible, and of course (unless there is reason to the

* No report of this communication has, so far as I know, hitherto appeared in print. [*Brit. Assoc. Report* 1883, p. 405, gives the title 'Gyrostatic Determination of the North and South Line and the Latitude of any Place.']

contrary in the shape of the framework) approximately perpendicular to this line; and, for convenience of putting on and off weights, hang, as in an ordinary balance, two very light pans by hooks on these edges in the usual way. Now, with the flywheel not running, adjust by weights in the pans if necessary, so that the framework rests in equilibrium in a certain marked position, with the axis of rotation inclined slightly to the horizontal in order that the axis of the flywheel, whether spinning or at rest, may always slip down so as to press on one and not on the other of the two end plates belonging to its two ends. Now, unhook the pans and take away the gyrostat and spin it; replace it on its knife edges, hang on the two pans, and find the weight required to balance it in the marked position with the flywheel now rotating rapidly. This weight, by an obvious formula which was placed before the Section at Southport, gives an accurate measure of the vertical component of the earth's rotation*.

Gyrostatic Model of the Dipping Needle.

I also showed at Southport that the gyrostatic balance described above, if modified by fixing the knife edges with their line passing as accurately as possible through the centre of gravity of the flywheel and framework, and with the faces of the knives so placed that they shall perform their function properly when the axis of the flywheel is parallel to the earth's axis of rotation, and the rotation of the flywheel in the same direction as the earth's, will act just as does an ordinary magnetic dipping needle; but showing latitude instead of dip, and dipping the South end of the axis downwards instead of the end that is towards the North as does the magnetic dipping needle. Thus, if the bearing of the knife edges be placed East and West, the gyrostat will balance with its axis parallel to the earth's axis, and therefore dipping with its South end downwards in northern latitudes and its North end downwards in southern latitudes. If displaced from this position and left to itself, it will oscillate according to precisely the same law as that by which the magnetic needle oscillates.

* The formula is $gw = a^{-1} W k^2 \omega \gamma \sin l$; where w denotes the balancing weight; gw the force of gravity upon it; a the arm on which this force acts; W the weight of the flywheel; k its radius of gyration; ω its angular velocity; γ the earth's angular velocity; and l the latitude of the place.

If the bearings be turned round in azimuth the position of equilibrium will follow the same law as does that of a magnetic dipping needle similarly dealt with. Thus, if the line of knife edges be North and South, the gyrostat will balance with the axis of the flywheel vertical, and if displaced from this position will oscillate still according to the same law; but with directive couple equal to the sine of the latitude into the directive couple experienced when the line of knife edges is East and West. Thus this piece of apparatus gives us the means of definitely measuring the direction of the earth's rotation, and the angular velocity of the rotation.

These experiments will, I believe, be very easily performed, although I have not myself hitherto found time to try them.

Gyrostatic Model of a Magnetic Compass.*

At Southport I showed that a gyrostat supported frictionlessly on a fixed vertical axis, with the axis of the flywheel horizontal or nearly so, will act just as does the magnetic compass, but with reference to "astronomical North" (that is to say rotational North) instead of "magnetic North." I also showed a method of mounting a gyrostat so as to leave it free to turn round a truly vertical axis, impeded by so little of frictional influence as not to prevent the realisation of the idea. The method, however, promised to be somewhat troublesome; and I have since found that the object of producing a gyrostatic model of the magnetic compass may, with a very remarkable dynamical modification, be much more simply attained by merely suspending the gyrostat by a very long fine wire or even by floating it with sufficient stability on a properly planned floater. To investigate the theory of this arrangement let us first suppose a gyrostat with the axis of its flywheel horizontal, to be hung by a very fine wire attached to its framework at a point, as far as can conveniently be arranged for, above the centre of gravity of flywheel and framework; and let the upper end of the wire be attached to a torsion head, capable of being turned round a fixed vertical axis as in a Coulomb's torsion balance. First, for simplicity, let us suppose the earth to be not rotating. The flywheel being set into rapid rotation, let the

* [Gyrostatic arrangements have been in recent times tried experimentally in the French and German navies, and are now in actual use.]

gyrostat be hung by the wire, and after being steadied as carefully as possible by hand, let it be left to itself. If it be observed to commence turning azimuthally in either direction, check this motion by the torsion head; that is to say, turn the torsion head gently in a direction opposite to the observed azimuthal motion until this motion ceases. Then do nothing to the torsion head, and observe if a reverse azimuthal motion supervenes. If it does, check this motion also by opposing it by torsion, but more gently than before. Go on until when the torsion head is left untouched the gyrostat remains at rest. The process gone through will have been undistinguishable from what would have had to be performed if, instead of the gyrostat with its rotating flywheel, a rigid body of the same weight, but with much greater moment of inertia about the vertical axis, had been in its place. The formula for the augmented moment of inertia is as follows. Denote by

W, the whole suspended weight of flywheel and framework,

K, the radius of gyration round the vertical through the centre of gravity of the whole mass regarded for a moment as one rigid body,

w, the mass of the flywheel,

k, the radius of gyration of the flywheel,

a, the distance of the point of attachment of the wire above the centre of gravity of flywheel and framework,

g, the force of gravity on unit mass,

ω, the angular velocity of the flywheel; the virtual moment of inertia round a vertical axis is

$$WK^2\left(1+\frac{w^2k^4\omega^2}{W^2K^2ag}\right)\qquad\qquad(1).$$

The proof is very easy. Here it is. Denote by

ϕ, the angle between a fixed vertical plane and the vertical plane containing the axis of the flywheel at any time t,

θ, the angle (supposed to be infinitely small and in the plane of ϕ) at which the line a is inclined to the vertical at time t,

H, the moment of the torque round the vertical axis exerted by the bearing wire on the suspended flywheel and framework.

By the law of generation of moment of momentum round an axis perpendicular to the axis of rotation requisite to turn the axis of rotation with an angular velocity $d\phi/dt$, we have

$$wk^2\omega\frac{d\phi}{dt}=gWa\theta \quad \text{.....................(2)},$$

because $gWa\theta$ is the moment of the couple in the vertical plane through the axis by which the angular motion $d\phi/dt$ in the horizontal plane is produced. Again by the same principle of generation of moment of momentum taken in connection with the elementary principle of acceleration of angular velocity, we have

$$wk^2\omega\frac{d\theta}{dt}+WK^2\frac{d^2\phi}{dt^2}=H \quad \text{.................(3)}.$$

Eliminating θ between these equations we find

$$\left(\frac{w^2k^4\omega^2}{gWa}+WK^2\right)\frac{d^2\phi}{dt^2}=H \quad \text{.................(4)},$$

which proves that the action of H in generating azimuthal motion is the same as it would be if a single rigid body of moment of inertia given by the formula (1), as said above, were substituted for the gyrostat.

Now to realise the gyrostatic model compass: arrange a gyrostat according to the preceding description with a very fine steel bearing wire, not less than 5 or 10 metres long (the longer the better; the loftiest sufficiently sheltered enclosure conveniently available should be chosen for the experiment). Proceed precisely as above to bring the gyrostat to rest by aid of the torsion head, attached to a beam of the roof or other convenient support sharing the earth's actual rotation. Suppose for a moment the locality of the experiment to be either the North or South pole, the operation to be performed to bring the gyrostat to rest will not be discoverably different from what it was, as we first imagined it when the earth was supposed to be not rotating. The only difference will be, that when the gyrostat hangs at rest relatively to the earth, θ will have a very small constant value; so small that the inclination of a to the vertical will be quite imperceptible, unless a were made so exceedingly small that the arrangement should give the result, to discover which was the object of the gyrostatic model balance described above, that is to say, to discover the vertical component of the earth's rotation. In reality we have made a as large as we conveniently can; and its inclination to the vertical will therefore be very small, when the moment of the tension of the wire round a horizontal axis perpendicular to the axis of rotation of the flywheel is just sufficient to cause the axis of the flywheel to turn round with the earth.

Let now the locality be anywhere except at the North or South pole; and now, instead of bringing the gyrostat to rest at random in any position, bring it to rest by successive trials in a position in which, judging by the torsion head and the position of the gyrostat, we see that there is no torsion of the wire. In this position the axis of the gyrostat will be in the North and South line, and, the equilibrium being stable, the direction of rotation of the flywheel must be the same as that of the component rotation of the earth round the North and South horizontal line, unless (which is a case to be avoided in practice) the torsional rigidity of the wire is so great as to convert into stability the instability which, with zero torsional rigidity, the rotational influence would produce, in respect to the equilibrium of the gyrostat with its axis reversed from the position of gyrostatic stability. It may be remarked, however, that even though the torsional rigidity were so great that there were two stable positions with no twist, the position of gyrostatic unstable equilibrium made stable by torsion would not be that arrived at: the position of stable gyrostatic equilibrium, rendered more stable by torsion, would be the position arrived at, by the natural process of turning the torsion head always in the direction of finding by trial a position of stable equilibrium with the wire untwisted by manipulation of the torsion head.

Now by manipulating the torsion head bring the gyrostat into equilibrium with its axis inclined, at any angle ϕ, to that position in which the bearing wire is untwisted; it will be found that the torque required to balance it in any oblique position will be proportional to $\sin\phi$.

The chief difficulty in realising this description results from the great augmentation of virtual moment of inertia, represented by the formula (1) above. The paper at present communicated to the section contains calculations on this subject, which throw light on many of the practical difficulties hitherto felt in any method of carrying out gyrostatic investigation of the earth's rotation, and which have led the author to fall back upon the method described by him at Southport, of which the essential characteristic is to constrain the frame of the gyrostat in such a manner as to leave it just one degree of freedom to move. The paper concludes with the description of a simplified manner of realising this condition for a gyrostatic compass—that is to say, a gyrostat free to move in a plane either rigorously or very approximately horizontal.

49. Gyrostatic Experiments.

[From *Proceedings of the Belfast Natural History and Philosophical Society*, April 16, 1889, pp. 89—91 (Abstract).]

...There are, however, other problems connected with gyrostatics which are far more difficult to solve. The lecturer next illustrated the rotatory stability of different forms of gyrostats, including prolate, oblate, and ordinary disc and gimbal-formed. He also gave an amusing solution of Columbus's problem how to make an egg stand on end. If the egg is hard-boiled it is practicable to spin it on end like a top, whereas the viscous fluid in the raw egg prevents its being treated in a similar manner.

He believed, if we are ever to solve the difficult problem of the elasticity of matter, it will be by the aid of the phenomena of rotation. Accepting the undulatory theory of light, we shall see that Faraday's brilliant discovery demonstrates the gyrostatic influence there. Nothing but the influence due to rotation could produce the effect which, as Faraday discovered, is produced by the magnet upon light passing through glass between the poles of the magnet. Some years ago we were all trying to find some kind of association between the vibrations of light and electricity, and a brilliant suggestion made by Professor FitzGerald (whom he was very pleased to see present) four years ago at Southport gave the key to the solution of the question. He suggested the employment of electric vibrators, and that suggestion had been realised in Hertz's splendid work within the past year, and the gap which we so desired to fill had been filled. It is almost impossible to go a step in the study of physics and dynamics without the aid of gyrostatics, and that is the reason the lecturer is interested in them—not merely because the phenomena they present are curious and interesting in themselves. But in studying and reconciling the laws of light, the laws of magnetism, the laws of electricity, and the laws of the elasticity of matter, gyrostatics play an undoubtedly important part.

Sir William concluded with a number of experiments tending to show the effect of gyrostatic domination in giving stability where instability exists, etc. One of the most interesting examples was the propulsion of smoke wreaths or rings, demonstrating the power of rotatory motion on so delicate a medium. Another curious effect obtained was the imparting of stability to water by means of rapid rotatory motion. In conclusion, he said that although the theory of which he had endeavoured to give some explanation is by no means complete, yet it will doubtless in time be rendered so, and meanwhile anything that tends to advance even a step towards the desired end is worthy of our attention.

50. On Some Test Cases for the Maxwell-Boltzmann Doctrine regarding Distribution of Energy*.

[From *Roy. Soc. Proc.* Vol. L. June 11, 1891, pp. 79—88; *Nature*, Vol. XLIV. pp. 355—358.]

1. MAXWELL, in his article (*Phil. Mag.* 1860) "On the Collision of Elastic Spheres," enunciates a very remarkable theorem, of primary importance in the kinetic theory of gases, to the effect that, in an assemblage of large numbers of mutually colliding spheres of two or of several different magnitudes, the mean kinetic energy is the same for equal numbers of the spheres irrespectively of their masses and diameters; or, in other words, the time-averages of the squares of the velocities of individual spheres are inversely as their masses. The mathematical investigation given as a proof of this theorem in that first article on the subject is quite unsatisfactory; but the mere enunciation of it, even if without proof, was a very valuable contribution to science. In a subsequent paper ("Dynamical Theory of Gases," *Phil. Trans.* for May, 1866) Maxwell finds in his equation (34) (*Collected Works*, p. 47), as a result of a thorough mathematical investigation, the same theorem extended to include collisions between Boscovich points with mutual forces according to any law of distance, provided only that not more than two points are in collision (that is to say, within the distances of their mutual influence) simul-

* [In a discussion ensuing on this paper, the position of Boltzmann and Maxwell was supported, among others, by Lord Rayleigh, *Phil. Mag.* Vol. XXXIII. 1892, pp. 356—9; *Scientific Papers*, Vol. III. pp. 554—7; cf. also *Phil. Mag.* Vol. XLIX. 1900, pp. 98—118; *Scientific Papers*, Vol. IV. pp. 433—451. The subject is discussed and largely amplified by Lord Kelvin in the second part of the Royal Institution Lecture, "Nineteenth Century Clouds over the Dynamical Theory of Light and Heat," as expanded February 2, 1901, reprinted as Appendix B of *Baltimore Lectures*, cf. pp. 493—527.]

taneously. Tait confirms Maxwell's original theorem for colliding spheres of different magnitudes in an interesting and important examination of the subject in §§ 19, 20, 21 of his paper "On the Foundations of the Kinetic Theory of Gases" (*Trans. R. S. E.* for May, 1866).

2. Boltzmann in his "Studien über das Gleichgewicht der lebendigen Kraft zwischen bewegten materiellen Punkten" (*Sitzb. K. Akad. Wien*, October 8, 1868), enunciated a large extension of this theorem, and Maxwell a still wider generalisation in his paper "On Boltzmann's Theorem on the Average Distribution of Energy in a System of Material Points" (*Cambridge Phil. Soc. Trans.*, May 6, 1878, republished in vol. II. of Maxwell's *Scientific Papers*, pp. 713—741), to the following effect (p. 716):—

"In the ultimate state of the system, the average kinetic energy of two given portions of the system must be in the ratio of the number of degrees of freedom of those portions."

Much disbelief and doubt has been felt as to the complete truth, or the extent of cases for which there is truth, of this proposition.

3. For a test case, differing as little as possible from Maxwell's original case of solid elastic spheres, consider a hollow spherical shell and a solid sphere—globule we shall call it for brevity—within the shell. I must first digress to remark that what has hitherto by Maxwell and Clausius and others before and after them been called for brevity an "elastic sphere," is not an elastic solid, capable of rotation and of elastic deformation; and therefore capable of an infinite number of modes of steady vibration, into which, of finer and finer degrees of nodal sub-division and shorter and shorter periods, all translational energy would, if the Boltzmann-Maxwell generalised proposition were true, be ultimately transformed by collisions. The "smooth elastic spheres" are really Boscovich point-atoms, with their translational inertia, and with, for law of force, zero force at every distance between two points exceeding the sum of the radii of the two balls, and infinite repulsion at exactly this distance. We may use Boscovich similarly for the hollow shell with globule in its interior, and so do away with all question as to vibrations due to elasticity of material, whether of the shell or of the globule. Let us simply suppose the mutual action between the shell and the globule to be nothing

except at an instant of collision, and then to be such that their relative component velocity along the radius through the point of contact is reversed by the collision, while the motion of their centre of inertia remains unchanged.

4. For brevity, we shall call the shell and interior globule of § 3, a double molecule, or sometimes, for more brevity, a doublet. The "smooth elastic sphere" of § 3 will be called simply an atom, or a single atom; and the radius or diameter or surface of the atom will mean the radius or diameter or surface of the corresponding sphere. (This explanation is necessary to avoid an ambiguity which might occur with reference to the common expression "sphere of action" of a Boscovich atom.)

5. Consider now a vast number of atoms and doublets, enclosed in a perfectly rigid fixed surface, having the property of reversing the normal component velocity of approach of any atom or shell or doublet at the instant of contact of surfaces, while leaving unchanged the absolute velocity of the centre of inertia of the two. Let any velocity or velocities in any direction or directions be given to any one or more of the atoms or of the shells or globules constituting the doublets. According to the Boltzmann-Maxwell doctrine, the motion will become distributed through the system, so that ultimately the time-average kinetic energy of each atom, each shell, and each globule shall be equal; and therefore that of each doublet double that of each atom. This is certainly a very marvellous conclusion; but I see no reason to doubt it on that account. After all, it is not obviously more marvellous than the seemingly well proved conclusion, that in a mixed assemblage of colliding single atoms, some of which have a million million times the mass of others, the smaller masses will ultimately average a million times the velocity of the larger. But it is not included in Maxwell's proof for single atoms of different masses [(34) of his "Dynamical Theory of Gases" referred to above]; and the condition that the globules enclosed in the shells are prevented by the shells from collisions with one another violates Tait's condition [(C) of § 18 of "Foundations of K.T. Gases"], "that there is perfectly free access for collision between each pair of particles whether of the same or of different systems." An independent investigation of such a simple and definite case as that of the atoms and doublets defined in §§ 3—5 is desirable

as a test, or would be interesting as an illustration were test not needed, for the exceedingly wide generalisation set forth in the Boltzmann-Maxwell doctrine.

6. Next, instead of only a single globule within the shell of § 4, let there be a vast number. To fix ideas let the mass of the shell be equal to a hundred times the sum of the masses of the globules, and let the number of the globules be a hundred million million. Let two such shells be connected by a push-and-pull massless spring. Let all be given at rest, with the spring stretched to any extent; and then left free. According to the Boltzmann-Maxwell doctrine, the motion produced initially by the spring will become distributed through the system, so that ultimately the sum of the kinetic energies of the globules within each shell will be a hundred million million times the average kinetic energy of the shell. The average velocity* of the shell will ultimately be a hundred-millionth of the average velocity of the globules. A corresponding proposition in the kinetic theory of gases is that, if two rigid shells each weighing 1 gram, and containing a centigram of monatomic gas, be attached to the two prongs of a massless perfectly elastic tuning fork, and set to vibrate, the gas will become heated in virtue of its viscous resistance to the vibration excited in it by the vibration of the shell, until nearly all the initial energy of the tuning fork is thus spent.

7. Going back to the double molecules of § 5, suppose the internal globule to be so connected by massless springs with the shell that the globule is urged towards the centre of the shell with a force simply proportional to the distance between the centres of the two. This arrangement, which I gave in my Baltimore Lectures, in 1884, as an illustration for vibratory molecules embedded in ether, would be equivalent to two masses connected by a massless spring, if we had only motions in one line to consider; but it has the advantage of being perfectly isotropic, and giving for all motions parallel to any fixed line exactly the same result as if there were no motion perpendicular to it. When a pair of masses connected by a spring strike a fixed obstacle or a movable body, with the line of their centres not exactly perpendicular to the

* The "average velocity of a particle," irrespectively of direction, is (in the kinetic theory of gases) a convenient expression for the square root of the time-average of the square of its velocity.

tangent plane of contact, it is caused to rotate. No such complication affects our isotropic doublet. An assemblage of such doublets being given moving about within a rigid enclosing surface, will the ultimate statistics be, for each doublet*, equal average kinetic energies of motion of centre of inertia, and of relative motion of the two constituents?

8. If we try to answer this question synthetically, we find a complex and troublesome problem in the details of all but the very simplest case of collision which can occur, which is direct collision between two not previously vibrating doublets, or any collision of one not previously vibrating doublet against a fixed plane. In this case, if the masses of globule and shell are equal, a complete collision consists of two impacts at an interval of time equal to half the period of free vibration of the doublet, and after the second impact there is separation without vibration, just as if we had had single spheres instead of the doublets. But in oblique collision between two not previously vibrating doublets, even if the masses of shell and globule are equal, we have a somewhat troublesome problem to find the interval between the two impacts,

* This implies equal average kinetic energies of the two constituents; and, conversely, equal average kinetic energies of the two constituents, except in the case of their masses being equal, implies the equality stated in the text. Let u, u' be absolute component velocities of two masses, m, m', perpendicular to a fixed plane; U the corresponding component velocity of their centre of inertia; and r that of their mutual relative motion. We have

$$u = U - \frac{m'r}{m+m'}, \qquad u' = U + \frac{mr}{m+m'} \quad \text{.......................(1);}$$

whence
$$mu^2 - m'u'^2 = (m-m')\left[U^2 - \frac{mm'r^2}{(m+m')^2}\right] - \frac{4mm'}{m+m'}Ur \quad \text{..............(2).}$$

Now suppose the time-average of Ur to be zero. In every case in which this is so we have, by (2),

$$\text{Time-av. } \{mu^2 - m'u'^2\} = (m-m') \times \text{Time-av. } \left\{U^2 - \frac{mm'r^2}{(m+m')^2}\right\} \quad \text{......(3).}$$

Hence in any case in which

$$\text{Time-av. } mu^2 = \text{Time-av. } m'u'^2 \quad \text{..........................(4)}$$

we have
$$(m-m') \times \text{Time-av. } \left\{U^2 - \frac{mm'r^2}{(m+m')^2}\right\} = 0 \quad \text{....................(5),}$$

and therefore, except when $m = m'$, we must have

$$\text{Time-av. } (m+m')U^2 = \text{Time-av. } \frac{mm'r^2}{m+m'} \quad \text{....................(6),}$$

which proves the proposition, because, as we readily see from (1), $\frac{1}{2}mm'r^2/(m+m')$ is, in every case, the kinetic energy of the relative motions, $u-U$, and $U-u'$.

when there are two, and to find the final resulting vibration. When the component relative motion parallel to the tangent plane of the first impact exceeds a certain value depending on the radius of the outer surface of the shell, the period of free vibration of the doublets, and the relative velocity of approach; there is no second impact, and the doublets separate with no relative velocity perpendicular to the tangent plane, but each with the energy of that component of its previous motion converted into vibrational energy. When the mass of the shell is much smaller than the mass of the interior globule, almost every collision will consist of a large number of impacts. It seems exceedingly difficult to find how to calculate true statistics of these chattering collisions, and arrive at sound conclusions as to the ultimate distribution of energy in any of the very simplest cases other than Maxwell's original case of 1860; but, if the Boltzmann-Maxwell generalised doctrine is true, we ought to be able to see its truth as essential, with special clearness in the simplest cases, even without going through the full problem presented by the details. I can find nothing in Maxwell's latest article on the subject (*Camb. Phil. Trans.* May 6, 1878), or in any of his previous papers, proving an affirmative answer to the question of § 7.

9. Going back to § 6, let the globules be initially distributed as nearly as may be homogeneously through the hollow; let each globule be connected with neighbours by massless springs; and let all the globules which are near the inner surface of the shell be connected with it also by massless springs. Or let any number of smaller shells be enclosed within our outer shell, and connected by massless springs as represented by the accompanying diagram, taken from a reprint of my Baltimore Lectures now in progress. Let two such outer shells, given at rest with their systems of globules in equilibrium within them, be connected by massless springs, and be started in motion, as were the shells of § 6. There will not now be the great loss of energy from the vibration of the shells which there was in § 6. On the contrary, the ultimate average kinetic energy of the whole two hundred million million globules will be certainly small in comparison with the ultimate average kinetic energy of the single shell. It may be because each globule of § 6 is free to wander that the energy

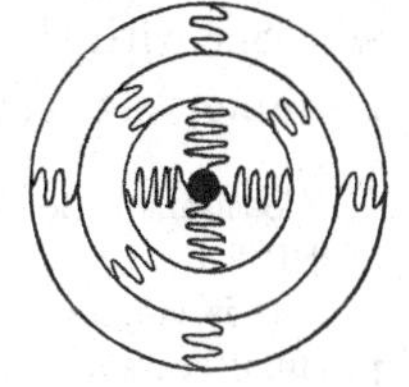

is lost from the shell in that case, and distributed among them. There is nothing vague in their motion allowing them to take more and more energy, now when they are connected by the massless springs. If we suppose the motions infinitesimal, or if, whatever their ranges may be, all forces are in simple proportion to displacements, the elementary dynamical theorem of *fundamental modes* shows how to find determinately each of the 600 million million and six simple harmonic vibrations of which the motion resulting from the prescribed initial circumstances is constituted. It tells us that the sum of the potential and kinetic energies of each mode remains always of constant value, and that the time-average of the changing kinetic energy during its period is half of this constant value. Without fully solving the problem for the 600 million million and six coordinates, it is easy to see that the gravest fundamental mode of the motion actually produced in the prescribed circumstances differs but little in period and energy from the single simple harmonic vibration which the two shells would take if the globules were rigidly connected to them, or were removed from within them, and the other initial circumstances were those of § 6. But this conclusion depends on the forces being *rigorously* in simple proportion to displacements.

10.* In no real case could they be so, and if there is any deviation from the simple proportionality of force to displacement, the independent superposition of motions does not hold good. We have still a theorem of fundamental modes, although, so far as I know, this theory has not yet been investigated†. For any stable system moving with a given sum, E, of potential and kinetic energies, there must in general be *at least as many fundamental modes of rigorously periodic motion as there are freedoms* (or independent variables). But the configuration of each fundamental mode is now not *generally* similar‡ for different values of E; and superposition of different fundamental modes, whether with the same or with different values of E, *has now no meaning*. It seems to me probable that every fundamental mode

* Sections 10 to 17 added July 10, 1891.

† [Cf. however Poincaré, *Mécanique Céleste*, or as quoted *infra*, p. 511.]

‡ It is similar for *adynamic* cases, that is to say, cases in which there is no potential energy, as, for example, a particle constrained to remain on a surface and moving along a geodetic line under the influence of no "applied" force.

is essentially unstable. It is so if Maxwell's fundamental assumption* "that the system if left to itself in its actual state of motion, will, sooner or later, pass through every phase which is consistent with the equation of energy" is true. It seems to me quite probable that this assumption *is* true, provided the "actual state of motion" is not exactly, as to position and velocity, a configuration of some one of the fundamental modes of rigorously periodic motion, and provided also that the "system" has not any exceptional character, such as those indicated by Maxwell for cases in which he warns† us that his assumption does not hold good.

11. But, conceding Maxwell's fundamental assumption, I do not see in the mathematical workings of his paper‡ any proof of his conclusion "that the average kinetic energy corresponding to any one of the variables is the same for every one of the variables of the system." Indeed, as a general proposition its meaning is not explained, and seems to me inexplicable. The reduction of the kinetic energy to a sum of squares§ leaves the several parts of the whole with no correspondence to any defined or definable set of independent variables. What, for example, can the meaning of the conclusion|| be for the case of a jointed pendulum? (a system of two rigid bodies, one supported on a fixed, horizontal axis and the other on a parallel axis fixed relatively to the first body, and both acted on only by gravity). The conclusion is quite intelligible, however (but is it true?), when the kinetic energy is expressible as a sum of squares of rates of change of single co-ordinates each multiplied by a function of all, or of some, of the co-ordinates. Consider, for example, the still easier case of these coefficients constant.

12. Consider more particularly the easiest case of all, motion of a single particle in a plane, that is the case of just two independent variables, say, x, y; and kinetic energy equal to $\frac{1}{2}(\dot{x}^2+\dot{y}^2)$. The equations of motion are

$$\frac{d^2x}{dt^2}=-\frac{dV}{dx}, \qquad \frac{d^2y}{dt^2}=-\frac{dV}{dy},$$

where V is the potential energy, which may be any function of

* *Scientific Papers*, Vol. II. p. 714.

† *Ibid.* pp. 714, 715.

‡ *Ibid.* pp. 716—726.

§ *Ibid.* p. 722.

|| Or of Maxwell's "b," in p. 723.

x, y, subject only to the condition (required for stability) that it is essentially positive (its least value being, for brevity, taken as zero). It is easily proved that, with any given value, E, for the sum of kinetic and potential energies there are two determinate modes of periodic motion; that is to say, there are two finite closed curves such that if m be projected from any point of either with velocity equal to $\sqrt{[2(E-V)]}$ in the direction, eitherwards, of the tangent to the curve, its path will be exactly that curve. In a very special class of cases there are only two such periodic motions, but it is obvious that there are more than two in other cases.

13. Take, for example,

$$V = \tfrac{1}{2}(\alpha^2 x^2 + \beta^2 y^2 + cx^2 y^2).$$

For all values of E we have

$$\left.\begin{array}{l} x = a\cos(\alpha t - e) \\ y = 0 \end{array}\right\} \text{ and } \left.\begin{array}{l} x = 0 \\ y = b\cos(\beta t - f) \end{array}\right\}$$

as two fundamental modes. When E is infinitely small we have only these two; but for any finite value of E we have clearly an infinite number of fundamental modes, and *every mode* differs infinitely little from being a fundamental mode. To see this let m be projected from any point N in OX, in a direction perpendicular to OX, with a velocity equal to $\sqrt{(2E - \alpha^2 ON^2)}$. After a sufficiently great number of crossings and re-crossings across the line $X'OX$, the particle will cross this line very nearly at right angles, at some point, N'. Vary the position of N very slightly in one direction or other, and re-project m from it perpendicularly and with proper velocity; till (by proper "trial and error" method) a path is found, which, after still the same number of crossings and re-crossings, crosses exactly at right angles at a point N'', very near the point N'. Let m continue its journey along this path and, after just as many more crossings and re-crossings, it will return *exactly* to N, and cross OX there, *exactly* at right angles. Thus the path from N to N'' is exactly half an orbit, and from N'' to N the remaining half.

14. When $cE/(\alpha^2\beta^2)$ is a small numeric, the part of the potential energy expressed by $\tfrac{1}{2}cx^2y^2$ is very small in comparison with the total energy, E. Hence the path is at every time very nearly the resultant of the two primary fundamental notes formu-

lated in § 13; and an interesting problem is presented, to find (by the method of the "variation of parameters") a, e, b, f, slowly varying functions of t, such that

$$x = a \ \sin(\alpha t - e), \qquad y = b \ \sin(\beta t - f),$$
$$\dot{x} = a\alpha \cos(\alpha t - e), \qquad \dot{y} = b\beta \cos(\beta t - f),$$

shall be the rigorous solution, or a practical approximation to it. Careful consideration of possibilities in respect to this case [$cE/(\alpha^2\beta^2)$ very small] seems thoroughly to confirm Maxwell's fundamental assumption quoted in § 10; and that it is correct whether $cE/(\alpha^2\beta^2)$ be small or large seems exceedingly probable, or quite certain.

15. But it seems also probable that Maxwell's *conclusion*, which for the case of a material point moving in a plane is

$$\text{Time-av. } \dot{x}^2 = \text{Time-av. } y^2 \quad \dots\dots\dots\dots\dots\dots(1)$$

is not true when α^2 differs from β^2. It is certainly not proved. No dynamical principle except the equation of energy,

$$\tfrac{1}{2}(\dot{x}^2 + \dot{y}^2) = E - V \quad \dots\dots\dots\dots\dots\dots(2),$$

is brought into the mathematical work of pp. 722—725, which is given by Maxwell as proof for it. Hence any arbitrarily drawn curve might be assumed for the path without violating the dynamics which enters into Maxwell's investigation; and we may draw curves for the path such as to satisfy (1), and curves not satisfying (1), but all traversing the whole space within the bounding curve

$$\tfrac{1}{2}(\alpha^2 x^2 + \beta^2 y^2 + c x^2 y^2) = E \quad \dots\dots\dots\dots\dots\dots(3),$$

and all satisfying Maxwell's fundamental assumption (§ 10).

16. The meaning of the question is illustrated by reducing it to a purely geometrical question regarding the path, thus:—calling θ the inclination to x of the tangent to the path at any point xy, and q the velocity in the path, we have

$$\dot{x} = q \cos\theta, \qquad \dot{y} = q \sin\theta \quad \dots\dots\dots\dots\dots\dots(4),$$

and therefore, by (2) $\quad q = \sqrt{\{2(E - V)\}} \quad \dots\dots\dots\dots\dots\dots(5).$

Hence, if we call s the total length of curve travelled,

$$\int \dot{x}^2 dt = \int q \cos^2\theta \, q dt = \int \sqrt{\{2(E - V)\}} \cos^2\theta \, ds \quad \dots\dots(6)$$

and the question of § 15 becomes, Is or is not

$$\frac{1}{S}\int_0^S ds \sqrt{\{2(E - V)\}} \cos^2\theta = \frac{1}{S}\int_0^S ds \sqrt{\{2(E - V)\}} \sin^2\theta \, ? \quad \dots(7),$$

where S denotes so great a length of path that it has passed a great number of times very near to every point within the boundary (3), very nearly in every direction.

17. Consider now separately the parts of the two members of (7) derived from portions of the path which cross an infinitesimal area $d\sigma$ having its centre at (x, y). They are respectively

$$\sqrt{\{2(E-V)\}}\, d\sigma \int_0^\pi N d\theta \cos^2\theta \quad \text{and} \quad \sqrt{\{2(E-V)\}}\, d\sigma \int_0^\pi N d\theta \sin^2\theta \qquad \text{.........(8)},$$

where $Nd\theta$ denotes the number of portions of the path, per unit distance in the direction inclined $\frac{1}{2}\pi + \theta$ to x, which pass eitherwards across the area in directions inclined to x at angles between the values $\theta - \frac{1}{2}d\theta$ and $\theta + \frac{1}{2}d\theta$. The most general possible expression for N is, according to Fourier,

$$\left.\begin{aligned} N = A_0 + A_1 \cos 2\theta + A_2 \cos 4\theta + \&c. \\ + B_1 \sin 2\theta + B_2 \sin 4\theta + \&c. \end{aligned}\right\} \qquad \text{.........(9)}.$$

Hence the two members of (8) become respectively

$$\sqrt{\{2(E-V)\}}\, d\sigma\, \tfrac{1}{2}\pi (A_0 + \tfrac{1}{2}A_1) \quad \text{and} \quad \sqrt{\{2(E-V)\}}\, d\sigma\, \tfrac{1}{2}\pi (A_0 - \tfrac{1}{2}A_1) \qquad \text{.........(10)}.$$

Remarking that A_0 and A_1 are functions of x, y, and taking $d\sigma = dx dy$, we find, from (10), for the two totals of (7) respectively

$$\left.\begin{aligned} &\tfrac{1}{2}\pi \iint dx dy (A_0 + \tfrac{1}{2}A_1) \sqrt{[2(E-V)]} \\ \text{and} \quad &\tfrac{1}{2}\pi \iint dx dy (A_0 - \tfrac{1}{2}A_1) \sqrt{[2(E-V)]} \end{aligned}\right\} \qquad \text{.........(11)},$$

where $\iint dx dy$ denotes integration over the whole space enclosed by (3). These quantities are equal if and only if $\iint dx dy A_1$ vanishes; it does so, clearly, if $\alpha = \beta$; but it seems improbable that, except when $\alpha = \beta$, it can vanish generally; and unless it does so, our present test case would disprove the Boltzmann-Maxwell general doctrine.

51. On a Decisive Test-case disproving the Maxwell-Boltzmann Doctrine regarding Distribution of Kinetic Energy*.

[From *Roy Soc. Proc.* Vol. LI. April 28, 1892, pp. 397—399; *Phil. Mag.* Vol. XXXIII. pp. 466, 467.]

The doctrine referred to is that stated by Maxwell in his paper "On the Average Distribution of Energy in a System of Material Points" (*Camb. Phil. Soc. Trans.* May 6, 1878, republished in Vol. II. of Maxwell's *Scientific Papers*) in the following words:—

"In the ultimate state of the system, the average kinetic energy of two given portions of the system must be in the ratio of the number of degrees of freedom of those portions."

Let the system consist of three bodies, A, B, C, all movable only in one straight line, KHL: B being a simple vibrator controlled by a spring so stiff that when, at any time, it has very nearly the whole energy of the system, its extreme excursions on each side of its position of equilibrium are small: C and A, equal masses: C, unacted on by force except when it strikes L, a fixed barrier, and when it strikes or is struck by B: A, unacted on by force except when it strikes or is struck by B, and when it is at less than a certain distance, HK, from a fixed repellent barrier, K, repelling with a force, F, varying according to any law, or constant, when A is between K and H, but becoming infinitely great when (if at any time) A reaches K, and goes infinitesimally beyond it.

* [The validity of the rejoinders made by Boltzmann, Poincaré, and Rayleigh is admitted in *Baltimore Lectures, loc. cit. supra*, p. 504.]

Suppose now A, B, C to be all moving to and fro. The collisions between B and the equal bodies A and C on its two sides must equalise, and keep equal, the average kinetic energy of A, immediately before and after these collisions, to the average kinetic energy of C. Hence, when the times of A being in the space between H and K are included in the average, the average of the *sum of the potential and kinetic energies of* A is equal to the average kinetic energy of C. But the potential energy of A at every point in the space HK is positive, because, according to our supposition, the velocity of A is diminished during every time of its motion from H towards K, and increased to the same value again during motion from K to H. Hence, the average kinetic energy of A is less than the average kinetic energy of C!

K

F

A

H

B

C

L

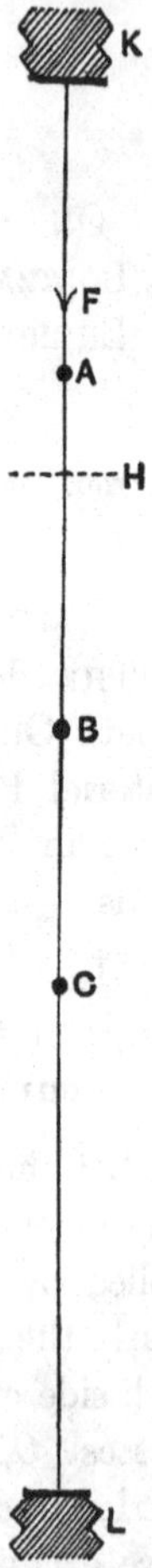

This is a test-case of a perfectly representative kind for the theory of temperature, and it effectually disposes of the assumption that the temperature of a solid or liquid is equal to its average kinetic energy per atom, which Maxwell pointed out as a consequence of the supposed theorem, and which, believed to be thus established, has been largely taught, and fallaciously used, as a fundamental proposition in thermodynamics.

It is in truth only for an approximately "perfect" gas, that is to say, an assemblage of molecules in which each molecule moves for comparatively long times in lines very approximately straight, and experiences changes of velocity and direction in comparatively very short times of collision, and it is only for the kinetic energy of the translatory motions of the "perfect gas," that the temperature is equal to the average kinetic energy per molecule, as first assumed by Waterston, and afterwards by Joule, and first proved by Maxwell.

52. ON PERIODIC MOTION OF A FINITE CONSERVATIVE SYSTEM.

[From *Philosophical Magazine*, Vol. XXXII. October, 1891, pp. 375—383, December, 1891, pp. 555—560.]

1. IN a recent communication to the Royal Society* I suggested an extension to stable systems in general of the well-known theory of "fundamental modes" for systems in which the potential energy is a quadratic function of coordinates and the kinetic energy a quadratic function of velocities, each with constant coefficients. This extension is the subject of the present communication to the British Association†.

2. In its title, "finite" means that the number of freedoms is finite and that the distance between no two points of the system can increase without limit. "Conservative" means that the kinetic energy is always altered by the same difference when the system passes from either to the other of any two configurations, whatever be the amount given to it when the system is projected from any configuration and left to move off undisturbed. By "path of a system" we shall understand, in generalized dynamics, the succession of configurations through which the system passes in any actual motion: or the group of single lines constituting the paths traversed by all points of the system. By an "orbit" we shall understand a circuital path; or a "path" of which every constituent line is a complete circuit, and all moving points are always at corresponding points of their circuits at the same time.

3. It will be convenient, though not necessary, to occasionally use the expression, "potential energy of the system in any configuration." When used at all it shall mean the difference by

* *Proceedings*, June, 1891 [*supra*, p. 490].

† §§ 1—10, and §§ 17—22 having been read before Section A of the British Association at its recent meeting in Cardiff.

which the kinetic energy is diminished when the system passes to the configuration considered, from *a* configuration or *the* configuration such that passage to any other permitted configuration involves diminution of the kinetic energy. By "total energy" of the system in any condition will be meant the sum of its kinetic and potential energies.

4. *Theorem of periodic motion.* For every given value, E, of the total energy, there is a fully determinate orbit such that if the system be set in motion along it, at any configuration of it, with the given total energy, E, it will circulate periodically in it.

5. To prove this theorem, suppose the number of freedoms to be i. Any configuration, Q, is fully specified by i given values for the i coordinates respectively. Suppose now the system to pass through some configuration, Q, at two times separated by an interval T, and to have the same velocities and directions of motion at those times. The path thus travelled in this interval is an orbit, and it is periodically travelled over in successive intervals each equal to T. To find how to procure fulfilment of our supposition, let the system be started from any configuration, Q, with any i values for the i velocity-components (or rates of variation per unit-time of the i coordinates). To cause it to return to Q after some unknown time T, we have $i-1$ equations to be satisfied: and to cause $i-1$ of its velocity-components to have the same values at the second as at the first passage through Q we have $i-1$ equations to satisfy; and, in virtue of the equation of energy, the remaining velocity-component also must have the same value at the two times. That the total energy may have the prescribed value, E, we have another equation. Thus we have in all $2i-1$ equations, among coordinates and velocity-components. Eliminate among these the i velocity-components, and there remain $i-1$ equations among the i coordinates which are the conditions necessary and sufficient to secure that Q is a configuration of an orbit of total energy E. Being $i-1$ equations among i coordinates, they leave only one freedom, that is to say they fully determine one path; of which, in the language of generalized analytical geometry, they are the equations. The or any path so determined is *an* orbit of total energy, E. Thus is proved the Theorem of § 4.

6. The solution of the determinate problem of finding an orbit whose total energy has the prescribed value E, is, in general, infinitely multiple, with different periods for the infinite number of different orbits determined by it.

7. A simple illustration with only two freedoms, will help to the full understanding of § 6 for every case, of any number of variables. Consider a jointed double pendulum consisting of two rigid bodies, A and B: one (A) supported on a fixed horizontal axis, I; the other (B) supported on a parallel axis, J, fixed relatively to A: and for simplicity let G, the centre of gravity of A, be in the plane of the two axes. Call H the centre of gravity of B. Let ϕ be the angle between the plane IJ and the vertical plane through I, which we shall call IV; and let ψ be the angle between the plane JH and the vertical. The coordinates and velocities of the system in any condition of motion are ϕ, ψ, $\dot{\phi}$, $\dot{\psi}$. The potential energy of the system, in kinetic units, will be gWz, where W denotes the sum of the masses, and z the height of their centre of gravity in any configuration of the system, above its lowest. Suppose now A to be placed in any particular position ϕ_0; and let it be required to find what must be the position, ψ_0, of B, and with what velocities, $\dot{\phi}_0$, $\dot{\psi}_0$ we must start A and B in motion, so that *the first time* ϕ has again the same value, ϕ_0, that is to say when A has made one complete turn in either direction, the system shall be wholly in the same position (ϕ_0, ψ_0) and moving with the same velocity ($\dot{\phi}_0$, $\dot{\psi}_0$) (in the same direction understood) as at the beginning. This implies only two equations, $\psi=\psi_0$, and either $\dot{\psi}=\dot{\psi}_0$ or $\dot{\phi}=\dot{\phi}_0$ (because either of these implies the other in virtue of the equation of energy). And we have just two disposables, ψ_0, and either $\dot{\psi}_0$ or $\dot{\phi}_0$ (the given total energy E determining either $\dot{\phi}_0$ or $\dot{\psi}_0$ when the other is known). The solution of this determinate problem is clearly possible, *unless E is too small*: but it is not generally unique. We may have solutions with the velocities of A and B started each in the positive direction, or each negative, or one negative and the other positive. If A is a flywheel of very great moment of inertia, and B a comparatively small pendulum hung on a crank-pin attached to it, and if for simplicity we suppose the crank to be counterpoised, so that the centre of gravity of A is in its axis, it is clear that, according to the greater or less value given for E, B may turn round and round many times before A comes again to its

primitive position ϕ_0. But it is clear that, though not generally unique, our problem of finding periodic motion with just one complete turn of A in its period has no real solution unless E is large enough; has many solutions for large enough values of E; but has not an infinite number of solutions for any finite value of E.

8. Again, let the condition be that, not the first time, but the *second* time A passes through its initial position, both co-ordinates and both velocities have their primitive values. When the given value of the total energy is not too great, the periodic motion which we now have will be purely vibratory; and the solution clearly duplex. But if E be great enough, A may still merely vibrate, while B may go round and round, first in one direction and then in the other, within the period of A's vibration. If the condition be that not at the first and not at the second, but at the third transit of A across its initial position, both co-ordinates and both velocities have their primitive values, we may, with sufficiently great total energy, have still wilder acrobatic performances, both bodies going round and round sometimes in one direction and sometimes in the other. Still with any finite value for E there is only a finite number of modes for the motion subject to the condition that the third transit of A through its initial position completes the first period. Wilder and wilder vagaries we have to think of if the first period is completed at the fourth transit of A; and so on.

9. This terrible Frankenstein of a problem is all involved in a very simple mathematical statement *not* including any declaration that it is the first, or the second, or the third, or other specified, transit of A that completes the first period. It will probably be convenient to arrange so as to find a transcendental equation which will have an infinite number of finite groups of roots equal to the periods of the modes of the periodic motions.

10. The case of no gravity presents a vastly simpler problem, of which the main solution has no doubt been many times found in terms of elliptic functions in the Cambridge Senate-house and Smith's Prize examinations. The character of the solution of this, as of all "adynamic" problems, is independent of the absolute value of the given energy, and of ϕ_0. It depends only on the value of the ratio $\dot{\psi}_0/\dot{\phi}_0$, which of course may be either positive

or negative. In the general solution $\psi - \phi$ is clearly a periodic function of the time; and our question of periodicity relatively to a *fixed* plane through I resolves itself into this:—During the period of the variation of $\psi - \phi$, is the change of ϕ either zero or a numeric commensurable with 2π? A corresponding question occurs for every case in which our "system" is free in space, without any fixed guides, and with no disturbing force from other bodies; as, for example, in the question of rigorous periodicity of the motion of three bodies such as the earth, moon, and sun, or of any finite number of mutually attracting bodies, such as the solar system, to be considered presently.

11. An (idealized) ordinary clock with weight, and pendulum, and dead-beat escapement, affords an interesting illustration. For simplicity let the cord be perfectly flexible and inextensible; let the cord-drum be rigidly fixed on the shaft of the escapement-wheel; let the escapement be rigidly fixed to the pendulum; and let the pendulum be a rigid body on perfect knife-edge bearings. Thus we have virtually two bodies, each with one freedom: A the escapement-wheel, cord, and weight; B the escapement and pendulum. Each impact of tooth on escapement is, in every clock and watch, followed by a mutual recoil. This recoil probably in almost all practical cases goes so far as to produce complete separation, followed by several more impacts and recoils before the tooth escapes, and the corresponding next tooth falls on the other side of the escapement. But there is a loss of energy by impacts and slipping, both on the non-working and on the working faces of the escapement. The loss on the working faces could be dispensed with: but the loss on the non-working faces is essential to the going of the clock. In our idealized clock we suppose each recoil to exactly reverse the relative motion of tooth and escapement in the direction perpendicular to the common tangent plane of the two surfaces at their point of impact; and we suppose the surfaces to be perfectly frictionless, so that the infinitely great mutual force at the instant of each impact is exactly in that direction. The jumping action thus produced would keep stopping the clock and letting it go on again: and would utterly prevent any regularity of going. Therefore I add the following arrangement of energy-receivers to annul the shocks on the non-working faces of the escapement:—Prolong the shaft of the escapement-wheel, and fix on it, in helical order,

sixty little arms each carrying at its end a disk, with its front face in a plane through the axis. Adjust the escapement-wheel on its shaft, so that when each of its thirty teeth strikes one or other of the two branches of the escapement, the lowest one of those sixty disks has its front face vertical. On a horizontal plane below the prolongation of the shaft place sixty little balls (energy-receivers) in such positions that each shall be struck by a disk *just* before the corresponding tooth touches the corresponding branch of the escapement. Let the mass of each ball be equal to the proper inertia-equivalent* of A, (the escapement-wheel, &c., and driving-weight). Each ball struck by it receives the whole kinetic energy which A had before the impact; and leaves A at rest, with a tooth of the escapement-wheel pressing on a non-working face of the escapement. Fix sixty rigid stops to prevent the balls from, in any circumstances, (§§ 13—16), going too far behind the positions in which they are initially placed. Each of these stops must be slotted, to allow the proper disk of the escapement-shaft to strike the ball and afterwards pass clear through with its carrying arm. For brevity these forked stops will be called the home-stops. Fix also sixty other stops (field-stops we shall call them) in such positions that the balls shall strike them all simultaneously and at exactly the instant (§ 12) when the weight strikes the bottom of the clock-case.

12. Suppose now the pendulum of our ideal clock, with its weight wound up very nearly to the top, to be started with sufficient range to let it keep going. For simplicity let this range be small enough to secure that when the weight is run down, the augmented range of vibration will still be within the limits allowed for proper action of the escapement mechanism. Let the bottom of the clock-case be a rigid horizontal plane fixed relatively to the framework bearing the wheel and pendulum in exactly such a position that when the weight, in running down, strikes it, the pendulum is at either end of its range. The weight jumps up after the impact, and the clock goes backwards, the energy-receivers return home from their field-stops at exactly

* Let r be the radius of the cord-drum: W the driving-weight: k the radius of gyration, and w the weight of the whole rotating body consisting of cord-drum, escapement-wheel, shaft, and 60 arms and disks: a the length of each arm reckoned from the axis to the point of its disk which strikes the ball. The "proper inertia-equivalent" is $(Wr^2+wk^2)/a^2$.

the right times, retracing exactly every step till the weight (wound up by the energy of the pendulum and of the returning energy-receivers) passes through its initial position. If it is allowed to go higher till it strikes against an unadjusted stop, the clock may be stopped for a time, with the pendulum vibrating through a moderately small range, and one tooth of the wheel chattering against one working facet of the escapement: but "sooner or later" (very soon) the tooth will escape; the clock will again go forward; the weight will run down, and again strike the bottom and jump from it, this time *not* when the pendulum is quite exactly at either end of its range.

13. Complicated but quite orderly action will follow; and "sooner or later" a tooth will be hooked up by the escapement and the clock will go backwards a beat or two; but after a very few beats, if more than one, it will go forward till the weight strikes the bottom again. "Sooner or later" the bottom will be struck at a time when the pendulum is *very nearly* at rest at either end of its range and when several energy-receivers are in such positions as to arrive home and strike disks at right times; and the clock will go backwards for a good many beats.

14. "Sooner or later," that is to say after some finite number of millions of millions of years, the weight will strike the bottom when the pendulum is *so very nearly* at rest at either end of its range and all the sixty balls so very nearly striking each its field-stop, that the clock will be driven back, winding up the weight till it again strikes the top stop, and immediately, or after a very few beats, begins again to go forward and let the weight run down. But our subject is not the fortuitous concourse of atoms. It is "periodic motion of a finite system."

15. Returning therefore to the end of § 12; let the top stop be so adjusted that it is struck by the weight at an instant when the pendulum is at one end of its range. The clock instantly begins to go forward; and goes on retracing every step, and repeating every one of the numerous impacts, of its first forward motion; till the weight strikes the bottom exactly when each of the sixty balls is striking its field-stop, and when the pendulum is at one end of its range, the same end of its range as when the weight struck the bottom the first time. Thus a *perfectly* periodic motion goes on for ever.

16. During any of the half-periods in which the clock is going forward, and the weight running down, any moderate disturbance, such as a slight blow on the pendulum, or a holding of the escapement-wheel stopped for some time, large or small, will make no noticeable difference in the subsequent motion: till the weight reaches the bottom of its range, when we find that the periodicity is lost, and the state of things described in §§ 13, 14 supervenes. But any such disturbance during a half-period when the clock is going backwards causes the backward motion to cease and regular forward motion to follow immediately, or after a few beats, a greater or less number according as the disturbance is exceedingly infinitesimal or but moderately small. This is a true dynamical illustration of the "dissipation of energy," and helps to show the vanity of attempts which have been made to found "Carnot's Principle," or "the Second Law of Thermodynamics," or theories of chemical action on Lagrange's generalized equations of motion.

17. Consider the "problem of the three bodies," in two varieties; first "the Lunar Theory," secondly "the Planetary Theory." One body (the Sun) is in each case vastly larger than either of the two others. In the first case the two others (the Earth and Moon) are so near one another in comparison with the Sun's distance from either that his force produces but a small disturbance of the relative motion of the Earth and Moon under their own mutual attraction. In the second case, two planets move each chiefly under the Sun's influence with comparatively small disturbance by their own mutual attraction. In each case we shall, for simplicity, neglect the motion of the Sun's centre of gravity, and consider him as an absolutely fixed "centre of force."

18. Taking first the lunar theory, suppose the centre of gravity, I, of Earth and Moon to move *very approximately* in a circle round the Sun. Now (without necessarily considering that the Moon is much smaller than the Earth) at an instant when the line MIE passes through S, give equal and opposite momentums to M and E in the line ME so as to annul their relative motion in this line if they had any, and to cause each to move exactly perpendicularly to it. If the next time their line passes through S they are again moving perpendicularly to ME,

their motion relatively to *SI* is rigorously periodic. This we see by considering that if both motions are reversed at any instant, *M* and *E* will exactly retrace their paths; and if such a reversal is made at an instant of perpendicularly crossing the line *SI*, the retraced paths are similar to the direct paths which are traced when there is no reversal.

19. Hence if the three bodies be given in line, *SME*, we secure rigorous periodicity of their motion if we project them in contrary directions perpendicular to this line with exactly such velocities that the next time *ME* is again in line with *S*, now *SEM*, their directions of motion are again perpendicular to *EM*. The problem of doing this has three solutions; in one of which the velocities of projection are so great that *M* and *E* are carried far away from one another, in opposite directions round the Sun till they again come near one another and in line on the far side of the Sun. Excluding this case we have certainly only two solutions left. In these *I* describes exceedingly nearly a circle round the Sun; while *M* and *E* move relatively to the point *I* and the line *IS*, somewhat approximately in circles, but to a second approximation in the ellipses corresponding to the lunar perturbation called the *variation*, and quite rigorously in two constant similar closed curves each differing very little from the variational ellipse. The centre of the variational ellipse is at *I*: its major axis is perpendicular to *SI* and exceeds the minor axis by approximately 1/179·6, being the square of the ratio (1/13·4) of the angular velocity of *SI* to the angular velocity of *ME*, each relative to an absolutely fixed direction. There are two solutions of this kind, in one of which (as in the actual case of Earth and Moon) *EM* turns samewards as, in the other contrary-wards to, *SI*.

20. If *ME* were two or three times as great as it is when the three bodies are in line, *SME*, and other dimensions the same, we should still have a solution for periodicity corresponding to that of § 19, but with the orbital curves of *M* and *E* round *I* differing very largely from circles and largely from ellipses. When *ME* exceeds a certain limit, this kind of solution becomes impossible. It would be not wholly uninteresting to follow the character of the orbital curves round *I* for increasing magnitudes of *ME* until they are lost. The solution referred to and rejected

in § 19 is still available and becomes now more interesting, but not so interesting as the corresponding solution in which M and E, now two planets, are projected so as to revolve in the same directions round the Sun.

21. *Rigorously periodic motion of two planets.* Given SVE in line, the Sun and two planets at distances such as those of Venus and the Earth:—it is required to project them with such velocities that the subsequent motion is rigorously periodic. A first solution is obtained by projecting them perpendicularly to SVE with such velocities that their periods of revolution round S are approximately equal; and exactly such that at the next time when VE is again in line with S, the motions are rigorously perpendicular to this line. The velocities which must be given to fulfil this condition must be such that the major axes of the ellipses approximately described are approximately equal. This solution, however, belongs rather to the Cometary than to the Planetary Theory.

22. Project the planets perpendicularly to SVE with such velocities that after some given number of times of their being in line with the Sun, their motions are, for the first time again, perpendicular to SVE. The determinate velocities which fulfil this condition must I think be such that the orbits are approximately ellipses of eccentricities not differing much from those required to make the major axes such that the periods have the proper commensurability to render the line of the three bodies at the second perpendicular crossing approximately coincident with their line at the initial perpendicular crossing.

On Instability of Periodic Motion, being a continuation of Article on Periodic Motion of a Finite Conservative System.

[From *Proc. Roy. Soc.* Nov. 26, 1891; *Phil. Mag.* Dec. 1891.]

23. Let $\psi, \phi, \chi, \vartheta, \ldots$ be generalized coordinates of a system; and let $A\ (\psi, \phi, \ldots \psi', \phi', \ldots)$ be the action in a path (§ 2 above) from the configuration $(\psi', \phi', \ldots)$ to the configuration $(\psi, \phi, \ldots)$ with kinetic energy $(E - V)$ with any given constant value for E, the total energy; V being the potential energy (§ 3 above), of which the value is given for every possible configuration of the

system. Let $\nu, \xi, \eta, \zeta \ldots$ and $\nu', \xi', \eta', \zeta' \ldots$ be the generalized component momentums of the system as it passes through the configurations $(\psi, \phi, \ldots)$ and $(\psi', \phi', \ldots)$ respectively. If by any means we have fully solved the problem of the motion of the system under the given forcive* (of which V is the potential energy), we know A for every given set of values of $\psi, \phi, \ldots \psi', \phi', \ldots$; that is to say, it is a known function of $(\psi, \phi, \ldots, \psi', \phi', \ldots)$. Then, by Hamilton's principle [Thomson and Tait's *Natural Philosophy*, § 330 (18)], we have

$$\left.\begin{array}{llll} \nu = \dfrac{dA}{d\psi}, & \xi = \dfrac{dA}{d\phi}, & \eta = \dfrac{dA}{d\chi}, & \zeta = \dfrac{dA}{d\vartheta}, \ldots \\ \nu' = -\dfrac{dA}{d\psi'}, & \xi' = -\dfrac{dA}{d\phi'}, & \eta' = -\dfrac{dA}{d\chi'}, & \zeta' = -\dfrac{dA}{d\vartheta'}, \ldots \end{array}\right\} \ldots (1).$$

24. Now let $P'P$ designate a particular path† from position $(\psi', \phi', \chi', \ldots)$ which for brevity we shall call P', to position $(\psi, \phi, \chi, \ldots)$ which we shall call P. Let ${}_0P'{}_0P$ be a part of a known periodic path, from which $P'P$ is infinitely little distant. But first, whether ${}_0P'{}_0P$ is periodic or not, provided it is infinitely near to $P'P$, and provided ${}_0P'$ and ${}_0P$ are infinitely near to P' and P, respectively, we have, by Taylor's theorem, and by (1),

$$\left.\begin{array}{l} A(\psi, \phi, \chi, \ldots, \psi', \phi', \chi', \ldots) \\ = A({}_0\psi, {}_0\phi, {}_0\chi, \ldots, {}_0\psi', {}_0\phi', {}_0\chi', \ldots) \\ + {}_0\nu(\psi - {}_0\psi) + {}_0\xi(\phi - {}_0\phi) + \ldots - {}_0\nu'(\psi' - {}_0\psi') - {}_0\xi'(\phi' - {}_0\phi') - \ldots \\ + \frac{1}{2}\left\{ {}_0\!\left(\dfrac{d^2A}{d\psi^2}\right)(\psi - {}_0\psi)^2 + {}_0\!\left(\dfrac{d^2A}{d\phi^2}\right)(\phi - {}_0\phi)^2 + \ldots \right. \\ \left. + 2\,{}_0\!\left(\dfrac{d^2A}{d\psi d\phi}\right)(\psi - {}_0\psi)(\phi - {}_0\phi) + \ldots \right\} \end{array}\right\} \ldots\ldots\ldots (2).$$

* This is a term introduced by my brother, Prof. James Thomson, to denote a force-system.

† For any given value of E, the total energy (§ 3 above), the problem of finding a path from any position P' to any position P is determinate. Its solution is, for each coordinate of the system, a determinate function of the coordinates which define P and P' and of t, the time reckoned from the instant of passing through P'. The solution is single for the case of a particle moving under the influence of no force; every path being an infinite straight line. For a single particle moving under the influence of a uniform force in parallel lines (as gravity in small-scale terrestrial ballistics) the solution is duplex or imaginary. For every constrainedly finite system the solution is infinitely multiple; as is virtually well known by every billiard player for the case of a Boscovichian atom flying about within an enclosing surface, and by every tennis player for the parabolas with which he is concerned, and their reflexions from walls or pavement.

25. Let us now simplify by choosing our coordinates so that the values of ϕ, χ, &c., are each zero for every position of the path ${}_0P'{}_0P$; and let ψ, for any position of this path, be the action along it reckoned from zero at ${}_0P'$. These assumptions, expressed in symbols, are as follows:—

$$\left.\begin{cases} \dfrac{dA}{d\phi}=0, \quad \dfrac{dA}{d\chi}=0,\ldots \quad \dfrac{dA}{d\psi}=1, \quad \dfrac{dA}{d\phi'}=0, \quad \dfrac{dA}{d\chi'}=0,\ldots \\ \text{for all values of } \psi \text{ and } \psi', \text{ if } \phi=0,\ \chi=0,\ldots;\ \phi'=0,\ \chi'=0,\ldots \end{cases}\right\} \quad \ldots\ldots\ldots(3).$$

26. Taking now

$$\psi'=0,\ \psi={}_0\psi,\ {}_0\phi=0,\ {}_0\chi=0,\ldots\ {}_0\psi'=0,\ {}_0\phi'=0,\ {}_0\chi'=0\ldots, \quad \ldots(4);$$

we have

$$A({}_0\psi, {}_0\phi, {}_0\chi, \ldots, {}_0\psi', {}_0\phi', {}_0\chi', \ldots) = A({}_0\psi, 0, 0, \ldots 0, 0, 0 \ldots) \quad \ldots(5)$$

and, in virtue of this and of (3) and (1), (2) becomes

$$\left.\begin{aligned} A(\psi, \phi, \chi, \ldots, 0, \phi', \chi', \ldots) = A({}_0\psi, 0, 0, \ldots\ 0, 0, 0) \\ + \tfrac{1}{2}\,[{}_{11}\phi^2 + {}_{22}\chi^2 + {}_{33}\vartheta^2 + {}_{44}\phi'^2 + {}_{55}\chi'^2 + {}_{66}\vartheta'^2 \\ + 2\,({}_{12}\phi\chi + {}_{13}\phi\vartheta + {}_{14}\phi\phi' + {}_{15}\phi\chi' + {}_{16}\phi\vartheta' \\ + {}_{23}\chi\vartheta + {}_{24}\chi\phi' + {}_{25}\chi\chi' + {}_{26}\chi\vartheta' \\ + {}_{34}\vartheta\phi' + {}_{35}\vartheta\chi' + {}_{36}\vartheta\vartheta' \\ + {}_{45}\phi'\chi' + {}_{46}\phi'\vartheta' \\ + {}_{56}\chi'\vartheta')] \end{aligned}\right\} \ldots(6),$$

where, merely for simplicity of notation, we suppose the total number of freedoms of the system, that is to say the total number of the coordinates ψ, ϕ, χ, ϑ, to be four; and for brevity put [printed by accident in (6) and (8) as subscripts]

$${}_0\!\left(\frac{d^2A}{d\phi^2}\right) = 11, \quad {}_0\!\left(\frac{d^2A}{d\phi\, d\chi}\right) = 12, \quad {}_0\!\left(\frac{d^2A}{d\chi^2}\right) = 22, \text{ \&c.} \quad \ldots(7).$$

27. From (6) we find, by (1),

$$\left.\begin{aligned} \xi &= {}_{11}\phi + {}_{12}\chi + {}_{13}\vartheta + {}_{14}\phi' + {}_{15}\chi' + {}_{16}\vartheta' \\ \eta &= {}_{21}\phi + {}_{22}\chi + {}_{23}\vartheta + {}_{24}\phi' + {}_{25}\chi' + {}_{26}\vartheta' \\ \zeta &= {}_{31}\phi + {}_{32}\chi + {}_{33}\vartheta + {}_{34}\phi' + {}_{35}\chi' + {}_{36}\vartheta' \\ -\xi' &= {}_{41}\phi + {}_{42}\chi + {}_{43}\vartheta + {}_{44}\phi' + {}_{45}\chi' + {}_{46}\vartheta' \\ -\eta' &= {}_{51}\phi + {}_{52}\chi + {}_{53}\vartheta + {}_{54}\phi' + {}_{55}\chi' + {}_{56}\vartheta' \\ -\zeta' &= {}_{61}\phi + {}_{62}\chi + {}_{63}\vartheta + {}_{64}\phi' + {}_{65}\chi' + {}_{66}\vartheta' \end{aligned}\right\} \quad \ldots\ldots\ldots(8).$$

These equations allow us to determine the three displacements, ϕ, χ, ϑ, and the three corresponding momentums, ξ, η, ζ, for any position on the path, in terms of the initial values ϕ', χ', ϑ', ξ', η', ζ', supposed known.

28. To introduce now our supposition (§ 24) that ${}_0P'{}_0P$ is part of a periodic path; let Q be a position on it between ${}_0P'$ and ${}_0P$; and let us now, to avoid ambiguity, call it ${}_0P'Q_0P$. Let ${}_0P'$ and ${}_0P$ now be taken to coincide in a position which we shall call O; in other words, let ${}_0P'Q_0P$, or OQO, be the complete periodic circuit, or orbit as we have called it (§ 2 above). Our path $P'P$ is now a path infinitely near to this orbit, and P' and P are two consecutive positions in it for which ψ* has the value zero. These two positions are infinitely near to one another and to O. We shall call them O_i, and O_{i+1}, considering them as the positions on our path in which ψ is zero for the ith time and for the $(i+1)$th time, from an earlier initial epoch than first passage through $\psi=0$ which we have been hitherto considering. It is accordingly convenient now to modify our notation as follows:—

$$\left.\begin{array}{llllll} \phi'=\phi_i, & \chi'=\chi_i, & \vartheta'=\vartheta_i; & \xi'=\xi_i, & \eta'=\eta_i, & \zeta'=\zeta_i \\ \phi=\phi_{i+1}, & \chi=\chi_{i+1}, & \vartheta=\vartheta_{i+1}; & \xi=\xi_{i+1}, & \eta=\eta_{i+1}, & \zeta=\zeta_{i+1} \end{array}\right\} \quad \text{.........(9)}.$$

Here ϕ_i, χ_i, ϑ_i are the generalized components of distance from O, at the ith transit through $\psi=0$, of the system pursuing its path infinitely near to the orbit; and ξ_i, η_i, ζ_i are the corresponding momentum-components. With the notation of (9), equations (8) become equations by which the values of these components for the $i+1$th time of transit through $\psi=0$ can be found from their values for the ith time. They are equations of finite differences, and are to be treated *secundum artem*, as follows.

29. Assume

$$\left.\begin{array}{lll} \phi_{i+1}=\rho\phi_i, & \chi_{i+1}=\rho\chi_i, & \vartheta_{i+1}=\rho\vartheta_i \\ \xi_{i+1}=\rho\xi_i, & \eta_{i+1}=\rho\eta_i, & \zeta_{i+1}=\rho\zeta_i \end{array}\right\} \quad \text{.........(10)}.$$

* [But the action continually increases along the path. What is meant is points which are equidistant, as regards action, by an amount which is equal to the action of one revolution in the periodic orbit. The subject may be elucidated by the simpler application to rays of light, where there are only two coordinates ϕ, χ, defining position in a plane transverse to the ray: cf. Thomson and Tait's *Nat. Phil.* § 334.]

Substituting accordingly in (8) modified by (9), and eliminating ξ_i, η_i, ζ_i, we find

$$\left.\begin{aligned}
&\left(11+\frac{14}{\rho}+41\rho+44\right)\phi+\left(12+\frac{15}{\rho}+42\rho+45\right)\chi \\
&\qquad\qquad+\left(13+\frac{16}{\rho}+43\rho+46\right)\vartheta=0 \\
&\left(21+\frac{24}{\rho}+51\rho+54\right)\phi+\left(22+\frac{25}{\rho}+52\rho+55\right)\chi \\
&\qquad\qquad+\left(23+\frac{26}{\rho}+53\rho+56\right)\vartheta=0 \\
&\left(31+\frac{34}{\rho}+61\rho+64\right)\phi+\left(32+\frac{35}{\rho}+62\rho+65\right)\chi \\
&\qquad\qquad+\left(33+\frac{36}{\rho}+63\rho+66\right)\vartheta=0
\end{aligned}\right\}\ldots(11).$$

Remarking that $41=14$, $12=21$, &c., we see that the determinant for the elimination of the ratios $\phi|\chi|\vartheta$ is symmetrical with reference to ρ and $1|\rho$. Hence it is

$$C_3(\rho^3+\rho^{-3})+C_2(\rho^2+\rho^{-2})+C_1(\rho+\rho^{-1})+2C_0 \quad \ldots(12),$$

where C_0, C_1, C_2, C_3 are coefficients of which the values in terms of 11, 12, &c. are easily written out. This determinant equated to zero gives an equation of the 6th degree for determining ρ, of which for each root there is another equal to its reciprocal. We reduce it to an equation of the third degree by putting

$$\rho+\rho^{-1}=2e \quad \ldots\ldots\ldots\ldots\ldots\ldots(13).$$

Let e_1, e_2, e_3 be the roots of the equation thus found. The corresponding values of ρ are

$$e_1 \pm \sqrt{(e_1^2-1)};\quad e_2 \pm \sqrt{(e_2^2-1)};\quad e_3 \pm \sqrt{(e_3^2-1)} \quad \ldots(14).$$

In the case of e having any real value between 1 and -1, it is convenient to put

$$\left.\begin{aligned}
&e=\cos\alpha, \\
\text{which gives}\qquad &\rho=\cos\alpha+\iota\sin\alpha \\
\text{and}\qquad &\rho^{-1}=\cos\alpha-\iota\sin\alpha
\end{aligned}\right\}\ldots\ldots\ldots\ldots\ldots(15).$$

30. Suppose now, for the first time of passing through $\psi = 0$, the three coordinates and three corresponding momenta, $\phi_1, \chi_1, \vartheta_1, \xi_1, \eta_1, \zeta_1$, to be all given; we find

$$\left.\begin{array}{l} \phi_{i+1} = A_1\rho_1{}^i + A_1'\rho_1{}^{-i} + A_2\rho_2{}^i + A_2'\rho_2{}^{-i} + A_3\rho_3{}^i + A_3'\rho_3{}^{-i} \\ \chi_{i+1} = B_1\rho_1{}^i + B_1'\rho_1{}^{-i} + B_2\rho_2{}^i + B_2'\rho_2{}^{-i} + B_3\rho_3{}^i + B_3'\rho_3{}^{-i} \\ \quad \cdots\cdots\cdots\cdots\cdots\cdots\cdots\cdots\cdots\cdots\cdots\cdots \\ \quad \cdots\cdots\cdots\cdots\cdots\cdots\cdots\cdots\cdots\cdots\cdots\cdots \\ \zeta_{i+1} = F_1\rho_1{}^i + F_1'\rho_1{}^{-i} + F_2\rho_2{}^i + F_2'\rho_2{}^{-i} + F_3\rho_3{}^i + F_3'\rho_3{}^{-i} \end{array}\right\} \ldots(16),$$

where $A_1, A_1', A_2, A_2' \ldots F_1, F_1', F_2, F_2'$ are 36 coefficients which are determined by the six equations (16), with $i = 0$: and the six equations (8), modified by (9); with i successively put $= 1, 2, 3, 4, 5$; with the given values substituted for $\phi_1, \chi_1, \vartheta_1, \xi_1, \eta_1, \zeta_1$, in them; and with for ϕ_2, χ_2, &c. their values by (16).

31. Our result proves that every path infinitely near to the orbit is unstable unless every root of the equation for e has a real value between 1 and -1. It does not prove that the motion *is* stable when this condition is fulfilled. Stability or instability for this case cannot be tested without going to higher orders of approximation in the consideration of paths very nearly coincident with an orbit.

POSTSCRIPT, *November* 10, 1891.

The subject of periodic motion and its stability has been treated with great power by M. Poincaré in a paper, "Sur le problème des trois corps et les équations de la dynamique," for which the prize of His Majesty the King of Sweden was awarded on the 21st of January, 1889. This paper, which has been published in Mittag-Leffler's *Acta Mathematica*, 13, 1 and 2 (270 4to pp.), Stockholm, 1890, only became known to me twelve days ago through Prof. Cayley. I am greatly interested to find in it much that bears upon the subject of my communication of last June to the Royal Society "On some Test Cases for the Maxwell-Boltzmann Doctrine regarding Distribution of Energy"; particularly in p. 239, the following paragraph:—"On peut démontrer que dans le voisinage d'une trajectoire fermée représentant une solution périodique, soit stable, soit instable, il passe une infinité d'autres trajectoires fermées. Cela ne suffit pas, en toute rigueur, pour conclure que toute région de l'espace, si petite qu'elle soit, est traversée par une infinité des trajectoires fermées mais cela suffit pour donner à cette hypothèse un haut caractère de vraisemblance *." This statement is exceedingly interesting in connexion with Maxwell's fundamental supposition quoted in § 10 of my paper, "that the system if left to itself in its actual state of motion, will, sooner or later, pass through every phase which is consistent with the equation of energy †"; an assumption which Maxwell gives not as a conclusion, but as a proposition which "we may with considerable confidence assert,...except for particular forms of the surface of the fixed obstacle." It will be seen that Poincaré's "hypothesis, having a high character of probability," does not go so far as Maxwell's, which asserts that every portion of space is traversed *in all directions* by *every trajectory*. The conclusion which I gave in § 13 ‡, as seeming to me quite certain, "that every mode differs infinitely little from being a fundamental mode," is clearly a necessary consequence of Maxwell's fundamental supposition; the truth of which still seems to me highly probable *provided exceptional cases are properly dealt with*.

I also find the following statement, pp. 100, 101:—"Il y aura donc en général n quantités α^2 distinctes. Nous les appellerons les *coefficients de stabilité* de la solution périodique considerée.

"Si ces n coefficients sont tous réels et négatifs, la solution périodique sera stable, car les quantités ξ_i et η_i resteront inférieures à une limite donnée.

"Il ne faut pas toutefois entendre ce mot de stabilité au sens absolu. En effet, nous avons négligé les carrés des ξ et des η, et rien ne prouve qu'en tenant compte de ces carrés le résultat ne serait pas changé. Mais nous pouvons dire au moins que les ξ et η, s'ils sont originairement très petits, resteront très petits pendant très longtemps. Nous pouvons exprimer ce fait en disant que la solution périodique jouit, sinon de la stabilité *séculaire*, du moins de la stabilité *temporaire*." Here the conclusion of § 31 of my present paper is perfectly anticipated and is expressed in a most interesting manner. M. Poincaré's investigation and mine are as different as two investigations of the same subject could well be, and it is very satisfactory to find perfect agreement in conclusions.

* The "trajectoire fermée" of M. Poincaré is what I called a "fundamental" mode of rigorously periodic motion," or "an orbit."

† *Scientific Papers*, Vol. II. p. 714. ‡ [*Supra*, p. 492.]

53. On a Theorem in Plane Kinetic Trigonometry suggested by Gauss's Theorem of Curvatura Integra.

[From *Philosophical Magazine*, Vol. xxxii. Nov. 1891, pp. 471—473.]

1. Albert Girard's beautiful theorem of the "Spherical Excess" in spherical trigonometry, published about 1637, and used practically 150 years later by General Roy in the trigonometrical survey of the British Isles, was splendidly extended by Gauss in his theorem* of the "Curvatura Integra." There must be a corresponding theorem in the "kinetic trigonometry" suggested

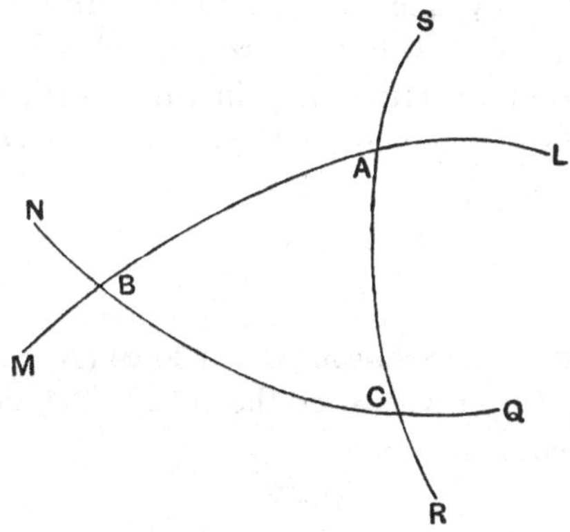

in the marginals of Thomson and Tait's *Natural Philosophy*, §§ 361 (*a*) (*b*) (*c*) (*d*), for the motion of the generalized conservative kinetic system of any number of variables. For the very simple case of a material point moving in a plane it is easily worked out, as I have found in endeavouring to write a continuation of my article (*Philosophical Magazine*, October) on the Periodic

* "Disquisitiones Generales circa Superficies Curvas; auctore Carolo Frederico Gauss; Societati Regiæ [Gottingensi] oblatæ D.VIII. Octobr. MDCCCXXVII." *Collected Works*, Vol. iv. Göttingen, 1873. Thomson and Tait's *Natural Philosophy*, §§ 131—138.

Motion of a Finite System, which I hope may be ready to appear in the December number. Here is the theorem meantime.

2. Let *LABM*, *NBCQ*, *RCAS* be three paths of a particle moving freely in a plane, under influence of a force $\left(\frac{-dV}{dx}, \frac{-dV}{dy}\right)$ and projected from any three places in any direction in the plane, with such velocities that the sum of the kinetic and potential energies has the same value (E) in each case. The sum of the three angles A, B, C exceeds two right angles by an amount which, reckoned in radians, is equal to the surface-integral of $\nabla^2 \log \sqrt{(E-V)}$, throughout the enclosed area ABC; ∇^2 denoting the Laplacian operation $\frac{d^2}{dx^2} + \frac{d^2}{dy^2}$.

3. To prove this; remark that

$$\iint dx\, dy \nabla^2 \psi = \int ds \frac{d\psi}{dn},$$

where ψ denotes any function of (x, y), $\iint dx dy$ surface-integration throughout any area, $\int ds$ line-integration all round its boundary, and $d\psi/dn$ rate of variation of ψ in the direction perpendicular to the boundary at any point. Hence the surface-integral mentioned in § 2 is equal to

$$\int ds \frac{-dV}{2(E-V)\,dn} \quad \text{.........................(1).}$$

But $-dV/dn$ is the normal-component force (N, we shall call it); and $2(E-V)$ is the square of the velocity (v^2, we shall call it). Hence (1) becomes

$$\int ds N/v^2 \quad \text{.........................(2).}$$

But N/v^2 is the curvature ($1/\rho$, we shall call it), at any point in any one of the three arcs AB, BC, CA. Hence, dividing $\int ds$ into the three parts belonging respectively to these three arcs, which we shall denote by $\int_A^B ds$, $\int_B^C ds$, $\int_C^A ds$, we find for (2),

$$\int_A^B \frac{ds}{\rho} + \int_B^C \frac{ds}{\rho} + \int_C^A \frac{ds}{\rho} \quad \text{.....................(3).}$$

But $\int_A^B \frac{ds}{\rho}$ is the change of direction in the arc AB, and similarly for the two others: hence the theorem.

54. On the Stability of Periodic Motion.

[From *British Association Report*, 1892, p. 638 (title only); *Nature*, Vol. XLVI. Aug. 18, 1892, p. 384.]

The mathematical investigation of this subject was illustrated by an experiment in which a simple harmonic vertical motion was given to the point of support of a pendulum. When the period of the superposed motion was one half of that of the natural motion of the pendulum, the equilibrium became unstable, and the slightest disturbance caused the vertical motion of the bob to be changed into transverse motion of increasing amplitude. If the superposed period were now lessened, the vertical motion again became stable. Similarly a rod poised vertically in unstable equilibrium could become stable by having its point of support moved with simple harmonic motion, of proper period, in a vertical line*.

Prof. Osborne Reynolds remarked that it was well known to practical engineers that a revolving shaft, when driven at a certain speed, began to bend, and might even break, though at higher speeds it would again become straight. Lord Kelvin had now explained this effect.

* [Cf. Lord Rayleigh, "On the Maintenance of Vibrations by Forces of Double Frequency,...," *Phil. Mag.* Vol. XXIV. 1887, pp. 145—159; *Scientific Papers*, Vol. III. pp. 1—14.]

55. On Graphic Solution of Dynamical Problems.

[From *British Association Report*, 1892, pp. 648—652; *Nature*, Vol. xlvi. Aug. 18, 1892, pp. 385, 386; *Phil. Mag.* Vol. xxxiv. pp. 443—448.]

The method of drawing meridianal curves of capillary surfaces of revolution, described in *Popular Lectures and Addresses*, Vol. i. 2nd edition, pp. 31—42, and illustrated by woodcuts made from large scale curves, worked out according to it with great care and success by Professor Perry when a student in the Natural Philosophy Class of Glasgow University, suggests a corresponding method for the solution of dynamical problems.

In dynamical problems regarding the motion of a single particle in a plane, it gives the following plan for drawing any possible path under the influence of a force of which the potential is given for every point of the plane. Suppose, for example, it is required to find the path of a particle projected, with any given velocity, in any given direction through any given point P_0 (fig. 1). Calculate the normal component force at this point; and divide the square of the velocity by this value, to find the radius of curvature of the path at that point. Taking this radius on the compasses, find the centre of curvature, C_0, in the line, P_0K, perpendicular to the given direction through P_0, and describe a small arc, $P_0P_1Q_1$, making P_1Q_1 equal to about half the length intended for the second arc. Calculate the altered velocity for the position Q_1, according to the potential law; and, as before for P_0, calculate a fresh radius of curvature for Q_1 by finding the normal component force for the altered direction of normal and for the velocity corresponding to the position of Q_1. With this radius, find the position of the centre of curvature, C_1, in P_1C_0L, the line of the radius through P_1. With this centre of curvature,

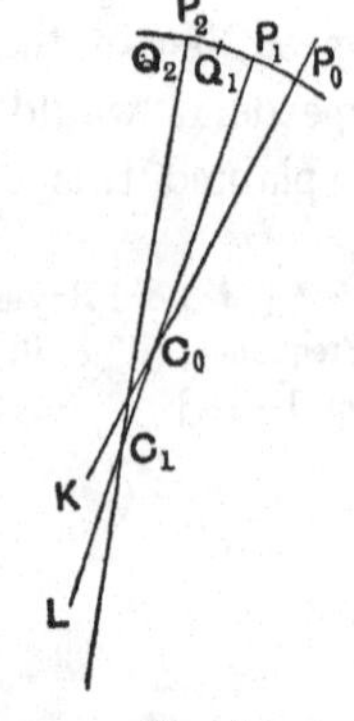

Fig. 1.

and the fresh radius of curvature, describe an arc $P_1P_2Q_2$ making P_2Q_2 equal to about half the length intended for the third arc; calculate radius of curvature for position Q_2; draw an arc $P_2P_3Q_3$; and continue the procedure. This process is well adapted for finding orbits by the "trial and error" method described in my article "On Some 'Test Cases' of the Maxwell-Boltzmann Doctrine regarding Distribution of Energy," § 13; *Proc. Roy. Soc.* June 11, 1891. [*Supra*, No. **50.**]

The accompanying curve (fig. 2) has been drawn with great care, and with very interesting success, in the "trial and error" method of finding the first and simplest orbit, by my secretary, Mr Thomas Carver, for the case of motion defined by the equations

$$\frac{d^2x}{dt^2} = -yx^2,$$

$$\frac{d^2y}{dt^2} = -xy^2.$$

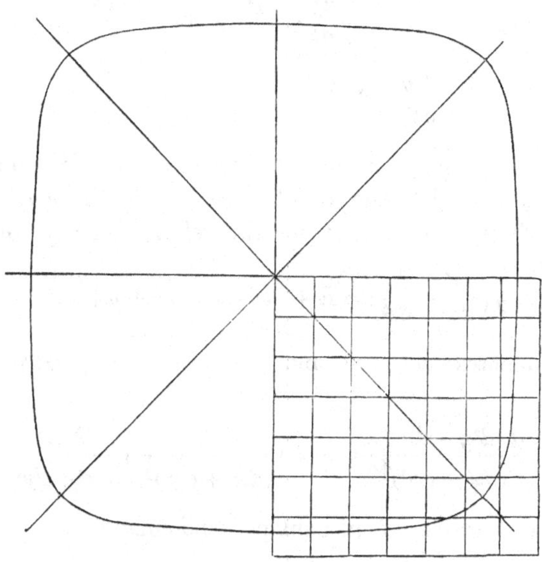

Fig. 2.

The initial point P_0 was taken on one of the lines cutting the axes of x and y at 45°, and at first at a random distance from the origin. A trial curve was worked according to the method described above, and was found to cut the axis of x at an oblique angle. Other trial curves, with unchanged energy-constant, were worked from initial points at greater or less distances from the origin, until a curve was found to cut the axis of x perpendicularly. This curve is one-eighth part of the orbit; and is shown in fig. 2 repeated eight times in order to complete the orbit, which is symmetrical on the two sides of the axis of x and y.

As an interesting case of motion related to the Lunar Theory, suppose the mass of the moon be infinitely small in comparison with the mass of the earth; and the earth and sun to have uniform motions in circles round their centre of gravity. Let (x, y) be coordinates of the moon relative to OX in line with the sun, outwards, and OY perpendicular to it in the direction of the earth's orbital motion. The well-known equation of motion relatively to revolving coordinates gives, for the equations of the moon's motion, if a denote the distance from O (the earth) of the centre of gravity of the sun and earth,

$$\frac{d^2x}{dt^2} - 2\omega\frac{dy}{dt} - \omega^2(a+x) = -\frac{dV}{dx} \quad \text{...........(1)},$$

$$\frac{d^2y}{dt^2} + 2\omega\frac{dx}{dt} - \omega^2 y = -\frac{dV}{dy} \quad \text{...........(2)},$$

where V is the potential of the attractions of the sun and earth on the moon, and ω the angular velocity of the earth's radius-vector. From this we find, for the relative-energy equation

$$\tfrac{1}{2}\left(\frac{dx^2}{dt^2} + \frac{dy^2}{dt^2}\right) = E + \tfrac{1}{2}\omega^2[(x+a)^2 + y^2] - V \quad \text{........(3)},$$

where E denotes a constant; and for the relative-curvature equation we find

$$\frac{dx\,d^2y - dy\,d^2x}{(dx^2+dy^2)^{\frac{3}{2}}} = -2\omega\frac{dt}{(dx^2+dy^2)^{\frac{1}{2}}} + \frac{N dt^2}{dx^2+dy^2} \quad \text{.....(4)},$$

where N denotes the component perpendicular to the path, of the resultant of (X, Y), with

$$X = \omega^2(x+a) - \frac{dV}{dx} \quad \text{...................(5)},$$

$$Y = \omega^2 y - \frac{dV}{dy} \quad \text{.........................(6)}.$$

Hence if q denote moon's velocity and ρ the radius of curvature of her path, relatively to the revolving plane XOY, we have

$$\tfrac{1}{2}q^2 = E + \tfrac{1}{2}\omega^2[(x+a)^2 + y^2] - V \quad\text{..........(7)},$$

and
$$\frac{1}{\rho} = \frac{-2\omega}{q} + \frac{N}{q^2} \quad\text{..........(8)}.$$

Calling S the sun's mass, and a his distance from the earth, and supposing the earth's mass infinitely small in comparison with the sun's, we have

$$S/a^2 = \omega^2 a \quad\text{..........(9)},$$

and therefore
$$-V = \frac{\omega^2 a^3}{[(a+x)^2 + y^2]^{\frac{1}{2}}} + \frac{m}{r} \quad\text{..........(10)},$$

where m denotes the earth's mass, and $r = \sqrt{(x^2 + y^2)}$.

Hence
$$-V \doteq \tfrac{1}{2}\omega^2(2a^2 - 2ax + 2x^2 - y^2) + \frac{m}{r} \quad\text{........(11)}.$$

With this, and with $\omega = 1$ and $m = b^3$, for simplicity in the numerical work which follows, we have

$$\frac{d^2x}{dt^2} - 2\frac{dy}{dt} = X = x\left(3 - \frac{b^3}{r^3}\right) \quad\text{..........(12)},$$

$$\frac{d^2y}{dt^2} + 2\frac{dx}{dt} = Y = -y\frac{b^3}{r^3} \quad\text{..........(13)},$$

$$q^2 = 2E + 3x^2 + \frac{2b^3}{r} \quad\text{..........(14)},$$

and
$$\rho = \frac{q^2}{N - 2q} \quad\text{..........(15)}.$$

From equations (12) and (13), G. W. Hill has, with four different values of E, found x and y explicitly in terms t, for the particular solution in each case which gives the simplest *orbit* (relatively to the revolving plane XOY); of which the one which presents the greatest deviation from the well-known "variational" oval of the elementary lunar theory is a symmetrical curve with two outwardly projecting cusps corresponding to the moon in quadratures. He supposed this to be the most extreme deviation from the variational oval possible for an orbit surrounding the earth. Poincaré, in his *Méthodes Nouvelles de la Mécanique*

Céleste, p. 109 (1892), admiring justly the manner in which Hill has thus "si magistralement" studied the subject of finite closed lunar orbits, points out that there are solutions corresponding to *looped* orbits, transcending Hill's, wrongly supposed extreme, cusped orbit. Mr Hill tells me that he accepts this criticism. The labour of working out a fairly accurate analytical solution for any of Poincaré's looped orbits, by Hill's method, would probably be very great. I have therefore thought it might interest others besides ourselves to apply my graphic method to the drawing of at least one of Poincaré's looped orbits, in our Physical (and Arithmetical) Laboratory in the University of Glasgow. Figure 3 represents a looped orbit, which has been worked out accordingly by Mr Magnus Maclean, Chief Official Assistant of the Professor of Natural Philosophy, from the equations (14) (15) above. The initial values used for obtaining the curve, were $x=2$; $y=0$; $b=10$; $2E=-130$; and therefore $q_0^2=882$ and $\rho_0=4{\cdot}8$.

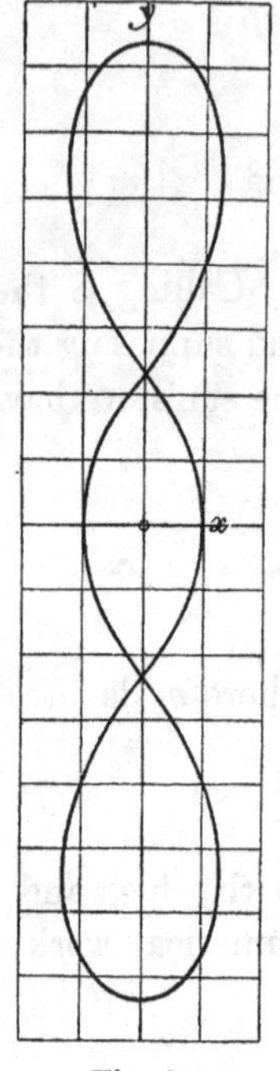

Fig. 3.

56. Reduction of every Problem of Two Freedoms in Conservative Dynamics to the Drawing of Geodetic Lines on a Surface of given Specific Curvature.

[From *British Association Report*, 1892, pp. 652, 653; *Nature*, Vol. XLVI. Aug. 18, 1892, p. 386.]

1. Any conservative case of two-freedom motion is proved to be reducible to a corresponding case of the motion of a material point in a plane.

2. In plane conservative dynamics, with any given value for the energy-constant, E, the resultant velocity, q, at any point (x, y) is a known function of (x, y), being given by the equation

$$q^2 = 2(E - V)$$

where V denotes the potential at (x, y); and every problem depends on drawing lines for which $\int q\,ds$ (the Maupertuis "action") is a minimum.

3. Considering any part, S, of the infinite plane, find a surface, S', such that any infinitesimal triangle $A'B'C'$ drawn on it has its sides q_0/q of those of a corresponding triangle ABC in the field, S, of our plane problem; q_0 denoting the value of q at any particular point (x_0, y_0) in the plane. By the principle of least action we see instantly that the lines on S', corresponding to paths on S, are geodetic. Thus the *adynamic* case of motion, of a particle on S', is found as a perfect and complete representative of the motion on the plane surface S, under force with any arbitrarily given function V for its potential, and any particular given value, E, for the total energy of the moving particle.

4. It is easily proved that the surface S', to be found according to § 3, exists; and that its specific curvature (Gauss's name for the product of its two principal curvatures) at any point is equal to*

$$\frac{q^2}{q_0^2} \nabla^2 \log q.$$

5. Examples are given of the finding of S'. As one example, illustrating the practical usefulness of this method in dynamics, the problem of the parabolic motion of an unresisted projectile is reduced to the drawing of geodetic lines on a certain figure of revolution of which the explicit equation is expressed in terms of elliptic functions.

* [Angles are not altered by the transformation from orbits on the plane to geodesics on the surface S'. For a kinetic triangle on the plane the excess of the angles over two right angles is the area multiplied by $\nabla^2 \log q$ (cf. *supra*, p. 514). On S' this excess is the area multiplied by the Gaussian curvature. By equating, the result in the text follows. On these kinetic transformations, cf. Larmor, *Proc. Lond. Math. Soc.* March, 1884; Darboux, *Théorie des Surfaces*, Vol. II. livre V. ch. VI. (1889).]

57. Generalization of "Mercator's" Projection performed by Aid of Electrical Instruments.

[From *Nature*, Vol. XLVI. Sep. 22, 1892, pp. 490, 491.]

The following mode of generalizing Mercator's Projection is merely an illustration of a communication to Section A of the British Association at its recent meeting in Edinburgh, entitled "Reduction of every Problem of Two Freedoms in Conservative Dynamics to the Drawing of Geodetic Lines on a Surface of given Specific Curvature." An abstract of this paper appeared in *Nature* for August 18.

In 1568, Gerhard Krämer, commonly known as "Mercator" (the Latin of his surname), gave to the world his chart, now of universal use in navigation. In it every island, every bay, every cape, every coast-line, if not extending over more than two or three degrees of longitude, or farther north and south than a distance equal to two or three degrees of longitude, is shown very approximately in its true shape: rigorously so if it extends over distances equal only to an infinitesimal difference of longitude. The angle between any two intersecting lines on the surface of the globe is reproduced rigorously without change in the corresponding angle on the chart.

Mercator's chart may be imagined as being made by coating the whole surface of a globe with a thin inextensible sheet of matter—sheet india-rubber for example (for simplicity, however, imagined to be perfectly extensible but inelastic)—cutting away two polar circles to be omitted from the chart; cutting the sheet through along a meridian, that of 180° longitude from Greenwich for example, stretching the sheet everywhere except along the equator so as to make all the circles of latitude equal in length to the circumference of the equator, and stretching the sheet in the direction of the meridian in the same ratio as the ratio in which the circles of latitude are stretched, while keeping at right angles the intersections between the meridians and the parallels. The sheet thus altered may be laid out flat or rolled up, as a paper chart.

What I call a generalized Mercator's chart for a body of any shape spherical or non-spherical, is a flat sheet showing for any intersecting lines that can be drawn on a part of the surface of the body, corresponding lines which intersect at the same angles. One Mercator chart of finite dimensions can only represent a part of the complete surface of a finite body, if the body be simply continuous; that is to say, if it has no hole or tunnel through it. The whole surface of an anchor ring can obviously be mercatorized on one chart. It is easily seen, for the case of the globe, that two charts suffice to mercatorize the whole surface; and it will be proved presently that two charts suffice for any simply continuous closed surface, however extremely it may deviate from the spherical form.

In *Liouville's Journal* for 1847, its editor, Liouville, gave an analytical investigation, according to which, if the equation of any surface whatever is given, a set of lines drawn on it can be found to fulfil the condition that the surface can be divided into infinitesimal squares, by these lines and the set of lines on the surface which cut them at right angles. Now it is clear that if we have any portion of a curved surface thus divided into infinitesimal square allotments, that is to say, divided into infinitesimal squares with the corners of four squares together, all through it, we can alter all these squares to one size and lay them down on a flat surface with each in contact with its four original neighbours; and thus the supposed portion of surface is mercatorized. Except for the case of a figure of revolution, or an ellipsoid, or virtually equivalent cases, Liouville's differential equations are of a very intractable kind. I have only recently noticed that we can solve the problem graphically (with any accuracy desired if the problem were a practical problem, which it is not) by aid of a voltmeter and a voltaic battery, or other means of producing electric currents, as follows:—

1. Construct the surface to be mercatorized in thin sheet metal of uniform thickness throughout. By thin I mean that the thickness is to be a small fraction of the smallest radius of curvature of any part of the surface.

2. Choose any two points of the surface, N, S, and apply the electrodes of a battery to it at these points.

3. By aid of movable electrodes of the voltmeter, trace an equipotential line, E, as close as may be around one electrode, and another equipotential line, F, as near as may be around the other electrode. Between these two equipotentials, E, F, trace a large number, n, of equidifferent equipotentials. Divide one of the equipotentials [E and F] into n equal parts; and through the divisional points draw lines cutting the whole series of equipotentials at right angles. These transverse lines and the equipotentials divide the whole surface between E and F into infinitesimal squares (Maxwell, *Electricity and Magnetism*, § 651).

4. Alter all the squares to one size and place them together, as explained above. Thus we have a Mercator chart of the whole surface between E and F.

N and S of our generalization correspond to the north and south poles of Mercator's chart of the world; and our generalized rule shows that a chart fulfilling the essential principle of similarity realized by Mercator may be constructed for a spherical surface by choosing for N, S any two points not necessarily the poles at the extremities of a diameter. If the points N, S are infinitely near one another, the resulting Mercator chart for the case of a spherical surface, is the stereographic projection of the surface on the tangent plane at the opposite end of the diameter through the point, C, midway between N and S. In this case the equipotentials and the stream-lines are circles on the spherical surface cutting NS at right angles, and touching it, respectively.

For a spherical or any other surface we may mercatorize any rectangular portion of it, $ABCD$, bounded by four curves, AB, BC, CD, DA, cutting one another at right angles as follows. Cut this part out of the complete metallic sheet; to two of its opposite edges, AB, DC, for instance, fix infinitely conductive borders. Apply the electrodes of a voltaic battery to these borders, and trace n equidifferent equipotential lines between AB and DC. Divide [the strip between consecutive equipotentials into squares], and through [n points each distant from the next by the same number of squares] draw curves cutting perpendicularly the whole series of equipotentials. These curves and the equipotentials divide the whole area into infinitesimal squares. Equalize the squares and lay them together on the flat as above.

If we have no mathematical instruments by which we can draw a system of curves at right angles to a system already drawn, we may dispense with mathematical instruments altogether, and complete the problem of dividing into squares by electrical instruments as follows. Remove the conducting borders from AB, DC; apply infinitely conductive borders to AD and BC, apply electrodes to these conducting borders, and as before draw n equidifferent equipotentials. This second set of equipotentials, and the first set, divide the whole area into squares.

58. To Draw a Mercator Chart on one Sheet representing the whole of any Complexly Continuous Closed Surface*.

[From *Nature*, Vol. xlvi. Oct. 6, 1892, pp. 541, 542.]

If a solid is not pierced by any perforation, its surface is called simply continuous, however complicated its shape may be. If a solid has one or more perforations, or tunnels†, its whole bounding surface is called "complexly continuous"; duplexly when there is only one perforation; $(n+1)$-plexly when there are n perforations. The whole surface of a group of n anchor-rings (or "toroids") cemented together in any relative positions, is a convenient and easily understood type of an $(n+1)$-plexly continuous closed surface.

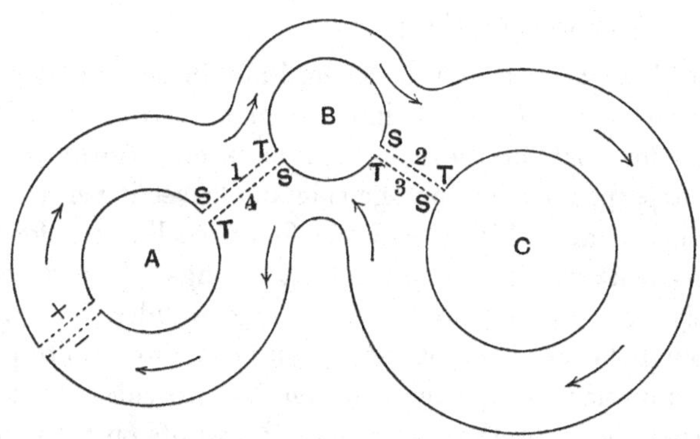

* [On the mathematics of the general problem, see Riemann, *Gesammelte Mathematische Werke*, "Theorie der Abel'schen Functionen," (2. Lehrsätze aus der Analysis Situs) and "Nachlass," Fragment aus der Analysis Situs; also Betti, *Annali di Matematica*, t. iv. (1870—1); also Forsyth's *Theory of Functions*, and other treatises.]

† A "hole" may mean a deep hollow, *not* through with two open ends. The word "tunnel" is inappropriate for the aperture of an anchor ring. Neither "hole" nor "tunnel" being unexceptionally available, I am compelled to use the longer word "perforation."

Let the diagram represent a quadruplexly continuous closed surface made of very thin sheet metal, uniform as to thickness and homogeneous as to quality throughout. To prepare for making a Mercator chart of it, cut it open between perforations C and B, B and A, A and outer space, in the manner indicated at $\frac{2}{3}$, $\frac{1}{4}$, and $\pm$. Apply infinitely conductive borders to the two lips separated by the cut at $\pm$, and apply the electrodes of a voltaic battery to these borders. By aid of movable electrodes of a voltmeter trace, on the metallic surface, a very large number $(n-1)$ of equi-different equipotential closed curves between the $+$ and $-$ borders. Divide any one of these equipotentials* into parts each equal to the infinitesimal distance perpendicularly across it to the next equipotential on either side of it; and through the divisional points draw curves, cutting the equipotentials at right angles. These curves are the stream-lines. They and the $(n+1)$ closed equipotentials (including the infinitely conductive borders) divide the whole surface into nm infinitesimal squares, if m be the number of divisions which we found in the equipotential. The arrows on the diagram show the general direction of the electric current in different parts of the complex circuit; each arrow representing it for the thin metal shell on either far or near side of the ideal section by the paper.

Considering carefully the stream-lines in the neighbourhoods of the four open lips marked in order of the stream 1, 2, 3, 4, we see that for each of these lips there is one stream-line which strikes it perpendicularly on one side and leaves it perpendicularly on the other, and which I call the flux-shed-line (or, for brevity, the flux-shed) for the lip to which it belongs. The stream-lines infinitely near to the flux-shed, on its two sides, pass infinitely close round the two sides of the lip, and come in infinitely near to the continuation of the flux-shed on its two sides. Let F_1, F_2, F_3, F_4 (not shown on the diagram) be the points on the $+$ terminal lip from which the flux-sheds of the lips 1, 2, 3, 4 proceed; and let G_1, G_2, G_3, G_4, be the points at which they fall on the $-$ lip. Let S_1, T_1, S_2, T_2, &c., denote the points on the four lips at which they are struck and left by their flux-shed lines.

* Two sentences of my previous article ("Generalization of Mercator's Projection"), in § 3, and in last paragraph but one, are manifestly wrong, and must be corrected to agree with the rule given for dividing into infinitesimal squares, in the present text. [Corrected in text, p. 525, *supra*.]

Let $p_1, l_1, p_2, l_2, p_3, l_3, p_4, l_4, p_5$ be the differences of potential from the + lip to S_1, from S_1 to T_1, T_1 to S_2, ... S_4 to T_4, and T_4 to the − lip. Measure these nine differences of potential. We are now ready to make the Mercator chart. We might indeed have done so without these elaborate considerations and measurements, simply by following the rule of my previous article; but the chart so obtained would have infinite contraction at eight points, the points corresponding to $S_1, T_1, \ldots S_4, T_4$. This fault is avoided, and a finite chart showing the whole surface on a finite scale in every part is obtained by the following process.

Take a long cylindric tube of thin sheet metal, of the same thickness and conductivity as that of our original surface; and on any circle H round it, mark four points, h_1, h_2, h_3, h_4, at consecutive distances along its circumference proportional respectively to the numbers of the m stream-lines which we find between F_1 and F_2, F_2 and F_3, F_3 and F_4, F_4 and F_1 on the + lip of our original surface. Through h_1, h_2, h_3, h_4 draw lines parallel to the axis of the cylinder.

Let now an electric current equal to the total current which we had from the + lip to the − lip through the original surface be maintained through our present cylinder by a voltaic battery with electrodes applied to places on the cylinder very far distant on the two sides of the circle H. Mark on the cylinder eight circles, $K_1, K_2 \ldots K_8$, at distances consecutively proportional to $l_1, p_2, l_2, p_3, l_3, p_4, l_4$, and absolutely such that l_1, p_1, &c., are equal to the differences of their potentials from one another in order.

Bore four small holes in the metal between the circles K_1 and K_2, K_3 and K_4, K_5 and K_6, K_7 and K_8 on the parallel straight lines through h_1, h_2, h_3, h_4, respectively. Enlarge these holes and alter their positions, so that the altered stream-lines through h_1, h_2, h_3, h_4 (these points supposed fixed and very distant) shall still be their flux-sheds. While always maintaining this condition, enlarge the holes and alter their positions until the extreme differences of potential in their lips become l_1, l_2, l_3, l_4, and the differences of potential between the lips in succession become p_2, p_3, p_4. In thus continuously changing the holes we might change their shapes arbitrarily; but to fix our ideas, we may suppose them to be always made circular. This makes the problem determinate, except the distance from the circle H of the hole nearest to it, which may be anything we please, provided it is very large in proportion to the diameter of the cylinder.

The determinate problem thus proposed is clearly possible, and the solution is clearly unique. It is of a highly transcendental character, viewed as a problem for mathematical analysis; but an obvious method of "trial and error" gives its solution by electric measurement, with quite a moderate amount of labour if moderate accuracy suffices.

When the holes have been finally adjusted to fulfil our conditions, draw by aid of the voltmeter and movable electrodes, the equipotentials, for p_1 above the greatest potential of lip 1, and for p_5 below the least potential of lip 4; and between these equipotentials, which we shall call f and g, draw $n-1$ equidifferent equipotentials. Draw the stream-lines, making infinitesimal squares with these according to the rule given above in the present article. It will be found that the number of the stream-lines is m, the same as on our original surface, and the whole number of infinitesimal squares on the cylinder between f and g is mn. Cut the cylinder through at f and g; cut it open by any stream-line from f to g, and open it out flat. We thus have a Mercator chart bounded by four curves cutting one another at right angles, and divided into mn infinitesimal squares, corresponding individually to the mn squares into which we divided the original surface by our first electric process. In this chart there are four circular blanks corresponding to the lips 1, 2, 3, 4 of our diagram; and there is exact correspondence of their flux-sheds and neighbouring stream-lines, and of the disturbances, which they produce in the equipotentials, with the analogous features at the lips of the original surface as cut for our process. The solution of this geometrical problem was a necessity for the dynamical problem with which I have been occupied, and this is my excuse for working it out; though it might be considered as devoid of interest in itself.

59. ISOPERIMETRICAL PROBLEMS.

[From *Roy. Institution Proc.* Vol. XIV. 1896, pp. 111—119 [May 12, 1893]; *Nature*, Vol. XLIX. March 29, 1894, pp. 515—518.

Reprinted in *Popular Lectures and Addresses*, Vol. II. pp. 571—592.]

60. NINETEENTH CENTURY CLOUDS OVER THE DYNAMICAL THEORY OF HEAT AND LIGHT.

[From *Roy. Institution Proc.* Vol. XVI. April 27, 1900, pp. 363—397; *Phil. Mag.* Vol. II. July, 1901, pp. 1—40.

Reprinted in *Baltimore Lectures*, Appendix B, pp. 486—527.]

ELASTIC PROPAGATION.

61. On Peristaltic Induction of Electric Currents in Submarine Telegraph Wires.

[From *British Association Report*, 1855, Pt. ii. p. 22.

Reprinted in *Math. and Phys. Papers*, Vol. ii. Art. lxxv. pp. 77, 78.]

62. On Signalling through Submarine Cables, illustrated by Signals Transmitted through a Model Submarine Cable, exhibited by Mirror Galvanometer and by Siphon Recorder.

[From *Inst. of Engineers in Scotland Trans.* Vol. xvi. March 18, 1873, pp. 119, 120.

Reprinted in *Math. and Phys. Papers*, Vol. ii. Art. lxxxv. pp. 168—172.]

63. Dynamical Illustrations of the Magnetic and the Helicoidal Rotatory Effects of Transparent Bodies on Polarized Light.

[From *Roy. Soc. Proc.* Vol. viii. June 12, 1856, pp. 150—158; *Phil. Mag.* Vol. xiii. March, 1857, pp. 198—204.

Reprinted in *Baltimore Lectures*, Appendix F, pp. 569—583.]

64. Vibrations and Waves in a Stretched Uniform Chain of Symmetrical Gyrostats*.

[From *London Math. Soc. Proc.* Vol. vi. April 8, 1875, pp. 190—194.]

A gyrostat is a rapidly rotating fly-wheel, frictionlessly pivoted on a stiff moveable framework or containing case. A symmetrical gyrostat is one in which not only the fly-wheel but the case is kinetically symmetrical round the axis of rotation of the wheel.

Let the chain consist of alternate gyrostats and massless connecting links, and let the connection be by universal flexure joints† at each end of each link. For simplicity, at present suppose the axes of the gyrostatic links to be all in one line when the chain is stretched straight. This line will be called the equilibrium axis.

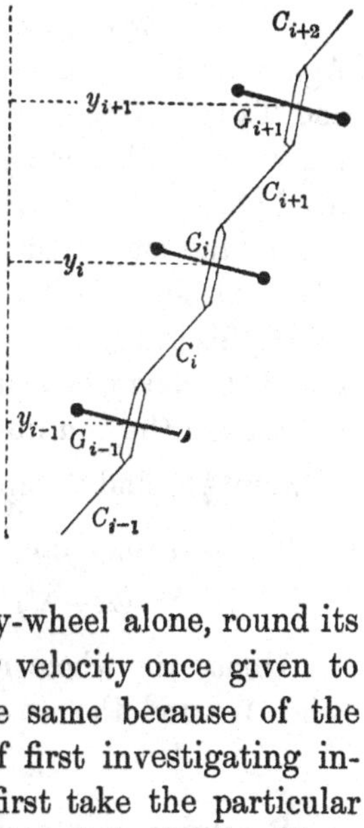

Let g and c be the lengths of the gyrostatic and connecting links respectively; m the mass, and λ and μ the moments of inertia round the axis of figure and a line perpendicular to it through centre of inertia, of a gyrostatic link, fly-wheel and case included; let λ' be the moment of inertia of each fly-wheel alone, round its axis of figure; and let ω be the angular velocity once given to each fly-wheel, and remaining always the same because of the frictionlessness of the pivots. Instead of first investigating infinitesimal motions in general, we shall first take the particular case of circular motions not limited to being infinitesimal. Then, taking them to be infinitesimal, by composition of circular

* [For developments, cf. Routh's *Advanced Rigid Dynamics*, § 419. Cf. also No. 63 *supra*.]

† Thomson and Tait's *Natural Philosophy*, § 109.

motions in similar or in contrary directions, and with different phases, we pass readily to the general solution of the problem of infinitesimal motions.

Problem.—Suppose a finite length of such a chain to be placed with its links forming an open plane polygon; and, the ends of the extreme links being held fixed by universal flexure joints, let each link be so set in motion perpendicularly to this plane that the polygon of axes moves as a rigid polygon rotating round the line joining the ends, with a given angular velocity n: required the form of the polygon, and the forces on the fixed ends, so that the chain, when left to itself, may continue revolving in the manner specified.

Call successive connecting and gyrostatic links ... C_i, G_i, C_{i+1}, G_{i+1}, ... Let ϑ_i and θ_i denote the inclinations of C_i and G_i to the line joining the ends, and let y_i be the distance of the centre of inertia of G_i from this line. We have the geometrical relation

$$y_{i+1} - y_i = \tfrac{1}{2} g (\sin \theta_{i+1} + \sin \theta_i) + c \sin \vartheta_{i+1} \quad \text{.....(1)}.$$

Again, by the geometry of the universal flexure joint (Thomson and Tait, § 109), each gyrostatic link moves as if its axis were produced to the line joining the fixed ends, and there joined to a fixed object by a universal flexure joint. Hence the instantaneous axis of the motion of G_i bisects the angle $\pi - \theta_i$ between the line of its axis and the line joining the fixed ends. Its angular velocity round this instantaneous axis is $2n \sin \frac{1}{2}\theta_i$. The components of this round the axis of G_i, and in a plane perpendicular to the axis of G_i through the equilibrium-axis, are

$$2n \sin^2 \tfrac{1}{2}\theta_i \text{ and } 2n \sin \tfrac{1}{2}\theta_i \cos \tfrac{1}{2}\theta_i, \text{ or } n(1 - \cos \theta_i) \text{ and } n \sin \theta_i.$$

The corresponding moments of momentum are

$$(\lambda - \lambda') \,.\, n(1 - \cos \theta_i) \text{ and } \mu \,.\, n \sin \theta_i.$$

Hence the whole moment of momentum of G_i (case and fly-wheel) round the axis of G_i is

$$(\lambda - \lambda') \,.\, n(1 - \cos \theta_i) + \lambda' \omega.$$

Resolving this and $\mu n \sin \theta_i$, round the equilibrium axis and a perpendicular to it in the plane of the chain, we have, for whole component moment of momentum round the last mentioned line*,

$$\{(\lambda - \lambda') n (1 - \cos \theta_i) + \lambda' \omega\} \sin \theta_i - \mu n \sin \theta_i \cos \theta_i.$$

* [The sign of μ should be changed henceforth.]

This line revolves with angular velocity n in a plane perpendicular to the equilibrium axis, and there must therefore be a couple equal to

$$n\left[\{(\lambda-\lambda')\,n\,(1-\cos\theta_i)+\lambda'\omega\}\sin\theta_i-\mu n\sin\theta_i\cos\theta_i\right],$$

acting on G_i round an axis perpendicular to the plane of the equilibrium axis and the axis of G_i. The direction of this couple, when positive, is such as to tend to increase the angle θ_i. We are now ready to write down the equations of motion (or kinetic equilibrium) of G_i. The component parallel to the equilibrium axis, of the pull in the connecting links, must be the same for all. (This *is* all the equations of motion parallel to the equilibrium line.) Let its amount be P: so that $P\sec\vartheta_i$ is the pull in the connecting link C_i. The applied forces on G_i are the pulls of C_i and C_{i+1} on its ends. Resolving them we have:—

Parallel to equilibrium axis.	Perpendicular to equilibrium axis.
$-P,$	$-P\tan\vartheta_i,$
$+P,$	$+P\tan\vartheta_{i+1};$

and, transposing to centre of inertia of G_i, we have finally

zero parallel to equilibrium axis;

$P(\tan\vartheta_{i+1}-\tan\vartheta_i)$ perpendicular to equilibrium axis;

and couple

$$-Pg\sin\theta_i+P(\tan\vartheta_{i+1}+\tan\vartheta_i)\,.\,\tfrac{1}{2}g\cos\vartheta_i$$

in plane through equilibrium axis and direction tending to increase θ_i. Hence, for motion of centre of inertia of G_i,

$$mn^2y_i+P(\tan\vartheta_{i+1}-\tan\vartheta_i)=0 \quad \text{...........(2)},$$

and, by couples,

$$n\left[\{(\lambda-\lambda')\,n\,(1-\cos\theta_i)+\lambda'\omega\}\sin\theta_i-\mu n\sin\theta_i\cos\theta_i\right]$$
$$=Pg\left\{-\sin\theta_i+\tfrac{1}{2}\cos\theta_i(\tan\vartheta_{i+1}+\tan\vartheta_i)\right\} \quad \text{...(3)}.$$

Equations (1), (2), (3), applied to each gyrostatic link, give as many equations as there are of unknown quantities ϑ_i, θ_i, y_i, if we suppose one end of the chain to be a gyrostatic and the other a connecting link, so that there be the same number of the two kinds of links.

When the displacements and inclinations are infinitely small, the "algorithm of finite differences," as applied by Lagrange to the transverse oscillations of a "linear system of bodies" (a case of what the present problem becomes when $\omega = 0$) is conveniently applicable. Equations (1), (2), and (3), when we can neglect the cubes of θ_i and ϑ_i, become

$$y_{i+1} - y_i = \tfrac{1}{2} g(\theta_{i+1} + \theta_i) + c\vartheta_{i+1} \quad\text{.............(4)},$$

$$mn^2 y_i + P(\vartheta_{i+1} - \vartheta_i) = 0 \quad\text{..............(5)},$$

$$n(\lambda'\omega - \mu n)\,\theta_i = Pg\{-\theta_i + \tfrac{1}{2}(\vartheta_{i+1} + \vartheta_i)\} \quad\text{........(6)}.$$

Let ρ denote a symbol of operation such that

$$\rho u_i = u_{i+1} \quad\text{.........................(7)},$$

where u_i is any function of i.

From (6) we have

$$\theta_i = \tfrac{1}{2}\frac{Pg(\rho + 1)\vartheta_i}{n(\lambda'\omega - \mu n) + Pg}.$$

Using this in (4), and eliminating ϑ_i between (4) and (5), we find, finally,

$$\left[P(\rho - 1)^2 + mn^2\left\{\tfrac{1}{4}\frac{Pg^2}{n(\lambda'\omega - \mu n) + Pg}(\rho + 1)^2 + c\rho\right\}\right] y_i = 0 \quad\text{......(8)};$$

and the same holds, of course, with θ_i or ϑ_i substituted for y_i. Hence to determine ρ we have the quadratic

$$\rho^2 - 2(1 - e)\rho + 1 = 0 \quad\text{....................(9)},$$

where

$$e = \tfrac{1}{2}\frac{\dfrac{Pg^2}{Pg + \lambda' n\omega - \mu n^2} + c}{\dfrac{P}{mn^2} + \tfrac{1}{4}\dfrac{Pg^2}{Pg + \lambda' n\omega - \mu n^2}} \quad\text{..............(10)}.$$

Put now $\quad 1 - e = \cos\alpha$, or $\sqrt{(\tfrac{1}{2}e)} = \sin\tfrac{1}{2}\alpha$ ....(11).

The solution of (9) becomes

$$\rho = \cos\alpha \pm \sin\alpha\sqrt{-1},$$

and therefore the general solution of (8) is

$$y_i = A\cos i\alpha + B\sin i\alpha \quad\text{................(12)};$$

and if x_i be the other coordinate of the centre of inertia of G_i (that of G_0 being taken as zero), we have

$$x_i = i(g + c) \quad\text{.......................(13)}.$$

Hence

$$y_i = A\cos\frac{\alpha x_i}{g + c} + B\sin\frac{\alpha x_i}{g + c} \quad\text{..........(14)};$$

that is to say, the centres of inertia of the links lie on a helix curve, whose wave length is

$$\frac{2\pi}{\alpha}(g+c) \text{ ....(15)},$$

where $2\pi/\alpha$ is the number of particles in the wave-length l.

The period is $$2\pi/n \text{ ....(16)};$$

and therefore, if V denote the velocity of propagation of the "circularly polarized" wave made up by proper superposition of our solutions, we have

$$V = n(g+c)/\alpha \text{ ....(17)};$$

and therefore, by (11),

$$V = \frac{\sin\frac{1}{2}\alpha}{\frac{1}{2}\alpha}\frac{n(g+c)}{\sqrt{2e}} \text{ ....(18)};$$

whence, by (10),

$$V = \frac{\sin\frac{1}{2}\alpha}{\frac{1}{2}\alpha}\left\{\frac{1+\frac{1}{4}\dfrac{mn^2g^2}{Pg+\lambda'n\omega-\mu n^2}}{1-\dfrac{g(\lambda'n\omega-\mu n^2)}{(g+c)(Pg+\lambda'n\omega-\mu n^2)}}\right\}^{\frac{1}{2}}\left\{\frac{P(g+c)}{m}\right\}^{\frac{1}{2}} \text{...(19)}.$$

The first two factors of this expression become each equal to unity when α is infinitely small, that is to say, when the wave length (15) is infinitely great in comparison with the distance from centre to centre of neighbouring molecules, and the expression becomes simply

$$\{P(g+c)/m\}^{\frac{1}{2}} \text{ ....(20)},$$

which is the known velocity of propagation of waves in a uniform stretched cord; $m/(g+c)$ being the mass per unit of length, and P the pull.

When α is not zero, but very small, we have approximately

$$\frac{\sin\frac{1}{2}\alpha}{\frac{1}{2}\alpha} = 1 - \frac{(\frac{1}{2}\alpha)^2}{6} = 1 - \frac{\pi^2}{6}\cdot\left(\frac{\alpha}{2\pi}\right)^2 = 1 - \frac{\pi^2}{6}\left(\frac{g+c}{l}\right)^2;$$

where l denotes the wave length. And by the approximate expression (20) for V, or $nl/2\pi$, $mn^2 = 4\pi^2P(g+c)/l^2$ approximately. Also, because each link is very small in all its linear dimensions in comparison with l, μ/ml^2 and λ'/ml^2 are each very small, and each comparable with g^2/l^2. Hence the second factor of (19) becomes approximately*

$$\left\{\frac{1+\pi^2\dfrac{g(g+c)}{l^2}}{1-4\pi^2\dfrac{\mu}{ml^2}\left(\dfrac{\lambda'\omega}{\mu n}-1\right)}\right\}^{\frac{1}{2}} \quad \text{or} \quad 1+\tfrac{1}{2}\frac{\pi^2}{l^2}\left\{g(g+c)+4\frac{\mu}{m}\left(\frac{\lambda'\omega}{\mu n}-1\right)\right\}.$$

And putting together the two factors, still approximately,

$$V=\left[1-\pi^2\left\{\tfrac{1}{6}\left(\frac{g+c}{l}\right)^2 - \tfrac{1}{2}\,\frac{g(g+c)+4\dfrac{\mu}{m}\left(\dfrac{\lambda'\omega}{\mu n}-1\right)}{l^2}\right\}\right]\sqrt{\frac{P(g+c)}{m}} \quad \text{......(21).}$$

[Take $\mu=0$, $(g+c)/l=0$, approximately. Then

$$V \doteq \left[1+\pi^2\frac{2\lambda'\omega}{l^2nm}\right]\sqrt{\frac{P(g+c)}{m}};$$

$$n=\frac{2\pi}{\tau}=\frac{2\pi V}{l}; \quad \therefore V \doteq \left(1+\pi\frac{\lambda'\omega}{mVl}\right)\sqrt{\frac{P(g+c)}{m}}.$$

Baltimore, *October* 5, 1884.]

* [The approximation now considered supposes $\dfrac{Pg}{\lambda' n\omega-\mu n^2}$ to be very great (Netherhall, Jan. 14, 1883).]

65. The Wave Theory of Light.

Lecture delivered in the Academy of Music, Philadelphia, under the auspices of the Franklin Institute, Sept. 29, 1884.

[From *Journal Franklin Institute*, Vol. LXXXVIII. Nov. 1884, pp. 321—341; *Nature*, Vol. XXXI. 1884, pp. 91—94, 115—118.

Reprinted in *Popular Lectures and Addresses*, Vol. I. pp. 300—348.]

66. On Cauchy's and Green's Doctrine of Extraneous Force to Explain Dynamically Fresnel's Kinematics of Double Refraction.

[From *Edin. Roy. Soc. Proc.* Vol. XV. Dec. 5, 1887, pp. 21—33; *Phil. Mag.* Vol. XXV. Feb. 1888, pp. 116—128.

Reprinted substantially in *Baltimore Lectures*, pp. 228—248.]

67. A SIMPLE HYPOTHESIS FOR ELECTRO-MAGNETIC INDUCTION OF INCOMPLETE CIRCUITS, WITH CONSEQUENT EQUATIONS OF ELECTRIC MOTION IN FIXED HOMOGENEOUS SOLID MATTER.

[From *British Association Report*, 1888, pp. 567—570; *Nature*, Vol. XXXVIII. pp. 569—571.]

1. To avoid mathematical formulas till needed for calculation consider three cases of liquid* motion, which for brevity I call Primary, Secondary, Tertiary, defined as follows: Half the velocity in the Secondary agrees numerically and directionally with the magnitude and axis of the molecular spin at the corresponding point of the Primary; or (short, but complete statement) *the velocity in the Secondary is twice the spin in the Primary;* and (similarly) *half the velocity in the Tertiary is the spin in the Secondary.*

2. In the Secondary and Tertiary the motion is essentially without change of density, and in each of them we naturally, therefore, take an incompressible fluid as the substance. The motion in the Primary we arbitrarily restrict by taking its fluid also as incompressible.

3. Helmholtz first solved the problem: Given the spin in any case of liquid motion, to find the motion. His solution consists in finding the potentials of three ideal distributions of gravitational matter having densities respectively equal to $1/4\pi$ of the rectangular components of the given spin; and, regarding for a moment these potentials as rectangular components of velocity in a case of liquid motion, taking the spin in this motion as the velocity in the required motion. Applying this solution to find the velocity in our Secondary from the velocity in our Tertiary, we see that the three velocity components in our Primary are the potentials of three ideal distributions of gravitational matter,

* I use "liquid" for brevity to signify incompressible fluid

having their densities respectively equal to $1/4\pi$ of the three velocity components of our Tertiary. This proposition is proved in a moment*, in § 5 below, by expressing the velocity components of our Tertiary in terms of those of our Secondary, and those of our Secondary in terms of those of our Primary, and then eliminating the velocity components of Secondary so as to have those of Tertiary directly in terms of those of Primary.

4. Consider now, in a fixed solid or solids of no magnetic susceptibility, any case of electric motion in which there is no change of electrification, and therefore no incomplete electric circuit; or, which is the same, any case of electric motion in which the distribution of electric current agrees with the distribution of velocity in a case of liquid motion. Let this case, with velocity of liquid numerically equal to 4π times the electric current density, be our Tertiary. The velocity in our corresponding Secondary is then the magnetic force of the electric current system†; and the velocity in our Primary is what Maxwell‡ has well called the "electro-magnetic momentum at any point" of the electric current system; and the rate of decrease per unit of time, of any component of this last velocity at any point, is the corresponding component of electro-motive force, due to electro-magnetic induction of the electric current system when it experiences any change. This electro-motive force, combined with the electrostatic force, if there is any, constitutes the whole electro-motive force at any point of the system. Hence by Ohm's law each component of electric current at any point is equal to the electric conductivity, multiplied into the sum of the corresponding component of electrostatic force and the rate of decrease per unit of time of the corresponding component of velocity of liquid in our Primary.

5. To express all this in symbols let (u_1, v_1, w_1), (u_2, v_2, w_2), and (u_3, v_3, w_3) denote rectangular components of the velocity at time t, and point (x, y, z) of our Primary, Secondary, and Tertiary. We have (§ 1)

$$u_2=\frac{dw_1}{dy}-\frac{dv_1}{dz},\quad v_2=\frac{du_1}{dz}-\frac{dw_1}{dx},\quad w_2=\frac{dv_1}{dx}-\frac{du_1}{dy}\text{......}(1),$$

* From Poisson's well-known elementary theorem $\nabla^2 V = -4\pi\rho$.

† *Electrostatics and Magnetism*, § 517 (postscript) (*c*).

‡ *Electricity and Magnetism*, § 604.

$$u_3 = \frac{dw_2}{dy} - \frac{dv_2}{dz}, \quad v_3 = \frac{du_2}{dz} - \frac{dw_2}{dx}, \quad w_3 = \frac{dv_2}{dx} - \frac{du_2}{dy} \quad \ldots\ldots(2).$$

Eliminating u_2, v_2, w_2 from (2) by (1), we find

$$u_3 = \frac{d}{dx}\left(\frac{du_1}{dx} + \frac{dv_1}{dy} + \frac{dw_1}{dz}\right) - \left(\frac{d^2u_1}{dx^2} + \frac{d^2u_1}{dy^2} + \frac{d^2u_1}{dz^2}\right), \text{ \&c.}\ldots(3).$$

But by our assumption (§ 2) of incompressibility in the Primary

$$\frac{du_1}{dx} + \frac{dv_1}{dy} + \frac{dw_1}{dz} = 0 \ldots\ldots\ldots\ldots\ldots\ldots\ldots\ldots(4).$$

Hence (3) becomes

$$u_3 = -\nabla^2 u_1, \quad v_3 = -\nabla^2 v_1, \quad w_3 = -\nabla^2 w_1 \ldots\ldots\ldots(5),$$

where, as in Article XXVII. (November, 1846) of my *Collected Mathematical and Physical Papers* (Vol. I.),

$$\nabla^2 = \frac{d^2}{dx^2} + \frac{d^2}{dy^2} + \frac{d^2}{dz^2} \quad \ldots\ldots\ldots\ldots\ldots\ldots(6)^*.$$

This (5) is the promised proof of § 3.

6. Let now u, v, w denote the components of electric current at (x, y, z) in the electric system of § 4; so that

$$4\pi u = u_3 = -\nabla^2 u_1; \quad 4\pi v = v_3 = -\nabla^2 v_1; \quad 4\pi w = w_3 = -\nabla^2 w_1 \ldots(7),$$

which, in virtue of (4), give

$$\frac{du}{dx} + \frac{dv}{dy} + \frac{dw}{dz} = 0 \ldots\ldots\ldots\ldots\ldots\ldots\ldots\ldots(8).$$

Hence the components of electro-motive force due to change of current being (§ 4)

$$-\frac{du_1}{dt}, \quad -\frac{dv_1}{dt}, \quad -\frac{dw_1}{dt},$$

are equal to

$$4\pi\nabla^{-2}\frac{du}{dt}, \quad 4\pi\nabla^{-2}\frac{dv}{dt}, \quad 4\pi\nabla^{-2}\frac{dw}{dt} \quad \ldots\ldots\ldots(9),$$

and therefore if Ψ denote electrostatic potential, we have, for the equations of the electric motion (§ 4)

$$u = \frac{1}{\kappa}\left(\nabla^{-2}\frac{du}{dt} - \frac{1}{4\pi}\frac{d\Psi}{dx}\right); \quad v = \frac{1}{\kappa}\left(\nabla^{-2}\frac{dv}{dt} - \frac{1}{4\pi}\frac{d\Psi}{dy}\right);$$

$$w = \frac{1}{\kappa}\left(\nabla^{-2}\frac{dw}{dt} - \frac{1}{4\pi}\frac{d\Psi}{dz}\right) \ldots\ldots(10),$$

where κ denotes $1/4\pi$ of the specific resistance.

* Maxwell, for quaternionic reasons, takes ∇^2 the negative of mine.

7. As Ψ is independent of t, according to § 4, we may, conveniently for a moment, put

$$u+\frac{d\Psi}{\kappa dx}=\alpha\,; \quad v+\frac{d\Psi}{\kappa dy}=\beta\,; \quad w+\frac{d\Psi}{\kappa dz}=\gamma \text{......(11)},$$

and so find, as equivalents to (10),

$$\frac{d\alpha}{dt}=\nabla^2(\kappa\alpha)\,; \quad \frac{d\beta}{dt}=\nabla^2(\kappa\beta)\,; \quad \frac{d\gamma}{dt}=\nabla^2(\kappa\gamma) \text{......(12)}.$$

The interpretation of this elimination of Ψ may be illustrated by considering, for example, a finite portion of *homogeneous* solid conductor, of any shape (a long thin wire with two ends, or a short thick wire, or a solid globe, or a lump, of any shape, of copper or other metal homogeneous throughout), with a constant flow of electricity maintained through it by electrodes from a voltaic battery or other source of electric energy, and with proper appliances over its whole boundary, so regulated as to keep any given constant potential at every point of the boundary; while currents are caused to circulate through the interior by varying currents in circuits exterior to it. There being no *changing electrification* by our supposition of § 4, Ψ can have no contribution from electrification within our conductor; and therefore throughout our field

$$\nabla^2\Psi=0 \text{.........................(13)},$$

which, with (8) and (11), gives

$$\frac{d\alpha}{dx}+\frac{d\beta}{dy}+\frac{d\gamma}{dz}=0 \text{.....................(14)}.$$

Between (12) and (14) we have four equations for three unknown quantities. These *in the case of homogeneousness* (κ constant) are equivalent to only three, because in this case (14) follows from (12) provided (14) is satisfied initially, and the proper surface condition is maintained to prevent any violation of it from supervening.

But unless κ is constant throughout our field, the four equations (12) and (14) are mutually inconsistent; from which it follows that our supposition of unchangingness of electrification (§ 4) is not generally true. An interesting and important practical conclusion is, that when currents are induced in any way, in a solid composed of parts having different electric conductivities (pieces of copper and lead, for example, fixed together in metallic contact), there must in general be changing electrification over every interface between these parts. This conclusion was not at first obvious to me; but it ought to be so by anyone approaching the subject with mind undisturbed by mathematical formulas.

8. Being thus warned off heterogeneousness until we come to consider changing electrification and incomplete circuits, let us apply (10) to an infinite homogeneous solid. As (8) holds through all space according to our supposition in § 4, and as κ is constant, (13) must now hold through all space, and therefore $\Psi = 0$, which reduces (10) to

$$u = \frac{1}{\kappa} \nabla^{-2} \frac{du}{dt}; \quad v = \frac{1}{\kappa} \nabla^{-2} \frac{dv}{dt}; \quad w = \frac{1}{\kappa} \nabla^{-2} \frac{dw}{dt} \quad \ldots(15).$$

These equations express simply the known law of *electro-magnetic* induction. Maxwell's equations (7) of § 783 of his *Electricity and Magnetism*, become in this case

$$\mu \left(4\pi C + K \frac{d}{dt} \right) \frac{du}{dt} = \nabla^2 u, \text{ \&c.} \quad \ldots\ldots\ldots\ldots(15'),$$

which cannot be right, I think, according to any conceivable hypothesis regarding electric conductivity, whether of metals, or stones, or gums, or resins, or wax, or shellac, or india-rubber, or gutta-percha, or glasses, or solid or liquid electrolytes; being, as seems to me, vitiated for complete circuits by the curious and ingenious, but as seems to me not wholly tenable hypothesis which he introduces, in § 610*, for incomplete circuits.

9. The hypothesis which I suggest for incomplete circuits, and consequently varying electrification, is simply that the components of the electro-motive force due to electro-magnetic induction are still $4\pi\nabla^{-2}du/dt$, &c. Thus, for the equations of motion, we have simply to keep equations (10) unchanged, while not imposing (8), but instead of it taking

$$4\pi \text{‘}v\text{’}^2 \left(\frac{du}{dx} + \frac{dv}{dy} + \frac{dw}{dz} \right) = \frac{d}{dt} \nabla^2 \Psi = -\frac{d\rho}{dt} \quad \ldots\ldots\ldots(16),$$

where ρ denotes 4π times the electric density at time t, and place (x, y, z), and ‘v’ denotes the number of electrostatic units in the

* [Namely, the fundamental postulate that rate of change of electric displacement operates as current, and so makes all currents flow effectively in complete circuits. The present paper, now only of historical interest, represents probably the result of a survey of Maxwell's scheme, prompted by Hertz's then recent discovery of electric waves.]

electro-magnetic unit of electric quantity. This equation expresses that the electrification of which Ψ is the potential, diminishes and increases in any place according as electricity flows more out than in or more in than out. We thus have four equations, (10) and (16), for our four unknowns, u, v, w, Ψ; and I find simple and natural solutions, with nothing vague or difficult to understand, or to believe when understood, by their application to practical problems, or to conceivable ideal problems, such as the transmission of ordinary or telephonic signals along submarine telegraph conductors and land lines, electric oscillations in a finite insulated conductor of any form, transference of electricity through an infinite solid, &c. &c. This, however, does not prove my hypothesis. Experiment is required for informing us as to the real electro-magnetic effects of incomplete circuits; and, as Helmholtz has remarked, it is not easy to imagine any kind of experiment which could decide between different hypotheses which may occur to anyone trying to evolve out of his inner consciousness a theory of the mutual force and induction between incomplete circuits.

68. On the Transference of Electricity within a Homogeneous Solid Conductor.

[From *British Association Report*, 1888, pp. 570, 571; *Nature*, Vol. xxxviii. p. 571.]

Adopting the notation and formulas of my previous paper [*supra*, p. 539], and taking ρ to denote 4π times the electric density at time t, and place (x, y, z), we have

$$-\rho = \nabla^2\Psi = 4\pi\, {}^{\backprime}v'^2 \int \left(\frac{du}{dx} + \frac{dv}{dy} + \frac{dw}{dz}\right) dt \quad \text{.........(17)},$$

and, eliminating u, v, w, Ψ by this from (10), we find, on the assumption of κ constant,

$$\kappa \frac{d}{dt} \nabla^2\rho = \frac{d^2\rho}{dt^2} - {}^{\backprime}v'^2 \nabla^2\rho \quad \text{.................(18)}.$$

The settlement of boundary conditions, when a finite piece of solid conductor is the subject, involves consideration of u, v, w, and for it, therefore, equations (17) and (12) must be taken into account; but when the subject is an infinite homogeneous solid, which, for simplicity, we now suppose it to be, (18) suffices. It is interesting and helpful to remark that this agrees with the equation for the density of a viscous elastic fluid, found from Stokes's equations for sound in air with viscosity taken into account; and that the values of u, v, w, given by (17) and (10), when ρ has been determined, agree with the velocity components of the elastic fluid if the simple and natural enough supposition be made that viscous resistance acts only against change of shape, and not against change of volume without change of shape.

For a type-solution assume

$$\rho = A\epsilon^{-qt} \cos\frac{2\pi x}{a} \cos\frac{2\pi y}{b} \cos\frac{2\pi z}{c} \quad \text{.........(19)},$$

and we find, by substitution in (18),

$$q^2 - \frac{\kappa}{L^2} q + \frac{\text{‘}v\text{'}^2}{L^2} = 0 \qquad \text{(20)},$$

where

$$L^{-2} = 4\pi^2 \left(\frac{1}{a^2} + \frac{1}{b^2} + \frac{1}{c^3}\right) \qquad \text{(21)}.$$

Hence, by solution of the quadratic (20) for q,

$$q = \tfrac{1}{2} \frac{\kappa}{L^2} \left\{1 \pm \sqrt{\left(1 - \frac{4\text{‘}v\text{'}^2 L^2}{\kappa^2}\right)}\right\} \qquad \text{(22)}.$$

[In the Communication to the Section numerical illustrations of non-oscillatory and of oscillatory discharge were given.]

69. Five Applications of Fourier's Law of Diffusion, illustrated by a Diagram of Curves with Absolute Numerical Values.

[From *British Association Report*, 1888, pp. 571—574; *Nature*, Vol. XXXVIII. pp. 571—573.

Reprinted in *Math. and Phys. Papers*, Vol. III. Art. xcviii. pp. 428—435.]

70. Discussion on Lightning Conductors at the British Association.

[From *British Association Report*, 1888, pp. 603—606; *Nature*, Vol. XXXVIII. Oct. 4, 1888, p. 546.]

71. On the Reflexion and Refraction of Light.

[From *Phil. Mag.* Vol. xxvi. pp. 414—425, Nov. 1888, and pp. 500, 501, Dec. 1888.

Extracts reprinted in *Baltimore Lectures*, pp. 174, 351—354, 407.]

72. Ether, Electricity, and Ponderable Matter.

[From *Inst. Elec. Engineers' Journal*, Vol. xviii. 1890, pp. 4—37 (Inaugural Address, Jan. 10, 1889).

Reprinted in *Math. and Phys. Papers*, Vol. iii. Art. cii. pp. 484—515.]

73. On a Mechanism for the Constitution of Ether.

[From *Edin. Roy. Soc. Proc.* Vol. xvii. March 17, 1890, pp. 127—132.

Reprinted in *Math. and Phys. Papers*, Vol. iii. Art. c. pp. 466—472.]

74. Motion of a Viscous Liquid; Equilibrium or Motion of an Elastic Solid; Equilibrium or Motion of an Ideal Substance called for Brevity "Ether"; Mechanical Representation of Magnetic Force.

[Written for *Math. and Phys. Papers*, Vol. iii. Art. xcix. May, 1890, pp. 436—465.]

75. Preliminary Experiments for Comparing the Discharge of a Leyden Jar through Different Branches of a Divided Channel. By Lord Kelvin and Alexander Galt.

[From *British Association Report*, 1894, pp. 555, 556.]

In these experiments the metallic part of the discharge channel was divided between two lines of conducting metal, each consisting in part of a test-wire, the other parts of the two lines being wires of different shape, material, and neighbourhood, of which the qualities in respect to facility of discharge through them are to be compared.

The two test-wires were, as nearly as we have been hitherto able to get them, equal and similar, and similarly mounted. Each test-wire was 51 cm. of platinum wire of ·006 cm. diameter and 12 ohms resistance, stretched straight between two metal terminals at the ends of a glass tube. One end of the platinum wire was soldered to a stiff solid brass mounting; the other was fixed to a fine spring carrying a light arm for multiplying the motion. The testing effect was the heat developed in the test-wire by the discharge, as shown by its elongation, the amount of which was judged from a curve traced, by the end of the multiplying arm, on sooted paper carried by a moving cylinder. Two of Lord Kelvin's vertical electrostatic voltmeters, suitable respectively for voltages of about 10,000 and 1,500, were kept constantly with their cases connected with the outer coatings of the leyden, and their insulated plates with the inside coatings of the leyden.

I. In the experiments hitherto made the two wires to be tested have generally been of the same length. When they were of the same material, but of different diameters, the testing elongation showed, as was to be expected, that the test-wire in the branch containing the thicker wire was more heated than the test-wire in the other branch. In a continuation of the experiments we hope to compare hollow and tubular wires of the same external diameter, and same length and same material.

II. With wires of different non-magnetic material—for example, copper and platinoid—of the same length, but of very different diameters, so as to have the same resistances, the testing elongations were very nearly equal.

III. In one series of experiments the tested conductors were two bare copper wires, each ·16 cm. diameter, 9 metres long, and resistance ·085 ohm, which, it will be observed, is very small in comparison with the 12 ohms in each of the platinum test-wires. One of the copper wires was coiled in a uniform helix of forty turns on a glass tube of 7 cm. diameter. The length of the helix was 35 cm., and the distance from centre to centre of neighbouring turns therefore $\frac{7}{8}$ cm. The middle of the other copper wire was hung by silk thread from the ceiling, and the two halves passed down through the air to the points of junction in the circuit. The elongation of the test-wire in this channel was more than twice as much as that of the test-wire in the channel of which the helix was part.

IV. One hundred and seventy-one varnished pieces of straight soft iron wire were placed within the glass tube, which was as many as it could take. This made the testing elongation ten times as great in the other channel.

V. The last comparison which we have made has been between iron wire and platinoid wire conductors. The length of each was 502·5 cm. The diameter of the iron wire was 034 cm., and its resistance 6·83 ohms. The diameter of the platinoid wire was ·058 cm., and its resistance 6·82 ohms. Each of these wires was supported by a silk thread from the ceiling, attached to its middle (as in III. and IV. for one of the tested conductors). Fourteen experiments were made, seven with the test-wires interchanged relatively to the branches in which they were placed for the first seven. The following table shows the means of the results thus obtained, with details regarding the electrostatic capacities of the leyden-jars and the voltages concerned in the results.

In each case four leyden-jars, connected to make virtually one of capacity ·02742 microfarad, were charged up to 9,000 volts, and discharged through divided channel. The energy, therefore, in the leyden before discharge was $11{\cdot}105 \times 10^6$ ergs. In each of the first three cases 1,450 volts were found remaining in the jars after discharge; in each of the last four 1,400.

Energy remaining in leyden after discharge	Energy used	Elongation of testing wires in cms.			
		In channel containing platinoid	means	In channel containing iron	means
$\cdot 29 \times 10^6$ ergs	$10{\cdot}82 \times 10^6$ ergs	·01794		01226	
„	„	·01861	·01829	·01247	·01239
„	„	·01832		·01244	
27×10^6 ergs	$10{\cdot}84 \times 10^6$ ergs	·01823		·01276	
„	„	·01828	·01836	·01280	·01261
„	„	·01828		·01244	
„	„	·01865		·01244	

The mode of measuring the elongation of the test-wires was, as may be understood from the preceding description, somewhat crude, but it is reassuring to see that the mean results in the cases of 10·82 and 10·84 megalergs of energy used are so nearly equal. The ratios for the two circuits are, in the two cases, respectively 1·48 and 1·46. The conclusion that the heating effect in the test-wire in series with the platinoid wire is nearly one-and-a-half times as great as that of the test-wire in series with the iron is certainly interesting, not only in itself, but in relation to Professor Oliver Lodge's exceedingly interesting and instructive experiments on alternative paths for the discharge of leyden-jars, described in his book on *Lightning Conductors and Lightning Guards*, which were not decisive in showing any general superiority of copper over iron of the same steady ohmic resistance, but even showed in some cases a seeming superiority of the iron for efficiency in the discharge of a leyden-jar. Our result is quite such as might have been expected from experiments made eight years ago by Lord Rayleigh and described in his paper "On the Self-induction and Resistance of Compound Conductors*."

* *Phil. Mag.* Vol. XXII. 1886, p. 469.

76. The Dynamical Theory of Refraction, Dispersion, and Anomalous Dispersion.

[From *British Association Report*, 1898, pp. 782, 783; *Nature*, Vol. LVIII. Oct. 6, 1898, pp. 546, 547.

Reprinted substantially in *Baltimore Lectures*, p. 148.]

77. Continuity in Undulatory Theory of Condensational-rarefactional Waves in Gases, Liquids, and Solids, of Distortional Waves in Solids, of Electric Waves in all Substances capable of Transmitting them, and of Radiant Heat, Visible Light, Ultra-violet Light.

[From *British Association Report*, 1898, pp. 783—787; *Nature*, Vol. LIX. Nov. 17, 1898, pp. 56, 57; *Phil. Mag.* Vol. XLVI. Nov. 1898, pp. 494—500.

Reprinted in *Baltimore Lectures*, pp. 148—162.]

78. On the Reflection and Refraction of Solitary Plane Waves at a Plane Interface between Two Isotropic Elastic Mediums—Fluid, Solid, or Ether.

[From *Edinb. Roy. Soc. Proc.* Vol. XXII. Dec. 19, 1898, pp. 366—378; *Phil. Mag.* Vol. XLVII. Feb. 1899, pp. 179—191.

Reprinted substantially in *Baltimore Lectures*, §§ 112—121.]

79. Application of Sellmeier's Dynamical Theory to the Dark Lines D_1, D_2 produced by Sodium-vapour.

[From *Edinb. Roy. Soc. Proc.* Vol. XXII. Feb. 6, 1899, pp. 523—531; *Phil. Mag.* Vol. XLVII. March, 1899, pp. 302—308.

Reprinted in *Baltimore Lectures*, pp. 176—184.]

80. On the Application of Force within a Limited Space, required to produce Spherical Solitary Waves, or Trains of Periodic Waves, of both Species, Equi-voluminal and Irrotational, in an Elastic Solid.

[From *Phil. Mag.* Vol. XLVII. May, 1899, pp. 480—493; Vol. XLVIII. August, 1899, pp. 227—236, Oct. 1899, pp. 388—393; also read as a Presidential Address before London Mathematical Society, cf. Vol. XXXI. June 8, 1899, p. 147.

Reprinted in *Baltimore Lectures*, pp. 190—219.]

81. On the Motion Produced in an Infinite Elastic Solid by the Motion through the Space occupied by it of a Body acting on it only by Attraction or Repulsion.

[From *Edin. Roy. Soc. Proc.* Vol. XXIII. July 16, 1900, pp. 218—235; *Phil. Mag.* Vol. L. Aug. 1900, pp. 181—198; *Congrès Internationale de Physique à l'Exposition de* 1900, Vol. II. pp. 1—22.

Reprinted in *Baltimore Lectures*, Appendix A, pp. 468—485.]

82. On the Duties of Ether for Electricity and Magnetism.

[From *Phil. Mag.* Vol. L. Sept. 1900, pp. 305—307.]

19.* In my paper published in the last number of the *Philosophical Magazine*, of which this is a continuation, I limited myself to a problem of mathematical dynamics; and merely suggested the possibility of finding in it an explanation of the fundamental difficulty in the Undulatory Theory of Light referred to in the first and last paragraphs (§§ 1, 18). The following communication is the substance of a supplementary statement relating to that paper given orally to the Congrès International de Physique at a meeting held in Paris last Wednesday (August 8).

20. I now cannot resist the temptation to speak of efforts which occupy me to find proper assumptions for including something of the allied subjects mentioned in the footnote on § 1.

21. For atoms of electricity, which, following Larmor, I at present call electrons, it inevitably occurs to suggest a special class of atoms not fulfilling the condition stated in lines 12—22 of § 5.

Thus a *positive electron*† would be an atom which by attraction condenses ether into the space occupied by its volume; and a negative electron would be an atom which, by repulsion, rarefies the ether remaining in the space occupied by its volume. The stress produced in the ether outside two such atoms by the attractions or repulsions which they exert on the ether within them, would cause apparent attraction between a positive and a

* [The numbering of the sections is continuous with that of No. 81.]

† It seems probable that this may be the *resinous* electrification, but it may possibly be the *vitreous*. It must be remembered that vitreous electrification has hitherto been called positive merely because it is it which is given by the "prime conductor" of the old ordinary electric machine.

negative electron; and apparent repulsion between two electrons both positive or both negative.

22. But these apparent attractions and repulsions would increase much more with diminished distance than according to the Newtonian law of the inverse square. This law, which we know from Coulomb and Cavendish to be true for electric attractions and repulsions, *cannot be explained by stress in ether* according to any known or hitherto imagined properties of elastic matter. But a very simple hypothesis, assuming action at distances, between different portions of ether, explains it perfectly. Consider two portions of ether occupying infinitesimal volumes V, V', at distance D asunder. My hypothesis is that they repel mutually with a force equal to

$$\frac{(\rho-1)\,V.(\rho'-1)\,V'}{D^2} \quad\ldots\ldots\ldots\ldots\ldots\ldots(16);$$

where ρ, ρ' denote the densities of the two portions of ether considered, and 1 is the natural density of undisturbed ether. This makes the force repulsion or attraction according as $(\rho-1)$, $(\rho'-1)$ are of the same or of opposite signs; and zero if either is zero, (which means that ether of undisturbed natural density experiences neither attraction nor repulsion from any other portion of ether far or near).

23. This closely resembles Aepinus' doctrine of the middle of the eighteenth century, commonly referred to as the "one-fluid theory of electricity"; but now, instead of electric fluid, we have "ether," an elastic solid pervading all space. According to our present hypothesis, similar electric atoms repel one another, and dissimilar attract; in virtue of *force* between each atom and the portion of ether within it, and mutual repulsion or attraction of these portions of ether with no contributive action of the ether in the space around them and between them.

24. Stress in ether, being thus freed from the *impossible task of transmitting both electrostatic and magnetic force*, is (we may well imagine) quite competent to perform the simpler duty of transmitting magnetic force alone.

25. Hitherto one seemingly insuperable obstacle against following up this idea to practical realization has been the greatness of the force in many well known cases of magnetic

attraction between iron poles, whether due to steel magnets or electromagnets. Considering that, in our most delicate experiments in various branches of science, ponderable bodies large and small are observed to be moved freely by forces of less than a thousandth of the heaviness* of a milligram, how can we conceive the ether through which they move to be capable of the stress required for the transmission of force between flat poles of an electromagnet amounting per square centimetre to more than two hundred† times the heaviness of a kilogram? This difficulty is annulled if we adopt the hypothesis which I have described to the Congrès (§ 2 above). We may now suppose the density of ether as great as we please, subject only to the limitation that it must not be so great as to disturb sensibly the proportionality of effective inertia to gravity in different kinds of matter, proved by Newton in his pendulum experiment, for lead, brass, glass, &c., and by his interpretation of Kepler's third law for the different planets of our system. Probably we might safely, if we wished it, assume the density of ether to be as much as 10^{-6}. I am content at present, however, to suggest 10^{-9}. This, with the velocity of light 300,000 kilometres per second, makes the rigidity (being density × square of velocity) equal to 9.10^{11} dynes per square centimetre, which is somewhat greater than the rigidity of steel (7.10^{11}). It is clearly not for want of strength that we need question the competence of ether to transmit magnetic force! I confess that I now feel hopeful of seeing solved some of the other formidable difficulties which meet every effort to explain electric insulation and conduction, and electromagnetic force, and the magnetic force of a steel magnet, by definite mechanical action of ether.

* I cannot without ambiguity use the simple word "weight" here; because this word means legally a mass, and is practically used more often to signify a mass than the gravitational heaviness of a mass.

† The most intense magnetic field hitherto measured is, I believe, that of Dubois (see his Report on Magnetism to this Congress [see under No. 81]) in which he found 76,000 C.G.S. between two small plane end-faces of soft iron poles of a powerful electromagnet. This makes the attraction per square centimetre of either face $(76,000)^2 \div 8\pi$, or approximately 23.10^7 dynes, or 230 kilograms.

83. A NEW SPECIFYING METHOD FOR STRESS AND STRAIN IN AN ELASTIC SOLID.

[From *Edin. Roy. Soc. Proc.* Vol. XXIV. Jan. 20, 1902, pp. 97—101; *Phil. Mag.* Vol. III. Jan. 1902, pp. 95—97, April, 1902, pp. 444—448.]

THE method for specifying stress and strain hitherto followed by all writers on elasticity has the great disadvantage that it essentially requires the strain to be infinitely small. As a notational method it has the inconvenience that the specifying elements are of two essentially different kinds (in the notation of Thomson and Tait e, f, g, simple elongations; a, b, c, shearings). Both these faults are avoided if we take the six lengths of the six edges of a tetrahedron of the solid, or what amounts to the same, though less simple, the three pairs of face-diagonals of a hexahedron*, as the specifying elements. This I have thought of for the last thirty years, but not till to-day (Dec. 16) have I seen how to make it conveniently practicable, especially for application to the generalized dynamics of a crystal.

1. We shall suppose the solid to be a homogeneous crystal of any possible character. Cut from it a tetrahedron $ABCD$ of any shape and orientation. Let the three non-intersecting pairs (AB, CD), (BC, AD), (CA, BD) of its six edges be denoted by

$$(3p, 3p'),\quad (3q, 3q'),\quad (3r, 3r') \qquad (1).$$

This notation gives

$$(p, p'),\quad (q, q'),\quad (r, r') \qquad (2)$$

for the six edges of a tetrahedron, similar to $ABCD$, formed by taking for its corners ($\alpha, \beta, \gamma, \delta$) the centres of gravity† of the four

* This name, signifying a figure bounded by three pairs of parallel planes, is admitted in crystallography; but the longer and less expressive "parallelepiped" is too frequently used instead of it by mathematical writers and teachers. A hexahedron with its angles acute and obtuse is what is commonly called, both in pure mathematics and crystallography, a rhombohedron. A right-angled hexahedron is a brick, for which no Greek or other learned name is hitherto to the front in usage A rectangular equilateral hexahedron is a cube.

† For brevity I shall henceforth call the centre of gravity of a triangle, or of a tetrahedron, simply *its centre*.

triangular faces BCD, CDA, DAB, ABC respectively, so that we have $p = \alpha\beta$, $q = \beta\gamma$, $r = \gamma\alpha$, $p' = \gamma\delta$, $q' = \alpha\delta$, $r' = \beta\delta$. Consider now, in advance, the amounts of work done by the six pairs of balancing forces constituting the six stress-components described in § 2, when the strain-components vary; for example, the balancing pulls P, parallel to AB, when $\alpha\beta$ increases from p to $p+dp$, all the other five lengths q, r, p', q', r' remaining constant. For the reckoning of work we may suppose the opposite forces, P, to be applied at α and β, instead of being equably distributed over the faces ADC, BDC. Hence the work which they do is Pdp; and other five pairs of balancing pulls, Q, R, P', Q', R', do no work.

2. Parallel to the edge AB apply to the faces ADC, BDC equal and opposite pulls, P, equally distributed over them. These two balancing pulls we shall call a stress or a stress-component. Similarly, parallel to each of the five other edges apply balancing pulls on the pair of faces cutting it. Thus we have in all six stress-components parallel to the six edges of the tetrahedron, denoted as follows:—

$$(P, P') \quad (Q, Q'), \quad (R, R') \quad \dots\dots\dots\dots\dots(3);$$

and we suppose that these forces, applied as they are to the surface of the solid, are balanced in virtue of the mutual forces between its particles, when its edges are of the lengths specified as in (1). Let p_0, p_0', q_0, q_0', r_0, r_0', be the values of the specifying elements in (2) when no forces are applied to the faces. Thus the differences from these values, of the six lengths shown in formula (2), represent the strain of the substance when under the stress represented by (3).

Let w be the work done when pulls upon the faces, each commencing at zero, are gradually increased to the values shown in (3). In the course of this process we have

$$dw = Pdp + P'dp' + Qdq + Q'dq' + Rdr + R'dr' \quad \dots(4).$$

3. Hence if we suppose w expressed as a function of p, p', q, q', r, r', we have

$$\frac{dw}{dp} = P, \quad \frac{dw}{dp'} = P', \quad \frac{dw}{dq} = Q, \quad \frac{dw}{dq'} = Q', \quad \frac{dw}{dr} = R, \quad \frac{dw}{dr'} = R' \quad \dots\dots\dots(4).$$

This completes the foundation of the molar dynamics of an elastic solid of the most general possible kind according to Green's theory, expressed in terms of the new mode of specifying stresses and strains.

4. To understand thoroughly the state of strain specified by (1) or (2), let the tetrahedron of reference, $A_0B_0C_0D_0$, for the condition of zero strain and stress, be equilateral (that is to say, according to the notation of § 2 (1) let $\frac{1}{3}$ of each edge

$$= p_0 = q_0 = r_0 = p_0' = q_0' = r_0').$$

In $A_0B_0C_0D_0$ inscribe a spherical surface touching each of the six edges. Its centre must be at K_0, the centre of the tetrahedron; and the points of contact must be the middle points of the edges. Alter the solid by homogeneous strain*, to the condition (p, q, r, p', q', r') in which $A_0B_0C_0D_0$ becomes $ABCD$. The inscribed spherical surface becomes an ellipsoid having its centre at K, the centre of $ABCD$, and touching its six edges at their middle points†. This ellipsoid shows fully and clearly the state of strain specified by p, q, r, p', q', r'. It is what is called the "strain ellipsoid."‡

5. Two ways of finding the ellipsoid touching the six edges of a tetrahedron are obvious. (1) Through AB and CD draw planes respectively parallel to CD and AB; and deal similarly with the two other pairs of non-intersecting edges. The three pairs of parallel planes thus found, constitute a hexahedron which contains the required ellipsoid touching the six faces at their centres; or (2) draw AK, BK, CK, DK, and produce to equal distances KA', KB', KC', KD' beyond K. We thus find four points, A', B', C', D', which, with A, B, C, D, are the eight corners of the hexahedron which we found by construction (1). A circumscribed hexahedron being thus given, the principal axes of the ellipsoid, and their orientation, are found by the solution of a cubic equation.

6. Another way of finding the strain-ellipsoid, which is in some respects simpler, and which has the advantage that in its

* Thomson and Tait's 'Natural Philosophy,' § 155; 'Elements,' § 136.

† Thus we have an interesting theorem in the geometry of the tetrahedron:—If an ellipsoid touching the edges of a tetrahedron has its centre at the centre of the tetrahedron, the points of contact are at the middles of the edges.

‡ Thomson and Tait's 'Natural Philosophy,' § 100; 'Elements,' § 141.

construction it does not take us outside the boundary of our fundamental tetrahedron, is as follows:—In the equilateral tetrahedron $A_0B_0C_0D_0$ describe, from its centre K_0, a spherical surface touching any three of its faces. It touches these faces at their centres; and it also touches the fourth face, and at its centre. Hence, if we solve the determinate, one-solutional, problem *to draw an ellipsoid touching at their centres any three of the four faces of any tetrahedron ABCD, and having its centre at K*, this ellipsoid touches at its centre the fourth face of the tetrahedron; and it is the strain-ellipsoid for the homogeneous strain by which an equilateral tetrahedron of solid is altered to the figure $ABCD$.

7. To bring our new method of specifying strain and stress into relation with the ordinary method for infinitesimal strains and the corresponding stresses:—Let λ denote the length of each edge of the equilateral tetrahedron of reference, $A_0B_0C_0D_0$; and let h be the edge of the cube of which A_0, B_0, C_0, D_0 are four corners (this cube being the hexahedron found by applying either of the constructions of § 5 to the tetrahedron $A_0B_0C_0D_0$). The twelve face-diagonals of this cube are each equal to λ, and therefore $\lambda = h\sqrt{2}$. Let now the cube be infinitesimally strained so that its edges become $h(1+e)$, $h(1+f)$, $h(1+g)$; and so that the angles in its three pairs of faces are altered from right angles to acute and obtuse angles differing respectively by a, b, c from right angles. This is the strain (e, f, g, a, b, c) in the notation of Thomson and Tait referred to in the introductory paragraph above. By the infinitesimal geometry of the affair, we easily find the corresponding alterations of the face-diagonals, which according to our present notations are $(p-1)\lambda$, $(p'-1)\lambda$, $(q-1)\lambda$, etc., and thus we have as follows:—

$$\left.\begin{aligned} p-1 &= \tfrac{1}{2}(f+g+a) \\ p'-1 &= \tfrac{1}{2}(f+g-a) \\ q-1 &= \tfrac{1}{2}(g+e+b) \\ q'-1 &= \tfrac{1}{2}(g+e-b) \\ r-1 &= \tfrac{1}{2}(e+f+c) \\ r'-1 &= \tfrac{1}{2}(e+f-c) \end{aligned}\right\} \quad \ldots\ldots\ldots\ldots\ldots\ldots(5)$$

for the relation between the two specifications of any infinitesimal strain. Adding these, and denoting $e+f+g$ by s, we find

$$p+p'+q+q'+r+r'-6 = 2s \quad \ldots\ldots\ldots\ldots\ldots(6).$$

And solving for a, b, c, e, f, g, in terms of p, q, r, p', q', r', we have

$$\left.\begin{array}{lll} a=p-p'; & b=q-q'; & c=r-r'; \\ e=s-p-p'+2; & f=s-q-q'+2; & g=s-r-r'+2 \end{array}\right\} \quad (7).$$

8. The work required to produce an infinitesimal strain e, f, g, a, b, c, in a homogeneous solid of cubic crystalline symmetry is expressed by the following formula:—

$$2w = \mathfrak{A}(e^2+f^2+g^2) + 2\mathfrak{B}(fg+ge+ef) + n(a^2+b^2+c^2). \ldots(8).$$

This may be conveniently modified by putting

$$k = \tfrac{1}{3}(\mathfrak{A}+2\mathfrak{B}); \quad n_1 = \tfrac{1}{2}(\mathfrak{A}-\mathfrak{B}) \quad \ldots\ldots\ldots..(9),$$

where k denotes the bulk modulus and n_1, n the two rigidity-moduluses. With this notation (8) becomes

$$2w = k(e+f+g)^2 + \tfrac{2}{3}n_1[(f-g)^2+(g-e)^2+(e-f)^2] + n(a^2+b^2+c^2) \quad \ldots\ldots(10).$$

The rigidity relative to shearings parallel to the pairs of planes of the cube, or, which is the same thing, changes of the angles of the corners of the square faces from right angles to acute or obtuse angles, is n_1. The rigidity relative to changes of the angles between the diagonals of the faces from right angles to acute or obtuse angles is n. The compressibility modulus is k. Using now (7) in (10) we have

$$2w = ks^2 + \tfrac{2}{3}n_1[(q+q'-r-r')^2 + (r+r'-p-p')^2 + (p+p'-q-q')^2] + n[(p-p')^2+(q-q')^2+(r-r')^2] \quad \ldots(11).$$

84. On the Electro-ethereal Theory of the Velocity of Light in Gases, Liquids, and Solids.

[From *British Association Report*, 1903, p. 535; *Phil. Mag.* Vol. vi. Oct. 1903, pp. 437—442.

Reprinted in *Baltimore Lectures*, xx. pp. 463—467.]

INDEX

CAMBRIDGE: PRINTED BY JOHN CLAY, M.A. AT THE UNIVERSITY PRESS.